Handbook of Proteomic Methods

Handbook
of Proteomic Methods

Edited by

P. Michael Conn

Oregon National Primate Research Center,
Oregon Health and Science University, Beaverton, OR

Humana Press ✳ **Totowa, New Jersey**

Library of Congress Cataloging in Publication Data

Handbook of proteomic methods / edited by P. Michael Conn.
   p. cm.
  Includes bibliographical references and index.
  ISBN 1-58829-340-8 (alk. paper); 1-59259-414-X (e-book)
   1. Proteomics—Handbooks, manuals, etc. I. Conn, P. Michael.

QP551.H33 2003
572'.6--dc21

2003040677

Preface

NIH Director Elias Zerhouni, speaking at the annual meeting of the Association of American Medical Colleges in late 2002, addressed the need for the development of new and revolutionary research tools to understand complex biological systems that can then be applied to cure diseases. "Among revolutionary methods of research, we need to accelerate proteomics and large-scale technology for the post-genomic era," he noted, reminding the audience that sequencing the genome is essentially only "getting the parts list."

It was in this spirit of enthusiasm that the groundwork for this volume was begun earlier that year, by selecting well-recognized authors, who have contributed mightily to the field of proteomics, and identifying areas of interest and potential growth that could lead to a useful methods handbook.

Proteomics, as a word and as a discipline, is new to most of us and we expect that many will find this volume a useful dictionary for understanding the work of others as well as a map for setting out on our own research programs. From the start, it was our goal to produce a volume that would be valuable both to the neophyte and the seasoned worker. We encouraged the authors to include hints and tricks that might not be obvious in those original publications that did not describe the procedure in the detail reported here.

This editor, who marveled at the speed that the human genome was sequenced, realized that he was looking only at the blueprint for the Taj Mahal, not seeing a photo, much less actually having an experience in Agra. The beauty of the detail is the study of Proteomics.

The editor wants to thank the authors for interrupting their busy schedules to participate in this important project and the publisher for recognizing the need for placing it on a high priority timetable.

P. Michael Conn

Contents

Contributors

HAIKE ANTELMANN • *Institut für Mikrobiologie und Molekularbiologie, Ernst-Moritz-Arndt-Universität Greifswald, Greifswald, Germany*

RON D. APPEL • *Proteome Informatics Group, Swiss Institute of Bioinformatics, Geneva, Switzerland*

NEBOJSA AVDALOVIC • *Research and Development, Dionex Corporation, Sunnyvale, CA*

GIOVANNI LUCA BERETTA • *Preclinical Chemotherapy and Pharmacology Unit, Experimental Oncology Department, Instituto Nazionale Tumori, Milan, Italy*

ROBERT J. BEYNON • *Protein Function Group, Faculty of Veterinary Science, University of Liverpool, Liverpool, UK*

JOEL R. BOCK • *Department of Bioengineering, University of California San Diego, La Jolla, CA*

DANAIL BONCHEV • *Program for Theory of Complex Natural Systems, Department of Marine Sciences, Fort Crockett Campus, Texas A&M University, Galveston, TX*

HANS-PETER BRAUN • *Institut für Angewandte Genetik, Universität Hannover, Hannover, Germany*

KERRY A. CHESTER • *Cancer Research UK Targeting and Imaging Group, Department of Oncology, Royal Free and University College Medical School, Royal Free Campus, London, UK*

CATHERINE A. COOPER • *Proteome Systems Ltd., North Ryde, New South Wales, Sydney, Australia*

BART DEVREESE • *Laboratory of Protein Biochemistry and Protein Engineering, Ghent University, Ghent, Belgium*

HENRY S. DUEWEL • *Department of Analytical Sciences, MDS Proteomics Inc., Toronto, Canada*

MICHAEL J. DUNN • *Department of Neuroscience, Institute of Psychiatry, London, UK*

MICHAEL R. EMMERT-BUCK • *Pathogenetics Unit, Laboratory of Pathology and Urologic Oncology Branch, National Cancer Institute, Bethesda, MD*

DELL FARNAN • *Dionex Corporation, Sunnyvale, CA*

DANIEL FIGEYS • *Department of Analytical Sciences, MDS Proteomics Inc., Toronto, Canada*

ERIC T. FUNG • *Ciphergen Biosystems Inc., Fremont, CA*

GIULIANO GALLI • *Biochemistry and Molecular Biology Unit, Chiron Srl, Siena, Italy*

IGNAZIO GARAGUSO • *Biochemistry and Molecular Biology Unit, Chiron Srl, Siena, Italy*

ELISABETH GASTEIGER • *Swiss Institute of Bioinformatics, Geneve, Switzerland*

LAURA GATTI • *Preclinical Chemotherapy and Pharmacology Unit, Experimental Oncology Department, Instituto Nazionale Tumori, Milan, Italy*

DAVID A. GOUGH • *Department of Bioengineering, University of California San Diego, La Jolla, CA*

GUIDO GRANDI • *Biochemistry and Molecular Biology Unit, Chiron Srl, Siena, Italy*

ROBIN GRAS • *Proteome Informatics Group, Swiss Institute of Bioinformatics, Geneva, Switzerland*

PAUL R. GRAVES • *Department of Pharmacology and Cancer Biology, Duke University, Durham, NC*

RENATA GRIFANTINI • *Biochemistry and Molecular Biology Unit, Chiron Srl, Siena, Italy*

MATHEW J. HARRISON • *Proteome Systems Ltd., North Ryde, New South Wales, Sydney, Australia*

TIMOTHY A. J. HAYSTEAD • *Department of Pharmacology and Cancer Biology, Duke University, Durham, NC*

MICHAEL HECKER • *Institut für Mikrobiologie und Molekularbiologie, Ernst-Moritz-Arndt-Universität Greifswald, Greifswald, Germany*

PATRICIA HERNANDEZ • *Proteome Informatics Group, Swiss Institute of Bioinformatics, Geneva, Switzerland*

VAN M. HOANG • *SAIC-Frederick Inc., National Cancer Institute, Frederick, MD*

HANSPETER HERZEL • *Innovationskolleg Theoretische Biologie, Humboldt University, Berlin, Germany*

SHENG-HE HUANG • *Division of Infectious Diseases, Childrens Hospital Los Angeles and Department of Pediatrics, University of Southern California School of Medicine, Los Angeles, CA*

ALEXANDRA HUHALOV • *Cancer Research UK Targeting and Imaging Group, Department of Oncology, Royal Free and University College Medical School, Royal Free Campus, London, UK*

EMILIA TIZIANA IACOBINI • *Biochemistry and Molecular Biology Unit, Chiron Srl, Siena, Italy*

PHILIP J. JOHNSON • *Division of Cancer Studies, University of Birmingham, Birmingham, UK*

AMBROSE JONG • *Division of Infectious Diseases, Childrens Hospital Los Angeles and Department of Pediatrics, University of Southern California School of Medicine, Los Angeles, CA*

HIREN J. JOSHI • *Proteome Systems Ltd., North Ryde, New South Wales, Sydney, Australia*

ELIAS KLEIN • *Core Proteomics Laboratory, Kidney Disease Program, Department of Medicine, University of Louisville, Louisville, KY*

JON B. KLEIN • *Core Proteomics Laboratory, Kidney Disease Program, Department of Medicine, University of Louisville, Louisville, KY*

VLADIMIR KNEZEVIC, *Research and Development, 20/20 GeneSystems Inc., Rockville, MD*

ELKE LECOCQ • *Laboratory of Protein Biochemistry and Protein Engineering, Ghent University, Ghent, Belgium*

WOLF D. LEHMANN • *Central Spectroscopy, German Cancer Research Center, Heidelberg, Germany*

SHANHUA LIN • *Research Proteomics, Ciphergen Biosystems Inc., Fremont, CA*

FRÉDÉRIQUE LISACEK • *Geneva Bioinformatics, Geneva, Switzerland and Génome and Informatique, Evry Cedex, France*

GARY D. LUKER • *Molecular Imaging Center, Mallinckrodt Institute of Radiology, Washington University School of Medicine, St. Louis, MO*

JAN MAARTEN VAN DIJL • *Department of Pharmaceutical Biology, University of Groningen, Groningen, The Netherlands*

KAREN J. MARTIN • *Proteomics Section, Molecular Probes, Inc., Eugene, OR*

CHRISTINA MIHR • *Institut National de la Recherche Agronomique, Avignon, France*

RALF MROWKA • *Johannes-Müller Institute for Physiology, Humboldt University, Berlin, Germany*

MARKUS MÜLLER • *Proteome Informatics Group, Swiss Institute of Bioinformatics, Geneva, Switzerland*

FRÉDÉRIC NIKITIN • *Geneva Bioinformatics, Geneva, Switzerland*

NATHALIE NORAIS • *Biochemistry and Molecular Biology Unit, Chiron Srl, Siena, Italy*

RENZO NOGAROTTO • *Biochemistry and Molecular Biology Unit, Chiron Srl, Siena, Italy*

NICOLLE H. PACKER • *Proteome Systems Ltd., North Ryde, New South Wales, Sydney, Australia*

WAYNE F. PATTON • *Proteomics Section, Molecular Probes Inc., Eugene, OR*

PAOLA PEREGO • *Preclinical Chemotherapy and Pharmacology Unit, Experimental Oncology Department, Istituto Nazionale Tumori, Milan, Italy*

DAVID PIWNICA-WORMS • *Molecular Imaging Center, Mallinckrodt Institute of Radiology, and Department of Molecular Biology and Pharmacology, Washington University School of Medicine, St. Louis, MO*

CHRIS POHL • *Dionex Corporation, Sunnyvale, CA*

TERENCE C. W. POON • *Department of Clinical Oncology at the Sir Y. K. Pao Centre for Cancer, Chinese University of Hong Kong, Shatin, Hong Kong*

MILAN RANDIĆ • *Laboratory for Chemometrics, National Institute of Chemistry, Ljubljana, Slovenia*

JOHANNES A. ROMIJN • *Department of Endocrinology and Metabolic Diseases, Leiden University Medical Center, Leiden, The Netherlands*

DAVID M. SCHIELTZ • *Department of Proteomics, Torrey Mesa Research Institute, Syngenta Research and Technology, San Diego, CA*

ANDREAS SCHLOSSER • *Institute of Medical Immunology, Charité, Humboldt University, Berlin, Germany*

BIRTE SCHULENBERG • *Proteomics Section, Molecular Probes, Inc., Eugene, OR*

VIJAY SHARMA • *Molecular Imaging Center, Mallinckrodt Institute of Radiology, Washington University School of Medicine, St. Louis, MO*

JOEL SMET • *Department of Pediatrics, Division of Pediatric Neurology and Metabolism, Ghent University Hospital, Ghent, Belgium*

JAN W. A. SMIT • *Department of Endocrinology and Metabolic Diseases, Leiden University Medical Center, Leiden, The Netherlands*

DANIEL I. R. SPENCER • *Cancer Research UK Targeting and Imaging Group, Department of Oncology, Royal Free and University College Medical School, Royal Free Campus, London, UK*

IAN I. STEWART • *Department of Analytical Sciences, MDS Proteomics Inc., Toronto, Canada*

JAMES T. SUMMERSGILL • *Infectious Diseases Laboratory, University of Louisville School of Medicine, Louisville, KY*

NING TANG • *Research Proteomics, Ciphergen Biosystems, Inc., Fremont, CA*

TY THOMSON • *Department of Analytical Sciences, MDS Proteomics Inc., Toronto, Canada*

VISITH THONGBOONKERD • *Core Proteomics Laboratory, Kidney Disease Program, Department of Medicine, University of Louisville, Louisville, KY*

AKIRA TSUGITA • *Proteomics Research Laboratory, NEC Laboratories, Tsukuba, Japan*

JOZEF VAN BEEUMEN • *Laboratory of Protein Biochemistry and Protein Engineering, Ghent University, Ghent, Belgium*

RUDY VAN COSTER • *Department of Pediatrics, Division of Pediatric Neurology and Metabolism, Ghent University Hospital, Ghent, Belgium*

FRANK VANROBAEYS • *Laboratory of Protein Biochemistry and Protein Engineering, Ghent University, Ghent, Belgium*

TIMOTHY D. VEENSTRA • *SAIC-Frederick Inc., National Cancer Institute, Frederick, MD*

MAUNO VIHINEN • *Institute of Medical Technology, University of Tampere, Tampere, Finland*

MICHAEL P. WASHBURN • *Department of Proteomics, Torrey Mesa Research Institute, Syngenta Research and Technology, San Diego, CA*

SCOT R. WEINBERGER • *Research Proteomics, Ciphergen Biosystems Inc., Fremont, CA*

JULES A. WESTBROOK • *Proteome Sciences plc, Institute of Psychiatry, London, UK*

MICHAEL WEITZHANDLER • *Dionex Corporation, Sunnyvale, CA*

MARC R. WILKINS • *Proteome Systems Ltd., North Ryde, New South Wales, Sydney, Australia*

LI-RONG YU • *SAIC-Frederick Inc., National Cancer Institute, Frederick, MD*

I

GENERAL TECHNIQUES

1

Proteomics and the Molecular Biologist

Paul R. Graves and Timothy A. J. Haystead

1. Introduction

This review is intended to provide the basics of proteomics for all scientists, regardless of their backgrounds, who are interested in using this approach in their research. We believe that proteomics when combined with pre-existing technologies like molecular biology can be a powerful tool in solving a wide variety of biological questions. However, for most who are not in the proteomics field, it is difficult to determine which proteomics approaches are feasible and which are not realistic. In this review, we highlight practical approaches that can be used by researchers without expertise in proteomics or mass spectrometry (MS) to gain insight into their system of choice.

1.1. Why Proteomics?

At any given time, a portion of the genome may be transcribed into proteins that make up the proteome of a cell. Thus, although genomics provides information about what proteins could potentially exist in a cell, only proteomics gives information about what proteins are actually present. In addition, the biological activity of proteins is often controlled by protein modifications or interactions with other proteins. Examples include proteolysis, phosphorylation, or formation of large protein complexes such as the nuclear pore (**Fig. 1**). These important events in the lifetime of a protein can only be determined through a proteomics approach, as they cannot be predicted from genomic data alone.

1.2. Types of Proteomics

1.2.1. Protein Expression Proteomics

The ability to measure protein expression levels accurately remains a fundamental goal in proteomics. However, considering that many other methods already exist to measure mRNA levels accurately such as DNA chip arrays, why would anyone want to measure protein levels? The reality is that the level of mRNA is not always a good reflection of the amount of protein present in the cell; in fact, several authors have shown a poor correlation between the two (*1,2*). Thus, protein levels will need to be measured directly to discern changes in expression patterns under different cellular conditions.

From: *Handbook of Proteomic Methods*
Edited by: P. Michael Conn © Humana Press Inc., Totowa, NJ

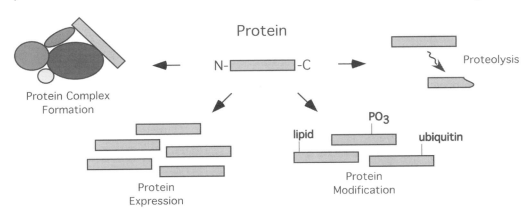

Fig. 1. Some possible fates of proteins in the cell. The proteome is dynamic because proteins are constantly being synthesized, modified, degraded, or formed into complexes. Proteomic approaches are necessary to determine the function of these events in the lifetime of a protein.

1.2.2. Cell Map Proteomics

Cell map proteomics strives to define all the proteins within particular organelles to gain insights into cellular architecture and protein function. An example of cell map proteomics is the characterization of all the proteins in rat mitochondria *(3)*.

1.2.3. Functional Proteomics

The goal of functional proteomics is to identify proteins in a cell, tissue, or organism that undergo changes in response to a specific biological condition. The types of changes could include protein modifications, proteolysis, subcellular localization, or interaction with other proteins *(4)*.

2. Proteomics Technologies

The two major developments that have catalyzed the growth of proteomics are the completion of genome sequencing projects and the development of mass spectrometry. A typical proteomics experiment can be divided into three main categories: protein preparation and analysis, mass spectrometry, and bioinformatics.

2.1. Protein Preparation and Analysis

The most important part of any proteomics experiment is the strategy for protein analysis. Surprisingly, it is also the most overlooked area in proteomics, partly because of the emphasis placed on MS and the development of high-throughput sequencing methods. However, of what value is high-throughput sequencing if the proteins are not of interest? The real challenge in proteomics is not speed of sample processing by MS but how to resolve the cellular proteome to allow for selection of proteins of interest. If small numbers of proteins are analyzed, this is a relatively simple task easily accomplished by one- or two-dimensional gel electrophoresis. But what about analysis of the entire proteome of a cell? Recently, it was reported that a typical eukaryotic cell may contain up to 1 billion protein molecules *(5)*. In addition, abundant cellular proteins can be present at 10 million copies per cell, whereas rare proteins may be present at only

1–10 copies per cell *(5)*. This results in a well-known bias in proteomics research favoring the identification of only the most abundant cellular proteins *(6)*. Currently, there are no systems available that can resolve and display all the proteins present in a typical cellular proteome. As a result, meaningful protein analysis is now where the bottleneck in proteomics occurs.

2.1.1. Electrophoresis-Based Proteome Analysis

Most protein separation in proteomics is still accomplished with polyacrylamide gels. This is because in comparison with other methods, gel-based strategies have the advantages of simplicity, reliability, and ready accessibility to researchers. In addition, gels perform an extremely important function—they allow for the visualization and selection of relevant proteins from a complex mixture. This focuses the analysis on specific proteins that have undergone changes within a particular system.

The two most common formats for polyacrylamide gels are one-dimensional (1-DGE) and two-dimensional (2-DGE) gel electrophoresis. The main advantage to 1-DGE is speed and the ability to resolve many different types of proteins regardless of size, hydrophobicity, or charge. However, 1-DGE, which resolves proteins on the basis of size, has limited resolving power. In contrast, 2-DGE has enormous resolving power, with the capacity to resolve thousands of proteins on a single gel *(7)*. As a result, it has been the cornerstone of proteomic studies for many years. In 2-DGE, proteins are resolved in the first dimension on the basis of their net charge and in the second dimension on the basis of size *(7)*. One disadvantage of 2-DGE is that membrane proteins, very large or small proteins, or highly basic or acidic proteins are not well resolved *(7)*.

As a result of these shortcomings, several strategies have been devised to enhance the ability of electrophoresis to resolve complex protein mixtures (**Fig. 2**). In one strategy, the complexity of a protein mixture is reduced by protein purification prior to analysis by 2-DGE (**Fig. 2A**). Examples include fractionation of cell lysates by ion-exchange chromatography *(8)*, affinity chromatography to isolate specific types of proteins *(9)*, or subcellular fractionation (**Fig. 2A**) *(3)*. Isolation of subproteomes can permit 2-DGE to be an effective tool for protein resolution *(10)*.

In another approach, proteins that cannot be resolved by 2-D gels (such as membrane proteins) are partially resolved by 1-D gels and then the proteins are excised and digested to peptides for further purification by high-performance liquid chromatography (HPLC; **Fig. 2B**). HPLC systems can be connected on-line with a mass spectrometer for continuous data analysis (**Fig. 2B**). This technique was used to characterize membrane proteins from human colon carcinoma cells *(11)*.

2.1.2. Nonelectrophoresis-Based Approaches

In an effort to analyze more of the proteome more rapidly, several methods have emerged that avoid the use of polyacrylamide gels alltogether. The common feature of these methods is that instead of manipulating proteins, the protein mixture is converted to peptides and the peptide mixture is resolved (**Fig. 2C**). Peptides are more manageable than large proteins and as a result, a much greater number of proteins can be represented in the analysis *(12)*. For example, although a hydrophobic membrane protein is difficult to analyze, the peptides from this protein may not be *(12)*.

The first step is to convert a complex mixture of proteins into peptides by protease digestion (**Fig. 2C**). The resultant peptide mixture is then resolved by different chro-

Protein Mixture

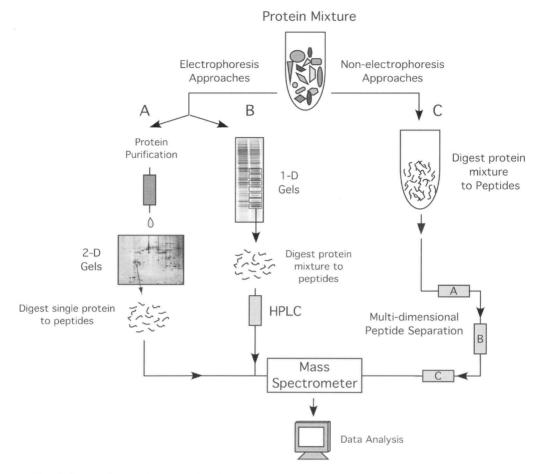

Fig. 2. Strategies for the analysis of protein mixtures for mass spectrometry. (**A**) Protein mixtures can be resolved by a combination of protein purification and two-dimensional (2-D) gel electrophoresis. Protein purification could include affinity chromatography, ion exchange, or subcellular fractionation. Proteins of interest can then be excised from the 2-D gel, digested to peptides, and analyzed by mass spectrometry. (**B**) Application of complex protein mixtures to a 1-D gel will result in only partial protein resolution. Thus, gel slices will contain multiple proteins, and the peptide mixture produced from their digest requires a further purification technique such as high-performance liquid chromatography (HPLC). (**C**) In a nonelectrophoresis approach, the entire protein mixture is digested to peptides and the peptide mixture is resolved by multi-dimensional chromatography.

matography steps and analyzed on-line by MS. An example of this strategy is multi-dimensional protein identification technology (MudPIT), developed by Yates and colleagues *(13)*. This technique has also been used to characterize protein modifications on a large scale without gels *(14)*. However, the major drawback to this approach is the loss of visual selection of relevant proteins afforded by polyacrylamide gels. Without such selection, every protein, or in this case, every peptide must be analyzed to detect for changes within a system. This results in the generation of an immense amount of irrelevant information to be analyzed by MS.

2.2. MS in Proteomics

In the last decade, the sensitivity of mass spectrometers has increased by several orders of magnitude *(15)*. It is now estimated that proteins can be identified in gels in the femtomolar range. Because MS is more sensitive, can tolerate protein mixtures, and is amenable to high-throughput operations, it has essentially replaced Edman sequencing as the protein identification tool of choice. Mass spectrometers resolve ions based on their mass-to-charge ratio in a vacuum. Therefore, proteins or peptides must be charged and dry. The two most common ways to ionize peptides for mass analysis are electrospray ionization (ESI) and matrix-assisted laser desorption/ionization (MALDI).

2.2.1. Electrospray Ionization

In ESI, a liquid form of the sample is sprayed into the mass spectrometer through a microcapillary tube under an electric field *(16)*. The spraying generates a fine mist of droplets, which evaporate into desolvated ions. Because ESI uses liquids, samples can be delivered using individual microcapillary tubes or from liquid chromatography systems in an automated fashion *(17)*.

2.2.2. Matrix-Assisted Laser Desorption/Ionization

In MALDI, the sample is mixed with a small energy-absorbing matrix molecule and spotted onto a 96-well metal plate. The plate is then systemically irradiated with a laser, promoting the formation of molecular ions *(18)*. The principle advantage to MALDI is speed, because many samples can be analyzed in an automated manner.

2.2.3. MALDI or ESI?

How does one decide whether to use MALDI or ESI to ionize samples for MS? The strategy used ultimately depends on how much sample is available. If the sample is limiting, then ESI is the best choice because it has superior sensitivity to MALDI and the mass spectrum produced can be manually interpreted to obtain *de novo* sequences. However, if the sample is not limiting, and many samples need to be analyzed, then MALDI should be used because it is more amenable to automation. Our general approach is to prepare samples in duplicate and use MALDI as the first attempt for protein identification. If a protein is not identified by MALDI, then it is submitted to ESI for a more detailed analysis. Although MALDI sample ionization is now fully automated, it should be noted that sample preparation for the mass spectrometer is not. Many of the gel spot-picking robot systems are not reliable, and samples are lost during preparation.

2.2.4. Protein Identification

Two general strategies for protein identification by MS are shown in **Fig. 3**. In each case, individual proteins are digested in-gel by trypsin to produce a mixture of specific peptides. If the peptides are analyzed by MALDI, then a characteristic mass spectrum is produced known as a peptide mass fingerprint **(Fig. 3A)**. The peptide mass fingerprint, which is essentially a list of masses for the peptides in the sample, is then searched against a database consisting of peptide mass maps produced from the theoretical digest of proteins with trypsin *(19)*. If enough peptides match in mass between the real protein digest and the theoretical one, then a protein can be identified **(Fig. 3A)**.

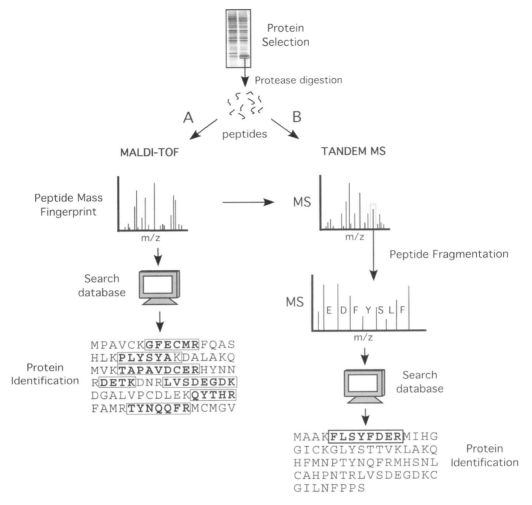

Fig. 3. Protein identification by mass spectrometry. (**A**) Protein identification by MALDI time-of-flight mass spectrometry (TOF MS). Ionization of peptides by MALDI produces information about the masses of the peptides and is known as a peptide mass fingerprint. The peptide masses obtained by MALDI are then searched against peptide mass maps generated from the theoretical digest of proteins in a database. If enough peptides match in mass, then a protein identification can be made. (**B**) In tandem MS, a peptide ion can be selected from an initial mass spectrum (obtained by either MALDI or ESI) and directed into the collision chamber in the mass spectrometer for fragmentation. Mass analysis of the peptide fragments generated (the second MS) can then be used to determine the amino acid sequence of the peptide, and this information can be used to search databases.

A much more accurate method of protein identification can be obtained by searching databases with peptide amino acid sequences. This can be obtained by a method known as tandem mass spectrometry (MS/MS; **Fig. 3B**). In this approach, a sample can be ionized by either MALDI or ESI and an initial mass spectrum obtained. From this mass spectrum, a specific peptide ion can be selected to undergo fragmentation in the collision chamber of the mass spectrometer, producing a series of peptide fragments. Under the

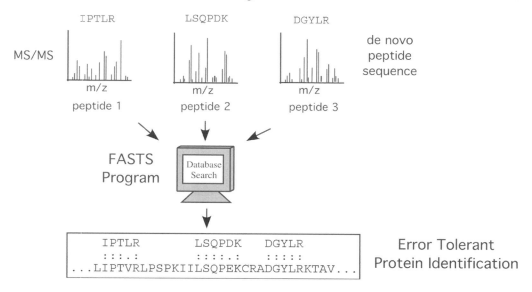

Fig. 4. Database searching by the FASTS algorithm. The FASTS program searches multiple peptide sequences (obtained by tandem mass spectrometry) against a specified database simultaneously. Because amino acid sequence is used, the search is error-tolerant and can allow for conservative amino acid substitutions, polymorphisms, or sequencing errors in the database. This method is also useful for identifying proteins from organisms whose genome has not yet been sequenced.

right conditions, peptide mass fragmentation can provide the amino acid sequence for the peptide, which is then used to search protein or DNA databases (**Fig. 3B**).

2.3. Bioinformatics

The method described in **Fig. 3** for protein identification is possible only if bioinformatics programs allow searching of databases with MS data. What is the best method for database searching? Although peptide mass fingerprint searching is the most rapid search method and is well suited for high-throughput sequencing operations, it is not as accurate as other methods that utilize peptide amino acid sequence *(20)*. In addition, peptide mass fingerprinting may fail if (1) more than one protein is in the sample, (2) the protein is modified post-translationally, or (3) the protein is from an organism whose database is not sequenced or fully annotated *(21)*.

Although difficult to obtain in some instances, peptide amino acid sequence is still the most accurate method to identify proteins. One of the most effective ways to identify proteins using amino acid sequence is with the FASTS program (**Fig. 4**). The main purpose of the FASTS program is to allow the identification of proteins from organisms whose genomes have not been sequenced or whose databases are not fully annotated *(22)*. It is able to accomplish this because the search program, by utilizing amino acid sequence and not peptide masses, is error-tolerant and can identify proteins that share homology between different species. Thus, it can accommodate sequencing errors, polymorphisms, and conservative amino acid substitutions that would invalidate a peptide mass fingerprint approach *(22)*. In the FASTS approach, *de novo* peptide sequence is obtained for several peptides by MS/MS and the sequence is searched simultaneously against a database of choice (**Fig. 4**). The FASTS program aligns peptides to sequence in the database and produces an expectation score to quantify the degree of matching *(22)*. The FASTS program is available on the world wide web at http://fasta.bioch. virginia.edu/.

3. Proteomics Applications

3.1. Protein Expression Profiling

One of the largest applications of proteomics is protein profiling. Although this is a worthy goal, it remains technically difficult. This is because it involves the comparison of protein levels on 2-D gels, which suffer from an inherent lack of reproducibility. In addition, it remains challenging to compare different 2-D gels accurately. Nonetheless, some progress has been made and new methods are emerging. One method to improve visualization and matching of proteins is known as differential in-gel electrophoresis (DIGE). In this strategy, two pools of proteins are labeled with two different fluorescent dyes, the proteins are mixed, and then the mixture is resolved by 2-D gels and proteins are visualized. This approach circumvents the need to match several different 2-D gels accurately because the protein mixtures can be separated and visualized on a single gel. A recent example of the use of DIGE technology was in the analysis of protein expression differences in esophageal carcinoma and normal epithelial cells *(23)*. Despite its drawbacks, 2-DGE currently remains the most practical method for protein expression comparison.

Protein expression profiling has also been performed without polyacrylamide gels in a procedure known as isotope-coded affinity tags (ICAT) *(24)*. The basis of this method is to use isotope tagging of proteins to allow their level of expression to be quantitated in the mass spectrometer *(25)*. However, this method requires that proteins be labeled with an isotopic reagent in vitro, and this turns out to be a major weakness of the approach *(24)*. The isotope tag is composed of a thiol-reactive group, a linker that incorporates deuterium to impart a mass change to the protein, and a biotin group to allow for purification of tagged peptides. However, because the isotopic reagent only reacts with cysteine residues in proteins, a large number of proteins are excluded from the analysis. In addition, of those proteins that do contain cysteine residues, the cysteines have to be located within appropriately spaced protease cleavage sites or the peptides will not be recovered. Indeed, recently ICAT was shown to be no more effective in the quantitation of proteins than 2-DGE *(26)*. As an alternative, a new approach was recently reported that involves the isotopic labeling of amino acids in tissue culture cells *(27)*. This strategy allows for the isotopic labeling of proteins without requiring an in vitro chemical reaction *(27)*.

3.2. Protein Modifications

One of the most important applications of proteomics is the characterization of protein modifications. It is estimated that more than 200 different types of protein modifications exist *(28)*. It was recently reported that one gene product results in an average of 10–15 spots on a 2-D gel as a result of post-translational protein modifications *(3)*.

3.2.1. Protein Phosphorylation Analysis

One of the most prevalent forms of protein modification is reversible protein phosphorylation. The cellular phosphoproteome can be defined as all the phosphoproteins in a cell *(29)*. Because protein phosphorylation is involved in signaling pathways, one aim of proteomics is to characterize changes in the phosphoproteome under different conditions. How can protein phosphorylation events be studied on a large scale? The most practical method for phosphoproteome analysis is the use of ^{32}P-labeling of proteins. Although the trend in most laboratories is to eliminate radioisotope use, none of the other methods approach the sensitivity of ^{32}P. In a typical experiment, cells are grown in the presence of ^{32}P-orthophosphate, proteins are resolved by 2-DGE, and changes in the phosphorylation state of proteins are detected by autoradiography. Proteins of interest are then excised from the gel and identified by mass spectrometry. One major limitation of this approach is that many phosphoproteins are present in low copy numbers relative to total cellular protein content and as a result are generally difficult to identify even by the most sensitive mass spectrometers. One solution to this problem is to enrich for phosphoproteins or phosphopeptides before analysis by electrophoresis.

3.2.1.1. PHOSPHOPROTEIN AND PHOSPHOPEPTIDE ENRICHMENT

A method to enrich effectively for all types of phosphoproteins from nonphosphoproteins would be of great value in the study of the phosphoproteome. Unfortunately, no such method currently exists. Although some success has been achieved in the enrichment of tyrosine-phosphorylated proteins with anti-phosphotyrosine antibodies *(30)*, much less progress has been obtained with anti-phosphoserine or threonine antibodies. Several chemical methods have been introduced to permit the enrichment of phosphoproteins from cellular lysates *(31,32)*. However, these methods rely on in vitro chemical reactions to convert phosphate groups in proteins to moieties that allow for their subsequent purification. As a result, these methods are limited by the poor yield of nonenzymatic reactions and are therefore inherently insensitive for low-copy phosphoproteins.

The enrichment of phosphopeptides can be achieved by the use of immobilized metal affinity chromatography (IMAC) *(33)*. However, in addition to phosphopeptides, IMAC is known to bind highly acidic peptides, rich in aspartate and glutamate residues. One solution to this problem is treat the peptides with methanolic HCl to convert acidic residues to their corresponding methyl esters *(34)*. This method was recently employed to identify phosphoproteins from *Saccharomyces cerevisiae* *(34)*.

3.2.1.2. MAPPING PHOSPHORYLATION SITES IN PROTEINS

Understanding protein phosphorylation in vivo requires identification of protein phosphorylation sites. However, despite advances in mass spectrometry, this can still be a formidable task. Therefore, in most cases, a variety of methods must be brought to bear on the problem including phosphopeptide enrichment, Edman degradation, and mass spectrometry. Site mapping can be challenging because most proteins are not phosphorylated to high stoichiometries in vivo. As a result, the phosphopeptides produced after digestion of a phosphoprotein represent only a small fraction of the total peptides.

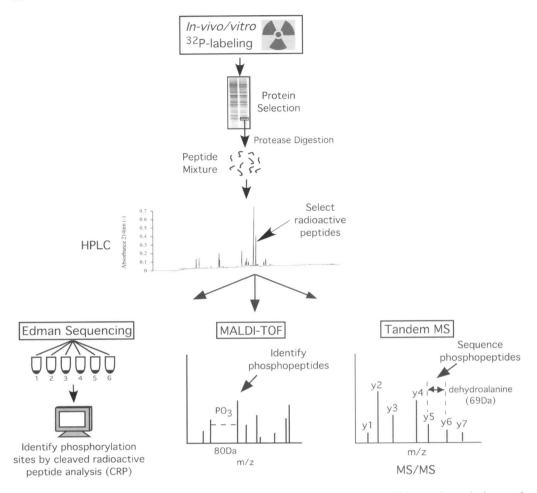

Fig. 5. Strategies for mapping phosphorylation sites in proteins by Edman degradation and mass spectrometry. Proteins that are labeled (either in vitro or in vivo) by ^{32}P can be resolved by electrophoresis and digested to peptides for site identification. In some cases, it is necessary to perform some peptide purification technique (such as HPLC) to enrich for the phosphopeptides in the sample. If phosphopeptides are sequenced by Edman degradation to monitor for ^{32}P release, the cleaved radioactive peptide program can assist in phosphorylation site identification *(35)*. Phosphopeptides can sometimes be identified using MALDI-TOF MS by detecting a mass change of 80 Daltons in a peptide corresponding to loss of a phosphate group. Once a phospho-peptide is identified, tandem MS can be used to identify which amino acid is phosphorylated.

Thus, the first task is to identify the phosphopeptides among a large excess of nonphosphopeptides.

One of the most effective methods to detect phosphopeptides is to label proteins with [32]P **(Fig. 5)**. ^{32}P-labeling permits phosphopeptides to be monitored during purification, and Edman sequencing can be used to identify phosphorylation sites by measuring ^{32}P release after each cleavage cycle. Edman sequencing can provide information about phosphopeptides that do not ionize well in the mass spectrometer **(Fig. 5)**. To aid the analysis of phosphopeptides by Edman sequencing, we developed a bioinformatics program called the Cleaved Radioactive Peptide (CRP) program *(35)*. This program requires that the protein sequence be known and predicts how many Edman cycles are required to obtain 100% coverage of all potential phosphorylation sites *(35)*.

In mass spectrometry, MALDI can be used to detect phosphopeptides from a mixture of nonphosphopeptides by scanning the time-of-flight spectrum for peptide ions that differ by multiples of 80 Daltons (80 Daltons per phosphate group) *(36)* (**Fig. 5**). Phosphopeptides can also be identified using electrospray and precursor ion scanning *(37)*. Once a peptide is identified, it can be sequenced by MS/MS to determine the site of phosphorylation (**Fig. 5**) *(36)*.

3.3. Protein Complexes

Proteomics can also be used to study the composition of protein complexes. Although the yeast two-hybrid method has been used extensively for this purpose, many of the results obtained by yeast two-hybrid have been questioned. A large-scale analysis of protein complexes in *S. cerevisiae* by MS showed a threefold higher success rate than the two-hybrid method at identifying known protein complexes *(38)*.

We believe that the best application of proteomics to study protein complexes is when the proper controls are conducted. For example, if a "pull-down" experiment is performed, only those proteins that are absent in the control should be sequenced. In some cases, further stringency is required and only those proteins that change compared with the control in response to some biological perturbation are sequenced.

3.4. Novel Proteomic Approaches: Proteome Mining

Proteome mining is a method to identify simultaneously the protein targets of drugs and identify drugs that interact with specific proteins. The basic principle of proteome mining is based on the assumption that all drug-like molecules selectively compete with a natural cellular ligand for a binding site on a protein target. Thus, if proteins can be immobilized by binding to their natural cellular ligand, then this can serve as a scaffold to measure protein–drug interactions. In the prototypical example, we isolated the purine binding proteome of cells using an affinity matrix in which the ligand, ATP, was immobilized via its γ-phosphate group *(9)*. This matrix is capable of binding protein and nonprotein kinases, dehydrogenases, and other miscellaneous ATP-utilizing enzymes *(9)*. Having isolated a subproteome, drugs are applied to the matrix and allowed to interact with the bound proteins. If a drug is able to compete with the natural ligand of the protein for binding, the protein is eluted from the matrix and recovered. Eluted proteins are then resolved by 1-DGE and identified by MS. Identification of the eluted proteins provides information about drug specificity. Thus, proteome mining can be used to identify the protein targets of drugs in various subproteomes. This approach was recently used to identify protein targets of the quinoline antimalarial drugs in the human purine binding proteome *(9)*.

4. Conclusions

Despite the difficulties in the large-scale analysis of proteins, proteomics is here to stay. This is because studying proteins makes sense, as proteins carry out the function of genes. Malfunctioning proteins can result in human diseases, and proteins are the targets of drugs. Currently, there are efforts in proteomics toward the development of high-throughput protein sequencing strategies. However, sequencing of the human proteome is not analogous to sequencing of the human genome. Considerably more effort is required to characterize a single protein than a single gene; as a result, it can often take 6–12 months to confirm that a single protein identified in a proteomics screen is a valid target. Thus, the rate-limiting step is not in the protein identification step, but rather in the characterization of proteins recovered. For these reasons, we believe that proteomics experiments should be designed to answer specific biological questions and only those proteins directly relevant to the analysis should be identified. By only sequencing those proteins involved in a specific response, protein identification can have a meaningful outcome, and this directs all subsequent investigation.

References

1. Gygi, S. P., Rochon, Y., Franza, B. R., and Aebersold, R. (1999) Correlation between protein and mRNA abundance in yeast. *Mol. Cell. Biol.* **19,** 1720–1730.
2. Anderson, L. and Seilhamer, J. (1997) A comparison of selected mRNA and protein abundances in human liver. *Electrophoresis* **18,** 533–537.
3. Fountoulakis, M., Berndt, P., Langen, H., and Suter, L. (2002) The rat liver mitochondrial proteins. *Electrophoresis* **23,** 311–328.
4. Graves, P. R. and Haystead, T. A. (2002) Molecular biologist's guide to proteomics. *Microbiol. Mol. Biol. Rev.* **66,** 39–63.
5. Kettman, J. R., Coleclough, C., Frey, J. R., and Lefkovits, I. (2002) Clonal proteomics: one gene-family of proteins. *Proteomics* **2,** 624–631.
6. Gygi, S. P., Corthals, G. L., Zhang, Y., Rochon, Y., and Aebersold, R. (2000) Evaluation of two-dimensional gel electrophoresis-based proteome analysis technology. *Proc. Natl. Acad. Sci. USA* **97,** 9390–9395.
7. Anderson, N. G. and Anderson, N. L. (1996) Twenty years of two-dimensional electrophoresis: past, present and future. *Electrophoresis* **17,** 443–453.
8. MacDonald, J. A., Borman, M. A., Muranyi, A., Somlyo, A. V., Hartshorne, D. J., and Haystead, T. A. (2001) Identification of the endogenous smooth muscle myosin phosphatase-associated kinase. *Proc. Natl. Acad. Sci. USA* **98,** 2419–2424.
9. Graves, P. R., Kwiek, J. J., Fadden, P., et al. (2002) Discovery of novel targets of quinoline drugs in the human purine binding proteome. *Mol. Pharmacol.* **62,** 1364–1372.
10. Cordwell, S. J., Nouwens, A. S., Verrills, N. M., Basseal, D. J., and Walsh, B. J. (2000) Subproteomics based upon protein cellular location and relative solubilities in conjunction with composite two-dimensional electrophoresis gels. *Electrophoresis* **21,** 1094–1103.
11. Simpson, R. J., Connolly, L. M., Eddes, J. S., Pereira, J. J., Moritz, R. L., and Reid, G. E. (2000) Proteomic analysis of the human colon carcinoma cell line (LIM 1215): development of a membrane protein database. *Electrophoresis* **21,** 1707–1732.
12. Goodlett, D. R. and Yi, E. C. (2002) Proteomics without polyacrylamide: qualitative and quantitative uses of tandem mass spectrometry in proteome analysis. *Funct. Integr. Genomics* **2,** 138–153.
13. Washburn, M. P., Wolters, D., and Yates, J. R. 3rd. (2001) Large-scale analysis of the yeast proteome by multidimensional protein identification technology. *Nat. Biotechnol.* **19,** 242–247.
14. MacCoss, M. J., McDonald, W. H., Saraf, A., et al. (2002) Shotgun identification of protein modifications from protein complexes and lens tissue. *Proc. Natl. Acad. Sci. USA* **99,** 7900–7905.
15. Pandey, A. and Mann, M. (2000) Proteomics to study genes and genomes. *Nature* **405,** 837–846.
16. Fenn, J. B., Mann, M., Meng, C. K., Wong, S. F., and Whitehouse, C. M. (1989) Electrospray ionization for mass spectrometry of large biomolecules. *Science* **246,** 64–71.

17. Davis, M. T. and Lee, T. D. (1998) Rapid protein identification using a microscale electrospray LC/MS system on an ion trap mass spectrometer. *J. Am. Soc. Mass Spectrom.* **9,** 194–201.

18. Karas, M. and Hillenkamp, F., Laser desorption ionization of proteins with molecular masses exceeding 10,000 daltons. *Anal. Chem.* **60,** 2299–2301.

19. Eng, J. K., McCormack, A. L., and Yates, J. R. (1994) An approach to correlate tandem mass-spectral data of peptides with amino-acid-sequences in a protein database. *J. Am. Soc. Mass Spectrom.* **5,** 976–989.

20. Wattenberg, A., Organ, A. J., Schneider, K., Tyldesley, R., Bordoli, R., and Bateman, R. H. (2002) Sequence dependent fragmentation of peptides generated by MALDI quadrupole time-of-flight (MALDI Q-TOF) mass spectrometry and its implications for protein identification. *J. Am. Soc. Mass Spectrom.* **13,** 772–783.

21. McDonald, W. H. and Yates, J. R. 3rd. (2000) Proteomic tools for cell biology. *Traffic* **1,** 747–754.

22. Mackey, A. J., Haystead, T. A. J., and Pearson, W. R. (2002) Getting more from less: algorithms for rapid protein identification with multiple short peptide sequences. *Mol. Cell. Proteomics* **1,** 139–147.

23. Zhou, G., Li, H., DeCamp, D., et al. (2002) 2D differential in-gel electrophoresis for the identification of esophageal scans cell cancer-specific protein markers. *Mol. Cell. Proteomics* **1,** 117–124.

24. Gygi, S. P., Rist, B., Gerber, S. A., Turecek, F., Gelb, M. H., and Aebersold, R. (1999) Quantitative analysis of complex protein mixtures using isotope-coded affinity tags. *Nat. Biotechnol.* **17,** 994–999.

25. Conrads, T. P., Issaq, H. J., and Veenstra, T. D. (2002) New tools for quantitative phospho-proteome analysis. *Biochem. Biophys. Res. Commun.* **290,** 885–890.

26. Patton, W., Schulenberg, B., and Steinberg, T. (2002) Two-dimensional gel electrophoresis; better than a poke in the ICAT? *Curr. Opin. Biotechnol.* **13,** 321.

27. Ong, S. E., Blagoev, B., Kratchmarova, I., et al. (2002) Stable isotope labeling by amino acids in cell culture, SILAC, as a simple and accurate approach to expression proteomics. *Mol. Cell. Proteomics* **1,** 376–386.

28. Krishna, R. G. and Wold, F. (1993) Post-translational modification of proteins. *Adv. Enzymol. Relat. Areas Mol. Biol.* **67,** 265–298.

29. Mann, M., Ong, S. E., Gronborg, M., Steen, H., Jensen, O. N., and Pandey, A. (2002) Analysis of protein phosphorylation using mass spectrometry: deciphering the phospho-proteome. *Trends Biotechnol.* **20,** 261–268.

30. Maguire, P. B., Wynne, K. J., Harney, D. F., O'Donoghue, N. M., Stephens, G., and Fitzgerald, D. J. (2002) Identification of the phosphotyrosine proteome from thrombin activated platelets. *Proteomics* **2,** 642–648.

31. Oda, Y., Nagasu, T., and Chait, B. T. (2001) Enrichment analysis of phosphorylated proteins as a tool for probing the phosphoproteome. *Nat. Biotechnol.* **19,** 379–382.

32. Zhou, H., Watts, J. D., and Aebersold, R. (2001) A systematic approach to the analysis of protein phosphorylation. *Nat. Biotechnol.* **19,** 375–378.

33. Porath, J. (1992) Immobilized metal ion affinity chromatography. *Protein Expr. Purif.* **3,** 263–281.

34. Ficarro, S. B., McCleland, M. L., Stukenberg, P. T., et al. (2002) Phosphoproteome analysis by mass spectrometry and its application to *Saccharomyces cerevisiae. Nat. Biotechnol.* **20,** 301–305.

35. MacDonald, J. A., Mackey, A. J., Pearson, W. R., and Haystead, T. A. (2002) A strategy for the rapid identification of phosphorylation sites in the phosphoproteome. *Mol. Cell. Proteomics* **1,** 314–322.

36. Bennett, K. L., Stensballe, A., Podtelejnikov, A. V., Moniatte, M., and Jensen, O. N. (2002) Phosphopeptide detection and sequencing by matrix-assisted laser desorption/ionization quadrupole time-of-flight tandem mass spectrometry. *J. Mass Spectrom.* **37**, 179–190.

37. Steen, H., Kuster, B., Fernandez, M., Pandey, A., and Mann, M. (2001) Detection of tyrosine phosphorylated peptides by precursor ion scanning quadrupole TOF mass spectrometry in positive ion mode. *Anal. Chem.* **73**, 1440–1448.

38. Ho, Y., Gruhler, A., Heilbut, A., et al. Systematic identification of protein complexes in *Saccharomyces cerevisiae* by mass spectrometry. *Nature* **415**, 180–183.

2

Protein Identification from 2-D Gels Using In Vitro Transcription Translation Products

Nathalie Norais, Renzo Nogarotto, Emilia Tiziana Iacobini, Ignazio Garaguso, Renata Grifantini, Giuliano Galli, and Guido Grandi

1. Introduction

The genomics revolution offers the opportunity to describe biological processes on the basis of global and quantitative gene expression patterns from cells or organisms representing different states. Although DNA microarrays are extremely powerful for measuring gene expression at the mRNA level (*1*), the nonpredictive correlation between mRNA and protein levels (*2,3*) and the discovery of post-translational mechanisms that either modify the structure and function of proteins or alter their half-life and rate of synthesis (*4*) indicate that direct measurement of protein expression is essential for a proper analysis of the biological processes.

Global analysis of proteins from cells and tissues is termed *proteomics*. Although a variety of alternative procedures are under study (*5,6*), at present the most popular approach for proteomics analysis requires the previous knowledge of the genome sequence of the organism under investigation and is based on the combination of two-dimensional gel electrophoresis (2-DGE) and mass spectrometry (*6*). According to this approach, proteins are separated by 2-DGE, stained, digested *in situ* with trypsin or other proteolytic enzymes, and finally subjected to matrix-assisted laser desorption/ionization time-of-flight (MALDI-TOF) mass spectrometry. The MALDI-TOF analysis provides peptide mass fingerprints, which lead to protein identification once they are compared with the theoretical *in silico* fingerprints generated from the available genome sequence. Usually, 80–85% of the analyzed spots give a mass fingerprint, which is in most cases sufficient for protein identification. A limited number of spots (approx 5%) further requires tandem mass spectrometry (MS/MS) analysis for unambiguous characterization.

The major drawback of this approach is that it requires expensive and sophisticated instruments, which need to be operated by well-trained and specialized scientists. In addition, the method presents some limitations in sensitivity, not because of mass spectrometers (which could analyze samples in the low fmole range), but to sample preparation procedures, which are usually inefficient, making the analysis feasible only when protein quantities greater than 0.1–0.2 pmol are available (*7*). Finally, although the approach is relatively rapid, especially when robotic stations are utilized to handle spot picking and processing, full characterization of a complex protein mixture may require a few weeks of intensive work.

We have recently described an alternative method for the characterization of proteomes, in particular bacterial proteomes, which may offer some advantages over the current proteomic approaches *(8)*. The method is based on in vitro transcription and translation of genes from the organism under investigation to obtain radioactive translation products. The proteins produced in vitro are then separated on 2-D gels, and the corresponding autoradiographs are superimposed by computer-assisted image acquisition and processing on the 2-D gel of the total proteins from the organism under investigation (sample gel). The matching of the radioactive products with spots of the sample gel allows immediate protein identification, demonstrating that the genes used in the transcription and translation experiments are in fact expressed in vivo.

Here, the method is applied to test rapidly the conservation of two membrane proteins among different clinical isolates of *Neisseria meningitidis* group B.

2. Materials and Methods

2.1. Reagents

1. Immobiline pH gradients (IPG) and IPG buffer, pH 3–10, nonlinear (NL) (Amersham Pharmacia Biotech, Piscataway, NJ).
2. Urea (enzyme grade, Life Technologies, Paisley, Scotland).
3. 3-[(3-Cholamidopropyl) dimethylammonio]-1-propanesulfonate (CHAPS), dithiothreitol (DTT), thiourea, L-glutamine, cocarboxylase, trypsin inhibitor, trypsinogen, carbonic anhydrase, myoglobin, and diethyl pyrocarbonate (DEPC) (Sigma, St. Louis, MO).
4. Acrylamide and agarose (Bio-Rad, Hercules, CA).
5. Glucose and benzonase (Merck, Darmstadt, Germany).
6. Sodium thiosulfate and tributyl phosphine (TBP) (Fluka, Buchs, Switzerland).
7. Amidosulfobetaine-14 (ASB14) *(9)* (a generous gift from Prof. C. Scolastico, University of Milan, Italy).
8. All other reagents were analytical grade.

2.2. Preparation of Protein Standards

1. Selected open reading frames (ORFs) from *Chlamydia pneumoniae* CWL029, *Streptococcus agalactiae* 2603 V/R, and *Neisseria meningitidis* serogroup B strain MC58 were cloned into expression vectors to produce histidine tag and glutathione-S-transferase (GST) fused proteins *(10)*.
2. Expression and purification of the recombinant proteins were as described by Montigiani et al. *(10)*, except that an ultimate step of purification by preparative electrophoresis was added. Gel pore size, length, and running conditions were selected according to the instrument manual (491 model Prep Cell, Bio-Rad, Hercules, CA).
3. Purified proteins were precipitated with 20% trichloroacetic acid, and the pellet was washed with acetone and redissolved in the reswelling solution [7 *M* urea, 2 *M* thiourea, 2% (w/v) CHAPS, 2% (w/v) ASB14, 2% (v/v) IPG buffer, pH 3–10, NL, 2 m*M* TBP, 65 m*M* DTT] to a final concentration of 3 mg/mL.

2.3. Preparation of Template for Transcription/Translation Reactions

1. The genes of interest were polymerase chain reaction (PCR)-amplified from *Neisseria meningitidis* strain MC58 chromosomal DNA using Pwo DNA polymerase (Roche Diagnostics, Mannheim, Germany) and appropriate synthetic forward and reverse primers carrying the *Nde*I and *Xho*I (or *Hind*III) restriction sites, respectively.

2. The amplified fragments were digested with either *NdeI/XhoI* or *NdeI/Hind*III enzyme mixtures and ligated to plasmid pET-21b+ (Novagen, Madison, WI) previously digested with the same enzymes. In pET-21b+ a stop codon was introduced upstream from the nucleotide sequence coding for the hexa-histidine tag to avoid the addition of any extra amino acid to the protein of interest.
3. Plasmid preparations were made using the Qiagen kit (Qiagen, Hilden, Germany).

2.4. In Vitro Transcription/Translation Reactions (TTRs)

1. TTR was carried out using circular plasmids as template (*11,12*).
2. S30 extracts were prepared from *Escherichia coli* BL21 containing endogenous T7 RNA polymerase (*13*) as described by Pratt (*14*).
3. Aliquots (250 µL) of S30 preparations (12–15 mg/mL, final protein concentration) were stored at –80°C.
4. In vitro coupled transcription-translation was carried out as described (*14*), with minor modifications. The reaction was performed in a 10-µL final volume containing 20–25 µg/mL of DNA, 0.42 mM ^{14}C-labeled L-leucine and L-lysine (11.7 and 12. 2 Gbq/mmol, respectively; Amersham Pharmacia Biotech), 0.42 mM of each of the other 18 unlabeled amino acids, and 34% (v/v) of S30 extract. All other reagents needed for the reaction were as indicated by Pratt (*14*).
5. The reaction was allowed to proceed for 3 h at 37°C.
6. Prior to electrophoretic analysis, the proteins were precipitated with cold ethanol (90%, v/v).

2.5. 2-D Electrophoresis and Autoradiography

1. *Neisseria meningitidis* membrane proteins were prepared as described by Molloy et al. (*15*). Proteins (125 µg) and TTRs (up to 20 µg) were mixed with 1 µg of each reference protein **(Table 1)** and brought to a final volume of 125 µL with reswelling solution.
2. The proteins were absorbed overnight onto an Immobiline DryStrip (7 cm, NL pH 3–10 gradient) using the Immobiline Dry-Strip Reswelling Tray (Amersham Pharmacia Biotech).
3. Proteins were then separated by 2-D electrophoresis. The first dimension was run using an IPGphor Isoelectric Focusing Unit (Amersham Pharmacia Biotech), applying sequentially 150 V for 35 min, 500 V for 35 min, 1000 V for 30 min, 2600 V for 10 min, 3500 V for 15 min, 4200 V for 15 min, and then 5000 V to reach 10 kVh. For the second dimension, the strips were equilibrated as described (*16*), and the proteins were separated on linear 9–16% polyacrylamide sodium dodecyl sulfate (SDS) gels (1.5 mm thick, 6 cm high) using the Mini Protean II Cell from Bio-Rad. Gels were stained with colloidal Coomassie (*17*).
4. Prior to autoradiography, gels were soaked with 125 mM salicylic acid in 40% methanol and dried on filter paper (Bio-Rad). Gels were autoradiographed overnight on BioMax MR-2 film (Eastman Kodak, Rochester, NY).

2.6. Mass Spectrometry Analysis

1. Protein spots were excised from the gel, washed with Milli-Q water and acetonitrile, and dried in a SpeedVac apparatus (Savant, Holbrook, NY) (*18*).
2. Gel pieces were rehydrated by adding 7–10 µL of 50 mM ammonium bicarbonate and 5 mM $CaCl_2$ containing 0.012 µg/µL sequencing grade trypsin (Roche Diagnostic, Mannheim, Germany).
3. Tryptic digestions were carried out at 37°C for 18 h, and peptides were eluted out by 30-min sonication in a sonicator bath after addition of 50% acetonitrile, 5% TFA (50 µL).
4. Peptide extraction was repeated once, extract solutions were pooled, and the volume was reduced to 10 µL in a SpeedVac apparatus.

Table 1
Standard Proteins Used for Gel Alignment

Name	Origin		MW (Kd)	pI
CPn0195	*Chlamydia pneumoniae*	recombinant GST fusion	84,0	6,78
CPn0197	*Chlamydia pneumoniae*	recombinant GST fusion	75,0	6,29
CPn0278	*Chlamydia pneumoniae*	recombinant GST fusion	55,0	6,67
CPn0324	*Chlamydia pneumoniae*	recombinant GST fusion	70,1	5,22
CPn0336	*Chlamydia pneumoniae*	recombinant GST fusion	60,2	6,21
CPn0420	*Chlamydia pneumoniae*	recombinant GST fusion	62,8	5,65
CPn1064	*Chlamydia pneumoniae*	recombinant GST fusion	32,0	5,00
GST	*Schistosoma japonicum*		27,4	6,40
CPn0278	*Chlamydia pneumoniae*	recombinant His Tag	29,6	7,01
CPn0764	*Chlamydia pneumoniae*	recombinant His Tag	47,9	7,28
CPn0399	*Chlamydia pneumoniae*	recombinant His Tag	23,3	8,07
CPn1034	*Chlamydia pneumoniae*	recombinant His Tag	29,0	6,05
CPn0113	*Chlamydia pneumoniae*	recombinant His Tag	41,0	5,92
CPn1067	*Chlamydia pneumoniae*	recombinant His Tag	21,8	6,26
CPn0321	*Chlamydia pneumoniae*	recombinant His Tag	40,9	5,57
CPn0273	*Chlamydia pneumoniae*	recombinant His Tag	22,9	6,63
CPn0297	*Chlamydia pneumoniae*	recombinant His Tag	34,7	5,67
CPn0624	*Chlamydia pneumoniae*	recombinant His Tag	37,7	6,6
SAg0681	*Group B Streptococcus*	recombinant GST fusion	65,7	8,62
SAg0079	*Group B Streptococcus*	recombinant His Tag	27,5	5,41
SAg0628	*Group B Streptococcus*	recombinant His Tag	50,8	4,89
NMB1119	*Neisseria meningitidis*	recombinant His Tag	21,8	4,95

GST, glutathione-S-transferase.

5. Samples were automatically prepared for mass spectrometry analysis using the MAP II system (Bruker, Bremen, Germany). The instrument was programmed to perform sample desalting with ZIP-TIP (C18, Millipore, Bedford, MA).

6. Peptides were eluted from ZIP-TIPs with a saturated solution of α-cyano-4-hydroxy-cinnamic acid in 50% acetonitrile/0,1% trifluoro-acetic acid and directly loaded onto the SCOUT 381 multiprobe (Bruker, Bremen, Germany).

7. The samples were then allowed to air dry at room temperature.

8. Spectra were acquired on a Bruker Biflex II MALDI-TOF (Bremen, Germany) equipped with the SCOUT 381 multiprobe ion source in a positive-ion reflector mode. The acceleration voltage was set to 19 kV, and the reflector voltage was set to 20 kV. Typically, about 100 laser shots were averaged per spectrum from a 337-nm N2 laser.

9. Spectra were calibrated externally using a combination of standards [angiotensin II (1046.54 Daltons), substance P (1347.74 Daltons), bombesin (1619.82 Daltons), and clipped human ACTH18-39 (2465.20 Daltons) located in spots adjacent to the samples.

10. Peptides were selected in the mass range of 700–3000 Daltons.

11. Monoisotopic peak values were used for Mascot search on private databases. All searches were performed using a 200-ppm constraint window.

2.7. Image Acquisition and Analysis

1. Autoradiographs and Coomassie Blue-stained gels were scanned with a Personal Densitometer SI (Amersham Pharmacia) at 8 bits and 50 µm per pixel. Prior to image acqui-

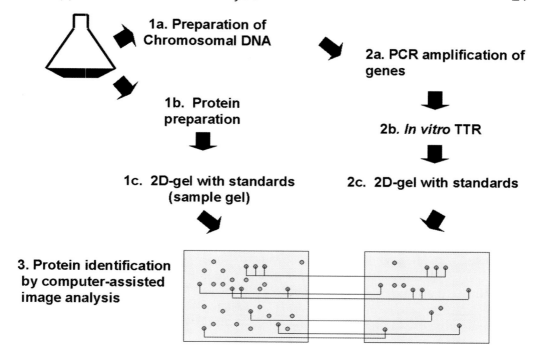

1a. Preparation of Chromosomal DNA

2a. PCR amplification of genes

1b. Protein preparation

2b. *In vitro* TTR

1c. 2D-gel with standards (sample gel)

2c. 2D-gel with standards

3. Protein identification by computer-assisted image analysis

Fig. 1. Schematic representation of the procedure for protein spot identification by in vitro transcription/translation reactions. A bacterial culture is used for chromosomal DNA and total protein preparation (steps 1a. and 1b.). Proteins are mixed with the set of protein standards and then separated on a 2-D gel (sample gel, step 2c.) whereas chromosomal DNA is used for gene amplification (step 2a.). The amplified genes are transcribed and translated in vitro in the presence of ^{14}C-labeled amino acids (step 2b.). TTRs are pooled, mixed with the set of protein standards, and then separated on 2-D gels (step 2c.). Finally, the 2-D gels are autoradiographed and superimposed on the sample gel using computer-assisted image analysis (step 3.). Spot matching allows the immediate identification of protein spots on the sample gel.

sition of autoradiographs, the positions of the reference proteins were manually reported on the films.
2. Image analysis was performed using Image Master 2D Elite software, version 3.10 (Amersham Pharmacia Biotech).

3. Results

3.1. Description of Strategy

The technology is represented schematically in **Fig. 1**. Total proteins and chromosomal DNA are prepared from the organism under investigation. Total proteins are mixed with selected standard proteins and resolved by 2-DGE to generate the "sample gel," whereas chromosomal DNA is used as template for gene amplification. The amplified genes are cloned in appropriate expression vectors for in vitro protein synthesis. TTRs are carried out in the presence of ^{14}C-labeled amino acids and resolved by 2-DGE in the presence of the same standard proteins used for the sample gel. The 2-D gels are first stained for standards localization and then dried and autoradiographed for TTR visualization. Finally, autoradiographs are superimposed *in silico* on the sample gel using the reference standards as common coordinates. Should the proteins encoded by the genes used in TTRs be present in the sample gel, they will match the TTR products, thus allowing their immediate identification.

3.2 Application of the Strategy

The potentiality of the procedure is illustrated here by analyzing two membrane proteins of *Neisseria meningitidis* group B (MenB), a bacterial pathogen under intense investigation in our laboratories for the development of an effective vaccine (19,20). In particular, we were interested in (1) determining whether the outer membrane protein OMP4 (a product of the NMB gene) and the chaperonin heat shock protein Hsp60 (a product of the NMB1972 gene) were present in the membrane protein preparation of the MC58 strain (19); and (2) the extent of conservation of these two proteins in different MenB clinical isolates.

To answer the first question, 125 μg of a membrane protein preparation of MenB cells were mixed with 20 standard proteins (*see* Materials and Methods and **Table 1**), resolved onto a 2-D gel, and subsequently stained with Coomassie Blue (**Fig. 2A**). The addition of the standard proteins to both the sample proteins and TTRs (*see* below) is a fundamental step of the entire procedure in that the standard proteins are subsequently utilized to superimpose the sample gel properly on the TTR gels. Standard proteins should be selected to have an even distribution throughout the gel. This in fact allows the software algorithm to compensate for the local gel distortions with high accuracy.

In a parallel experiment, plasmid pET-0382 and pET-1972 carrying the NMB0382 and NMB1972 genes, respectively, were used for in vitro transcription and translation reactions. The TTR products were then mixed with the standard proteins and resolved on 2-D gel. Finally the gel was autoradiographed. As shown in **Fig. 2B**, the main com-

---

Fig. 2. *(see facing page)* Matching analysis of TTR 2-D gel spots with sample 2-D gel spots. **(A)** Sample gel. A membrane protein preparation of MenB (125 μg) was loaded onto a 2-D gel in the presence of the 22 standard proteins (1 μg each, spots assigned with green arrows labeled ST) and stained with Coomassie Blue. **(B)** Autoradiograph of a TTR 2-D gel. Radiolabeled NMB 1972 and NMB 0382 were synthesized in vitro. Aliquots (20 μg of total proteins) of each reaction were mixed together, combined with the 22 standard proteins (1 μg each), and loaded onto a 2-D gel. The gel was first stained with Coomassie Blue to visualize the standard proteins and then dried and autoradiographed. The TTR products are circled in red and boxed on the autoradiograph. The standard position (solid blue spots) was determined by superimposing the autoradiograph on the corresponding Coomassie Blue-stained gel. **(C)** Superimposition of the TTR autoradiograph (B) on the sample gel (A) using Image Master 2D Elite, 3.1 software. The software creates the combined gel shown in C in which the TTR gel is superimposed on the sample gel using the reference proteins (solid blue spots). The combined gel is then scanned for TTR spots located within a fixed area (vector box) from any spot of the sample gel. If a TTR spot is within the selected vector box range from one protein of the sample gel, the two spots are automatically matched (boxed spots). **(D)** Details of the matching areas are highlighted in C. The panel shows the extent of matching accuracy between TTR spots (circled in red) and the corresponding spots on the sample gels (circled in blue). The "comet-like" spots of the NMB 1972 gene product synthesized in vitro match spot numbers 56–61 of the sample gel, identified as NMB 1972 protein by mass spectrometry, within a vector box size of 12. Similarly, spots 206 and 209, identified as two different NMB 0382 isoforms by mass spectrometry, match the two autoradiographic spots obtained by TTR of the NMB 0382 gene, within a vector box size of 13. ST, standard.

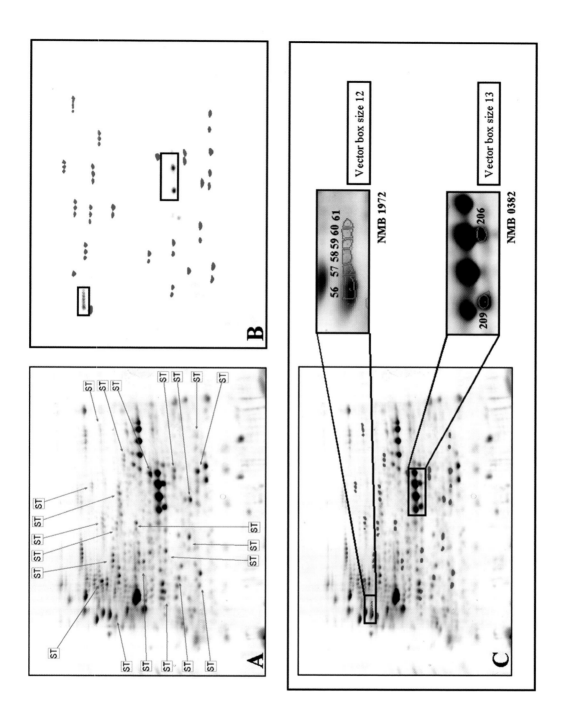

ponents of the NMB 1972 and NMB 0382 gene products were resolved on the autoradiograph in "comet-like" spots and two spots, respectively, corresponding to spots having different pI but identical molecular weight. Only minor intermediates, most likely represented by prematurely interrupted TTR products, were visible on the autoradiograph.

To determine whether the proteins were present in the MenB MC58 membrane preparation, the autoradiograph was superimposed *in silico* on the sample gel using the Image Master 2D Elite 3.1 program. Using the reference proteins, the software first creates a combined gel in which the TTR gel is superimposed on the sample gel and then scans the combined gel for TTR spots located within a fixed area (vector box) from any spot of the sample gel. If a TTR spot is within the selected vector box range from one protein of the sample gel, the two spots are automatically matched. As shown in **Fig. 2C**, the two NMB 0382 TTR spots matched two spots on the sample gel when a vector box size of 13 was selected. Similarly, the NMB 1972 "comet-like" TTR spots matched the corresponding protein spots 56–61 using a vector box size of 12. The details of the matching areas are given in **Fig. 2C**, showing the extent of accuracy of the matches between the TTR spots (circled in red) and the spots of the sample gel (circled in blue).

Having demonstrated that both OMP4 and Hsp60 are present in the membrane protein preparation of the MC58 strain, we then asked whether the two proteins were also conserved in six different MenB isolates. To this aim, their membrane proteins were resolved on 2-D gels (**Fig. 3A**), and the gels were matched with the TTR autoradiograph. The matching analysis revealed that the two proteins were highly conserved in all six strains. **Figure 3B** gives a detailed analysis of the matching experiment with the MenB 1000 strain 2-D map. In this strain, the matching spots were subjected to mass spectrometry analysis *(8)*, which confirmed the identity of the spots (data not shown).

4. Discussion

The growing importance of proteomics analysis in the understanding of biological processes is pushing research toward the identification of new, more efficient methods for protein identification. Although it is well recognized that new concepts in protein separation must be applied for complete proteome characterization *(21)*, the methods based on 2-DGE will remain the first choice until alternative, innovative approaches have reached the mature state of general applicability. Once a pool of proteins have been separated on a 2-D gel, each protein spot must be processed and eventually identified. Although spot processing can be carried out either *in situ*, after spot picking, or on a membrane, after electroblotting *(22)*, the method for spot identification is mass spectrometry. However, as already pointed out, the approach is hampered by problems such as high costs, complexity of the analysis and relatively low sensitivity. Here, we propose an alternative approach, which may offer some advantages over the current methods, at least for some specific applications. The method is based on 2-D gel separation

---

Fig. 3. *(see facing page)* Matching analysis of TTR 2-D gel spots with 2-D maps of six different MenB isolates. **(A)** Membrane preparations of six MenB strains *(20)* were separated on 2-D gels, and the gels were matched with the TTR autoradiograph. Red boxes indicate the spots matching the OMP4 TTR product, whereas blue boxes highlight the Hsp60 matches. **(B)** Detailed matching analysis of the membrane preparation of MenB 1000 strains. For details, *see* **Fig. 2**.

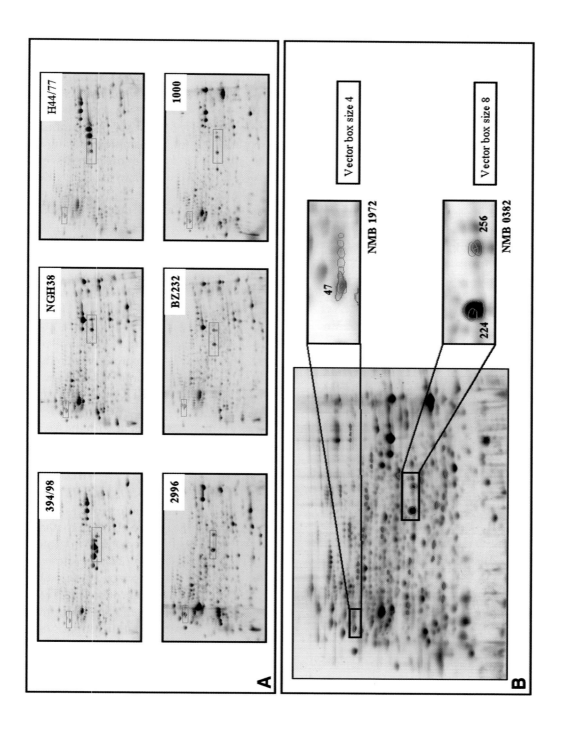

of radiolabeled synthetic proteins derived from TTRs of PCR-amplified genes. The gel is autoradiographed and the autoradiograph is superimposed on the sample gel whose protein spots have to be identified. Matching between spots from the autoradiograph and the sample gel immediately allows protein identification. The method is sufficiently accurate and sensitive and, when compared with other procedures, relatively simple, rapid, and inexpensive.

As far as accuracy is concerned, we have previously shown that the method is highly accurate in terms of its capacity to establish with a high degree of confidence whether a given protein is present in the protein mixture under investigation. In fact, mass spectrometry analysis of the sample gel spots matching the TTR spots revealed that all matching spots corresponded to the expected proteins *(8)*. The most serious problem, which weakens the degree of accuracy of this procedure, is the resolution power of 2-D electrophoresis. Proteins having similar molecular weight and isoelectric point are often hardly resolved on 2-D gels, and protein comigration may lead to misinterpretations during the comparative analysis of TTR gels with sample gels. Although in our experience comigration events are relatively rare (in the course of a MALDI-TOF analysis of bacterial protein 2-D maps obtained on mini-gels, we found that only 3 of 254 spots were constituted by more than one protein species) and should not affect the interpretation of proteins resolved in more than one isoform (the probability that two isoforms of the same protein comigrate with two protein species is expected to be very low), the possibility that TTR products superimpose on unrelated proteins cannot be ruled out.

With regard to the sensitivity of the proposed procedure, in general, the lowest sensitivity limit of all proteomic approaches based on 2-D gel separation is the sensitivity limit of the staining system. In other words, for a protein to be characterized, it must be visible on the gel. However, although mass spectrometry-based methods often fail to assign a name unambiguously to a poorly visible spot, in the case of our procedure the sensitivity of the staining system is truly the sensitivity limit of the procedure. The more sensitive the staining system used for protein detection, the more sensitive the procedure proposed here is.

Finally, one of the advantages of the procedure proposed here is that it can be utilized by any molecular biologist with relatively simple instrumentation. Gene amplification and TTRs can be completed in 1 wk, and, if a simple robotic station is available, up to 100 reactions can be carried out simultaneously. If appropriate TTR pools are run on the same gel, 50–100 TTRs can be analyzed in 2 wk by a single scientist. Finally, gel autoradiography usually takes few hours, and the available software packages for image acquisition and analysis allow for processing of several gels per day *(7)*.

The most critical aspect of the procedure is to guarantee a consistently accurate superimposition of the sample gel on the TTR gels. This can only be achieved if good sets of standard proteins and advanced software for image analysis are available. Standard proteins are indispensable, representing the hallmark of the software used for gel normalization. Because gel distortions and aberrant migrations of proteins have a local and unpredictable behavior, it is fundamental for the subsequent analysis to use several standard proteins, which are evenly distributed in the gel.

In addition to standard proteins, the availability of proper software packages for gel analysis is also critical. Although software distributors are working on more and more sophisticated packages, we found some of the products already available on the market to be more than adequate. In this study, we have utilized a software from Pharmacia Biotech, but a preliminary analysis with Melanine *(22)* has provided comparable satisfactory results.

We consider this procedure particularly amenable for bacterial proteomics analysis in consideration of the fact that post-translational modifications, which may alter protein pI, are relatively limited. Obviously, substantial protein modifications such as endoprotease digestion cannot be detected with this procedure, whose main deliverable property is to establish whether a protein with the molecular weight and pI of the TTR gene product is present in the protein pool under investigation.

Although this procedure can be utilized for the elucidation of whole bacterial proteomes, we believe that the method is particularly useful for the analysis of subsets of specific proteins. In fact, whereas with 2-D gel/mass spectrometry the presence or absence of specific proteins can be established only after scanning a large portion of the gel, with this procedure their identification is much simpler since only their coding genes need to be amplified and used for the analysis. In the particular example presented here, we were interested in knowing whether two specific proteins were present and conserved in the MenB membrane protein subproteome of different clinical isolates. The superimposition of the autoradiograph of TTR products on the 2-D maps of the membrane fractions of the different isolates allowed us to demonstrate rapidly the conservation of the two antigens in all the strains analyzed. This kind of analysis will be invaluable to the identification of new vaccine candidates against MenB.

Acknowledgments

We are deeply grateful to Prof. C. Scolastico (University of Milan, Italy) for the chemical synthesis of ABS14. We also wish to thank Dr. S. Toma and Dr. G. Ratti for helpful discussions and Mrs. A. Maiorino for her expert secretarial assistance.

References

1. Lockhart, D. J. and Winzeler, E. A. (2000) Genomics, gene expression and DNA arrays. *Nature* **405,** 827–836.
2. Futcher, B., Latter, G. I., Monardo, P., McLaughlin, C. S., and Garrels, J. I. (1999) A sampling of the yeast proteome. *Mol. Cell. Biol.* **19,** 7357–7368.
3. Gygi, S. P., Rochon, Y., Franza, B. R., and Aebersold, R. (1999) Correlation between protein and mRNA abundance in yeast. *Mol. Cell. Biol.* **19,** 1720–1730.
4. Varshavsky, A. (1996) The N-end rule: functions, mysteries, uses. *Proc. Natl. Acad. Sci. USA* **93,** 12142–12149.
5. Washburn, M. P. and Yates, J. R. III. (2000) Analysis of the microbial proteome. *Curr. Opin. Microbiol.* **3,** 292–297.
6. Pandey, A. and Mann, M. (2000) Proteomics to study genes and genomes. *Nature* **405,** 837–846.
7. Lopez, M. F. (2000) Better approaches to finding the needle in a haystack: optimizing proteome analysis through automation. *Electrophoresis* **21,** 1082–1093.
8. Norais, N., Nogarotto, R., Lacobini, E. T., Garaguso, I., Grifantini, R., Galli, G., and Grandi, G. (2001) Combined automated PCR cloning, in vitro transcription/translation and two-dimensional electrophoresis for bacterial proteome analysis. *Proteomics* **1,** 1378–1389.
9. Chevallet, M., Santoni, V., Poinas, A., et al. (1998) New zwitterionic detergents improve the analysis of membrane proteins by two-dimensional electrophoresis. *Electrophoresis* **19,** 1901–1909.

10. Montigiani, S., Falugi, F., Scarselli, M., et al. (2002) Genomic approach for analysis of surface proteins in *Chlamydia pneumoniae*. *Infect. Immun.* **70,** 368–379.

11. Gygi, S. P., Corthals, G. L., Zhang, Y., Rochon, Y., and Aebersold, R. (2000) Evaluation of two-dimensional gel electrophoresis-based proteome analysis technology. *Proc. Natl. Acad. Sci. USA* **97,** 9390–9395.

12. Grandi, G. (2001) Antibacterial vaccine design using genomics and proteomics. *Trends Biotechnol.* **19,** 181–188.

13. Tabor, S. and Richardson, C. C. (1985) A bacteriophage T7 RNA polymerase/promoter system for controlled exclusive expression of specific genes. *Proc. Natl. Acad. Sci. USA* **82,** 1074–1078.

14. Pratt, J. M. (1984) Coupled transcription-translation in prokaryotic cell-free systems, in *Transcription and Translation: A Practical Approach* (Hames, B. D. and Higgins, S. J., eds.), IRL Press, New York, pp. 179–209.

15. Molloy M. P., Herbert B. R., Slade M. B., Rabilloud T., Nouwens A. S., Williams K. L., and Gooley A. A. (2000) Proteomic analysis of the *Escherichia coli* outer membrane. *Eur. J. Biochem.* **267,** 2871–288.

16. Herbert, B. R., Molloy, M. P., Gooley, A. A., Walsh, B. J., Bryson, W. G., and Williams, K. L. (1998) Improved protein solubility in two-dimensional electrophoresis using tributyl phosphine as reducing agent. *Electrophoresis* **19,** 845–851.

17. Doherty, N. S., Littman, B. H., Reilly, K., Swindell, A. C., Buss, J. M., and Anderson, N. L. (1998) Analysis of changes in acute-phase plasma proteins in an acute inflammatory response and in rheumatoid arthritis using two-dimensional gel electrophoresis. *Electrophoresis* **19,** 355–363.

18. Wilm, M., Shevchenko, A., Houthaeve, T., Breit, S., Schweigerer, L., Fotsis, T., and Mann, M. (1996) Femtomole sequencing of proteins from polyacrylamide gels by nano-electrospray mass spectrometry. *Nature* **379,** 466–469.

19. Tettelin, H., Saunders, N. J., Heidelberg, J., et al. (2000) Complete genome sequence of *Neisseria meningitidis* serogroup B strain MC58. *Science* **287,** 1809–1815.

20. Pizza, M., Scarlato, V., Masignani, V., et al. (2000) Identification of vaccine candidates against serogroup B meningococcus by whole-genome sequencing. *Science* **287,** 1816–1820.

21. Binz, P. A., Muller, M., Walther, D., et al. (1999) A molecular scanner to automate proteomic research and to display proteome images. *Anal. Chem.* **71,** 4981–4988.

22. Hoogland, C., Sanchez, J. C., Walther, D., et al. (1999) Two-dimensional electrophoresis resources available from ExPASy. *Electrophoresis* **20,** 3568–3571.

3

Selective Chemical Cleavage Methods in Proteomics, Including C-Terminal Successive Degradation

Akira Tsugita

1. Introduction

The addition of trifluoroacetic acid (TFA) to HCl, in both liquid *(1)* and vapor phases *(2)*, increases the hydrolysis efficiency of peptide bonds, suggesting a cleavage type other than conventional HCl hydrolysis. Partial acid hydrolysis using high concentrations of perflouric acid have indicated a novel cleavage specificity of peptide bonds, different from those of either Sanger's high-concentration acid hydrolysis *(3)* or the conventional dilute-acid hydrolysis *(4)*. This chapter summarizes this high-concentration perfluoric acid, which essentially allows C-terminal protein sequencing as well as specific site cleavages at the C-side of Asp and the N-side of Ser/Thr. Akabori et al. *(5)* used anhydrous hydrazine for protein C-terminal determination. In this reaction, internal peptide bonds are hydrazinolyzed, yielding the hydrazides of the constituent amino acids and peptides and allowing the residual C-terminal amino acid residue to be identified *(6)*. The similar reaction has been widely used for the deglycosylation of glycoprotein *(7)*. Recently, milder conditions, including the use of hydrazine hydrate, were developed for the deblocking of N-blocked protein and for the specific cleavage of Asn residues *(8,9)*. This chapter also includes a discussion of this Asn C-side peptide bond cleavage by hydrazine and hydrazine-hydrate.

2. C-Terminal Sequencing and Successive Degradation

Exposure of the peptides and proteins to vapors of high-concentration aqueous perfluoric acid at elevated temperature caused successive C-terminal degradation *(10)*. In general, the vapor reaction was manually carried out by the following double-tube system: a small glass tube (6 × 40 mm), containing dried peptide or protein, was placed in a large glass tube (13 × 100 mm), which contained 100 µL of reagent, and flame-sealed under vacuum. The tube was placed in a bath of a given temperature for a given period. After the reaction, the tube was opened and the small tube was placed in a vacuum desiccator to remove traces of the reagent.

2.1. Vapors of High-Concentration Pentafluoropropionic Acid (PFPA) and Heptafluorobutyric Acid (HFBA) Aqueous Solution

A hexapeptide, LWMRFA, was reacted by the double-tube system using 90% aqueous TFA, PFPA, or HFBA containing 5% dithiothreitol (DTT) in the vapor phase at

From: *Handbook of Proteomic Methods*
Edited by: P. Michael Conn © Humana Press Inc., Totowa, NJ

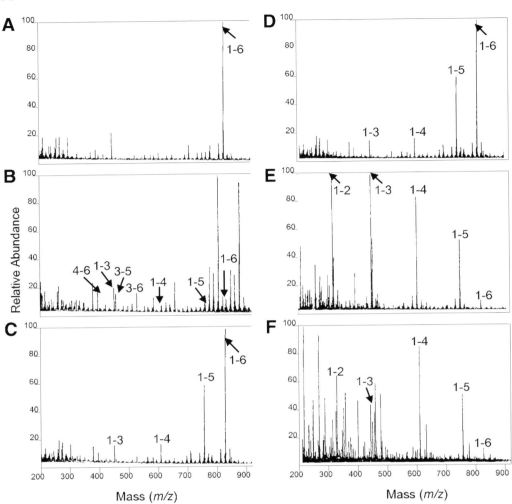

Fig. 1. FAB-MS of the truncated products of a hexapeptide, LWMRFA. The peptid (200 pmol) was truncated with vapors of various 90% perfloric acid containing 5% DTT at 90°C. (**A**) 0 time. (**B**) TFA for 4 h. (**C**) PFPA for 4 h. (**D**) HFBA for 4 h. (**E**) PFPA for 24 h. (**F**) HFBA for 24 h.

90°C, and the product was analyzed by fast-atom-bombardment mass spectroscopy (FAB MS). As shown in **Fig. 1B**, TFA vapor resulted in truncated molecular ions such as 1–5, 1–4, and 1–3, but other modified ions as well as internally cleaved ions (4–6, 3–5, and 3–6) were clearly observed, indicating that TFA may not be an adequate reagent. However, the other two acids, vapors of both PFPA (**Fig. 1C** and **E**) and HFBA (**Fig. 1D** and **F**), mainly resulted in C-terminal-truncated molecular ions. From the observation of –18u ions accompanied by the respective truncated peptide ions, the successive degradation first forms an oxazolone, a five-membered ring, at the C-terminal amino acid, followed by hydrolysis of the oxazolon.

This mechanism is unfavorable for the Pro residue because Pro is unable to make an oxazolone ring. A nonapeptide, RVYIHPFHL reacted by HFBA for 4 h, resulted in a three-residue truncation showing resistance at the Pro residue, but 24-h reaction resulted in a six-residue truncation beyond Pro, as shown in **Fig. 2B** and **C**. The peptide, composed of 23 amino acid residues, was reacted with PFPA for 2 h, and the product was analyzed by electrospray ionization (ESI) mass spectrometry (MS). The data are shown in **Fig. 3**; partial cleavage at the Ser-N side was observed. Also, Asp-C-side cleavage and occasionally Gly cleavages on both sides were observed in addition to the C-terminal degradation *(11)*. These are discussed in the next section.

2.2. C-Terminal Sequencing at Multiple Sites of Proteins

In the previous discussion, we demonstrated that the vapor generated from concentrated perfluoric acid aqueous solution at high temperature results in a C-terminal truncation reaction. However, this reaction simultaneously facilitates the specific internal peptide bond cleavages at Asp-C and Ser/Thr-N *(10)*. The conditions were carefully re-examined. Both the temperature and the reagent concentration may not be changed, but it is better to vary the reaction time from protein to protein and also better to use several reaction times for one protein *(12)*. **Figure 4** shows matrix-assisted laser desorption ionization time-of-flight (MALDI-TOF) MS of bovine carbonic anhydrase II exposed to a vapor of 90% PFPA aqueous solution containing 5% DTT at 90°C for 2, 4, and 8 h. Also, **Table 1** shows the summary of C-terminal truncation data of human glucagon, egg white lysozyme (pyridylethylated), and carbonic anhydrase II, where F was obtained by FAB MS and M by MALDI-TOF MS. A software named DST-RUNC was developed to assist in the analysis of mass spectrometic data *(10)*, in addition to the use of conventional software.

2.3. Dilute HCl for Successive C-Terminal Degradation of Peptide and Protein

Diluted HCl (10% w/v, 2.74 N) was reacted on peptides and proteins at 25°C for 14 and 30 d, causing successive C-terminal degradation and deamination of the C-terminal α-amide and partially of internal acidic amino acid amides *(13)*. **Figure 5** shows the results after two peptides exposed to a 10% HCl aqeous solution at room temperature (25°C) for 2 wk (14 d) and 1 mo (30 d). The reaction on the first peptide angiotensin, DRVYIHPFHL, clearly truncated three C-terminal residues, where the formation of oxazolone at the C-termini of each truncated peptides is shown by −18u ions. The reaction on the second peptide, the tyrosine protein kinase substrate, RRLIEDAEYAARG, demonstrated C-terminal four-residue successive truncation and accompanied −18u ions, which was owing to cyclization of Asp^6 or Glu^5/Glu^8. For more practical use, the reaction temperature was raised to 50°C. The same two peptides were exposed to 50°C for 16 h, but essentially similar truncation data were obtained with additional internal peptide bond cleavages.

Substance P, RPKPQQFFGLM-NH$_2$, was tested at 50°C for 1 and 16 h (**Fig. 6A** and **B**). The α-carboxyl amide was hydrolyzed, revealing truncation of four amino acids from the C-terminus. Liberation of the last amino acid, Phe, was achieved by a small, 1-h reaction but was expanded by 16 h of reaction. The two Gln^5 and Gln^6 residues were deaminated (+2u). Bombesin, pGlu-QRLGNQWAGHLM-NH$_2$, was also exposed to a 50°C truncation reaction for 1 and 4 h (**Fig. 6C** and **D**). The results again showed hydrolysis of the α-carboxyl amide, Gln^2, Asn^6, and Gln^7 (+4u) and confirmation of the truncation C-terminal amino acid residues. The N-terminal pyrrolidon-

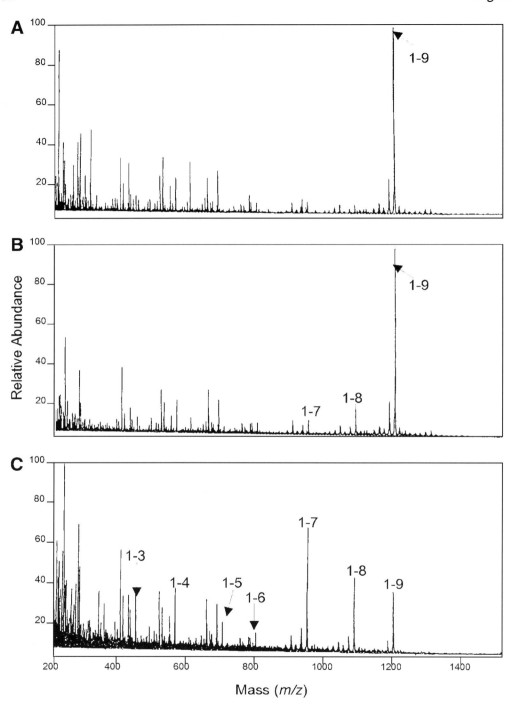

Fig. 2. FAB MS of the truncated products of a Pro-containing peptide, RVYIPFHL. The peptide (200 pmol) was truncated with a vapor of HFBA 90% containing 5% DTT at 90°C for (**A**) 0 time, (**B**) 4 h, and (**C**) 24 h.

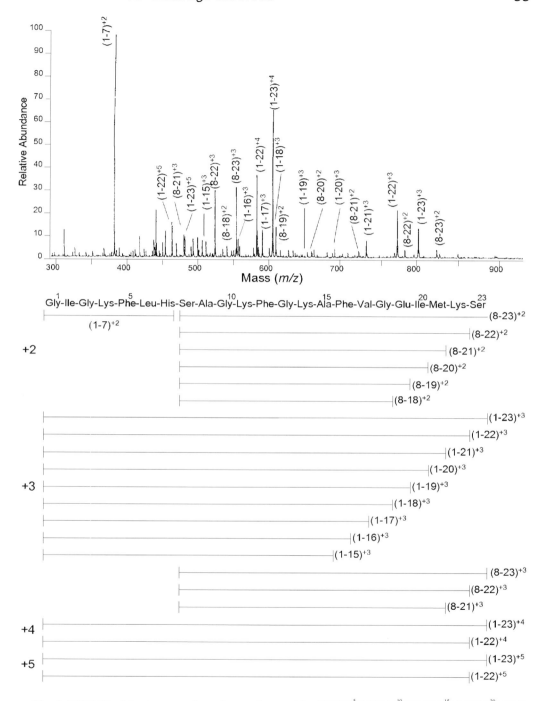

Fig. 3. ESI MS of the truncated products of a peptide, GIGKFLHSAGKFGKAFVGWIMKS. The peptide (125 pmol) was truncated with a vapor of 90% PFPA in the presence of 50 mg DTT at 90°C for 2 h.

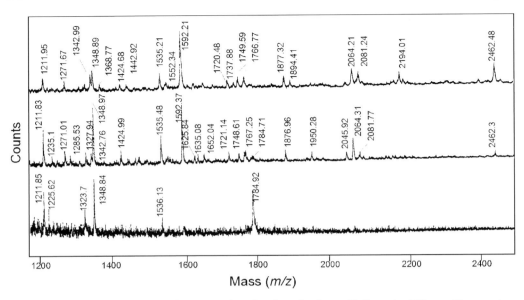

Fig. 4. MALDI-TOF MS of truncated carbonic anhydrase II. Protein (30 pmol) was electrophoresed on 12.5% SDS-PAGE, electroblotted on an Immobilon-CD membrane, negatively stained, excised and exposed to a vapor of 90% PFPA (5% DTT) at 90°C for (**top**) 2 h, (**middle**) 4 h, and (**bottom**) 8 h.

carboxylic acid was hydrated (+18u). Trp[8] is known as an acid-labile amino acid, but such an effect was prevented in the reaction. Glucagon, composed of 29 amino acid residues, was subjected to a 10% HCl, 50°C, 16-h reaction. As shown in **Fig. 7**, the internal peptide bonds, Gly-Thr[5], Phe-Thr[5], Tyr-Ser[11], and Asp-Ser[16], were cleaved. Having these cleaved sites as the N-termini, the C-terminal truncations were observed at a maximum of five residues, which stopped at Trp[25].

This 10% HCl method is similar to the perfluoric acid C-terminal truncation reaction, although these two methods differ in several points. This method decomposes C-terminal and internal amides, whereas the perfluoric acid reaction does not. This difference is potentially useful when studying C-terminal amidated peptides, however truncation reaction is more clearly conducted by perfluoric acid reaction. In HCl method the C-terminal successive degradation, is stopped at Pro-X bond and reduced by aromatic amino acid residues. Internal peptide bond cleavages were observed at the C sides of Asp, N sides of Ser/Thr and both sides of Gly.

2.4. C-Terminal Successive Degradation with Perfluoric Anhydrides

In the mass spectrum of C-terminal successive degradation with perfluoric acid (*see* **Subheading 2.2.1.**) –18u peaks were mostly seen in the respective degraded peptides suggesting the formation of oxazolone at the respective C-termini as the intermediate *(14)*. The acid anhydride is known to be a more effective reagent to produce oxazolone. When a vapor of perfluoric anhydrides was reacted on a peptide at –15°C for only 10 min, a series of successive degraded peptide ions was observed accompanied by both –18u and –45u ions. These observations suggest the following reaction mechanism. First the C-terminal amino acid loses H_2O (–18u), converting to the oxazolone, as the reaction intermediate; then this oxazolone-containing peptide loses CO_2 (–45u) and forms a compound that is not involved in the further successive degradation. A dodecapeptide, ARGIKGIRGFSG, was exposed to a vapor generated from 10% pentafluoropropionic anhydride (PFPAA) acetonitrile for 10 and 30 min at –15°C. The 10-min reaction did not result in acylation, but the C-terminal truncated eight residues of the peptide (1–4) **(Fig. 8B)**. The 30-min reaction resulted in acylation and almost complete degradation into the dipeptide (1–2), as shown in **Fig. 8C**. In both figures the asterisks stand for –18u and the plus marks for –45u.

This reaction was studied further in detail *(15)*. Reagents selected were PFPAA or hexafluorobutylic anhydride (HFBAA) vapors but not trifluoroacetic anhydride (TFAA) vapor, because the TFAA vapor gave several side reaction products. Also the N-acylation reaction of TFAA increases by +96u, PFPAA by +146u, and HFBAA by +196u, with these increments being similar to natural amino acid residual weights (97 for the Pro residue and 147 for the Phe residue, but no similar residual weight for 196). Therefore the HFBAA vapor seemed to be the best reagent except for poor volatility. The reaction temperature selected was –20°C, and the reaction times were between 30 min and 1 h. The 10-min reaction gave partial acylation, and the extension of reaction time to 2 or 5 h did not increase the extent of truncation. In the 0°C reaction, a stronger acylation was observed, and dehydration of Ser to dehydroalanine and internal peptide bond cleavages were occasionally observed **(Fig. 9A)**. The reagent was the vapor of 10% PFPAAA acetonitrile solution. This acetonitrile is not essential, but it is convenient, and the acetonitrile also absorbs the trace of water in the reagent. After the reaction, the product was dried in vacuum and treated with a vapor of 10% aqueous pyridine at 100°C for 30 min. This treatment reduced the complexity of the results owing to a shift in oxazolone peaks (–18u) to the principle truncated molecular ions and partial disappearance of the acylated peaks (probably O-acylation) **(Fig. 9B and C)**. **Figure 10** shows possible schemes of byproducts in the perfluoracyl anhydride vapor reaction, in addition to the oxyzolone formation (–18u) and C-terminal decarboxylation (–45u) previously mentioned.

2.5. Application of the Successive Degradation Method to Proteins

It is difficult to apply the perfluoric anhydride method to protein C-terminal sequencing, because protein sizes are large for direct analysis by MALDI-TOF MS. ESI MS may measure protein molecules directly, although analysis of a sample containing a mixture of the products and the other similar molecular masses is not always easy. Therefore a protein is first fragmented and the C-terminal peptide is selectively isolated to detect the C-terminal sequence. The following two methods were developed.

2.5.1. Cyanogen Bromide Cleavage and Perfluoric Anhydride Degradation

A classical cyanogen bromide cleavage for the Met C-side was used *(16)*. The cleaved fragments with homoserine lactone at the C-termini were covalently bound to N-(2-aminoethyl)-3-aminopropyl glass (APG; LKB Biochrom), leaving the C-terminal peptide. The C-terminal peptide was exposed to an HFBAA vapor and analyzed by MS. Both a dodecapeptide, YGGFMRRVGRPE, and sheep myoglobin were

Table 1
Summary of Multiple C-Terminal Truncation Data

M/Z	Residue	Sequence	Time (h)						Cleavage	
			1	2	4	6	8	16	N	C
Human glucagon			(HFBA)		(HFBA)					
980	1–9	HSQGTFTD–s	F						—	DS
865	1–8	HSQGTFTS	F		F					
778	1–7	HSQGTFT	F		F					
677	1–6	HSQGTF	F		F					
530	1–5	HSQGT	F		F					
1753	16–29	d-SRRAQDFVQWLMNT	F						DS	—
1652	16–28	d-SRRAQDFVQWLMN	F		F					
1538	16–27	d-SRRAQDFVQWLM	F		F				DS	DF
733	16–21	d-SRRAQD–f	F							
618	16–20	d-SRRAQ	F							
490	16–19	d-SRRA	F							
418	16–18	d-SRR	F							
Egg white lysozyme (pyridylethylated)										
2002	1–18	KVFGRCELAAAMKRHGLD–n	PFPA	PFPA	PFPA	PFPA	PFPA	PFPA	—	DN
1887	1–17	KVFGRCELAAAMKRHGL			F					
1774	1–16	KVFGRCELAAAMKRHG			F					
1717	1–15	KVFGRCELAAAMKRH			F					
1580	1–14	KVFGRCELAAAMKR			F					
1308	120–129	d-VQAWIRGCRL			F				DV	—
1194	120–128	d-VQAWIRGCR			F					
1038	120–127	d-VQAWIRGC			F					
830	120–126	d-VQAWIRG			F					
Bovine carbonic anhydorase II										
1442	41–54	d-PALKPLALVYGEAT–s	F	F	M	M			DP	TS
1341	41–53	d-PALKPLALVYGEA	F	F	M	M				

ID	Position	Peptide	1	2	3	4	5	6	7	8	Mut
1270	41–52	d-PALKPLALVYGE	F	F	M	F	M	F			
1141	41–51	d-PALKPLALVYG	F	F	M	F	M	F		F	
1084	41–50	d-PALKPLALVY	F	F	M	F	M	F		F	
921	41–49	d-PALKPLALV	F	F		F		F		F	
822	41–48	d-PALKPLAL	F			F		F		F	
709	41–47	d-PALKPLA				F					
638	41–46	d-PALKPL	F			F		F			
673	75–80	d-KAVLKD–g									DK DG
558	75–79	d-KAVLK	F	F		F		F		F	
1591	86–97	g-TYRLVQFHFHWG–s	F	F	M	F	M	F		F	GT GS
1534	86–96	g-TYRLVQFHFHW		F	M	F	M	F		F	
1348	86–95	g-TYRLVQFHFH			M		M	F	M	F	
1211	86–94	g-TYRLVQFHF			M		M	F	M	F	
1063	86–93	g-TYRLVQFH			M		M	F		F	
1767	110–123	d-RKKYAAELHLVHWN–t	F	F	M		M				DR NT
1652	110–122	d-RKKYAAELHLVHW		F	M		M				
1463	110–121	d-RKKYAAELHLVH					M				
1328	110–120	d-RKKYAAELHLV					M				
1273	138–150	d-GLAVVGVFLKVGD–a	F	F		F					DG DA
1158	138–149	d-GLAVVGVFLKVG	F	F		F					
1101	138–148	d-GLAVVGVFLKV	F	F		F					
1002	138–147	d-GLAVVGVFLK	F	F		F					
875	138–146	d-GLAVVGVFL	F	F		F					
761	138–145	d-GLAVVGVF	F	F		F					
1368	151–163	d-ANPALOQKVLDALD–s	F	F		F					DA DS
1253	151–162	d-ANPALOQKVLDAL	F	F		F					
1068	151–160	d-ANPALOQKVLD	F	F		F					
952	151–159	d-ANPALOQKVL	F			F					
840	151–158	d-ANPALOQKV				F					
1024	179–188	d-PGSLLPNVLD–y	F								DP DY
909	179–187	d-PGSLLPNVL	F								
796	179–186	d-PGSLLPNV	F								

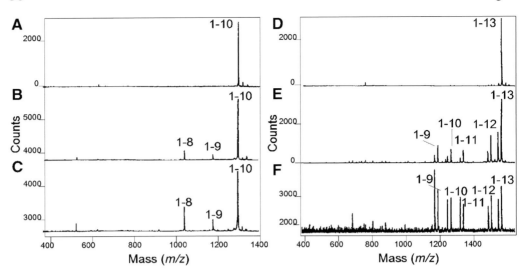

Fig. 5. MALDI-TOF MS of 10% HCl reaction on two peptides (each 5 pmol) at 25°C. DRVYIHPFHL at (**A**) 0 time, (**B**) 14 d, and (**C**) 30 d; RRLIEDAYARG at (**D**) 0 time, (**E**)14 d, and (**F**) 30 d. (**E**) and (**F**) accompanied the –18u peaks.

dissolved in 10 μL of 70% formic acid and reacted with equal weights of cyanogen bromide at 20°C for 16 h. The mixtures were vacuum-evaporated, and 5 μL of anhydrous TFA was added. Each mixture was kept for 30 min at 20 °C and evaporated to ensure the formation of homoserine lactone at the cleaved C-termini. The products were disolved in 10 μL of water, and then 30 μL of dimethylformamide (DMF) was added. The solution was mixed with 5 mg of APG, equilibrated with a 2% triethyl amine in DMF solution, at 45°C for 2 h. The APG was washed with two 25-μL portions of DMF and 100 m*M* pyridine collidine acetate buffer, pH 8.2. The eluted and washed solutions were collected and evaporated in small test tubes (6 × 40 mm). The tubes were then placed in reaction vessels (19 × 100 mm; Pierce), which contained 15% (v/v) HFBAA acetonitrile solution (50 μL). After evacuation (10^{-2} Torr) under cooling by liquid nitrogen, the reactions were carried out at –20°C for 1 h. The reaction products were evaporated to dry and treated with a vapor of 10% aqueous pyridine at 100°C for 30 min.

The final products were analyzed by FAB MS, as shown in **Fig. 11A** and **B**. The dodecapeptide was first cleaved into two fragments 1–5 and 6–12; the N-terminal fragment disappeared, and the acetylated C-terminal peptide (6–12) was completely truncated, as shown in **Fig. 11A**. The truncated peaks 6–12, 6–10, and 6–8 accompanied –18u peaks, owing to dehydration, and two peaks, 6–8 and 6–7, accompanied the decarboxylation –45u peaks. Sheep myoglobin clearly showed the C-terminal peptide isolation (143–153), and four-residue C-terminal truncation was seen, as shown in **Fig. 11B**. The dehydrations seemed to be on both Gln[145] and Gln[152].

2.5.2. Acetylation, Perfluoric Acid Degradation, and Trypsin Digestion

Another similar line of approach was recently taken (*17*). The peptide was treated with a vapor of 5% acetic acid in acetic anhydride at 50°C for 2 h. This completely acetylates the N-terminal and internal amino groups. The truncation was made with a vapor

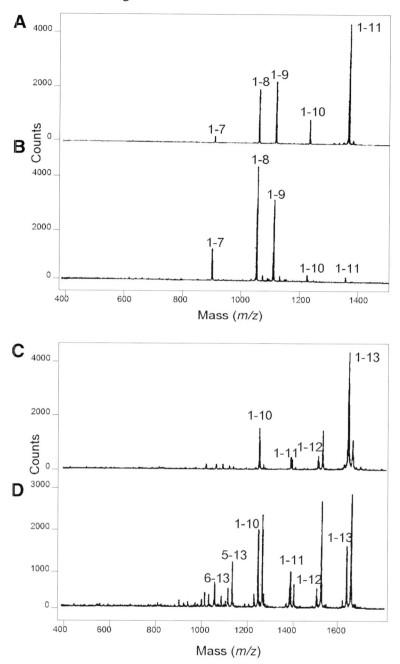

Fig. 6. MALDI-TOF MS of 10% HCl reaction on two peptides (each 5 pmol) at 50 °C. RPKPQQFFGLM-NH₂ at (**A**) 1 h and (**B**) 16 h; p-EQRLGNQWAGHLM-NH₂ at (**C**) 1 h and (**D**) 4 h. The data show dehydration of N-pyrrolidon carboxylic acid (+18u) and deamidation of C-NH₂, Asn and Gln residues (+4u).

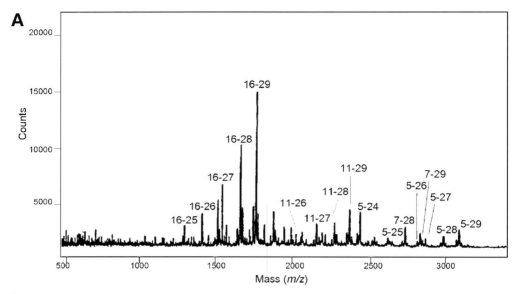

Fig. 7. MALDI-TOF MS of 10%HCl reaction products on human glucagon. (A) Glucagon (5 pmol) was reacted with 10% HCl at 50°C for 16 h. (B) The cleavage and molecular ions are schematically illustrated.

of 5% TFA in acetic anhydride at 40°C for 16 h. The truncated products were treated with a vapor of 10% pyridine aqueous solution at 100°C for 30 min and then subjected to MALDI-TOF MS analysis.

MALDI-TOF MS of angiotensin (10 pmol), DRVYIHPFHLVIH, resulted in 1 h four major peaks, 1687.9, 1550.8, 1437.8, and 1338.7, which correspond to

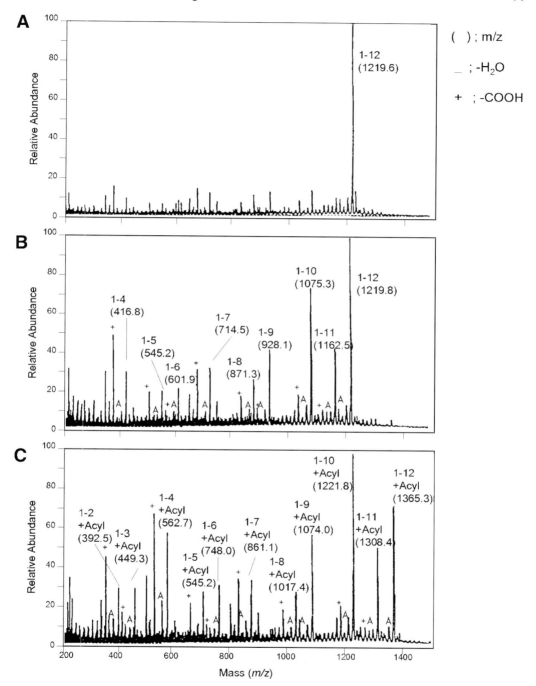

Fig. 8. FAB MS of truncated products with PFPAA vapor on a peptide, ARGIKGIRGFSG. The peptide was truncated with a vapor from 10% PFPA acetonitril solution at –15°C for (**A**) 0 time, (**B**) 10 min, and (**C**) 30 min.

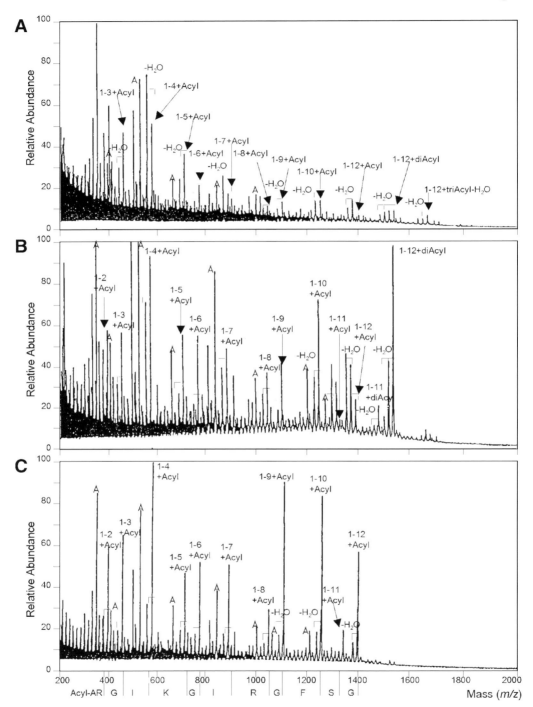

Fig. 9. FAB MS of truncated products of a peptide, ARGIKGIRGFSG. The peptide (200 pmol) was truncated with a vapor of 10% PFPA acetonitril solution for 1 h (**A**) at 0°C, (**B**) at –20°C, and (**C**) at –20°C and further treated with a vapor of 10% aqueous pyridine for 30 min at 100°C.

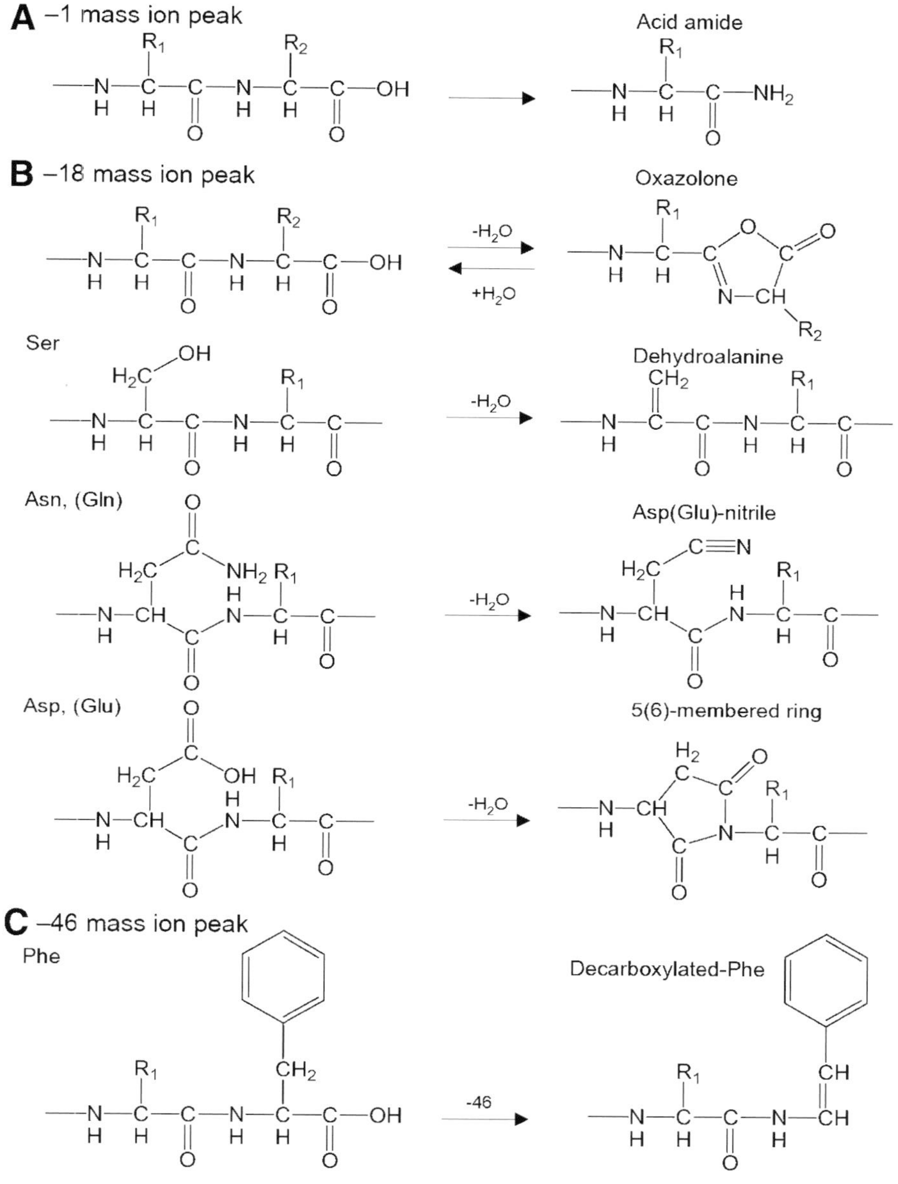

Fig. 10. (**A–C**) Possible schemes in the reaction of perfluoroacyl anhydride vapor on peptides.

Ac-DRVYIHPFHLVIH, Ac-DRVYIHPFHLVI, Ac-DRVYIHPFHL and Ac-DRVYIH PFHL, Ac-DRVYIHPF, and Ac-DRVYIHP. A diacetylated product was seen with +42u peaks possibly because of acetylation on the Tyr residue (**Fig. 12**).

For proteins (50 pmol), the acetylation reaction process was the same as for peptides. However, truncation was performed with a vapor from 1% HFBA in acetic anhydride at 40°C for 3 h, and the product was treated with a vapor of 10% dimethylaminoethanol (DMAE) aqueous solution at 60°C for 1 h. The product was dissolved in 0.5 *M* pyridine acetate buffer, pH 7.0, and digested with trypsin that was half the weight of the starting protein at 37°C for 8 h. The digests were evaporated and analyzed

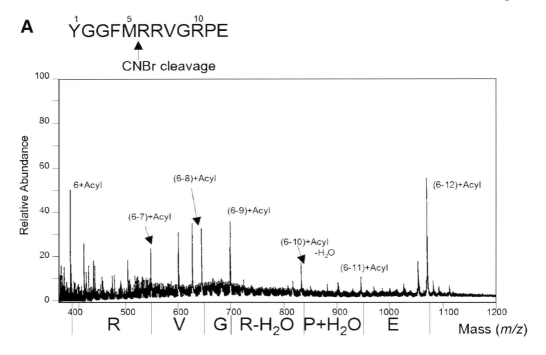

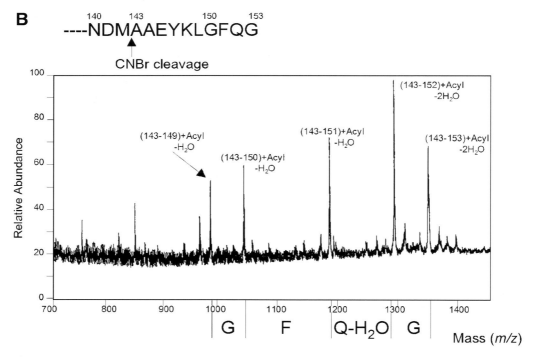

Fig. 11. FAB MS of successively truncated C-terminal peptides of (**A**) a dodecapeptide, YGGFMRRVGRPE, and (**B**) sheep myoglobin.

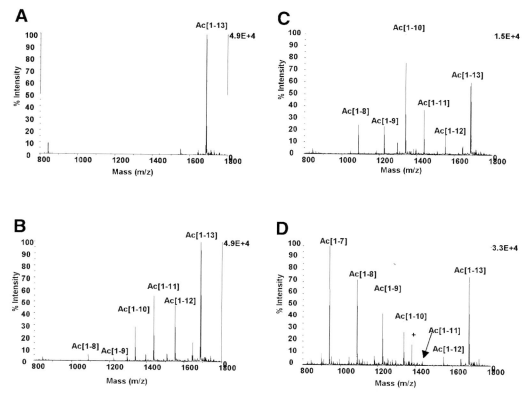

Fig. 12. MALDI-TOF MS of truncated products of a tridecapeptide. A peptide, DRVYIH-PFHLVIH (10 pmol), was acetylated with a vapor of 5% acetic acid in acetic anhydride at 50°C for 2 h. The dried product was truncated with a vapor of 5% TFA in acetic anhydride at 40°C for specified times. The dried product was then hydrated with a vapor of 10% DMAE aqueous solution at 60°C for 1 h. Truncation time are (**A**) 0 time after acetylation, (**B**) 1 h, (**C**) 3 h, and (**D**) 16 h.

by MALDI-TOF MS in negative mode. Acetylation of the amino groups of the N-terminus and internal Lys residues was almost complete. In the truncation reaction, TFA-acetic anhydride vapor produced side reactions; therefore HFBA-acetic anhydride vapor was used. Hydration of the truncated product with 10% pyridine vapor was not enough; therefore DMAE vapor was chosen.

Trypsin cleaves at the C-terminal sides of both the Lys and Arg residues of proteins; in this reaction, Lys residues were acetylated, and the Lys peptide bond became uncleavable. Thus all tryptic peptides have Arg at their C-termini but the protein C-terminal peptide. Mass spectrometry by the negative ion mode selected the C-terminal truncated peptides, carrying the C-terminal sequence information from the other tryptic positively charged peptides ending with Arg. This method can be used for the isolated protein spot on the polyacrylamide gel of 2-DGE with a few modifications (Miyazaki and Tsugita, unpublished data). **Figure 13** shows the negative mode MS of the C-terminal truncated myoglobin. As seen in this figure, four residues, Gly, Gln, Phe, and Gly, were sequenced from differences of each peak. This C-terminal fragment has two acetyl Lys residues.

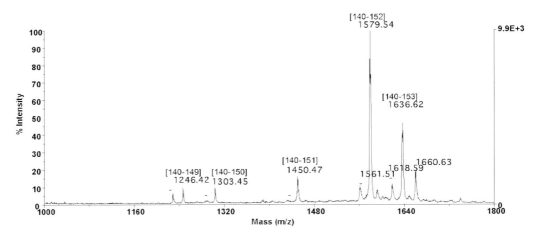

Fig. 13. MALDI-TOF negative mode MS of C-terminal truncation products of sheep myoglobin. Myoglobin (10 pmol) was acetylated with a vapor from 5% acetic acid in acetic anhydride at 50°C for 2 h. The dried product was truncated with a vapor of 1% HFBA in acetic anhydride at 40°C for 3 h. The dried product was hydrated with a vapor of 10% DMAE at 60°C for 1 h. The products were then digested with trypsin (10 pmol) in 0.5 *M* pyridine acetate buffer, pH 7.0, for 8 h. The product was dried in vacuum and subjected to MS. The C-terminal tryptic peptide of myoglobin is [140]NDIAAKYKELGFQE[153] (1636.6u).

3. Bond-Specific Chemical Cleavage of Protein

In 1953, Sanger and Thompson *(3)* first reported Ser/Thr and Gly peptide bond cleavage in concentrated HCl at room temperature. It was proposed that this cleavage was caused by an N→O shift of the hydroxy groups. In 1950, Partridge and Davis *(18)* observed that Asp was the only amino acid released when proteins were heated in weak acid solution. This chapter summarizes data on peptide bond cleavage of common amino acids such as the N-sides of Ser/Thr, the C-sides of Asp, and the C-sides of Asn peptide bonds.

3.1. Asp-C Cleavage

Vapors generated from various concentrations of HFBA aqueous solutions containing 5% DTT were reacted at both 50°C and 90°C for 24 h on a hexadecapeptide, EADKADVNVLTKALSQ *(11)*. The cleavage yields for Ser-N and Asp-C were estimated from the relative heights of molecular ions of FAB MS **(Fig. 14)**. From these experiments a vapor of 75% aqueous HFBA at 50°C was chosen for Ser/Thr-N cleavage, and a vapor of 0.2% aqueous HFBA at 90°C was chosen for Asp-C cleavage. **Figure 15** shows Asp-C cleavage. The same hexadecapeptide was exposed to the vapor of 0.2% aqueous HFBA containing 5% DTT at 90°C for 4 h **(Fig. 15A)**. A tridecapeptide, RRLIEDAEYAARG, was exposed to the HFBA vapor for 4 h **(Fig. 15B)**, 8 h **(Fig. 15C)**, and 24 h **(Fig. 15D)**. The reaction cleaved Asp-C; this Asp residue was activated and cleaved the Asp-N bond, resulting in liberation of this Asp **(Fig. 15C and D)**.

Among the Asp peptide bonds, the Asp-Pro bond is known to be an extremely cleavable bond. Several peptides and proteins containing Asp-Pro sequences were exposed to the same vapor of 0.2% aqueous HFBA (5% DTT) but at 50°C for 16 h; the results showed specific cleavages for the Asp-Pro bond but not the other Asp-C peptide bonds.

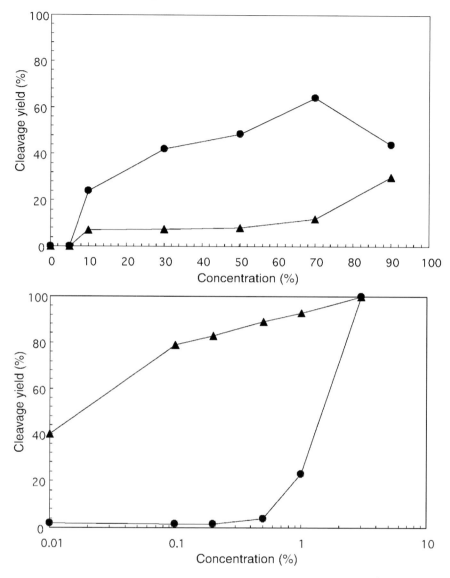

Fig. 14. Ser-N and Asp-C cleavage reaction on a hexadecapeptide, EADKÅDVNVĹTKAKŜQ. The peptide was exposed to vapors from various concentrations of aqueous HFBA containg 5% DTT for 24 h at 50°C **(top)** and at 90°C **(bottom)**. The products were analyzed by FAB MS and cleavage yields for Ser-N (●) and AspC (▲) were estimated from the relative heights of molecular ions.

3.2.1. Ser/Thr-N Cleavage by Acid Vapor

The concentrated HFBA aqueous vapor at 50°C was observed to cleave selectively the N-side of the Ser residue in the peptide *(11)* **(Fig. 14A)**. The peptides with Ser residues were exposed to a vapor from 75% aqueous HFBA containing 5% DTT at 50°C for 24 h and the products analyzed by FAB MS. The results are shown in **Fig. 16**. A peptide, KRPP(OH)ĠFSPFR̊, was cleaved into two fragments (1–6 and 7–10) by cleav-

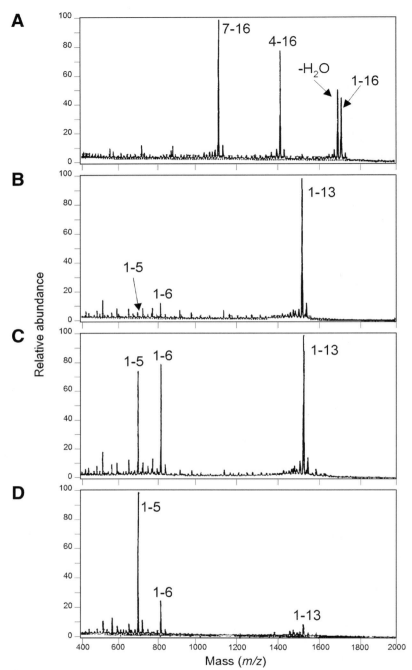

Fig. 15. FAB MS of Asp-C cleavage products. (**A**) A peptide EADKĂDVNVL̆TKAKŜQ, was exposed to a vapor of 0.2% HFBA aqeuous solution (5% DTT) at 90°C for 4 h. Another peptide, RRLIĔDAEYĂARG, was exposed to the same vapor at 90°C for (**B**) 4 h, (**C**) 8 h, and (**D**) 24 h. The yields for Asp-C peptide (1–6) for (**B**), (**C**), and (**D**) are estimated at 9, 31, and 18% and those for Asp-N peptide (1–5) for (**B**), (**C**), and (**D**) are each estimated at 2, 29, and 77%.

age of the Phe-Ser[7] peptide bond (**Fig. 16A**). Another peptide, KILGN[8]QGSFL[10]TKGP[15]SKL, was cleaved at the Gly-Ser peptide bond and very slightly at the Pro-Ser[15] peptide bond, but not at the N-side of the Thr bond (**Fig. 16B**). The other peptide, p-EKRP[5]SQR[10]SKYL, was cleaved at the Arg-Ser peptide bond but not at the Pro-Ser[5] bond. Pro-Ser cleavage has rarely been observed under the present conditions (**Fig. 16C**).

Quite a few proteins whose sequences are known were examined for this type of Ser/Thr N cleavage reaction, but under the milder conditions. The vapor was the same as before, 75% HFBA, but reaction temperature was 30°C instead of 50°C, and the reaction times were 16–24 h. These conditions may avoid unnecessary side reactions. The reaction failed to cleave Pro-Ser and Ser-Ser peptide bonds, whereas Gly-Ser, Thr-Ser, Asp-Ser, Ala-Ser, Val-Ser, Leu-Ser, Phe-Ser, Tyr-Ser, and N-acetyl-Ser were cleaved. No X-Thr was observed to be cleaved except Gly-Thr. The cleavage of Gly-Thr was found to be unexpectedly higher than that of Gly-Ser, which is also known to be an acid-labile peptide bond.

3.2.2. Ser/Thr-N Cleavage by S-Ethyltrifluorothioacetate Vapor

A vapor of S-ethyltrifluorothioacetate (ETFTA) was found to cleave the N sides of Ser and Thr peptide bonds specifically (*19*). The cleavage reactions were carried out at 50°C for 6–24 h or at 30°C for 24 h. When vapors were generated from ETFTA solutions of several organic solvents, the cleavage reactions were reduced or stopped, or the side reactions took place. When the reagent vapor contained water components, the cleavages at Gly residues were enhanced. This reagent did not oxidize any amino acid residues such as Cys, Met, and Trp. This cleavage was also effective on proteins that were membrane blotted or electroblotted from polyacrylamide gels. The acetyl-Ser bond, which was often observed in mature proteins, was exposed to the ETFTA vapor at 50°C for 6 h (**Fig. 17A**) and at 30°C for 24 h (**Fig. 17B**). The Ser-N bond was clearly cleaved off and deacetylated (–42u). A peptide, motilin (FVPIF[5]TYGEL[10]QRM Q[15]EKERN[20]KGQ) was exposed to ETFTA vapor at 30°C for 24 h. FAB MS (**Fig. 18A**) shows the cleavage at the N-side of Thr, Phe-Thr[6], which is the only susceptible residue in this peptide, resulting in two peptide fragments (1–5 and 6–22). The other peptide, α-endorphin (YGGF[5]MTSEK[10]SQTPL[15]VT), was subjected to ETFTA at 50°C for 6 h (**Fig. 18B**) and at 30°C for 24 h (**Fig. 18C**). The 50°C, 6-h conditions showed cleavages of Gln-Thr[12] and Lys-Ser[10], and the 30°C, 24-h conditions showed essentially the same two cleavages.

These vapor cleavage reactions were also performed on proteins on a membrane. Proteins from the spots of 2-DGE or from PAGE bands were electroblotted onto an immobilon CD membrance (Millipore), negatively stained, excised, and cut into 1×1-mm pieces; the pieces were placed in the small glass tube. After the reaction, the peptide fragments was extracted with 20 µL of 2.5% TFA in 30% acetonitrile and 20 µL of 2.5% TFA in 60% acetonitrile aqueous solution; they were then combined and subjected to MS analysis. Several proteins were subjected to ETFTA vapor reactions; the cleavage efficiencies of the N-sides of Ser, Thr, and Gly are summarized in **Table 2**. The ETFTA vapor method has a wider specificity, such as Thr-N cleavages and Gly-N cleavages, than the HFBA vapor. The cleavage was found only in the Gly-Thr bond by the HFBA vapor, but the ETFTA vapor cleavage showed high efficiency

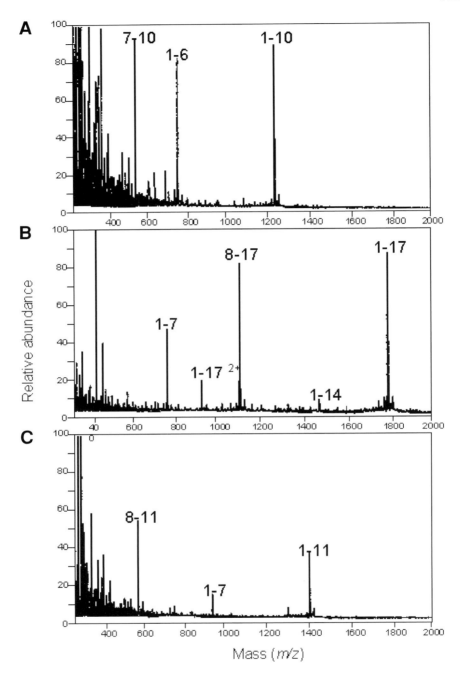

Fig. 16. FAB MS of Ser-N cleavage products of three peptides. The reactions were carried out at 50°C for 24 h with a vapor of 75% HFBA aqueous solution (5% DTT) on (**A**) KRPP(OH)GFSPFR, (**B**) KILGNQGSFLTKGPSKL, and (**C**) pEKRPSQRSKYL, where P(OH) and pE stand for hyproxyproline and pyroglutamic acid, respectively.

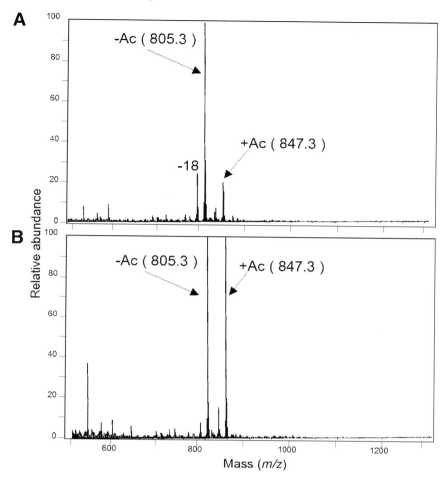

Fig. 17. FAB MS of the cleavage products of a peptide with ETFTA vapor. A peptide Ac-SQNPVV (110 pmol) was exposed to the ETFTA vapor at 50°C for 6 h (**A**), and at 30°C for 24 h (**B**).

for Arg-Thr, Lys-Thr, Ala-Thr, and Ser-Thr in addition to the Gly-Thr bond. The Ser-N cleavage was clearly observed for Gln-Ser, Asn-Ser, Ala-Ser, Pro-Ser, Leu-Ser, Phe-Ser, Tyr-Ser, His-Ser, and Lys-Ser peptide bonds.

3.3. N-Terminal Deblocking and Asn-C and Gly-C Side Cleavages with Anhydrous Hydrazine Vapor and Hydrazine-Hydrate Vapors

Many mature proteins have blocked N-termini, and the blockage typically occurs by acylation of an α-amino group (formylation or acetylation) or the N-terminal pyrrolidone carboxylic acid formed by cyclization of Gln or Glu (*8,9,20*). When anhydrous hydrazine vapor was reacted on N-formyl proteins at –5°C for 8 h, the formyl group

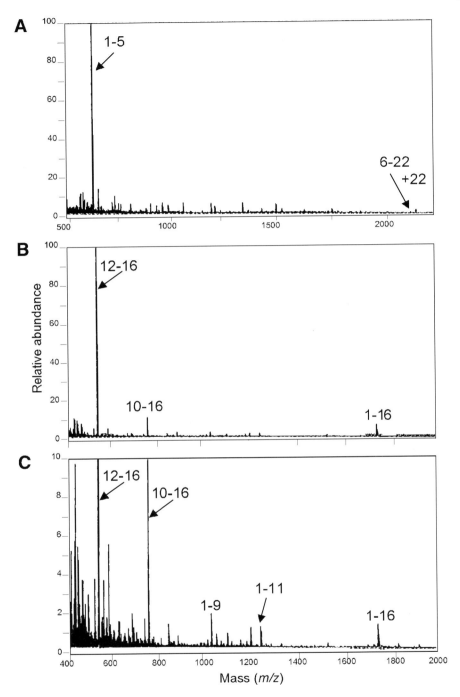

Fig. 18. FAB MS of the cleavage products from two peptides with ETFTA vapor. (**A**) FVPIFTYGELQRMQEKERNKGQ (100 pmol) at 30°C for 24 h. (**B**) YGGFMTSEKSQTPLVT (120 pmol) at 50°C for 6 h and (**C**) at 30°C for 24 h.

Table 2
Cleavage Frequency of Peptide Bonds by ETFTA Vapor[a]

	Xaa-Ser			Xaa-Thr			Xaa-Gly		
	No. of residues	No. of cleavages	Rate (%)	No. of residues	No. of cleavages	Rate (%)	No. of residues	No. of cleavages	Rate (%)
Ala	12	9	75	5	3	60	11	3	27
Asn	2	1	50	–	–	–	4	1	25
Asp	4	1	25	4	1	25	3	1	33
Arg	1	0	0	3	3	100	4	1	25
Cys	1	0	0	–	–	–	1	1	100
Gly	6	1	17	5	4	80	6	0	0
Gln	1	1	100	4	1	25	4	2	50
Glu	9	2	22	2	1	50	3	1	33
His	2	1	50	1	0	0	5	1	20
Ile	6	1	17	4	1	25	4	0	0
Leu	6	4	67	3	0	0	6	0	0
Lys	4	2	50	1	1	100	6	1	17
Met	3	0	0	–	–	–	4	0	0
Phe	4	2	50	1	0	0	3	0	0
Pro	4	3	75	1	0	0	2	0	0
Ser	8	3	38	5	3	60	6	0	0
Thr	5	0	0	3	0	0	–	–	–
Trp	1	0	0	1	0	0	–	–	–
Tyr	2	1	50	1	0	0	–	–	–
Val	6	1	17	2	1	50	9	0	0
Total	87	33	38	46	19	41	81	12	15

[a]Values are the total numbers from the coat protein of cucumber green mottle mosaic virus, whale sperm myoglobin, ovalbumin, and yeast alcohol dehydrogenase 1.

was removed (–28u), and the residual peptides or proteins were not cleaved or modified. Deblocking of N-terminal pyrrlidone carboxylate residue by conversion to γ-hydrazidyl Glu (+32u) was performed by exposure to an anhydrous hydrazine vapor at 20°C for 4 h (*8*).

These conditions cause partial modification of Asn and Gln residues to the corresponding hydrazides (+15u) and conversion of the Arg residue to ornithine (–42u) (*9*). When exposure was lengthened to 16 h at 20°C, the reactions of Asn and Gln to their β- and γ-hydrazides increased. Then the β-hydrazidyl Asp cyclized to an anhydous hydrazide six-member ring, acylaminosuccinyl hydrazide, that caused an Asn-C side bond cleavage. Other cleavages, Gly-C side cleavages including Gly-Gly bond cleavages, were also observed (*9*). **Table 3** summarizes the cleavage data of peptide bonds exposed to anhydrous hydrazine vapor at 20°C for 16 h for 20 peptides and 7 proteins. The C-terminal α-amide was also converted to α-hydrazide (+15u).

Anhydrous hydrazinolysis has been successfully used for deglycosylation of glycoprotein (*7*), where the proteins were considerably damaged. The milder conditions

Table 3
Stability of Peptide Bonds (A–B) to Hydrazine Treatment[a]

A/B	G	A	V	I	L	S	T	D	E	N	Q	C	M	F	Y	P	H	K	R	W
G	X	X	O	O	O	X		X	O		O	X		O		X			O	
A	O	O	O			O		O	O	O			O	O				O	O	
V	O			O	O	O	O		O	O				O	O	O	O			
I	O		O	O	O	O		O	O		O		O				O	O	O	
L	O		O	O	O	O		O						O	O	O			O	
S	O	O			O	O	O	O	O	O	O	O		O	O		O			
T						O	O	O	O	O	O			O			O	O	O	
D	X	O	O	O	O	O			O	O						O	O	O	O	
E		O		O				O		O		O			O	O	O			
N	X			X	X	X		X				X		X	X	X		X	X	
Q	X	O		O				O	O	O						O				
C								O							O	O		O		
M								O									O			
F		O		O	O	O		O	O			O	O			O				O
Y	O	O	O	O			O		O							O				
P	O		O	O	O		O		O					O	O			O	O	
H	X			O	O		O		O							O	O	O		O
K	O	O			O	O			O					O			O	O		
R	O		O		O				O							O			O	
W		O			O												O			

[a]X, cleaved peptide bond; O, stable peptide bond.

were introduced *(20)*. Aqueous hydrazine vapors (NH$_2$NH$_2$ • nH$_2$O, n = 1 or 4) were reacted at 30°C for 4 or 16 h. These conditions provided Asn peptide bond cleavage and partial conversions of internal Asn and Gln to their hydrazides and Arg residues to ornithine residues. The N-formyl residue was cleaved off, and the C-terminal α-amide was converted to the hydrazide, but conversion of the N-pyroglutamyl residue to the γ-hydrazidyl Glu residue was less extensive. N- and O-glycosyl groups were cleaved off by these conditions using model compounds and proteins, and the residual peptides seemed to be less damaged. Asn-C cleavages were observed on Asn-Cys, Asn-Asn, Asn-Gln, Asn-Asp, Asn-Glu, Asn-Gly, Asn-Ala, Asn-Val, Asn-Ile, Asn-Phe, Asn-Tyr, and Asn-Met peptide bonds, whereas Asn-Lys and Asn-Arg were not cleaved, and the Asn residues remained as their β-hydrazides. The peptide bond Gly-Gly was cleaved but not the other Gly bonds. In general, the cleavages at 30°C by NH$_2$ • H$_2$O vapor for 4 h were similar but a little less extensive than that by NH$_2$NH$_2$ • 4H$_2$O vapor for 16 h, at 30°C *(20)*.

4. An Application of Chemical Selective Cleavage to Analyze Post-Translational Modification of Proteins

Three specific chemical cleavage reactions, the Asp-C-side, Asn-C-side, and Ser/Thr-N-side peptide bonds, simultaneously act on several post-translationally modified

Table 4
Reaction in Modified Peptides with Three Chemical Cleavage Methods

Modification	M/Z	Peptide	Yield (%)		
			Ser/Thr-N	Asp-C	Asn-C
N-terminal blockade					
N-formyl	−28	Formyl-MLFK	0	40	40
N-acetyl Ser	−42	Ac-SYSMEHFRWGKPV-NH$_2$	100	0	0
N-acetyl X	−42	Ac-MDKVLNRY	0	0	0
		Ac-AADISQWAGPL	0	0	0
N-myristoyl	−240	Myristoyl-RKRTLRRL	0	0	0
Pyroglutamyl	+32	Pyr-LYENKPRRPYIL	0	0	40
	+32	Pyr-HWSYGLRPG-NH$_2$	0	0	40
C-terminal amide	+1/+15	Pyr-HWSRGLRPG-NH$_2$	30	0	80
	+1/+15	CYFQNCPRG-NH$_2$	30	–	60
Internal acetylation					
O-acetyl Ser	−42	YAFGYPS(O-Ac)-NH$_2$	100	5	100
Phosphorylation					
H$_2$PO$_3$-Ser	−80	CLNRQLS(PO$_3$H$_2$) SGVSEIR	0	30	0
H$_2$PO$_3$-Tyr	−80	Y(PO$_3$H$_2$)VPML	0	100	0
	−80	NPEY(PO$_3$H$_2$)	0	100	0
	−80	ENDY(PO$_3$H$_2$)INASL	0	70	0
	−80	DADPY(PO$_3$H$_2$)L-NH$_2$	0	100	0
Sulfonylation					
HSO$_3$-Tyr	−80	DY(SO$_3$ H)MGWMDF-NH$_2$	75	0	0
	−80	NY(SO$_3$ H)Y(SO$_3$ H)GWMDF-NH$_2$	75	0	0
Glycosylation					
N-glycosylation		Fmoc-Asn[GLcNAc(Ac)$_3$]	0	0	100
O-glycosylation		Fmoc-Ser[GLcNAc(Ac)$_3$]	0	0	100

groups in peptides or proteins *(21)*. Also, these amino acids, Asp, Asn, and Ser/Thr, are involved in modification sites such as acetylation, phoshorylation, and glycosylation sites of protein, and the modified amino acid residues themselves are not cleaved by these chemical reactions. The specific reactions of Ser/Thr-N-side bonds, with a vapor of ETFTA at 30°C or 50°C for 24 h *(19)*, cleaved off N-acetyl groups (−42u) from N-acetyl-Ser of proteins and partially cleaved off sulphonate groups (−80u) from O-sulfo-Tyr. The specific reaction of Asp-C-side bonds, with a vapor generated from 0.2% PFPA aqueous solution containing 5% DTT at 90°C for 4–8 h *(12)*, cleaved off an N-formyl group (−28u) and partially cleaved off phosphate groups (−80u) from phospho-Ser and phospho-Tyr residues in protein. This cleavage, of course, may reveals the deamination sites (+1u) of Asn in the protein. The specific reaction Asn-C-side bonds, with NH$_2$NH$_2$ • H$_2$O vapor at 30°C for 4 h or with NH$_2$NH$_2$ • 4H$_2$O vapor at 30°C for 16 h *(19)*, cleaved off an N-formyl group (−28u), deblocked an N-pyroglutamyl group (+32u), converted the C-terminal amide to its hydrazide (+15u), and cleaved off the glycosyl residue from the N-glycosyl Asn residue and the O-glycosyl Ser/Thr residue in protein.

These data are summarized in **Table 4**. These three chemical cleavage reactions may be useful for analyzing post-translational modification because they cleave specific peptide bonds and at the same time react on modified groups. These methods are mild and have been used for microanalysis. Furthermore, these cleavage reactions have a direct connection to the key amino acids in post-translational modifications. The vapor reactions of both Asp-C and Asn-C can also be performed in liquid phases.

References

1. Tsugita, A. and Scheffler, J. J. (1982) Rapid method for hydrolysis of protein. *Eur. J. Biochem.* **124,** 585–588.
2. Tsugita, A., Uchida T., Mews, H. W., and Ataka, T. (1987) A rapid vapor-phase acid hydrolysis of peptide. *J. Biochem.* **102,** 243–246.
3. Sanger, J. and Thompson, E. O. P (1953) Sequence of glycyl chain of insulin: partial hydrolysates. *Biochem. J.* **53,** 353–366.
4. Inglis, A. S. (1983) Cleavage at aspartic acid. *Methods Enzymol.* **91,** 324–332.
5. Akabori, S., Ohno, K., and Narita, K. (1952) Hydrazinolysis of protein C-terminal amino acid. *Bull. Chem. Soc. Jpn.* **25,** 214–218.
6. Akabori, S., Ohno, K., Ikenaka, T., Okada, Y., Haruna, H. L., and Tsugita, A. (1956) Hydrazinolysis of peptide. *Bull. Chem. Soc. Jpn.* **29,** 507–518.
7. Takasaki, S., Mizuochi, T., and Kobata, A. (1982) Hyrazinolysis of asparagines-linked sugar chains. *Methods Enzymol.* **83,** 263–268.
8. Miyatake, N., Kamo, M., Satake, K., Uchiyama, Y., and Tsugita, A. (1993) Removal of N-terminal formyl groups and deblocking of pyrrolidon carboxylic acid with anhydrous hydrazine. *Eur. J. Biochem.* **212,** 785–789.
9. Miyatake, N., Satake, K., Kamo, M., and Tsugita, A. (1994) A specific chemical cleavage of asparaginyl bonds in peptides. *J. Biochem.* **115,** 208–212.
10. Tsugita, A., Takamoto, K., Kamo, M., and Iwadate, H. (1992) C-terminal sequencing of protein. *Eur. J. Biochem.* **206,** 691–696.
11. Miyazaki, K., Shen, R., Takayama, M., Kamo, M., Kawakami, T., and Tsugita, A. (1998) C-terminal sequencing at multiple sites of proteins. *Res. Commun. Biochem, Cell. Mol. Biol.* **2,** 175–194.
12. Kawakami, T., Kamo, M., Takamoto, K., Miyazaki, K., Chow, L. P., Ueno, Y., and Tsugita, A. (1997) Specific cleavage of peptide, amino side of serine and carboxyl side of asparatic acid. *J. Biochem.* **121,** 68–76.
13. Takayama, M., Matsui, T., Sakai, T., and Tsugita, A. (1999) Peptide successive C-terminal degradation using dilute HCl. *J. Biomol. Tech.* **10,** 194–198.
14. Tsugita, A., Takamoto, K., and Satake, K. (1992) A carboxy peptidase mimetic degradation. *Chem. Lett.* **1992,** 235–238.
15. Takamoto, K., Kamo, M., Kubota, K., Satake, K., and Tsugita, A. (1995) C-terminal degradation of protein with perfluoroacyl anhydrides. *Eur. J. Biochem.* **228,** 362–372.
16. Nabuchi, Y., Yano, H., Kamo, M., Takamoto, K., Satake, K., and Tsugita, A. (1994) C-terminal sequencing of peptides and proteins by successive degradation. *Chem. Lett.* **1994,** 757–760.
17. Miyazaki, K. and Tsugita, A. (2002) C-terminal sequencing by vapor reaction in acetic anhydride, in *AOHUPO Conference Proceedings*, p. 23.
18. Partridge, S. M. and Davis, H. F. (1950) Release of aspartic acid during partial acid hydrolysates. *Nature* **165,** 62–63.

19. Kamo, M. and Tsugita, A. (1998) Specific cleavage of N side of serine and threonine with S-ethyltrifluoroactate vapor. *Eur. J. Biochem.* **255,** 162–171.
20. Nabetani, T., Miyazaki, K., and Tsugita, A. (2000) Chemical cleavage of asparaginyl peptide bond with aqueous hydrazine vapor. *Res. Commun. Biochem. Cell. Mol. Biol.* **4,** 205–220.
21. Tsugita, A., Miyazaki, K., Nabetani, T., Nozawa, T., Kamo, M., and Kawakami, T. (2001) Application of chemical cleavage to analyze post-translational modification. *Proteomics* **1,** 1082–1091.

4

Means of Hydrolyzing Proteins Isolated upon ProteinChip® Array Surfaces

Chemical and Enzymatic Approaches

Shanhua Lin, Ning Tang, and Scot R. Weinberger

1. Introduction

1.1. Overview of SELDI ProteinChip Array Technology

ProteinChip® Array, surface-enhanced laser desorption ionization time-of-flight mass spectrometry (SELDI-TOF MS) is an advanced analytical technique providing facile protein analysis of complex biological mixtures. ProteinChip Array surfaces function as solid-phase extraction media that support on-probe isolation and clean-up of analytes prior to mass spectrometric investigation. Two fundamental types of arrays are available: general chromatographic arrays that function to extract biocompounds using quasi-specific levels of affinity, often employed for differential protein expression studies; and preactivated surfaces capable of covalently binding target compounds for specific biomolecular interaction studies (**Fig. 1**). The advancement of ProteinChip Array technology has been championed by researchers at Ciphergen Biosystems and is embodied in Ciphergen's Protein Biology System platforms.

When used for differential expression studies, ProteinChip Array profiling allows the rapid creation of phenotypic fingerprints often elucidating biomarkers of a particular physiological event or disease state (*1–6*). Invariably, biomarker discovery leads to a need for biomarker identification, and the process of on-chip proteolysis and MS analysis ensues (*7*). Alternatively, preactivated arrays may be used to immobilize target biologicals for interaction studies, as is often the case when searching for ligands against orphan receptors or utilizing antibodies to extract target antigens. Often, *in situ* proteolytic digestion followed by mass spectrometric studies are employed to identify captured species, as well as to elaborate specific ligand-binding domains or antigenic determinants for these respective biomolecular interactions (*8*).

1.2. Protein Identification by Mass Spectrometry

Protein identification by MS is best accomplished by fragmenting peptides, although some sequence information can be obtained by direct fragmentation of intact proteins (*9,10*). Therefore, protein digestion is a critical step in protein identification. The three major protein profiling and identification techniques are sodium dodecyl sulfate poly-

From: *Handbook of Proteomic Methods*
Edited by: P. Michael Conn © Humana Press Inc., Totowa, NJ

Preactivated Surfaces for Specific Protein Interaction Studies

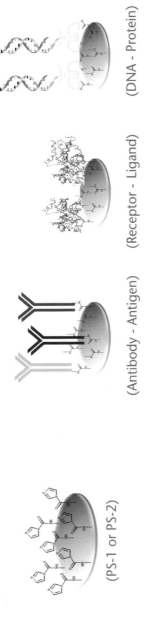

(PS-1 or PS-2)　　(Antibody - Antigen)　　(Receptor - Ligand)　　(DNA - Protein)

Chromatographic Surfaces for General Profiling

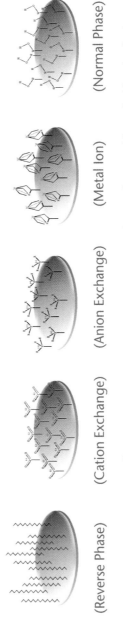

(Reverse Phase)　(Cation Exchange)　(Anion Exchange)　(Metal Ion)　(Normal Phase)

Fig. 1. Various ProteinChip® Array surfaces. Preactivated surfaces contain reactive epoxides or imidazoles for immobilizing biomolecules using covalent attachment to primary amines or alcohols. Immobilized biomolecules are then used for specific interaction studies. Chromatographic surfaces function to extract proteins via hydrophobic, electrostatic, hydrogen binding, or coordinate covalent binding mechanisms.

acrylamide gel electrophoresis (SDS-PAGE) separation followed by in-gel digestion; global in solution digestion of protein mixtures and separation by multidimensional chromatography; and protein biochip array purification using MS profiling, ultimately followed by on-array digestion. The first two methods of protein profiling and their respective digestion strategies have been well documented *(11–14)* and are currently widely used in the field of proteomic research. On the other hand, on-array or on-target proteolysis has rarely been done. Patterson and coworkers *(15)* have been using carboxypeptidase Y to digest peptides on matrix-assisted laser desorption/ ionization (MALDI) plates. Exoglycosidase digestion on MALDI targets has also been demonstrated to be effective for structure characterization of glycoconjugates and oligosaccharides at picomole levels *(16,17)*.

Although the performance of either chemical or enzymatic proteolysis in free solution or gels has become a relatively straightforward routine, on-chip proteolysis often presents a comparatively greater challenge. For a variety of reasons, proteins immobilized upon ProteinChip Array surfaces are, at times, refractory to conventional enzymolysis. ProteinChip Array surfaces are generally stratified, employing various surface chemical modifications. In some instances, such modification may bind the applied enzyme in an effectively nonreversible manner or may sterically inhibit enzyme access to bound proteins. Additionally, proteins bound to array surfaces may become denatured, and when enzyme-containing aqueous solutions are applied, these denatured proteins often become sequestered by an interfacial plane of surface tension, excluding the protein from entering the solution phase where protease activity must occur. The total amount of protein captured on ProteinChip Arrays is also limited. Finally, during protein profiling or biomolecular interaction analysis, strong detergents are often used under stringent wash conditions. Residual amounts of these detergents remaining on the array surface may compromise proteolytic activity. To address these problems, we have investigated the use of limited acid hydrolysis and limited enzymolysis as opposed to conventional proteolysis regimes.

1.3. Overview of Limted Proteolysis

Limited proteolysis, also known as partial proteolysis or controlled proteolysis, has been used to provide low-resolution protein structure *(18)*, to probe protein conformational changes *(19)*, and to dissect protein domains *(20)*. Limited proteolysis was rarely used for protein identification purposes, for it routinely created an insufficient number of peptides to use for database mining. Furthermore, when multiple proteins are simultaneously digested, a heterogeneous peptide pool is frequently created, and successful database mining requires not only high mass accuracy, but primary sequence information that is embedded in tandem MS/MS data. Until recently, the only MS/MS approach available for laser desorption-based analyses was postsource decay analysis (PSD), which has demonstrated low fidelity in generating useful peptide fragments at trace levels of starting material. The recent development of a laser desorption/ionization quadruple time-of-flight mass spectrometer (TOF MS) has made high-efficiency, collision-induced dissociation (CID) MS/MS analysis possible for SELDI and MALDI approaches. The high mass accuracy of both MS and MS/MS provided by such devices has allowed identification of a previously sequenced protein using only one or two peptides.

Recently, three alternative methods of enzymatic digestion for protein identification by MS have been reported. The first method, vapor-phase acid hydrolysis, which was reported by Gobom et al. *(21)*, uses pentafluoropropionic acid (PFPA) to produce peptide fragments for protein identification. This method produces sequence ladders as well as three different types of cleavages. Fragmentation is reported to occur on either side of glycine and on the C-terminal side of aspartic acid. The second method, developed by Li et al. *(22)*, is based on chemical cleavage by formic acid. It cleaves efficiently and specifically at aspartyl residues, thus allowing database searches. Protein identification using in-gel cleavage by formic acid has also been performed using this method. The third method, published by our group *(23)*, showed that limited acid hydrolysis of proteins using trifluoroacetic acid (TFA) produces similar cleavage patterns to those reported by Gobom et al. *(21)*. The examples and detailed discussions given in our previous publication *(23)* demonstrate the effectiveness of this method for protein ID in free solution as well as on-array. In addition, we have developed a protocol that can be applied to gel-purified proteins (unpublished results).

This chapter presents a number of proven strategies to facilitate on-array or on-target proteolytic studies of captured proteins. Several different limited enzymatic and acid hydrolysis protocols and representative results are presented.

2. Materials

2.1. Reagents

1. Dithiothreitol (DTT) and TFA (Sigma, St. Louis, MO).
2. α-Cyano-4-hydroxycinnamic acid (CHCA) and sinapinic acid (SA) (Ciphergen Biosystems, Fremont, CA).
3. Horse heart apo-myoglobin and hen egg white lysozyme (Sigma). Prior to acid hydrolysis, both proteins were confirmed to be free of degradation by SELDI-TOF analysis.
4. Cytochrome C (horse), fetal calf serum, and Triton X-100 (Sigma).
5. Carcinoembryonic antigen (CEA) and monoclonal anti-CEA antibody (BioDesign International, Saco, Maine).
6. Phosphate-buffered saline (pH 7.4) and HEPES buffer (1 *M*, pH 7.4) (Invitrogen, Chicago, IL).

3. Methods

3.1. Mass Spectrometry

3.1.1. Single MS Analysis

1. A saturated solution of CHCA matrix was used for analysis of acid hydrolysis products. The matrix solvent was 50%/50% H_2O/ACN (v/v) and 0.5% TFA.
2. Spectra were acquired in the positive-ion mode on a Ciphergen PBS II System (Fremont, CA), a time lag focusing, linear, laser desorption/ionization TOF MS. Time lag focusing delay time was set to 400 ns.
3. Ions were extracted using a 3-kV ion extraction pulse and accelerated to final velocity using 20 kV of acceleration potential.
4. The system employed a pulsed nitrogen laser at repetition rates varying from 2 to 5 pulses/s. Typical laser fluence varied from 30 to 150 µJ/mm^2.
5. An automated analytical protocol was used to control the data acquisition process in most of the sample analyses. Each spectrum was an average of at least 50 laser shots and was externally calibrated against a mixture of known peptides. Peptide sequences could be directly derived from the mass spectrum when peptide ladders were generated. Protein sequences of our model systems were retrieved using NCBI database and Prowl software (http://prowl1.rockefeller.edu/prowl/proteininfo.html).

3.1.2. Tandem MS Analysis

1. For MS/MS experiments, spectra were acquired on a PE-Sciex (Concord, Canada) Q-STAR tandem Qq-TOF MS equipped with a Ciphergen ProteinChip® Array interface (ProteinChip Qq-TOF).
2. Ions were created using a pulsed nitrogen laser (VSL 337 NDS, Laser Science, Franklin, MA) operated at 30 pulses/s and delivering an average pulse fluence of 130 μJ/mm^2.
3. Nitrogen gas, at 10 mtorr of pressure, was used for collisional cooling of formed ions, and argon was used as a collision gas for all low-energy collision-induced dissociation experiments.
4. The previously described CHCA matrix system was used for tandem analysis of all hydrolysis products.
5. Applied collision energy in general followed the rule of 50 eV/kDa, and each acquisition was typically the sum of 1 min of spectra at a TOF scan rate of 7 kHz.
6. Protein identification was carried out using the UCSF ProteinProspector MS-Tag program (http://prospector.ucsf.edu/ucsfhtml3.4/mstagfd.htm).

3.2. Limited Acid Hydrolysis

3.2.1. Limited Acid Hydrolysis Cleavage Rules

The rules and principles of acid hydrolysis describe a total of six cleavage sites as well as the formation of peptide ladders (*7*). The preferred cleavage sites for acid hydrolysis are at either side of glycine, at the C-terminal side of aspartic acid and glutamic acid, and at the C-terminal side of asparagine and glutamine, the latter two being spontaneously deamidated before cleavage. Peptide ladders with C-terminal peptide cleavage sites are produced from the N-terminal portion of a peptide or protein, while peptide ladders with N-terminal peptide cleavage sites are produced from the C-terminal portion of the respective peptide or protein.

3.2.2. In-Gel Acid Hydrolysis

The following is a recommended protocol for in-gel acid hydrolysis. Representative results are provided for didactic purposes.

1. Protein samples should be dissolved in 10 μL of sample loading buffer, loaded onto a 4–20% Tris-glycine gel, and stained by colloidal blue stain.
2. The gel band containing the desired protein (about 10 μg) should be cut out and transferred to a 0.5-mL microcentrifuge tube.
3. The gel slices should be destained and SDS should be removed by incubating on a shaker with 0.4 mL of 50% methanol and 10% acetic acid.
4. After 1 h, this procedure should be repeated.
5. The methanol/acetic acid solution should be removed and the gel band incubated in 0.4 mL of 50% ACN solution for 1 h.
6. Finally, the gel band should be dehydrated in 50 μL of ACN for 15 min and dried in a vacuum oven.
7. Acid hydrolysis solution, containing 6% TFA and 1% DTT, should be added to the microcentrifuge tube to cover the gel band (about 50 μL).
8. The microcentrifuge tube should be incubated at 65°C for 2 h.
9. Multiple sample spots should be prepared on the MALDI or SELDI target.
10. For each spot, 1 μL of hydrolysis product and 1 μL of matrix should be deposited.

After following the above-recommended protocol, the MS obtained in a positive-ion analysis of the acid hydrolysis products of apo-myoglobin (apo-Mb) shows a series of

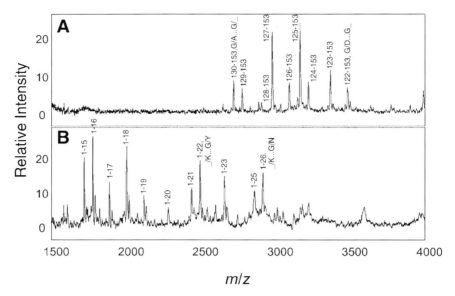

Fig. 2. Positive-ion mass spectra of peptide products resulting from in-gel acid hydrolysis of proteins, as analyzed by the PBS II. (**A**) apo-Mb. (**B**) Lysozyme.

ion peaks in the mass range of 2500–4000 amu (**Fig. 2A**). These peaks represent a series of N-terminal peptide ladders for the C-terminal peptide fragment [122–153 (G/D . . . G/_), 123–153 (D/F . . . G/_), 124–153 (F/G . . . G/_), 125–153 (G/A . . . G/_), 126–153 (A/D . . . G/_), 127–153 (D/A . . . G/_), 128–153 (A/Q . . . G/_), 129–153(Q/G . . . G/_), and 130–153 (G/A . . . G/_)]. Similarly, in-gel acid hydrolysis of lysozyme produced a series of C-terminal peptide ladders for the N-terminal peptide fragment [1–26 (_K/ . . . G/N), 1–25 (_/K . . . L/G), 1–23 (_/K . . . Y/S), 1–22 (_/K . . . G/Y), 1–21 (_/K . . . R/G), 1–20 (_/K . . . Y/R), 1–19 (_/K . . . N/Y), 1–18 (_/K . . . D/N), 1–17 (_/K . . . L/D), 1–16 (_/K . . . G/L), and 1–15 (_/K . . . H/G); **Fig. 2B**]. Furthermore, high-accuracy MS and MS/MS measurements revealed that deamidation of glutamine and asparagines were frequently observed.

3.2.3. On-Array Acid Hydrolysis

The following is a recommended protocol for on-array acid hydrolysis for SELDI analysis. Similar steps can be taken for performing on-target MALDI proteolysis. A representative set of results from SELDI experiments is presented for demonstrative purposes.

1. Typically, 1–10 pmol of proteins should be deposited on each feature of a ProteinChip Array and allowed to air dry.
2. In these experiments, eight-spot, mixed-mode arrays were used. Mixed-mode arrays demonstrate mostly a hydrophobic binding nature with some hydrophilic character.
3. After drying, 2 μL of 0.6% TFA (with 1% DTT) should be added directly to each spot.
4. Arrays should be immediately put into a sealed humidity chamber (a plastic container employing a liquid reservoir) and allowed to incubate for 2–4 h. The bottom of the humidity chamber is filled with water, and all arrays should be placed on a rack, suspended above the water surface.

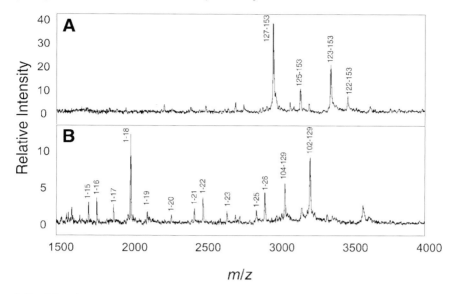

Fig. 3. Positive-ion mass spectra of peptide products resulting from on-chip acid hydrolysis of proteins, as analyzed by the PBS II. (**A**) apo-Mb. (**B**) Lysozyme.

5. The humidity chamber should be placed into a 65°C oven.
6. After incubation, arrays should be taken out and all spots allowed to air-dry prior to the addition of the CHCA matrix solution.

Results of on-array acid hydrolysis experiments for apo-Mb and lysozyme are depicted in **Fig. 3**. The hydrolysis products of apo-Mb (**Fig. 3A**) demonstrate preferred cleavage on the C-terminus of aspartic acid and glycine [(e.g., fragments 122–153 (G/D . . . G/_), 123–153 (D/F . . . G/_), 125–153 (G/A . . . G/_), and 127–153 (D/A . . . G/_)]. Similar hydrolytic patterns are observed for lysozyme [e.g., fragments 102–129 (D/G . . . L/_), 104–129 (N/G . . . L/_), and 1–18 (_/K . . . D/V)]. In addition, all the C-terminal peptide ladders (1–15 to 1–26) are detected (**Fig. 3B**). Deamidation of residue 103 converted asparagine to aspartic acid, and therefore it became a preferred cleavage site.

3.3. On-Array Enzyme Digestion

3.3.1. Commonly Used Enzymes for On-Array Digestion

The choice of appropriate protease depends on the analytical goals of the experiment. Trypsin is the most commonly used protease for protein identification purposes. Trypsin specifically cleaves at the C-terminus of Lys and Arg and creates reasonably sized peptide pools for MS analysis. Commercial sequencing-grade trypsin is modified to produce minimum autolysis products. Other frequently used specific proteases for on-array digestion are Lys-C, Asp-N, and Glu-C. On the other hand, aggressive, nonspecific proteases, which function under denaturing conditions, have been found to be highly successful for on-array digestion. For example, pepsin has an optimum pH of 1–6, and thermolysin functions at elevated temperature. These proteases can effectively digest proteins without introducing additional denaturants, which are not directly compatible with MS analysis. Another very useful protease for on-array digestion is carboxypeptidase Y. This protease cleaves successively from the C-terminus of proteins and is

Table 1
Commonly Used Proteases and Typical Conditions for On-Array Digestion

Protease	Type	Cleavage preferences	Typical digestion conditions
Trypsin	Serine protease	C-terminus of Lys and Arg	100 mM ammonium bicarbonate, pH 8.0, at 37°C for 2 h
Lys-C	Serine protease	C-terminus of Lys	In 100 mM ammonium bicarbonate, pH 8.0, at 37°C for 2 h
Asp-N	Metallo protease	N-terminus of Asp	100 mM ammonium bicarbonate, pH 8.0, at 37°C for 2 h
Glu-C	Serine protease	C-terminus of Glu (in ammonium bicarbonate, pH 7.8, or ammonium acetate buffer, pH 4.0) or Glu and Asp (in phosphate buffer, pH 7.8)	100 mM ammonium bicarbonate, pH 8.0, at 37°C for 2 h
Pepsin	Aspartic protease	C-terminus of aromatic and hydrophobic amino acids (especially Phe and Leu); it will not cleave bonds containing Val, Ala, or Gly residues	0.5% TFA at 37°C for 2 h
Thermolysin	Metallo protease	N-terminal side of Leu, Phe, Ile, and Val, but it could also cleave other residues at a lower rate, such as Met, Tyr, Ala, Asn, Ser, Thr, Gly, Lys, and Glu.	100 mM ammonium bicarbonate, pH 8.0, at 65°C for 2 h
Carboxy-peptidase Y	Serine protease	Successively release amino acid from C-terminal	100 mM ammonium acetate, pH 6.0, at room temperature

especially useful during epitope mapping experiments, when attempting to qualify true limit epitopes.

Some commonly used proteases, their type, cleavage rules, and typical on-array digestion conditions are listed in **Table 1**. Volatile buffers have been chosen to avoid introducing Na^+, K^+, or other MS troublesome cations. The concentration of applied protease (usually ranging from 1 ng to 100 ng/µL) depends on the estimated amount of proteins on the array. Because some enzyme may be lost owing to binding to the array surface, the optimal enzyme-to-substrate ratio is usually higher than is suggested for in-solution digestion. The volume of on-array digestion reaction is generally kept between 2 and 5 µL. An important step during all digestion experiments is to include

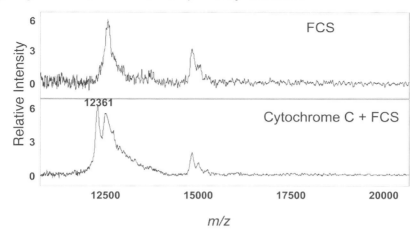

Fig. 4. Positive-ion mass spectra of differential protein profiles of fetal calf serum (FCS) and cytochrome C spiked into FCS obtained on PBS II.

parallel digestion of a control sample. This control sample can be a blank spot or a sample of the same origin as the real sample.

3.3.2. Typical Protocol for On-Array Digestion

The following represents a typical protocol for on-array digestion. In this example, trypsin is used. Other enzymes may be employed recognizing salient differences in optimal pH and recommended enzyme substrate stoichiometry.

1. A model system was created by spiking 10 µL of cytochrome C (80 µg/mL = 6.5 nmol/mL) into 40 µL of 10% fetal calf serum (FCS) in PBS (6 mg/mL).
2. Five microliters of this sample and 5 µL of 8% FCS were spotted on different NP20 ProteinChip Arrays.
3. Samples were bound to the NP20 ProteinChip Arrays by incubation in the previously described humidity chamber for 15 min.
4. All arrays were then washed with 5 mM HEPES, pH 7.4, for 5 min. The wash was repeated two more times with fresh buffer.
5. One of the arrays was air-dried.
6. One microliter of sinapinic acid (SPA) matrix (in 50% acetonitrile/0.5% aq. TFA) was loaded on one of the ProteinChip Arrays, and this array was read in a PBS II as previously described **(Fig. 4)**. **Figure 4** shows the PBS II mass spectra and resultant protein profiles for the sample (cytochrome C spiked-in FCS) and control (FCS). A peak uniquely appearing in the sample is marked (12,361 *m/z*).
7. The rest of the NP20 ProteinChip Arrays were loaded with 2 µL of trypsin in 100 mM NH_4HCO_3, pH 8.0. They were then incubated in a humid chamber at 37°C for 2 h. One microliter of CHCA matrix (in 50% acetonitrile/0.1% aq. TFA) was added on the arrays at the end of the digestion.
8. These arrays were analyzed using the previously described PBS II and SELDI-QqTOF systems to identify peptides unique to the cytochrome C spiked system (the sample) compared with the control FCS system. Unique peptides were analyzed using low-energy CID on the SELDI-QqTOF system. Protein identification was done using MS-Tag (http://prospec tor.ucsf.edu).

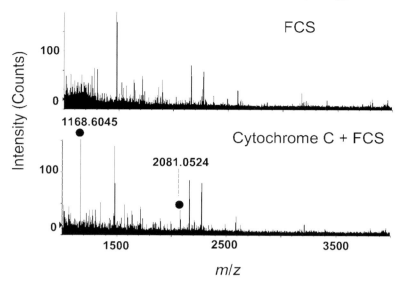

Fig. 5. Positive-ion mass spectra of on-array trypsin digestion of fetal calf serum (FCS) and cytochrome C spiked into FCS obtained on a SELDI-QqTOF. The labeled peaks are unique in the cytochrome C + FCS sample.

Figure 5 illustrates MS spectra for sample and control arrays acquired on the SELDI-QqTOF after on-array digestion with trypsin. The spectrum at the top shows the control and the spectrum at the bottom shows the sample. Peptides that are uniquely present in the sample $((M+H)^+ = m/z$ 1168.6045, 2081.0524) are labeled. The peptide at 1168 m/z was then selected for CID and MS/MS analysis, with the resulting fragment spectrum shown in **Fig. 6**. Peptide fragment masses were submitted to MS-Tag, and cytochrome C was positively identified. These results demonstrate that cytochrome C can be identified directly as a differentially displayed protein **(Fig. 4)** and can also be rapidly identified based on the differential display of a constituent peptide following on-array proteolytic digestion.

3.3.3. Capturing of Carcinoembryonic Antigen by Its Antibody Followed by On-Array Pepsin Digestion

This example demonstrates the use of on-array digestion for protein identification after antibody capture upon preactivated array surfaces. CEA is a glycoprotein that is expressed in a variety of secretory tissues. Elevated serum levels of CEA are associated with several malignant states, and immunoassays for CEA have been used for several years in monitoring malignancy. CEA is difficult to detect in MALDI owing to its glycoprotein heterogeneity. CEA is also very difficult to digest with specific proteases under native conditions. On-array pepsin digestion converts the proteins to peptides and facilitates identification of the resultant CEA peptides.

1. Two microliters of 1 m*M* protein G solution was applied on all the spots of a Ciphergen PS20 ProteinChip Array. PS20 ProteinChip Arrays employ an epoxy surface that covalently reacts with amine and thiol groups, covalently binding protein G to the array's surface.
2. Each array was incubated in the previously described humid chamber at 4°C for 16 h.

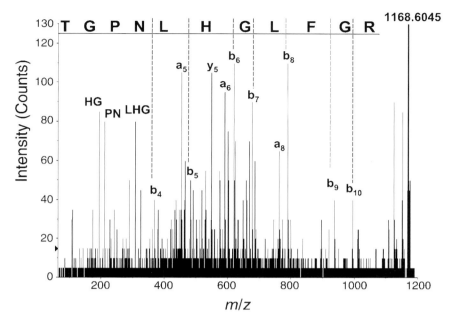

Fig. 6. MS/MS spectra of $(M+H)^+ = m/z$ 1168.6045.

3. Residual active sites were later blocked by placing the array in a conical 15-mL tube with 8 mL of blocking buffer (0.5 M ethanolamine in PBS, pH 8.0).

4. The tube was agitated using a rotating platform for 15 min at room temperature. After blocking, the array was washed with 0.5% Triton X-100 in PBS buffer for 15 min and then with PBS three times.

4. The array was air-dried, and 2 μL of anti-CEA antibody (2.3 mg/mL) was applied to each feature.

5. The array was incubated in the humid chamber at room temperature for 2 h and then washed with 0.5% Triton X-100 in PBS for 15 min and PBS three times.

6. Two microliters of antigen at 2 pmol/μL was applied to each feature, and the array was incubated in a humid chamber at room temperature for 2 h.

7. The array was then washed with 0.5% Triton X-100 in PBS for 15 min three times, followed by a PBS wash three times.

8. After the array was air-dried, 2 μL of pepsin at 0.01 mg/mL in 0.5% TFA was applied to all features, excluding negative controls, and the array was incubated in a humid chamber at 37°C for 2 h. At the end of the digestion, 1 μL of CHCA matrix was applied on the digested features and 1 μL of SPA on the undigested spots. The array was first read in PBS II and then on a SELDI-QqTOF system to obtain MS/MS spectra.

TOF-MS spectra of pepsin-digested sample spots are shown in **Fig. 7**. The top row shows the spectrum from the array having protein G covalently bound ("Protein G"); the middle row provides the spectrum from the array further binding anti-CEA monoclonal antibody ("Protein G + anti-CEA"); and the bottom row shows the spectrum from the array further binding CEA ("Protein G + anti-CEA + CEA"). We found that anti-CEA antibody was also digested by pepsin (**Fig. 7**, row 2). This spectrum was used as the control. After comparing the digestion pattern of the control (**Fig. 6**, row 2) against the CEA-containing array (row 3), we observed one major difference at m/z 1896. This

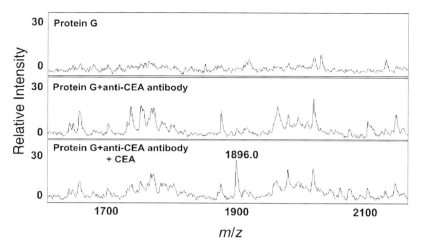

Fig. 7. Positive-ion mass spectra of on-array pepsin digestion of protein G, protein G + anti-carcinoembryonic antigen (CEA) antibody, and protein G + anti-CEA antibody + CEA.

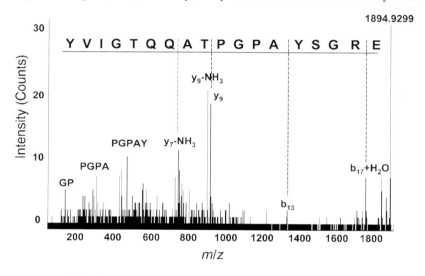

Fig. 8. MS/MS spectra of (M+H)$^+$ = *m/z* 1894.9299.

peak was further analyzed by CID on SELDI-QqTOF. **Figure 8** shows the MS/MS spectrum of (M+H)$^+$ = 1894.9299. Fragment masses were submitted to MS-Tag for protein identification using the least stringent searching parameters (molecular weight range: all; species: all; enzyme: none; parent ion: 20 ppm; fragment ions: 50 ppm; 640,428 entries). This peptide was identified as peptide YVIGTQQATPGPAYSGRE from CEA. Peptide fragments arising from amide bond cleavage were observed corresponding to charge retention on the N-terminus (b ions), C-terminus (y ions), and internal fragments (labeled according to their sequence).

On-array digestion of CEA has also been proven to be sensitive and quantitative. Serial dilutions of CEA from 800 fmol–8 fmol were loaded on the antibody array. After on-array digestion, the resulting peptide at *m/z* 1894.9 was observed when as little as

20 fmol CEA was loaded. The measured intensity of this peptide peak was found to be quantitative from 20 fmol –80 fmol of total CEA.

References

1. Fang, L., Zhang, R., Williams, E. R., and Zayre, R. N. (1994) On-line time-of-flight mass spectrometric analysis of peptides separated by capillary electrophoresis. *Anal. Chem.* **66,** 126–133.
2. Fung, E. T., Wright, G. L., Jr., and Dalmasso, E. A. (2000) Proteomic strategies for biomarker identification: progress and challenges. *Curr. Opin. Mol. Ther.* **2,** 643–650.
3. Fung, E. T., Thulasiraman, V., Weinberger, S. R., and Dalmasso, E. A. (2001) Protein biochips for differential profiling. *Curr. Opin. Biotechnol.* **12,** 65–69.
4. Merchant, M. and Weinberger, S. R. (2000) Recent advancements in surface-enhanced laser desorption/ionization-time of flight-mass spectrometry. *Electrophoresis Field* **21,** 1164–1177.
5. Weinberger, S. R., Morris, T. S., and Pawlak, M. (2000) Recent trends in protein biochip technology. *Pharmacogenomics* **1,** 395–416.
6. Rubin, R. B. and Merchant, M. (2000) A rapid protein profiling system that speeds study of cancer and other diseases. *Am. Clin. Lab.* **19,** 28–29.
7. Lin, S., Tornatore, P., Weinberger, S. R., King, D., and Orlando, R. (2001) Limited acid hydrolysis as a means of fragmenting proteins isolated upon ProteinChip® array surfaces. *Euro. J. Mass Spectrom.* **7,** 131–141.
8. Howard, J. C., Heinemann, C., Thatcher, B. J., Martin, B., Gan, B. S., and Reid, G. (2000) Identification of collagen-binding proteins in *Lactobacillus* spp. with surface-enhanced laser desorption/ionization-time of flight ProteinChip technology. *Appl. Environ. Microbiol.* **66,** 4396–4400.
9. Loo, J. A., Edmonds, C. G., and Smith, R. D. (1991) Tandem mass spectrometry of very large molecules: serum albumin sequence information from multiply charged ions formed by electrospray ionization. *Anal. Chem.* **63,** 2488–2499.
10. Ge, Y., Lawhorn, B. G., El Naggar, M., et al. (2002) Top down characterization of larger proteins (45 kDa) by electron capture dissociation mass spectrometry. *J. Am. Chem. Soc.* **124,** 672–678.
11. Lee, T. D. and Shively, J. E. (1990) Enzymatic and chemical digestion of proteins for mass spectrometry. *Methods Enzymol.* **193,** 361–374.
12. Courchesne, P. L., Luethy, R., and Patterson, S. D. (1997) Comparison of in-gel and on-membrane digestion methods at low to sub-pmol level for subsequent peptide and fragment-ion mass analysis using matrix-assisted laser-desorption/ionization mass spectrometry. *Electrophoresis* **18,** 369–381.
13. Medzihradszky, K. F., Leffler, H., Baldwin, M. A., and Burlingame, A. L. (2001) Protein identification by in-gel digestion, high-performance liquid chromatography, and mass spectrometry: peptide analysis by complementary ionization techniques. *J. Am. Soc. Mass Spectrom.* **12,** 215–221.
14. Wolters, D. A., Washburn, M. P., and Yates, J. R. 3rd. (2001) An automated multi-dimensional protein identification technology for shotgun proteomics. *Anal. Chem.* **73,** 5683–5690.
15. Patterson, D. H., Tarr, G. E., Regnier, F. E., and Martin, S. A. (1995) C-terminal ladder sequencing via matrix-assisted laser desorption mass spectrometry coupled with carboxypeptidase Y time-dependent and concentration-dependent digestions. *Anal. Chem.* **67,** 3971–3978.
16. Geyer, H., Schmitt, S., Wuhrer, M., and Geyer, R. (1999) Structural analysis of glycoconjugates by on-target enzymatic digestion and MALDI-TOF-MS. *Anal. Chem.* **71,** 476–482.

17. Colangelo, J. and Orlando, R. (2001) On-target endoglycosidase digestion matrix-assisted laser desorption/ionization mass spectrometry of glycopeptides. *Rapid Commun. Mass Spectrom.* **15,** 2284–2289.

18. Crabb, J. W., Gaur, V. P., Garwin, G. G., et al. (1991) Topological and epitope mapping of the cellular retinaldehyde-binding protein from retina. *J. Biol. Chem.* **266,** 16,674–16,683.

19. Spolaore, B., Bermejo, R., Zambonin, M., and Fontana, A. (2001) Protein interactions leading to conformational changes monitored by limited proteolysis: apo form and fragments of horse cytochrome C. *Biochemistry* **40,** 9460–9468.

20. Aubert-Foucher, E. and Font, B. (1990) Limited proteolysis of synapsin I. Identification of the region of the molecule responsible for its association with microtubules. *Biochemistry* **29,** 5351–5357.

21. Gobom, J., Mirgorodskaya, E., Nordhoff, E., Hojrup, P., and Roepstorff, P. (1999) Use of vapor-phase acid hydrolysis for mass spectrometric peptide mapping and protein identification. *Anal. Chem.* **71,** 919–927.

22. Li, A., Sowder, R. C., Henderson, L. E., Moore, S. P., Garfinkel, D. J., and Fisher, R. J. (2001) Chemical cleavage at aspartyl residues for protein identification. *Anal. Chem.* **73,** 5395–5402.

23. Lin, S., Tornatore, P., King, D., Orlando, R., and Weinberger, S. R. (2001) Limited acid hydrolysis as a means of fragmenting proteins isolated upon ProteinChip array surfaces. *Proteomics* **1,** 1172–1184.

5

A Combined Radiolabeling and Silver Staining Technique for Improved Visualization and Localization of Proteins on Two-Dimensional Gels

Jules A. Westbrook and Michael J. Dunn

1. Introduction

Two-dimensional gel electrophoresis (2-DGE) remains the method of choice for the separation of complex mixtures of proteins, whereas mass spectrometry (MS) is rapidly becoming the key tool for protein identification. When combined, 2-DGE and MS form the current operating paradigm for classical proteomics. One of the key challenges of proteome research is that of detecting and identifying all the elements (proteins) of a proteome. Silver staining and radiolabeling, e.g., using [^{35}S]methionine ([^{35}S]met), represent two sensitive methods that are used to visualize many of the constitutive and synthesized elements of a proteome, respectively. The latter method allows a very low total protein loading on a 2-D gel and, consequently, challenges protein identification using current MS-based technology. That is, identifying 2-D gel-resolved radiolabeled proteins is problematic because the amounts of radiolabeled protein present on the gel are insufficient for excision, digestion, and identification using current MS technology. Therefore, it is necessary to refer to and locate a radiolabeled spot's equivalent on a preparatively loaded silver- or dye-stained gel (or dye-stained Western blot) and to use that protein spot for identification. Unfortunately, the images of autoradiographs and preparative gels or blots, even of the same sample, are often dissimilar, making it difficult to locate and select spots of interest accurately by visual comparison. A technique that permits the visualization and unambiguous localization of radiolabeled proteins to their silver-stained (MS-compatible) equivalents on the same 2-D gel is described in this chapter. Subsequent protein identification of superimposed spots by, e.g., peptide mass profiling using matrix-assisted laser desorption/ionization time-of-flight (MALDI-TOF) MS and peptide sequencing using tandem MS by hybrid quadrupole/orthogonal acceleration time-of-flight (Q-TOF) MS is possible.

2. Materials

1. Tissue media, dithiothreitol (DTT), iodoacetamide, bovine serum albumin (BSA), and bromophenol blue (Sigma-Aldrich, Gillingham, Dorset, UK).
2. Radiolabel, [^{35}S]methionine (Redivue), Pharmalyte, immobilized pH gradient strips (IPGs), Plus One silver stain kit, and all electrophoresis equipment (reswelling trays, Multiphor

From: *Handbook of Proteomic Methods*
Edited by: P. Michael Conn © Humana Press Inc., Totowa, NJ

electrophoresis units, ISO-DALT electrophoresis tanks, and other equipment) (Amersham Biosciences, Little Chalfont, Buckinghamshire, UK).
3. Glycine, sodium dodecyl sulfate (SDS) and Tris base (Genomic Solutions, Huntingdon, Cambridgeshire, UK).
4. Urea (GibCo/Invitrogen, Paisley, UK).
5. CHAPS (Calbiochem; CN Biosciences, Beeston, Nottingham, UK).
6. Acetone, glycerol, acetic acid, hydrochloric acid, methanol, phosphate-buffered saline (PBS), sucrose, silica gel sachets, trichloroacetic acid (TCA), Kodak MR x-ray film, and general labware (VWR International, Poole, Dorset, UK).
7. Glass fiber filter pads and 3MM paper (Whatman, Maidstone, Kent, UK).
8. Liquid scintillant counting cocktail (Filter-Count; Packard BioScience; Pangbourne, Berkshire, UK).
9. Human heart endomyocardial biopsy material was obtained from patients recruited to an ethically approved study in accordance with the Helsinki Declaration of 1975.

3. Methods
3.1. Sample Tissue Radiolabeling Conditions

1. Endomyocardial biopsy tissue was obtained "fresh" from the patient undergoing scheduled monitoring as part of postoperative care for heart transplantation. The biopsy sample was transported from the theater to the laboratory in an Eppendorf tube containing a small aliquot of Hanks' balanced salts solution tissue media.
2. In the laboratory, the biopsy was incubated, at 37°C, in methionine-free Hanks' balanced salts solution tissue media containing 225 µCi [^{35}S]methionine radiolabel for 20–24 h.
3. Following incubation, and immediately prior to storage (at –20 or –80°C) or homogenization, the biopsy was rinsed in 0.35 M sucrose to remove culture media salts, which are found to interfere with isoelectric focusing.

3.2. Sample Preparation for 2-DGE
3.2.1. Preparation of Radiolabeled Proteins

The radiolabeled biopsy was homogenized in 200 µL lysis buffer (9.5 M urea, 1% DTT, 2% CHAPS, 0.8% Pharmalyte, pH 3–10) (*1*) and centrifuged for 15–30 min at 10,000g (at 17–20°C) to remove insoluble material. The radioactive content of the supernatant was determined by scintillation counting.

3.2.2. Isolation of Radiolabeled Proteins for Scintillation Counting

1. A 5-µL aliquot of (radiolabeled) supernatant was added to 100 µL BSA (1 mg/mL) in a 1.5-mL Eppendorf tube.
2. One milliliter of ice-cold TCA (10% w/v) was added and, after brief vortexing, the tube was placed on ice for 1 h.
3. Radiolabeled proteins were isolated on glass fiber filter pads using a Millipore 1225 sampling manifold vacuum filtration device (Millipore, Watford, UK).
4. Prior to use, filter pads were blocked by soaking in a solution of BSA (5 mg/mL, prepared in PBS) for 1 h (minimum), and the filtration device was cooled by packing ice in the liquid collection compartment.
5. The entire contents of the ice-incubated Eppendorf tube were dispensed onto a filter pad under vacuum.
6. The pad was then washed twice with ice-cold TCA (10% w/v) followed by one rinse with ice-cold acetone (100% v/v).

7. The filter pad was removed from the vacuum filtration apparatus and allowed to air dry.
8. The dried pad was placed in a scintillation counting tube, and 3 mL scintillant cocktail fluid was added.
9. The counting tube was capped, and the pad was dissolved by vigorous shaking for 30 s.
10. The radioactivity was then counted using a Packard Tri-Carb liquid scintillation counter set to ^{35}S mode and allowing 3 min for each count. The radioactivity of the sample was then resolved in units of counts per minute (cpm).

3.2.3. Preparation of Nonradiolabeled Proteins

A piece of nonradiolabeled heart material (left ventricle) of approx 100 mg was homogenized in 1 mL lysis buffer (9.5 *M* urea, 1% DTT, 2% CHAPS, 0.8% Pharmalyte, pH 3–10) *(1)* and, following centrifugation (15–30 min at 10,000*g*, 17–20°C), the protein content was determined using a modified dye-binding protein assay *(2)*.

3.3. Two-DGE

3.3.1. Isoelectric Focusing (First-Dimension Separation)

1. 2-DGE was carried out essentially according to Gorg et al. *(1)*.
2. A pH 3–10 nonlinear (NL; 18 cm) IPG strip was rehydrated in a solution containing 1×10^6 cpm radiolabeled material and 400 µg nonradiolabeled protein made up to a final volume of 450 µL with reswelling solution (8 *M* urea, 0.2% DTT, 0.5% CHAPS, 0.2% Pharmalyte, pH 3–10) *(1)*.
3. Following overnight rehydration by the in-gel method *(3,4)*, isoelectric focusing was performed for 65–75 kVh using a Multiphor II platform.

3.3.2. Second-Dimension Separation

1. Prior to second-dimension separation, the IPG strip was equilibrated in buffer (1.5 *M*, pH Tris-HCl buffer solution, 8.8, 6 *M* urea, 30% glycerol, 2% SDS, 0.01% bromophenol blue) containing 1% DTT for 15 min, followed by another 15 min in the same buffer containing 4.8% iodoacetamide *(1)*.
2. Following equilibration, SDS-polyacrylamide gel electrophoresis (PAGE) was performed overnight at 20 mA/gel, 10°C, using 12% T/2.6% C (200 × 200 × 1-mm) separating gels.

3.4. MS-Compatible Silver Staining

1. Following SDS-PAGE, the gel was fixed overnight (10% acetic acid, 50% methanol) prior to staining using a modified version of the Plus-One silver stain kit (Amersham Biosciences) according to Yan et al. *(5)*. No quenching of the radioactive signal is observed using this stain.
2. The stained gel was scanned at 100-µm, 12 bit resolution using a Personal Densitometer SI (Amersham Biosciences).

3.5. Gel Drying

1. Prior to drying, the stained gel was soaked overnight in a solution containing 3% glycerol and 30% methanol (400 mL total) *(6)*. The solution renders polyacrylamide gels (10–15%) amenable to drying with reduced incidences of cracking.
2. The gel was placed on a sheet of 3MM (Whatman) paper cut to a size just slightly greater than the dimensions of the gel. It was found that wetting the 3MM sheet in the gel drying solution just before placing the gel on top reduced the likelihood of the gel cracking during the drying procedure.

3. The gel/wetted 3MM paper was placed on top of two dry sheets of 3MM paper on the drying table of a Savant SpeedGel (SG210D) integrated gel dryer (Savant Instruments, Holbrook, NY).
4. The gel was then dried for 1 h at 62°C followed by 1 h at 66°C.
5. Following vacuum-assisted drying, the gel was allowed to cool at room temperature and air dry by placing the gel in a thin cardboard box (to prevent the gel from rolling-up on itself).
6. The box was then placed in an opaque plastic bag containing a silica gel sachet overnight. It was found that this extra drying step removed any traces of moisture and prevented the gel from sticking to the x-ray film during autoradiography.

3.6. Autoradiography and Film Scanning

1. The dried gel was then exposed to suitable x-ray film (e.g., BioMax-MR, Kodak), typically for 6–8 d (1×10^6 cpm).
2. The film was developed using an automated Agfa film processor and scanned as described as above.
3. When scanning x-ray films it was found to be useful to wet the scanning tray with water and then to place the film on top, carefully ensuring a uniform seal between the glass and the film. Air bubbles can be "squeezed out" by gentle pressure using soft tissue (to prevent the film being scratched). This extra step prevented the appearance of refraction lines when the scanned image was viewed at high contrast.

3.7. Image Alignment

Once the x-ray film had been developed, it was possible to overlay the film on the dried (silver-stained) gel for optimal alignment to allow accurate matching and localization of radiolabeled protein spots to the nonradiolabeled protein spot profile. Following rehydration of the dried gel (*see* **Subheading 3.8.**), protein spots of interest were excised and identified by peptide mass profiling using MS *(7)*.

3.8. Rehydration of the Silver-Stained 2-DGE

In order to excise any protein spots of interest for identification by MS-based methods, it was necessary to rehydrate the dried gel. **Figure 1** illustrates the steps involved. We found it difficult to cut any spot of interest accurately and carefully from the dried gel owing to the tough and sometimes brittle nature of such a gel. Therefore, the gel was rehydrated by immersing it fully gel-side-down in a tray of high-quality water (approx 400 mL). In our hands, the gel quickly dissociates from the 3MM paper and remains intact (provided there are no cracks already present on the gel). The excess 3MM paper that remains initially bound to the gel can be removed by gently rubbing it away using one's fingers (wearing gloves). It is best to remove the excess 3MM residue before the gel has reswollen to its original size since it naturally becomes more fragile with rehydration. Furthermore, it is nearly impossible to remove the excess 3MM totally; however, the trace film of 3MM that remains was found to not interfere with subsequent MS *(7)*. Once the gel had rehydrated to the original size, it was possible to excise the spots of interest for MS.

4. Final Comments

The technique described here is useful for comparing the display of radiolabeled proteins (i.e., proteins that are synthesized during the incubation period and conditions of

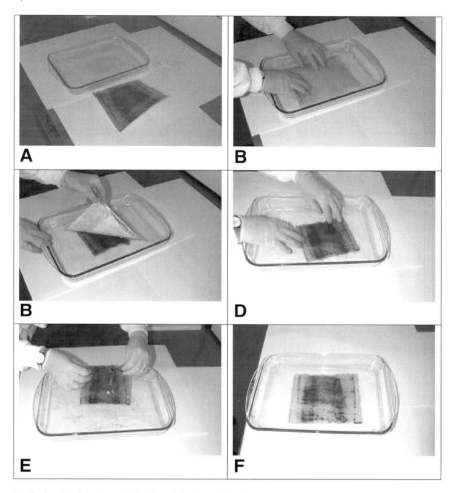

Fig. 1. Gel rehydration. **(A)** The dried gel is shown next to a glass dish containing approx 400 mL ddH$_2$O. **(B,C)** The dried gel is placed gel-face-down in the dish, which then dissociates from the backing paper with a vigorous shrinking action within approx 1 min. **(D,E)** Residual gel-bound filter paper can be removed by careful rubbing. **(F)** The gel will rehydrate to its regular size within approx 45 min in ddH$_2$O.

an experiment) with the display of proteins that are constitutively expressed in a tissue and are revealed by silver staining. Ultimately, the method permits such comparisons on one gel and has the added facility of permitting subsequent identifications to be made. Being able to compare protein profiles on one gel removes the problem of inter-gel variation caused by many factors both experimental (e.g., batch variation in sample buffers and gels) and sample dependent. Any synthesized (radiolabeled) protein spots of interest can then be localized accurately to the constitutive (silver-stained) version and cut from the gel (following gel rehydration) for identification by MS.

Acknowledgments

J. A. W. gratefully acknowledges Dr. Jun Yan for invaluable assistance in the development of the methodology. Mass spectrometry was originally performed by Dr. Robin

Wait at the Kennedy Institute of Rheumatology Division, Imperial College School of Medicine, London, UK *(7)*.

References

1. Gorg, A., Obermaier, C., Boguth, G., et al. (2000) The current state of two-dimensional electrophoresis with immobilized pH gradients. *Electrophoresis* **21,** 1037–1053.
2. Ramagli, L. S. and Rodriguez, L. V. (1985) Quantitation of microgram amounts of protein in two-dimensional polyacrylamide gel electrophoresis sample buffer. *Electrophoresis* **6,** 559–563.
3. Rabilloud, T., Valette, C., and Lawrence, J. J. (1994) Sample application by in-gel rehydration improves the resolution of 2-dimensional electrophoresis with immobilised pH gradients in the first dimension. *Electrophoresis* **15,** 1552–1558.
4. Sanchez, J.-C., Rouge, V., Pisteur, M., et al. (1997) Improved and simplified in-gel sample application using reswelling of dry immobilised pH gradients. *Electrophoresis* **18,** 322–327.
5. Yan, J. X., Wait, R., Berkelman, T., et al. (2000) A modified silver staining protocol for visualization of proteins compatible with matrix-assisted laser desorption/ionization and electrospray ionization-mass spectrometry. *Electrophoresis* **21,** 3666–3672.
6. Doyle, K. (ed.) (1996) Protein analysis, in *Protocols and Application Guide*, 3rd ed., Promega, Madison, WI, pp. 285–286.
7. Westbrook, J. A., Yan, J. X., Wait, R., and Dunn, M. J. (2001) A combined radiolabelling and silver staining technique for improved visualisation, localisation, and identification of proteins separated by two-dimensional gel electrophoresis. *Proteomics* **1,** 370–376.

6

Qualitative and Quantitative Proteomic Analyses via Multidimensional Protein Identification Technology

Michael P. Washburn and David M. Schieltz

1. Introduction

The integration of multidimensional chromatography and mass spectrometry (MS) as a proteomic tool has developed into a mature technology that is impacting biological discovery. Early descriptions of coupled two-dimensional chromatography and MS systems focused on protein separation and hinted at the potential of the concepts but failed to identify large numbers of proteins in a sample (reviewed in **ref. 1**). Link et al. *(2)* reported a dramatic improvement in the integration of multidimensional chromatography and MS by coupling two-dimensional chromatography of peptides, rather than proteins, and tandem mass spectrometry (MS/MS) in a method named direct analysis of large protein complexes (DALPC). In a tandem mass spectrometer, peptides can be fragmented in a predictable fashion, which allows for the computational determination of the peptide sequence and therefore the identification of the protein from which the peptide was derived *(3)*. In DALPC, 80 proteins were found in a *Saccharomyces cerevisiae* ribosomal preparation and 189 total proteins from a *S. cerevisiae* soluble protein extract *(2)*. DALPC is now commonly referred to as multidimensional protein identification (MudPIT).

MudPIT couples a biphasic microcapillary column packed with reversed phase (RP) and strong cation exchange (SCX) high-performance liquid chromatography (HPLC) grade materials to a tandem mass spectrometer (**Fig. 1**). In MudPIT, a complex peptide mixture is generated and loaded onto the biphasic microcapillary column. The biphasic microcapillary column is placed in-line with an HPLC and also acts as the ion source for the tandem mass spectrometer. Peptides are directly eluted off of the biphasic microcapillary column, ionized, and analyzed in the tandem mass spectrometer. The data are searched using the SEQUEST algorithm, which interprets the MS/MS generated and identifies the peptide sequence from which it was generated, resulting in the determination of the protein content of the original sample *(3)*.

Modifications to MudPIT *(4)* resulted in the detection and identification of 1484 proteins from the proteome of *S. cerevisiae (5)*. Proteins with extremes in MW, pI, hydophobicity (integral membrane proteins), and predicted abundance (proteins with low Codon Adaptation Index values) were detected and identified by MudPIT *(5)*. These results demonstrated the superior capabilities of MudPIT compared with two-

From: *Handbook of Proteomic Methods*
Edited by: P. Michael Conn © Humana Press Inc., Totowa, NJ

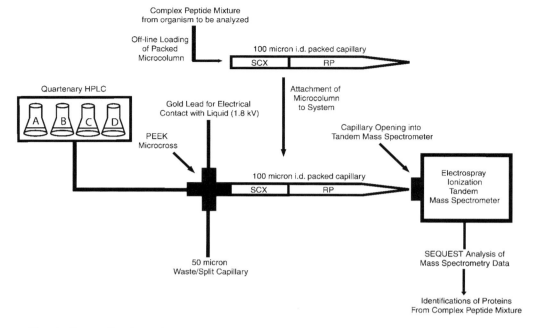

Fig. 1. Generalized MudPIT system setup and schema. Upon the preparation of a biphasic microcapillary column packed with reverse phase (RP) and strong cation exchange (SCX) packing materials, a complex peptide mixture from *S. cerevisiae*, for example, is loaded off-line onto the column. The microcapillary column is then integrated into a PEEK microcross, which couples the column, a gold lead for the electrospray voltage, the flow splitter to obtain flow rates of 0.15–0.5 µL/min, and HPLC inlet line. Xcaliber software, HPLC, and the mass spectrometer were controlled simultaneously via the user interface of the mass spectrometer. The HPLC is a quaternary system containing four different buffers consisting of buffer A (5% ACN/0.5% acetic acid/0.012% HFBA), buffer B (80% ACN/0.5% acetic acid/0.012% HFBA), buffer C (250 m*M* ammonium acetate/5% ACN/0.5% acetic acid/0.012% HFBA), and buffer D (500 m*M* ammonium acetate/5% ACN/0.5% acetic acid/0.012% HFBA). Peptides directly elute into the tandem mass spectrometer because a voltage (kV) supply is directly interfaced with the microcapillary column. The tandem mass spectra generated by the ion trap mass spectrometer are correlated to theoretical mass spectra generated from protein or DNA databases by the SEQUEST algorithm *(3)*.

dimensional polyacrylamide gel electrophoresis (2D-PAGE), which has a limited dynamic range *(6)* and access to membrane proteins *(7)*. In a more complex organism, that of *Orysa sativa* (rice), 2528 proteins were detected and identified when both MudPIT and 2D-PAGE were used in a parallel analysis to detect and identify proteins from three separate rice tissues *(8)*. A total of 391 proteins were identified by both methods, 165 were uniquely identified by 2D-PAGE, and 1972 were uniquely identified by MudPIT, demonstrating the complementary nature of these two technologies *(8)*. As a result, 2D-PAGE is worth considering when complete coverage of a proteome is needed, but it is a more time-consuming and expensive process.

MudPIT may be used for both qualitative proteomic analyses *(5,8,9)* and quantitative proteomic analyses *(10)*. In a quantitative proteomic analysis, the relative abundance of a protein from two separate cell states is determined. Cells are grown under at least two different conditions that result in altered protein expression levels, and the proteins from each cell growth condition are labeled with a "heavy" or "light" tag, thereby separating the masses of identical protein from each sample. This allows for the relative abundance of any given peptide from any given protein to be calculated by MS. The purpose of this chapter is to provide detailed protocols for qualitative and quantitative proteomic analyses via MudPIT using *S. cerevisiae* as the model organism.

2. Materials

1. Urea, ammonium acetate, ammonium bicarbonate (AmBic), Tris(carboxyethyl)phosphine (TCEP), iodoacetamide (IAM), EDTA, dibasic sodium phosphate (Na_2HPO_4), dibasic potassium phosphate (KH_2PO_4), and sodium carbonate (Na_2CO_3) (Sigma, St. Louis, MO).
2. Ammonium-^{15}N sulfate (99 atom %), ammonium-^{14}N sulfate (99.99 atom %), sodium vanadate ($NaVO_3$), sodium fluoride (NaF), sodium pyrophosphate ($Na_4P_2O_7$), HPLC grade acetonitrile (ACN), HPLC grade methanol, HPLC grade water, glacial acetic acid, formic acid, and cyanogen bromide (CNBr) (Aldrich, Milwaukee, WI).
3. DIFCO dextrose and yeast nitrogen base without amino acids or ammonium sulfate (Becton Dickinson Microbiology Systems, Sparks, MD).
4. Poroszyme™ bulk immobilized trypsin (Applied Biosystems, Framingham, MA).
5. Endoproteinase Lys-C (Roche Diagnostics, Indianapolis, IN).
6. Heptafluorobutyric acid (HFBA), BCA or Micro BCA protein assay kits (Pierce, Rockford, IL).
7. SPEC-PLUS PTC18 solid phase extraction pipet tip (Ansys Diagnostics, Lake Forest, CA).
8. Nano-LC ion-source, LCQ family of mass spectrometers (includes the LCQ Classic, Deca, or DecaXP Plus) (ThermoFinnigan, San Jose, CA).
9. QTOF1 or QTOF2 mass spectrometers (Micromass, Beverly, MA).
10. Quaternary HP 1100 series HPLC pump, fused silica capillary (100 µm inner diameter by 365 µm outer diameter and 50 µm inner diameter by 365 µm outer diameter) and Zorbax™ XDB 5 µm C$_{18}$ reverse phase packing material (Agilent Technologies, Palo Alto, CA).
11. P-2000 laser puller (Sutter Instruments, Novato, CA).
12. Partisphere SCX strong cation exchange resin (Whatman, Clifton, NJ).
13. PEEK™ MicroCross and Microtight tubing sleeves (UpChurch Scientific, Oak Harbor, WA).
14. Gold wire (Scientific Instrument Services, Ringoes, NJ).

3. Qualitative Analysis of Proteomes by MudPIT
3.1. Interfering Substances

Any organism may be qualitatively analyzed by MudPIT. Examples in the literature include analyses of the *S. cerevisiae* proteome *(5)*, soluble extracts of *O. sativa* tissue proteomes *(8)*, and life cycle stages of the malaria parasite, *Plasmodium falciparum (9)*. The sample preparation in nearly every case involves lysing a cell type in a buffer devoid of protease inhibitors, detergents, glycerol, and sucrose, for example. These chemicals interfere with nano-LC electrospray MS/MS.

Generally, if a detergent is present in solution, the mass spectrometer will only see the detergent because detergents ionize much more readily than peptides. Also, once a detergent comes into contact with the reversed phase material it will continue to leach off the column in the following runs, polluting other samples. So it is best if the detergent is removed from buffers before it is introduced to the mass spectrometer.

No protease inhibitors are used. Because the methodology looks at peptides derived from enzymatic cleavage of proteins, one needs to be able to digest the proteins with enzymes. Specifically, trypsin is used so the buffer in which the protein mixture is in needs to be at a pH of approx 8.5 and free of serine protease inhibitors.

No sucrose or glycerol is used. The viscosity of these solutions will plug the microcapillary column even if these are present in small amounts.

No EDTA or EGTA is used. Trypsin depends on a small amount of Ca^{2+} for its activity, and EDTA will chelate the calcium ions, preventing efficient digestion.

3.2. Digestion of Soluble Protein Extracts for MudPIT

1. Adjust the pH of the protein extract to 8.5 with 1 *M* ammonium bicarbonate. Determine the protein concentration by a protein assay.
2. Add solid urea to the protein solution to a final concentration of 8 *M* urea.
3. Add TCEP to the protein solution to a final concentration of 1 m*M* and incubate the solution for 20 min at room temperature.
4. Add iodoacetamide to the reduced protein solution to a final concentration of 10 m*M* and incubate the solution for 20 min at room temperature in the dark.
5. Add endoproteinase Lys-C to a final substrate to enzyme ratio of 100:1 and incubate the reaction overnight at 37°C while mixing.
6. On the following day, dilute the sample to 2 *M* urea with 100 m*M* ammonium bicarbonate, pH 8.5, and add $CaCl_2$ to a final concentration of 2 m*M*.
7. Add approx 1 μL of Porosyzme immobilized trypsin slurry to every 50 μg of protein starting material. Incubate the reaction overnight at 37°C while rotating.
8. On the following day centrifuge the reaction at maximum speed for 6 min in a microfuge to remove the trypsin beads. Transfer the supernatant to a fresh microfuge tube.

3.3. Digestion of Insoluble Protein Fractions for MudPIT Analysis

Proteins associated with insoluble particulate matter from a whole cell lysate or carefully prepared membrane samples can also be analyzed using MudPIT. However, preliminary steps are required to prepare a usable complex protein mixture from these types of samples. Because of the toxicity of CNBr, it is very important that these steps be carried out in a fume hood exercising proper safety precautions.

1. In a 1.7-mL microfuge tube, lyophilize the insoluble fraction.
2. Add 100 μL of 90% formic acid to the lyophilized pellet and incubate the tube for 5 min at room temperature.
3. Add 50 mg of CNBr to the protein, mix the solution well, and incubate it overnight at room temperature in the dark (wrap the tube in aluminum foil).
4. Transfer the solution to a 15-mL falcon tube. Add saturated ammonium bicarbonate to the tube until the pH reaches 8.5. Add the ammonium bicarbonate solution very slowly, because a large amount of bubbling occurs during this step. Use pH paper to check the pH of the sample.
5. Concentrate the fraction to approx 200 μL by lyophilization in a vacuum concentrator.
6. Proceed with **step 1**.

3.4. Concentration of Sample

Peptide samples can be solubilized in any number of reagents, including Tris-HCl, ammonium bicarbonate, acetic acid, formic acid, and urea. However, peptide samples are typically the product of a digested protein or protein mixture, in which a variety of reagents may be present including some that will interfere with the performance of the reversed-phase column and the mass spectrometer. Peptide sample volumes range from several microliters to a milliliter or more. The bed volume of the micro-column is approx 1.5 µL, which allows samples up to 50 µL to be loaded directly onto the column. For sample volumes greater than 50 µL, concentration of the sample is necessary.

1. Wet the SPEC Plus PT C18 solid-phase extraction pipet tip with 1 mL of methanol as follows. Push approximately half of the methanol through the disk, and then wait for 15–30 s to allow the disk to activate. Push the remainder of the methanol through, but do not push air through the disk.
2. Equilibrate the disk with 1 mL of 5% ACN/0.5% acetic acid (HoAc) by pushing it through the disk.
3. Pull the peptide solution into the pipet tip and then push the solution back out. This can be repeated two to three times. Peptides remain in the tip and are concentrated onto the disk. The flowthrough can be discarded.
4. Elute the peptides with approx 100 µL of 90% ACN/0.5% acetic acid. Push the eluting solvent through the disk, pull the solution back up through the disk, and then push it through one final time, to give three passes across the disk.
5. Remove the acetonitrile using a vacuum concentrator, until the peptides are nearly dry. Resuspend the peptides in approx 10–15 µL of 5% ACN/5% formic acid. The sample is ready to load onto the MudPIT column.

3.6. Quantitative Proteomic Analysis by MudPIT

A variety of chromatography-based quantitative proteomic analytical methods have been described in the past few years. A promising method is metabolic labeling, in which an organism is grown in a media in such a fashion as to replace the naturally abundant isotope of an atom with that of a heavy isotope. Metabolic labeling by growth of cells in media with either ^{14}N or ^{15}N as the sole nitrogen source has been demonstrated as a potential method for quantitative proteomics in *S. cerevisiae (10,11)*, *E. coli (12)*, *D. radiodurans*, and mammalian tissue culture *(13)*. For a quantitative proteomic analysis by metabolic labeling, it is important to determine the extent of the enrichment of the isotope into the sample before beginning an analysis. The extent of the isotopic enrichment in a sample is an important factor in determining the relative abundance of peptides prior to carrying out a comprehensive analysis. The following sections will present first an example of preparing a metabolically labeled sample, that of *S. cerevisiae* and then a method by which the isotopic enrichment may be determined. Finally, the details of carrying out a MudPIT analysis on a metabolically labeled sample will be described.

3.6.1. Growth and Lysis of S. cerevisiae

An *S. cerevisiae* strain, like S288C, can be used to prepare a metabolically labeled sample because it is a strain of yeast capable of growing in minimal media without supplementation of amino acids whose metabolic pathways have been genetically disrupted.

1. For the best enrichment of *S. cerevisiae* with ^{15}N, first carry out an overnight growth at 30°C of a 10 µL liquid culture inoculation into 5 mL of ^{14}N or ^{15}N minimal media (1.7 g yeast nitrogen base without amino acids and ammonium sulfate, 20 g dextrose, and 5 g of either ammonium sulfate per L).
2. On the following day, inoculate 250 mL of identical media with the 5 mL overnight growth at 30°C and grow to mid log phase (O.D. 0.6).

3. After they have reached mid log phase, the cells should be pelleted by centrifugation at 1000*g*.
4. Wash the pellets three times with 1X phosphate-buffered saline (PBS, 1.4 m*M* NaCl, 0.27 m*M* KCl, 1 m*M* Na$_2$HPO$_4$, 0.18 m*M* KH$_2$PO$_4$, pH 7.4) by vigorous resuspension in 1× PBS followed by centrifugation.
5. Add lysis buffer (0.1 *M* Na$_2$CO$_3$, 310 m*M* NaF, 3.45 m*M* NaVO$_3$, 12 m*M* EDTA, 250 m*M* NaCl) to the cells and lyse in a mortar and pestle frozen in liquid nitrogen by grinding the frozen cell pellet for several minutes.
6. After lysis, thaw the lysed cells and solution and place them on ice for 30 min. After incubation, spin the lysed cells at 10,000*g* for 10 min at 4°C in an Eppendorf microfuge and remove the supernatant.

3.6.2. Preparation and Analysis of Ratio Mixtures

Depending on the amount of cells that one has and the expected protein concentration, either a BCA or Micro BCA protein assay from Pierce should be run on the soluble portion of the proteome from each of the samples. The BCA assay takes longer than the Bradford assay, but the BCA assay is more compatible with contaminants and generally gives more reliable results.

Any number of different types of samples can be prepared depending on the experiment that is being run. However, each sample to be analyzed by the mass spectrometer is a ratio mixture of 1× ^{14}N proteins to 1× ^{15}N proteins sample. In an experimental setting one control sample of untreated cells grown in both medias should be prepared. However, before committing resources to a biological experiment, it is best to first do a cell growth and sample preparation and determine the extent of the isotopic enrichment (*see* **Subheading 3.6.3.**).

Each of the three samples were then prepared for MudPIT analysis by denaturation in urea, digestion with TCEP, modification with IAM, sequential digestion with Endoproteinase Lys-C and trypsin beads, and buffer exchange/concentration with SPEC Plus PTC18 SPE tips as described in **Subheading 3.4.**

3.6.3. Isotope Enrichment Analysis by High-Resolution Mass Spectrometry

A variety of instruments may be used to determine the isotope enrichment of a sample, but in each case the instrument should be of sufficient resolution to determine the difference between a peptide containing differences of as little as one ^{14}N or ^{15}N in a given peptide. An example of an experimental setup to determine the isotopic enrichment of a sample is given below.

3.6.3.1. SINGLE-DIMENSION REVERSE-PHASE MIXTURE ANALYSIS VIA THE MICROMASS, Q-TOFII™

1. A 1× ^{14}N:1× ^{15}N sample is appropriate to analyze on a Q-TOFII, which is a quadrapole time-of-flight instrument. The Q-TOFII is coupled to an Agilent quaternary HP 1100 series HPLC pump.
2. A Q-TOFII is equipped with a Z-spray™ nanoelectrospray source fitted with the glass capillary probe option.
3. A fused silica microcapillary column (100 µm i.d. × 365 µm o.d.) is pulled with a Model P-2000 laser puller as described *(14)*.

4. The pulled microcapillary column is packed with 10 cm of 5 µm Zorbax Eclipse XDB-C$_{18}$ and affixed to a PEEK finger tight microcross electrospray interface described previously *(14)*.

5. The capillary/corona voltage line is attached directly to the PEEK finger tight microcross electrospray interface through a gold rod.

6. The microcapillary column packed with RP material is threaded through the opening to the glass capillary option stage of the Q-TOFII tail end first, after which it is connected to the PEEK finger tight microcross.

7. The Q-TOFII is run according to the manufacturer's instructions for usage of the instrument with an HPLC pump and the glass capillary option of the Z-spray nanoelectrospray source.

8. The electrospray capillary voltage should be set to 2.2 kV, the cone voltage set to 40 V, and the source temperature set to 80oC. The MS survey scan is *m/z* 400–1400 with a scan time of 0.42 s and an interscan time of 0.08 s. When the intensity of a peak rises above a threshold of 8 counts, tandem mass spectra are acquired. Normalized collision energies for peptide fragmentation is set using the charge state recognition files for +1, +2, and +3 peptide ions. The scan range for MS/MS acquisition is from *m/z* 50 to 1900 with a scan time of 1.92 s and an interscan time of 0.08 s. After fragmentation, an *m/z* value is dynamically excluded for 4 min.

9. Upon the completion of a run, the MS/MS data are extracted from *.RAW files to *.DAT files according to the manufacturer's instructions and analyzed via the SEQUEST *(3)* algorithm. The SEQUEST *(3)* algorithm is run as described below for analysis of both the ^{14}N and ^{15}N peptides with two separate SEQUEST parameters files. The only specialized alteration to the SEQUEST parameters file for analysis of data from a Q-TOFII is the setting of the parent and fragment mass types to monoisotopic. Xcorr and ΔCn criteria are applied to the results as described in **Subheading 3.9.**

3.7. System Setup for MudPIT Analysis

There are several emerging choices of systems to carry out MudPIT type analyses. Systems are becoming available for purchase from a variety of instrument manufacturers like the ProteomeX from ThermoFinnigan, a Q-TOF-CapLC system may be purchased from Micromass/Waters (Beverly, MA), and others are available. Essentially every HPLC company sells a 2-D HPLC system that can be coupled to a mass spectrometer. For researchers who wish to pursue this route, it is essential that the 2-D HPLC and MS systems be compatible. Unified HPLC/MS systems like the ProteomeX are popular for this reason. These are options to the researcher, but the authors of this chapter have not evaluated them, and, therefore, we do not endorse any of these products.

The MudPIT system we use has been extensively described in the literature and is a combination of a HP1100 quartenary pump (Agilent Technologies), an LCQ^{Deca} mass spectrometer from ThermoFinnigan, and a home-built stage to hold the 2-D packed microcapillary column, which forms the ion source that couples the two *(14)*. A description of the setup is given below.

3.7.1. Preparation of a 2-D Nano-LC Column

Note: If the researcher prefers not to purchase a laser puller and to construct individual columns, packed 2-D fused silica microcapillary columns (Biphasic PicoFrit™) may be purchased from New Objective (Woburn, MA).

1. Make a window in the center of approx 50 cm of a 100×365-μm fused silica capillary by holding it over an alcohol flame until the polyimide coating has been charred. Remove the charred material by wiping the capillary with a tissue soaked in methanol.

2. To pull a needle, place the window portion of the capillary into the P-2000 laser puller. Position the exposed window on the capillary in the mirrored chamber of the puller. Arms on each side of the mirror have grooves and small vises, which properly align the fused silica and hold it in place. A typical four-step parameter setup for pulling approx 3-μm tips from a 100×365-μm capillary is as follows, with all other values set to zero:

   Heat = 320, velocity = 40, and delay = 200
   Heat = 310, velocity = 30, and delay = 200
   Heat = 300, velocity = 25, and delay = 200
   Heat = 290, velocity = 20, and delay = 200

3. Pack the pulled needle capillary with C_{18} reverse-phase packing material and SCX packing material using a stainless steel pressurization bomb.

4. Place approx 20 mg of XDB-C_{18} RP packing material (Agilent Technologies) into a 1.7-mL microfuge tube, add 1 mL of MeOH, and place the tube into the bomb. Secure the bomb lid by tightening the six bolts.

5. The lid has a Swagelok® fitting containing a Teflon ferrule. Feed the fused silica, pulled needle capillary down through the ferrule until the end of the capillary reaches the bottom of the microfuge tube. Tighten the ferrule to secure the pulled needle capillary.

6. Apply pressure to the bomb by first setting the regulator on the gas cylinder to approx 400–800 psi and then opening a valve on the bomb to pressurize it. The packing material will begin filling the pulled needle capillary. Pack the capillary with 8–10 cm of RP packing material.

7. Slowly release the pressure from the bomb so as to not cause the packed RP material to unpack. Open the stainless steel pressurization bomb and remove the microfuge tube containing the RP material in MeOH.

8. Place approx 20 mg of Whatman Paritshpere SCX packing material into a 1.7-mL microfuge tube, add approx 1 mL of MeOH, and place the tube into the bomb. Secure the bomb as described in **step 4**.

9. Apply pressure as described in **step 6** in order to pack the SCX material behind the RP material. Pack the capillary with 3–5 cm of the SCX material and then slowly release the pressure as in **step 7**.

10. Place a 1.7-mL microfuge tube containing 1 mL of 5% ACN/0.5% HoAc into the stainless steel bomb and seal the bomb as in **step 4**.

11. Apply pressure as in **step 6** and allow several column volumes of 5% ACN and 0.5% HoAc to flow through the packed 2-D microcapillary column. This is an equilibration step needed to remove the MeOH from the column to allow for proper binding of the peptide mixture.

3.7.2. Loading the Sample onto the C_{18} Column

1. Spin the samples for 10 min at $10,000g$ in a microfuge to pellet any solid material.

2. If any pellet is visible, transfer the supernatant (peptide sample) to a fresh microfuge tube. Even minute amounts of solid material will plug the microcolumn. Place the tube containing the peptide sample into the pressurization bomb, and tighten the lid to the bomb.

3. Feed the RP capillary microcolumn down through a Teflon ferrule until the end of the capillary reaches the bottom of the microfuge tube. Tighten the ferrule to secure the capillary microcolumn.

4. Set the regulator on the gas cylinder to 400–800 psi, and then pressurize the bomb by opening the valve on the bomb that connects to the gas cylinder. The peptide sample will begin to flow into the column.

5. Measure the loaded volume at the tip of the needle using a graduated glass capillary.
6. When the sample is loaded, release the pressure, remove the column from the bomb, and place it back into the UpChurch MicroCross.

3.7.3. Ion-Source Setup

1. Attach the needle to the HPLC system through an UpChurch PEEK™ MicroCross.
2. One connection point of the cross contains the transfer line from the HPLC pump. This consists of a 50×365-μm fused silica capillary, with a length sufficient to reach from the HPLC pump to the ion source.
3. A second connection point contains a length of fused silica capillary that is used as a split/waste line. This split line allows a majority of the flow to exit through the split; therefore, very low flow rates can be achieved through the packed capillary microcolumn. The size and length of this section of capillary depend on the flow rate from the pump and the length of the microcolumn. A good starting point is to use a 12-inch section of 100×365-μm for the split line.
4. A third connection contains a section of gold wire. This is to allow the solvent entering the needle to be energized to approx 1800 V, thus allowing electrospray to occur.
5. A fourth connection point is for the packed capillary microcolumn.
6. Place the UpChurch MicroCross, with the connections into a stage, which in this case is designed for the ThermoFinnigan LCQ series mass spectrometer. This stage performs a threefold purpose:
   a. to support the MicroCross and hold it in place along with the connections.
   b. to insulate the MicroCross electrically from contact with its surroundings when it is held at high voltage potential.
   c. to allow for fine position adjustment of the microcolumn with respect to the entrance of the mass spectrometer (heated capillary) by using an XYZ manipulator.
7. Position the microcolumn using the XYZ manipulator so that the needle tip is within 5 mm from the orifice of the mass spectrometer's heated capillary.
8. Measure the flow from the tip of the capillary microcolumn, using graduated glass capillaries. To do this, turn on the flow rate of the HP1100 from the controller. Adjust the flow rate as necessary on the panel. The target flow rate at the tip should be approx 100–300 nL/min. If the flow rate is above this, cut off a portion of the split line capillary. This will cause more of the flow to exit out of the split and cause less flow through the microcolumn. If the flow is less than 100 nL/min, a longer piece of capillary or a section with a smaller inner diameter can be used to force more flow through the microcolumn. Measuring the flow rate and adjusting the split line may have to be repeated a number of times until the target flow rate is reached.

3.7.4. Instrument Method Design Description

Data-dependent acquisition of tandem mass spectra during the HPLC gradient is also programmed through the LCQ Xcalibur™ software. Here we provide guidance for setting the parameters for data-dependent acquisition using the ThermoFinnigan LCQ system. The following settings are for a typical data-dependent MS/MS acquisition analysis. The method consists of a continual cycle beginning with one scan of MS (scan one), which records all the *m/z* values of the ions present at that moment in the gradient, followed by two rounds of MS/MS. The initial MS/MS scan is of the first most intense ion recorded from the MS scan. The second MS/MS scan is of the second most intense ion recorded from scan one. Dynamic exclusion is activated to improve the protein identification capacity during the analysis.

1. In the main Xcalibur™ software page, select Instrumental Setup.
2. In the next window select the button labeled "Data Dependent MS/MS."
3. The next window will contain the design parameters for the acquisition setup. At the top, input the time of the acquisition. The time of the acquisition is the same as the duration of the HPLC gradient.
4. The "Segments Setting" is generally left at 1 and the "Start Delay" for the acquisition is set to 0.
5. The "Scan Event" is set to 4. The first scan event will be a full scan of MS, and the next three scan events will be MS/MS. This will allow the mass spectrometer to perform one round of MS followed by three rounds of MS/MS, of the first, second, and third most intense peaks.
6. Highlight the bar showing "Scan Event 1." Below this check "Normal Mass Range," "Scan Mode," as MS, and "Scan Type" as full. Set the "m/z range" to 400–1400, the "Polarity" to positive, and the "Data Type" as centroid.
7. The "Tune Method" box specifies the path for a file containing the parameters for the electrostatic lenses and ion trap. These parameters are established through the "LCQ Tune" as described in the ThermoFinnigan LCQ Getting Started manual.
8. Highlight the bar labeled as "Scan Event 2," check the box next to "Dependent Scan," and then click the "Settings" button. A window will come up in which the parameters can be set for all of the MS/MS scan events (Global and/or Segment) and for individual MS/MS scan events.
9. At the side of the window, press the "Global" button. These are parameters that are common among all segments and scan events. Several tabs will be visible. The first, "Global" tab, sets the values for both the MS and the MSn data-dependent masses and is left at the default settings. In the next tab, "Mass Widths," for high-throughput protein identification, these values are all left at the default settings. The "Dynamic Exclusion" tab contains several parameters that are important for effective data-dependent tandem mass analysis. Dynamic exclusion prevents an abundant peptide ion, with a broad elution profile, from being continually selected for MS/MS. The intense ions prevent other lower abundance peptides from being selected. Ions that are selected for MS/MS are placed onto a list. While on this list they will be excluded for a period of time so that other ions may be selected. The "Enable" box is checked and typical settings include a "Repeat Count" of 2, a "Repeat Duration" of 0.5 min, and an "Exclusion Duration" of 10 min. The "Exclusion Mass Width" is set for "by mass" and a width of 3 Daltons is set around the peak, 0.8 Daltons on the low mass side, and 2.2 Daltons on the high mass side.
10. Again on the left side is the "Segment" button. Press this to bring up the parameters for the current segment (in this case there is only one segment containing three scan events). Under the "Current Segment" tab leave the "Parent Masses" and the "Reject Masses" blank; therefore, leave the box next to "Most Intense If No Parent Mass Found" unchecked. The "Normalized Collision Energy" is set at 35%. This is the amount of energy delivered to the peptide to cause fragmentation. The "Activation Q" and the "Activation Time" are both left at their default settings of 0.250 and 30, respectively. The "Default Charge State" is set to 2, which makes the assumption that all peptides entering the mass spectrometer are doubly charged. The "Min MS Signal (10^4 counts)" is set to 10, which is the signal intensity needed for a peak recorded in an MS scan to trigger the mass spectrometer to select this peak for MS/MS. The "Min MSn Signal (10^4 counts)" is set to 0.5, which is the threshold to trigger the mass spectrometer to perform higher order tandem mass spectrometry. In this type of analysis, no higher order MS/MS is programmed into the segments so this value does not get utilized. The last value is the "Isolation Width," which is the window around the ion in the MS scan that is select for MS/MS. This typically is set to 2.5 Daltons. The "Add/Sub" tab is not enabled.

11. Pressing the "Scan Event" button accesses the last set of values. This defines parameters for this specific scan event, which is "Scan Event 2." This allows the user to select the ranking of ion abundances per MS scan (scan event 1) to be selected for MS/MS. In this case the "Nth Most Intense Ion" is checked and set to 1 (this is the first most intense ion). After pressing the okay button, the process is repeated by going back to the "method page" and pressing the "Scan Event 3" bar, checking "Dependent Scan" and pressing the "Settings" button. Then the "Scan Event" button is selected. "Use Previous List of Ions" is checked so the selection for MS/MS is from Scan 1 and is set to select the second most intense ion. Repeat the process for "Scan Event 4" to select the third most intense ion. Press okay to accept.

12. Under the "Timed Events" window of the instrument setup, the following sequence should be inserted

    Time (min) = 0.00, settings = Contact 1, value = open
    Time (min) = 0.01, settings = Contact 1, value = closed
    Time (min) = 0.05, settings = Contact 1, value = open

13. In the "Gradient Program" window of the instrument setup the following sequence is a possible first step that can be used or adjusted by a researcher:

Time (min)	Flow (mL/min)	A (%)	B (%)	C (%)	D (%)
0.00	0.150	100.0	0.0	0.0	0.0
5.00	0.150	100.0	0.0	0.0	0.0
5.10	0.150	90.0	0.0	10.0	0.0
7.00	0.150	90.0	0.0	10.0	0.0
7.10	0.150	100.0	0.0	0.0	0.0
12.00	0.150	100.0	0.0	0.0	0.0
12.10	0.150	100.0	0.0	0.0	0.0
21.00	0.150	75.0	25.0	0.0	0.0
80.00	0.150	60.0	40.0	0.0	0.0
110.00	0.150	40.0	60.0	0.0	0.0

The solvents A–D correspond to the following solvents below:
    Solvent A: 5% ACN/0.5% acetic acid/0.012% HFBA
    Solvent B: 80% ACN/0.5% acetic acid/0.012% HFBA
    Solvent C: 250 mM ammonium acetate/5% ACN/0.5% acetic acid/0.012% HFBA
    Solvent D: 500 mM ammonium acetate/5% ACN/0.5% acetic acid/0.012% HFBA

14. This file should then be saved as the 10% salt step method file. It can then be altered in terms of %C in each successive MuDPIT step in order to build up a series of methods files, which will then be used in a "Sequence Setup" to run a MuDPIT analysis.

15. An example of a MudPIT analysis would include the 10% Buffer C step described above and an additional 12 steps with the methods files written to accomplish the following. The next nine methods would be identical to that of the first excepting that the 2-min buffer C would be as follows: method 2–20%, method 3–30%, method 4–40%, method 5–50%, method 6–60%, method 7–70%, method 8–80%, method 9–90%, and method 10–100%. Method 11 contains a 20-min 100% buffer C wash but is otherwise identical to method 1. Method 12 is identical to method 11 except that the 20-min salt wash was with 100% buffer D, and method 13 is identical to method 12 except that the last 130 min of the method is from 10–100% buffer B.

3.7.5. Sequence Setup and Running Sequence

1. In the Xcalibur software main page, check the sample list button. In the "Sequence Setup" window fill in the "Filename," "Path," and "Inst Method" lines with the appropriate meth-

ods made previously. Save the Sequence to the same directory of the paths for the *.RAW files.

2. Once everything is set up and the needle is positioned correctly in front of the heated capillary on the mass spectrometer hit "Actions"-"Run Sequence." A box will come up that should have the HP1100 and LCQ Deca MS listed, and "Yes" should be typed under the "Start Instrument" box of the HP1100. The "start when ready" boxed should be checked as well as the "After Sequence Set System to –Standby." Do not enable the "Run Synchronously" boxes. Hit "O.K." and the run will begin.

3. A message will probably come up saying that there are devices that need to be turned on; hit "O.K."

4. Upon the completion of a run, the *.RAW files that have been accumulated by the mass spectrometer need to be converted to *.DAT files for SEQUEST analysis. To do this, go to the "Home Page" and hit the "Tools-File Converter" buttons. Follow the instructions to select the jobs that need to be converted to *.DAT files and hit "O.K."

3.8. SEQUEST Analysis of MS/MS Datasets

Several articles in the literature can be referred to for additional information regarding the method of visually assessing SEQUEST results *(2,3,5,15,16)*. One set of SEQUEST scoring criteria that has been reported in the literature is described below *(2)*. However, in general, individuals come up with their own set of criteria for accepting or rejecting SEQUEST results.

1. Peptides entering the ion trap mass spectrometer and identified by MudPIT in general have three different charge states, +1, +2, or +3. Each of these three charge states on a peptide provides for a unique spectrum, and their respective SEQUEST results must be analyzed accordingly.

2. In nearly every case, an accepted SEQUEST result has to have a ΔCn score of at least 0.1 regardless of charge state. In rare instances SEQUEST outputs were accepted when the ΔCn score was less than 0.1.

3. Peptides with a +1 charge state are accepted if they are fully tryptic and have a cross-correlation (Xcorr) of at least 1.9.

4. Peptides with a +2 charge state are accepted if they are fully tryptic or partially tryptic between the Xcorr ranges of 2.2 and 3.0. Partially tryptic peptides are especially relevant in samples in which CNBr is used to digest a protein. CNBr cleaves at the C-terminus of methionine, and many peptides will be identified with a modified methionine on one end. Peptides with a +2 charge state above an Xcorr above 3.0 are accepted regardless of their tryptic nature.

5. Peptides with a +3 charge state are only accepted if they are fully or partially tryptic and have an Xcorr above 3.75.

6. In general, when more than four peptides for a protein are identified by the preceding charge state-specific ΔCn and Xcorr criteria, the presence of protein is ensured.

7. In cases when three or fewer peptides using the criteria described are identified to a protein at least one of the spectra is assessed manually.

3.8.1. Manual Assessment Criteria

The MS/MS must be of good quality with fragment ions clearly above baseline noise. Sometimes a spectrum will consist purely of "sticks." In general, these are poor quality spectra. There should be some noise that allows one to define what the baseline noise is.

There must be some continuity to the *b* or *y* ion series. Of ions that are matched by SEQUEST to the original tandem mass spectrum, the matched ions should correspond to signal rather than noise. Furthermore, better matches contain strings of ions rather than isolated ions here and there throughout the sequence.

The intensity of a *b* or *y* ion resulting from a proline should be far greater than the other ion in the spectrum. If there are multiple prolines present in a peptide, one may see internal fragment ions corresponding to the mass of the portion of the peptide between each of the two prolines.

After a visual assessment as described above, either reject or accept the spectrum.

3.9. SEQUEST Analysis of Quantitative MudPIT Datasets

1. The SEQUEST algorithm is run two separate times on each of the three datasets against the yeast_orfs.fasta database from the National Center for Biotechnology Information.
2. The *.DAT files from each have to be run twice to detect and identify peptides separately from the ^{14}N minimal media sample and from the ^{15}N minimal media sample by using two separate SEQUEST parameters files in which the masses of each amino acid are set to the corresponding growth conditions and nitrogen content therein. Specifically, to search the dataset for peptides where all of the amino acids were fully ^{15}N labeled, the mass of each amino acid needs to be statically modified to an increased mass depending on the number of nitrogen atoms in that amino acid.
3. Using the previously described SEQUEST analysis criteria, two independent sets of results are generated depending on the nitrogen content of the samples (^{14}N or ^{15}N).
4. Next, a single separate file is generated for each of the three ratio samples. Each file contains both the ^{14}N and ^{15}N peptides corresponding to particular proteins from each of the samples. Included in these files for each of the peptides in the file is the parent ion mass, the charge state of the identified peptides, and the MudPIT step information. Each file is then submitted to a software program that determines the relative abundance of each peptide in the chromatographic runs of MudPIT. When only one of the peptides from a peak pair generated a tandem mass spectrum of sufficient quality to be included in the list, the software calculates the expected *m/z* of the missing partner assuming 100% replacement of ^{14}N in every amino acid in the identified peptide with ^{15}N or vice versa. Next, the MS scans used are searched to confirm that no charge state of any other peptide, either ^{14}N or ^{15}N, in the same scan range has an *m/z* within 2.0 amu of the predicted pair. Then the chromatographic elution peak area for the missing partner is determined. Relative abundance ratios for individual peptides are generated and when multiple hits to a protein exist, the software determines the average ratio using all the successful ratio determinations. The software then reports back a new data file containing the peptides detected and identified from each protein and the relative abundance of each of these peptides.

References

1. Washburn, M. P. and Yates, J. R. (2000) Novel methods of proteome analysis: multidimensional chromatography and mass spectrometry. *Proteomics Curr. Trends Suppl.* 28–32.
2. Link, A. J., Eng, J., Schieltz, D. M., et al. (1999) Direct analysis of protein complexes using mass spectrometry. *Nat. Biotechnol.* **17,** 676–682.
3. Eng, J. K., McCormack, A. L., and Yates, J. R. I. (1994) An approach to correlate tandem mass spectral data of peptides with amino acid sequences in a protein database. *J. Am. Soc. Mass Spectrom.* **5,** 976–989.
4. Wolters, D. A., Washburn, M. P., and Yates, J. R. 3rd (2001) An automated multidimensional protein identification technology for shotgun proteomics. *Anal. Chem.* **73,** 5683–5690.

5. Washburn, M. P., Wolters, D., and Yates, J. R. 3rd (2001) Large-scale analysis of the yeast proteome by multidimensional protein identification technology. *Nat. Biotechnol.* **19,** 242–247.
6. Gygi, S. P., Corthals, G. L., Zhang, Y., Rochon, Y., and Aebersold, R. (2000) Evaluation of two-dimensional gel electrophoresis-based proteome analysis technology. *Proc. Natl. Acad. Sci. USA* **97,** 9390–9395.
7. Santoni, V., Molloy, M., and Rabilloud, T. (2000) Membrane proteins and proteomics: un amour impossible? *Electrophoresis* **21,** 1054–1070.
8. Koller, A., Washburn, M. P., Lange, B. M., et al. (2002) Proteomic survey of metabolic pathways in rice. *Proc. Natl. Acad. Sci. USA* **99,** 11,969–11,974.
9. Florens, L., Washburn, M. P., Raine, J. D., et al. (2002) A proteomic view of the *Plasmodium falciparum* life cycle. *Nature* **419,** 520–526.
10. Washburn, M. P., Ulaszek, R., Deciu, C., Schieltz, D. M., and Yates, J. R. 3rd (2002) Analysis of quantitative proteomic data generated via multidimensional protein identification technology. *Anal. Chem.* **74,** 1650–1657.
11. Oda, Y., Huang, K., Cross, F. R., Cowburn, D., and Chait, B. T. (1999) Accurate quantitation of protein expression and site-specific phosphorylation. *Proc. Natl. Acad. Sci. USA* **96,** 6591–6596.
12. Pasa-Tolic, L., Jensen, P. K., Anderson, G. A., et al. (1999) High throughput proteome-wide precision measurements of protein expression using mass spectrometry. *J. Am. Chem. Soc.* **121,** 7949–7950.
13. Conrads, T. P., Alving, K., Veenstra, T. D., et al. (2001) Quantitative analysis of bacterial and mammalian proteomes using a combination of cysteine affinity tags and ^{15}N-metabolic labeling. *Anal. Chem.* **73,** 2132–2139.
14. Gatlin, C. L., Kleemann, G. R., Hays, L. G., Link, A. J. and Yates, J. R. 3rd (1998) Protein identification at the low femtomole level from silver-stained gels using a new fritless electrospray interface for liquid chromatography-microspray and nanospray mass spectrometry. *Anal. Biochem.* **263,** 93–101.
15. Yates, J. R. 3rd, Morgan, S. F., Gatlin, C. L., Griffin, P. R., and Eng, J. K. (1998) Method to compare collision-induced dissociation spectra of peptides: potential for library searching and subtractive analysis. *Anal. Chem.* **70,** 3557–3565.
16. Yates, J. R. 3rd, Carmack, E., Hays, L., Link, A. J., and Eng, J. K. (1999) Automated protein identification using microcolumn liquid chromatography-tandem mass spectrometry. *Methods Mol. Biol.* **112,** 553–569.

7

Di- and Tri-Chromatic Fluorescence Detection on Western Blots

Karen J. Martin and Wayne F. Patton

1. Introduction

The standard protocol for investigators striving to visualize concurrently the total protein profile and a specific antigen by immunodetection is to run replicate gels, using one for general protein staining and the other for immunoblotting. Similar procedures are often employed to visualize total protein and glycoprotein profiles by lectin blotting. However, the multiple manipulations required for electrophoresis and blotting often result in an inability to align bands or spots on the gel with bands or spots from the immunoblot. Shrinking, swelling, and other incidental side effects of electrophoresis, immunoblotting, and staining result in gel-to-gel and/or gel-to-blot variations, which make definitive band identification at best uncertain. The possibility of registration errors arising from the cited dimensional changes are compounded when addressing identification of specific proteins in two-dimensional (2-D) gel profiles, in which the immunoblot may have very few landmarks to aid in registration with the complex spot pattern.

To address the issue of spot and band registration, several methods have been devised over the years for serial detection of total protein profiles followed by immunodetection, or vice versa, on a single blot. These methods utilize dyes, such as SYPRO® Ruby protein blot stain, that reversibly stain proteins on blots (*1*). Serial detection methods avoid the potential for interference by the colorimetric or fluorescent dye labels with epitope recognition by immunodetection reagents. Although some of these methods have proved sensitive and provide more information about total protein profiles, there can still be difficulties in alignment that can be more easily resolved by methods of *simultaneous* detection performed on one blot. Some methods described for simultaneous detection involve conventional, colorimetric, or chemiluminescent immunoblotting procedures performed after visualizing total protein profiles on blots by colloidal gold staining or followed by Coomassie Blue dye staining. In addition, a method for the detection of proteins electroblotted to polyvinylidene fluoride (PVDF) membranes was developed, exploiting direct fluorescence detection of 2-methoxy-2,4-diphenyl-2(2H)-furanone (MDPF)-labeled proteins on PVDF membrane using an ultraviolet (UV) transilluminator, followed by chemiluminescent immunodetection using photographic film (*2,3*). The following section describes three methods for simultaneous

From: *Handbook of Proteomic Methods*
Edited by: P. Michael Conn © Humana Press Inc., Totowa, NJ

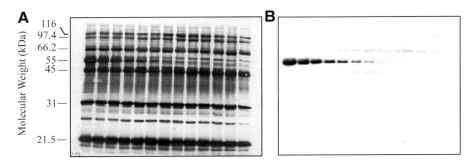

Fig. 1. Simultaneous dichromatic detection of total protein and a specific target protein on a PVDF electroblot generated from an SDS-polyacrylamide mini-gel. (**A**) Total protein detection with the red fluorophore BODIPY TR-X, SE dye. (**B**) and specific antigen detection on the same membrane, using mouse monoclonal anti-α-tubulin antibody, alkaline phosphatase-conjugated secondary antibody, and the alkaline phosphatase substrate, ELF 39 phosphate, which forms the green-fluorescent ELF 39 alcohol upon cleavage.

detection of total protein profiles and specific antigens on a single blot utilizing reactive fluorescent dyes to label covalently the total protein on the membrane coupled with fluorescent substrates for detection of alkaline phosphatase or horseradish peroxidase enzyme conjugates used in immunodetection.

The simultaneous detection methods described in **Subheadings 3.1.** and **3.2.** are dichromatic systems utilizing two fluorophores, one for total protein detection and the other for specific protein identification, which can be spectrally separated by their absorbance and emission maxima using the appropriate light sources and filter sets. **Figure 1** illustrates one of these dichromatic systems. Both systems rely on the use of a succinimidyl ester fluorophore to covalently label the primary amines of proteins (**Fig. 2B**) coupled with an alkaline phosphatase substrate, either DDAO phosphate or ELF® 39 phosphate, which become fluorescent upon cleavage by alkaline phosphatase-conjugated reporter molecules *(4,5)*. In both systems, the BODIPY® succinimidyl ester fluorophores do not interfere with and remain bound to proteins throughout the immunodetection process. Each system offers particular advantages for its users. The green/red system exploits a green BODIPY FL-X, SE fluorophore, and the red fluorescent DDAO (product of DDAO phosphate cleavage), both of which have absorbance maxima in the UV and visible regions of the spectrum and both of which can be imaged on a variety of gel-based image acquisition devises. The red/green system employs a red BODIPY TR-X, SE fluorophore and the green fluorogenic ELF 39 phosphate, which forms ELF 39 alcohol upon cleavage. Only the BODIPY TR-X, SE fluorophore of this pair has both UV and visible absorbance maxima; ELF 39 alcohol is only UV excitable. Although ELF 39 alcohol can only be imaged on image analysis platforms utilizing UV light sources, it can be advantageous compared with DDAO in that it forms an actual precipitate as a result of enzymatic action and forms at the site of cleavage. It is, therefore, not easily disturbed, and blots can be extensively washed, unlike those stained with DDAO phosphate. The hydrolysis product DDAO can be removed by washing and sometimes smears at protein bands with very high antigen loads.

Subheading 3.3. describes a simultaneous trichromatic system for target detection on blots, in which a fluorescent total protein label is used in conjunction with two flu-

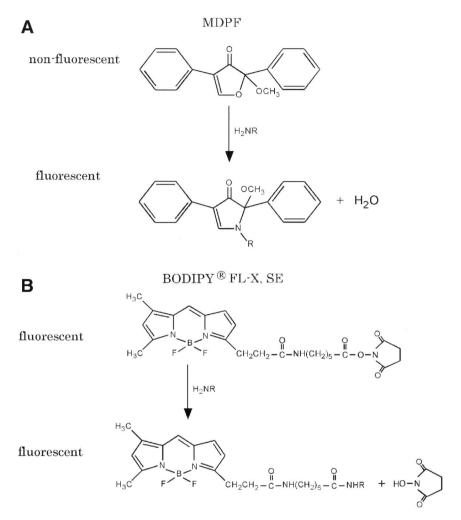

Fig. 2. Mechanisms of fluorescent total protein labeling with 2-methoxy-2,4-diphenyl-2(2H)-furanone (MDPF) and 6-((4,4-difluoro-5,7-dimethyl-4-bora-3a,4a-diaza-s-indacene-3-propionyl)amino)hexanoic acid, succinimidyl ester (BODIPY FL-X, SE dye). **(A)** MDPF reacts with any free amino group in a protein. **(B)** BODIPY FL-X, SE dye reacts with the primary amines of proteins to form amide bonds.

orogenic enzyme substrates, allowing the visualization of two specific antigens at one time. MDPF is utilized to covalently label the free amines of proteins **(Fig. 2A)**. This labeling is followed by immunodetection, in which a solution containing both fluorogenic substrates for alkaline phosphatase (DDAO phosphate) and horseradish perodixase (Amplex® Gold reagent) conjugates are employed to detect and determine the location of two separate antigens on a single immunoblot. Thus, this system can be used to identify two different specific proteins on one blot when primary antibodies raised in different host animals are available for each protein, i.e., using a mouse monoclonal antibody and a rabbit polyclonal antibody. DDAO and Amplex Gold dye both have UV and visible absorbance maxima, whereas MDPF is a blue fluorophore with an emission maximum at 485 nm and an absorbance maximum at 330/435 nm. During immuno-detection both primary antibodies are added to the blot at the same time, and then, following a wash step, both secondary conjugates are added to the blot at the same time. Detection of the enzymes is also performed with one solution, in which both substrates are present.

Choosing among the three methods presented depends on the particular advantages of each system relative to the experimental design constraints of the investigator. All three systems are detailed below.

2. Materials

1. ELF 39 alcohol.
2. Sodium borate buffer: 0.01 M sodium borate. The solution is brought to pH 9.5 with sodium hydroxide.
3. Wash buffer: 0.05 M Trizma base, 0.15 M NaCl, pH 7.5.
4. Blocking buffer: 0.05 M Trizma base, 0.15 M NaCl, 0.2% Tween 20, 0.25% Mowiol® 4-88, 0.5% bovine serum albumin, pH 7.5 (*see* **Note 1**).

2.1. Green/Red: Dual Detection with BODIPY FL-X, SE Dye, and DDAO Phosphate

1. BODIPY FL-X, SE total protein stain: 0.01 M 6-[(4,4-difluoro-5,7-dimethyl-4-bora-3a, 4a-diaza-s-indacene-3-propionyl)]aminohexanoic acid, succinimidyl ester (BODIPY FL-X, SE, Molecular Probes, cat. no. D-6102) in dimethylformamide stock solution. The stock solution is then diluted 1000-fold in sodium borate buffer to a 1× concentration of 0.01 mM to make the total protein stain. The 0.01 M dye stock is stable at –20°C for 2–3 mo. The dye is not stable in aqueous solution; the staining solution should be made up immediately prior to use.
2. DDAO phosphate stock solution: 1.25 mg/mL 9H-(1,3-dichloro-9,9-dimethylacridin-2-one-7-yl) phosphate, diammonium salt (DDAO phosphate) in dimethylformamide. The 1.25 mg/mL dye stock is stable at –20°C for 2–3 mo.
3. DDAO phosphate staining solution: 0.01 M Trizma base, 0.001 M MgCl$_2$, pH 9.5. Dilute the DDAO phosphate stock solution 1000-fold to a final concentration of 1.25 µg/mL in the staining solution. The dye is not stable in aqueous solution; the staining solution should be made up immediately prior to use.

All detection reagents are available packaged as convenient kits from Molecular Probes: DyeChrome™ Western Blot Stain Kit no. 1 *with goat anti-mouse IgG (cat. no. D-21881), DDAO phosphate, and BODIPY FL-X, SE*, DyeChrome Western Blot Stain Kit no. 2 *with goat anti-rabbit IgG, DDAO phosphate, and BODIPY FL-X, SE* (cat. no. D-21882), DyeChrome Western Blot Stain Kit no. 3 *with streptavidin, DDAO phosphate, and BODIPY FL-X, SE* (cat. no. D-21883).

2.2. Red/Green: Dual Detection with BODIPY TR-X, SE Dye, and ELF 39 Phosphate

1. BODIPY TR-X, SE total protein stain: 0.01 M 6-[(4-4,4-difluoro-5-2-thienyl-4- bora-3a,4a-diaza-s-indacene-3-yl)]phenoxyacetylaminohexanoic acid, succinimidyl ester (BODIPY TR-X, SE, Molecular Probes, cat. no. D-6116) in dimethylformamide stock solution. The stock solution is then diluted 1000-fold in sodium borate buffer to a 1× concentration of 0.01 mM to make the total protein stain. The 0.01 M dye stock is stable at –20°C

for 2–3 mo. The dye is not stable in aqueous solution; the staining solution should be made up immediately prior to use.

2. ELF 39 phosphate stock solution: 10 mg/mL ELF 39 phosphate (Molecular Probes, cat. no. D-21884, D-21885, or D-21886, DyeChrome Western Blot Stain Kit no. 4, no. 5, or no. 6) in dimethylformamide. The 10 mg/mL dye stock solution is stable at $-20°C$ for 2–3 mo.

3. ELF 39 phosphate staining solution: 0.01 M Trizma base, 0.001 M $MgCl_2$, pH 9.5. Dilute the ELF 39 phosphate stock solution 1000-fold to a final concentration of 10 µg/mL in the staining solution. The dye is not stable in aqueous solution; the staining solution should be made up immediately prior to use.

All detection reagents are available packaged as convenient kits from Molecular Probes: DyeChrome Western Blot Stain Kit no. 4 *with goat anti-mouse IgG, ELF 39 phosphate, and BODIPY TR-X, SE (cat. no. D-21884), DyeChrome Western Blot Stain Kit no. 5 *with goat anti-rabbit IgG, ELF 39 phosphate, and BODIPY TR-X, SE* (cat. no. D-21885), DyeChrome Western Blot Stain Kit no. 6 *with streptavidin, ELF 39 phosphate, and BODIPY TR-X, SE* (cat. no. D-21886).

2.3. Blue/Yellow/Red: Multicolor Detection with MDPF, Amplex Gold Reagent, and DDAO Phosphate

1. MDPF total protein stain: 0.035 M 2-methoxy-2,4-diphenyl-3(2H)-furanone (MDPF) in dimethysulfoxide stock solution. The stock solution is then diluted 200-fold in sodium borate buffer to a 1× concentration of 0.175 mM to make the total protein stain. The 0.035 M dye stock is stable at $-20°C$ for 2–3 mo. The dye is not stable in aqueous solution; the staining solution should be made up immediately prior to use.

2. DDAO phosphate stock solution (*see* **Note 2**): 1.25 mg/mL 9H-(1,3-dichloro-9, 9-dimethylacridin-2-one-7-yl) phosphate, diammonium salt (DDAO phosphate) in dimethylformamide. Dye stock is stable at $-20°C$ for 2–3 mo.

3. Amplex Gold dye stock solution (*see* **Note 2**): 2.08 mg/mL Amplex Gold reagent (Molecular Probes, cat. no. D-21887) or as described in kit, in dimethysulfoxide. Dye stock is stable at $-20°C$ for 2–3 mo.

4. DDAO phosphate and Amplex Gold reagent staining solution: 0.01 M Trizma base, 0.001 M $MgCl_2$, 0.001 M $ZnCl_2$, pH 7.5. Dilute the DDAO phosphate stock solution 1000-fold to a final concentration of 1.25 µg/mL and the Amplex Gold dye 200-fold to a final concentration of 10.4 µg/mL together into the staining solution. The dye is not stable in aqueous solution; the staining solution should be made up immediately prior to use.

All detection reagents are available packaged as a convenient kit from Molecular Probes: DyeChrome Double Western Blot Stain Kit *for mouse IgG, rabbit IgG, and total protein detection* (cat. no. D-21887).

2.4. Equipment

1. Forceps and gloves for handling blots.
2. Rotary shaker or rotisserie.
3. Image documentation system (one or more of the following):
   a. Hand-held or other UV epi-illumination light source in conjunction with Polaroid MP4+ camera system and appropriate cellophane filters (DyeChrome Red/Green Photographic Filter Set *two filters*, cat. no. D-24771, Molecular Probes).
   b. Computerized CCD camera-based image analysis system.
   c. Laser-based gel scanner.

3. Methods

3.1. Dual Detection with BODIPY FL-X, SE Dye, and DDAO Phosphate Alkaline Phosphatase Substrate

The following method is optimized for staining one 6-cm × 9-cm blot. For larger blots, the volumes should be scaled up proportionally. All wash and incubation steps require continuous, gentle agitation (e.g., on an orbital shaker at 50 rpm) unless stated otherwise. Do not use glass containers.

1. After electroblotting, equilibrate the blot by placing it in 100 mL of sodium borate buffer for 10 min. Incubate at room temperature without agitation. Repeat this step once.
2. Dilute 20 µL of the BODIPY FL-X, SE stock solution 1000-fold into 20 mL of sodium borate buffer. The final dye concentration will be of 0.01 mM. The staining solution must be used immediately, as the reactive dye is not stable in aqueous solution.
3. Pour off the sodium borate buffer used to equilibrate the blot and add 20 mL of the BODIPY FL-X, SE staining solution. Incubate the blot in staining solution at room temperature for 30 min.
4. Remove the staining solution and wash the blot in sodium borate buffer for 2 min. Repeat this wash step.
5. Remove the sodium borate buffer and wash the blot in 100% methanol for 10 min. Repeat the methanol wash twice for a total of three washes. Rinse the blot briefly in water and allow it to dry. The total protein stain may be visualized at this time (*see* **Subheading 3.4.**), or you may proceed directly to immunoblotting (**step 6**).
6. After staining with BODIPY FL-X, SE dye, wash the blot in wash buffer at room temperature for 10 min. Repeat the wash step for a total of three washes.
7. Incubate the blot in blocking buffer at room temperature for 1–2 h, or until the blot is completely wet. Additionally, this step can be performed overnight.
8. Dilute the primary antibody into blocking buffer. The optimal concentration of primary antibody must be determined empirically by the investigator.
9. Remove the blocking buffer and incubate the blot with the diluted primary antibody at room temperature for at least 1 h. This step may be performed overnight as well. If incubating overnight with primary antibody, specificity of antibody recognition may increase by incubation at 4°C. If you are incubating overnight, a small amount of sodium azide can be added to prevent bacterial growth. The final concentration of sodium azide should be 2 mM.
10. Remove the solution of primary antibody and wash the blot in blocking buffer at room temperature for 10 min. Repeat the wash step for a total of three washes.
11. Dilute the secondary antibody alkaline phosphatase conjugate or streptavidin alkaline phosphatase conjugate into 10 mL of blocking buffer. The optimal concentration of secondary antibody to be used must be determined by the investigator.
12. Incubate the blot with the diluted conjugate at room temperature for 30–45 min.
13. Remove the solution of conjugate and wash the blot in blocking buffer at room temperature for 10 min. Repeat the wash step for a total of three washes.
14. Perform two final washes in wash buffer at room temperature for 5 min each.
15. Dilute the DDAO phosphate stock solution 1000-fold into 10 mM Tris, 1 mM MgCl$_2$, pH 9.5, for a final concentration of 1.25 µg/mL. Approximately 1 mL of the DDAO phosphate staining solution is needed per 6-cm × 9-cm blot (*see* **Note 3**).
16. The staining step may be performed either on the imaging surface of an instrument or at the bench depending on the instrumentation available to the user. If an instrument is available and depending on the configuration of the imaging instrumentation, the blot can be placed face up or face down in substrate (*see* **Note 4**). For face-up staining at the bench, place the blot on the plastic wrap and pipet 1 mL of the DDAO phosphate staining solu-

tion onto the blot. For face-up staining on an imaging surface, place the blot directly on the imaging surface and pipet 1 mL of the DDAO phosphate staining solution onto the blot. For face-down staining on an imaging surface, pipet 1 mL of the DDAO phosphate staining solution directly onto the imaging surface of the instrument and lay the blot face down onto the solution, being careful not to trap any air bubbles.

17. Incubate the blot in the substrate for 5–20 min; do not agitate (*see* **Note 5**). The time required for optimal staining must be determined empirically because the substrate turnover rate depends on the amount of protein on the blot and the quality and quantity of the antibodies used (*see* **Note 6**).

18. Do not wash the blot after the reaction is complete (*see* **Note 5**). Image the blot wet or dry according to the imaging instructions below. Drying typically increases the fluorescent signal, as well as the background.

3.2. Dual Detection with BODIPY TR-X, SE Dye and ELF 39 Phosphate Alkaline Phosphatase Substrate

The following method is optimized for staining one 6-cm × 9-cm blot. For larger blots, the volumes should be scaled up proportionally. All wash and incubation steps require continuous, gentle agitation (e.g., on an orbital shaker at 50 rpm) unless stated otherwise. Do not use glass containers.

1. After electroblotting, equilibrate the blot by placing it in 100 mL of sodium borate buffer for 10 min. Incubate at room temperature without agitation. Repeat this step once.

2. Dilute 20 µL of the BODIPY TR-X, SE stock solution 1000-fold into 20 mL of sodium borate buffer. The final dye concentration will be of 0.01 mM. The staining solution must be used immediately, as the reactive dye is not stable in aqueous solution.

3. Pour off the sodium borate buffer used to equilibrate the blot and add 20 mL of the BODIPY TR-X, SE staining solution. Incubate the blot in staining solution at room temperature for 30 min.

4. Remove the staining solution and wash the blot in sodium borate buffer for 2 min. Repeat this wash step.

5. Remove the sodium borate buffer and wash the blot in 100% methanol for 10 min. Repeat the methanol wash twice for a total of three washes. Rinse the blot briefly in water and allow it to dry. The total protein stain may be visualized at this time (*see* **Subheading 3.4.**), or you may proceed directly to Western blotting (**step 6**).

6. After staining with BODIPY TR-X, SE dye, wash the blot in wash buffer at room temperature for 10 min. Repeat the wash step for a total of three washes.

7. Incubate the blot in blocking buffer at room temperature for 1–2 h, or until the blot is completely wet. Additionally, this step can be performed overnight.

8. Dilute the primary antibody into blocking buffer. The optimal concentration of primary antibody must be determined empirically by the investigator.

9. Remove the blocking buffer and incubate the blot with the diluted primary antibody at room temperature for at least 1 h. This step may be performed overnight as well. If incubating overnight with primary antibody, specificity of antibody recognition may increase by incubation at 4°C. If incubating overnight, sodium azide can be added to prevent bacterial growth. The final concentration of sodium azide should be 2 mM.

10. Remove the solution of primary antibody and wash the blot in blocking buffer at room temperature for 10 min. Repeat the wash step for a total of three washes.

11. Dilute the secondary antibody alkaline phosphatase conjugate or streptavidin alkaline phosphatase conjugate into 10 mL of blocking buffer. The optimal concentration of secondary antibody to be used must be determined by the investigator.

12. Incubate the blot with the diluted conjugate at room temperature for 30–45 min.

13. Remove the solution of the conjugate and wash the blot in blocking buffer at room temperature for 10 min. Repeat the wash step for a total of three washes.
14. Perform two final washes in wash buffer at room temperature for 5 min each.
15. Dilute the ELF 39 phosphate stock solution 1000-fold into 10 mM Tris-HCl, 1 mM MgCl$_2$, pH 9.5, for a final concentration of 10 µg/mL. Approximately 1 mL of the ELF 39 phosphate staining solution is needed per 6-cm × 9-cm blot (*see* **Note 7**).
16. The staining step may be performed either on the imaging surface of an instrument or at the bench depending on the instrumentation available to the user. If an instrument is available and depending on the configuration of the imaging instrumentation, the blot can be placed face up or face down in substrate (*see* **Note 4**). For face-up staining at the bench, place the blot on the plastic wrap and pipet 1 mL of the ELF 39 phosphate staining solution onto the blot. For face-up staining on an imaging surface, place the blot directly on the imaging surface and pipet 1 mL of the ELF 39 phosphate staining solution onto the blot. For face-down staining on an imaging surface, pipet 1 mL of the ELF 39 phosphate staining solution directly onto the imaging surface of the instrument and lay the blot face down onto the solution, being careful not to trap any air bubbles.
17. Incubate the blot in the substrate for 5–20 min; do not agitate (*see* **Note 8**). The time required for optimal staining must be determined empirically because the substrate turnover rate depends on the amount of protein on the blot and the quality and quantity of the antibodies used (*see* **Note 6**).
18. The blot can be washed briefly at this step (*see* **Note 8**). Image the blot wet or dry according to the imaging instructions below. Drying typically increases the fluorescent signal, as well as the background.

3.3. Simultaneous Multicolor Detection with MDPF, Amplex Gold Reagent as a Horseradish Peroxidase Substrate, and DDAO Phosphate Alkaline Phosphatase Substrate

1. After electroblotting, equilibrate the blot by placing it in 100 mL of sodium borate buffer for 10 min. Incubate at room temperature without agitation. Repeat this step once.
2. Dilute 200 µL of the MDPF stock solution 200-fold into 40 mL of sodium borate buffer. The final dye concentration will be 0.175 mM. The staining solution must be used immediately, as the reactive dye is not stable in aqueous solution.
3. Pour off the sodium borate buffer used to equilibrate the blot and add 20 mL of the MDPF staining solution. Incubate the blot in staining solution at room temperature for 10 min.
4. Remove the staining solution and wash the blot in sodium borate buffer for 2 min. Repeat this wash step.
5. Remove the sodium borate buffer and wash the blot in 100% methanol for 5 min. Repeat the methanol wash once for a total of two washes. Rinse the blot briefly in water and allow it to dry. The total protein stain may be visualized at this time (*see* **Subheading 3.4.**), or you may proceed directly to Western blotting (**step 6**).
6. After staining with MDPF, wash the blot in wash buffer at room temperature for 10 min. Repeat the wash step for a total of three washes.
7. Incubate the blot in blocking buffer at room temperature for 1–2 h, or until the blot is completely wet. Optionally, this step can be performed overnight.
8. Dilute both primary antibodies into blocking buffer. The optimal concentration of primary antibody must be determined empirically by the investigator.
9. Remove the blocking buffer and incubate the blot with the diluted primary antibodies at room temperature for at least 1 h. This step may be performed overnight as well. Specificity of antibody recognition may increase by incubation at 4°C. If you are incubating overnight,

sodium azide can be added to prevent bacterial growth. The final concentration of sodium azide should be 2 mM.

10. Remove the solution of primary antibodies and wash the blot in blocking buffer at room temperature for 10 min. Repeat the wash step for a total of three washes.

11. Dilute the secondary antibody alkaline phosphatase conjugate and secondary antibody horseradish peroxidase conjugate into 10 mL of blocking buffer. The optimal concentration of secondary antibody to be used must be determined by the investigator.

12. Incubate the blot with the diluted conjugates at room temperature for 30–45 min.

13. Remove the solution of conjugates and wash the blot in blocking buffer at room temperature for 10 min. Repeat the wash step for a total of three washes.

14. Perform two final washes in wash buffer at room temperature for 5 min each.

15. Dilute the DDAO phosphate stock solution 1000-fold and the Amplex Gold dye stock solution 200-fold into 10 mM Tris, 1 mM MgCl$_2$, 1 mM ZnCl$_2$, pH 7.5. The final concentration of DDAO phosphate will be 1.25 µg/mL and the final concentration of Amplex Gold dye will be 10.4 µg/mL. Approximately 1 mL of the substrate solution is needed per 6-cm × 9-cm blot (*see* **Note 9**).

16. The staining step may be performed either on the imaging surface of an instrument or at the bench depending on instrumentation available to the user. If an instrument is available and depending on the configuration of the imaging instrumentation, the blot can be placed face up or face down in substrate (*see* **Note 4**). For face-up staining at the bench, place the blot on the plastic wrap and pipet 1 mL of the staining solution onto the blot. For face-up staining on an imaging surface, place the blot directly on the imaging surface and pipet 1 mL of the staining solution onto the blot. For face-down staining on an imaging surface, pipet 1 mL of the staining solution directly onto the imaging surface of the instrument and lay the blot face down onto the solution, being careful not to trap any air bubbles.

17. Incubate the blot in the substrate for 5–20 min; do not agitate (*see* **Note 10**). The time required for optimal staining must be determined empirically because the substrate turnover rate depends on the amount of protein on the blot and the quality and quantity of the antibodies used (*see* **Note 6**).

18. Do not wash the blot after reaction is complete (*see* **Note 10**). Image the blot wet or dry according to imaging instructions below. Drying typically increases the fluorescent signal, as well as the background (*see* **Note 11**).

3.4. Visualization

3.4.1. Dual Detection with BODIPY FL-X, SE Dye, and DDAO Phosphate Alkaline Phosphatase Substrate

Both fluorescent stains can be visualized using either UV epi-illumination and a Polaroid or digital camera or using a visible light source, such as those available in laser scanning instruments. A standard UV transillumination light source is not suitable for this application because it is difficult for light to penetrate the blot.

3.4.1.1. BODIPY FL-X, SE Dye

The green fluorescent BODIPY FL-X, SE dye has absorbance/emission maxima at 365 and 505/515 nm. The green fluorescence is best documented with epi-illumination using a fixed overhead or a hand-held UV light source. It is not recommended to use UV transillumination. If using a Polaroid camera and Polaroid 667 black-and-white print film for documentation, a Wratten 61 filter can be used to separate the green fluorescence of the BODIPY FL-X, SE dye from the red-fluorescent DDAO signal. A cellophane filter set for imaging of both BODIPY FL-X dye-labeled proteins and DDAO is also available (DyeChrome Red/Green Photographic Filter Set, Molecular Probes). For other types of cameras or image documentation systems, choose a filter that matches the emission wavelengths of the dye. A bandpass filter will separate the green fluorescence from the red fluorescence, whereas a longpass filter will show both fluorescent colors simultaneously.

3.4.1.2. DDAO

The red-fluorescent DDAO phosphate reaction product (DDAO) has absorption maxima at 275 and 646 nm and an emission maximum at 659 nm. The dye can be visualized and documented using either UV or visible light. If using UV epi-illumination, a Polaroid camera and Polaroid 667 black-and-white print film for documentation, the red signal can be separated from the green signals using a Wratten 92 filter. A cellophane filter set for imaging of both BODIPY FL-X dye-labeled proteins and DDAO is also available (DyeChrome Red/Green Photographic Filter Set, Molecular Probes). For other types of cameras, image documentation systems, or laser scanners, choose a light source and a filter that match the emission of the dye.

3.4.2. Dual Detection with BODIPY TR-X, SE Dye, and ELF 39 Phosphate Alkaline Phosphatase Substrate

Both fluorescent stains can be visualized using UV epi-illumination and a Polaroid or digital camera. The two fluorescent signals can be photographed separately using filter sets matched to the spectral properties of the dyes. A standard UV transillumination light source is not suitable for this application because it is difficult for light to penetrate the blot.

3.4.2.1. BODIPY TR-X, SE DYE

The red fluorescent BODIPY TR-X, SE dye has absorbance/emission maxima at 300 and 590/615 nm. The red fluorescence is best documented with epi-illumination using a fixed overhead or a hand-held UV light source. It is not recommended to use UV transillumination. If using a Polaroid camera and Polaroid 667 black-and-white print film for documentation, a Wratten 92 filter can be used to separate the red fluorescence of the BODIPY TR-X, SE dye from the green-fluorescent ELF 39 dye signal. A cellophane filter set for imaging of both BODIPY TR-X dye-labeled proteins and ELF 39 dye is also available (DyeChrome Red/Green Photographic Filter Set, Molecular Probes). For other types of cameras or image documentation systems, choose a filter that matches the emission wavelengths of the dye. A bandpass filter will separate the red fluorescence from the green fluorescence, whereas a longpass filter will show both fluorescent colors simultaneously.

3.4.2.2. ELF 39 DYE

The green-fluorescent ELF 39 phosphate reaction product (ELF 39 alcohol) has absorption maxima at 345 and an emission maximum at 495 nm; therefore, the dye can be visualized and documented using only a UV light source. If using a Polaroid camera and Polaroid 667 black-and-white print film for documentation, the green signal can be separated from the red signals using a Wratten 61 filter. A cellophane filter set for imaging of both BODIPY TR-X dye labeled proteins and ELF 39 dye is also available (DyeChrome Red/Green Photographic Filter Set, Molecular Probes). For other types of cameras or image documentation systems, choose a filter that will match the emission of the dye. Because ELF 39 alcohol has only UV excitation, a laser-based scanner cannot be used to image ELF 39 alcohol.

3.4.3. Simultaneous Multicolor Detection with MDPF, Amplex Gold Reagent as a Horseradish Peroxidase Substrate, and DDAO Phosphate Alkaline Phosphatase Substrate

All three fluorescent stains can be visualized using UV epi-illumination and a Polaroid or digital camera. However, for best results with each dye, document the fluorescent signals separately, using visible or UV light sources and filters matched to the absorbance and emission maxima for each dye. A standard UV transillumination light source is not suitable for this application because it is difficult for light to penetrate the blot.

3.4.3.1. MDPF

The blue-fluorescent MDPF dye has absorbance/emission maxima at 330/435 and 485 nm. The blue fluorescence is best documented with epi-illumination using a fixed overhead or a hand-held UV light source. It is not recommended to use a UV transillumination light source. If using a Polaroid camera and Polaroid 667 black-and-white print film for documentation, a Wratten 98 filter can be used to separate the blue fluorescence of the MDPF from the other fluorescent signals. For other types of cameras or image documentation systems, choose a filter that matches the emission wavelengths of the dye. A bandpass filter will separate the blue fluorescence from the yellow and red fluorescence, whereas a longpass filter will show all three fluorescent colors simultaneously.

3.4.3.2. AMPLEX GOLD DYE

The yellow-fluorescent Amplex Gold dye reaction product has absorbance/emission maxima at 515/535 nm, with some absorbance in the UV as well. The dye can be visualized and documented using either UV or visible light. As with MDPF, it is possible to use UV epi-illumination to excite the dye. Again, transillumination is not recommended. If using a Polaroid camera and Polaroid 667 black-and-white print film for documentation, the yellow signal can be separated from the blue and red signals using a combination of Wratten 15 or 16 filter and Wratten 61 filters stacked together. A Wratten 99 can also be used, by itself, to image the blue signal. For other types of cameras, image documentation systems, or laser scanners, choose a light source and a filter that match the emission of the dye.

3.4.3.3. DDAO

The red-fluorescent DDAO phosphate reaction product (DDAO) has absorption maxima at 275 and 646 nm and an emission maximum at 659 nm. The dye can be visualized and documented using either UV or visible light. If using UV epi-illumination, a Polaroid camera and Polaroid 667 black-and-white print film for documentation, the red signal can be separated from the blue and yellow signals using a Wratten 92 filter. For other types of cameras, image documentation systems or laser scanners, choose a light source and a filter that match the emission of the dye.

4. Notes

1. Other protein components can be added to the blocker, such as casein or gelatin, or can be left out all together.
2. DDAO phosphate is a substrate for alkaline phosphatase that will produce a red-fluorescent product when cleaved. Amplex Gold dye is a substrate for horseradish peroxidase that will produce a yellow-fluorescent product when oxidized.
3. DDAO phosphate is unstable when stored at room temperature as an aqueous solution. The DDAO phosphate dye stock should be added to the buffer to make a staining solution immediately prior to use.
4. The face of the membrane refers to the surface of the membrane on which the proteins have been immobilized. If using UV epi-illumination or a laser scanner (visible light excited dyes only) with a light source that illuminates from above the bed of the image surface, stain the blot face up. If you are staining the blot at the bench or on another flat, hard surface, stain the blot face up. If using a laser scanner (visible light excited dyes only) with a light source that illuminates the blot from below a glass bed, stain the blot faced down. If using an imaging tray or another imaging surface, the staining solution can be pipeted directly onto the imaging surface. After the reaction is complete, wash the surface with H_2O followed by ethanol.
5. The product of the DDAO phosphate reaction with alkaline phosphatase is the nonprecipitating red-fluorescent DDAO. Because the product does not adhere to the blotting membrane, it can diffuse from the site of the reaction. Therefore, it is important to minimize the amount of solution in which the reaction takes place and to avoid washing the blot after the reaction is complete. It is also necessary to avoid agitation or disturbing the membrane during incubation in the substrate for two other reasons: (1) agitation may cause greater diffusion of the product across the membrane; and (2) agitation can break the surface tension holding the small volume of substrate to the blot, causing it to leave the membrane and reducing the amount of substrate available to the enzyme.
6. To find the optimal staining time, visualize the blot immediately after exposing it to the substrate staining solution and then every 10–15 min up to 1 h. Visualization after 1 h will most likely not improve sensitivity, as the background will also increase. If it is desirable to allow the reaction to continue for longer than 1 h, it may be necessary to add more substrate to the membrane, as it is likely that the small volume of substrate added initially will evaporate or dry out after 1 h.
7. ELF 39 phosphate is unstable when stored at room temperature as an aqueous solution. The ELF 39 phosphate dye stock solution should be added to the buffer to make a staining solution immediately prior to use.
8. The product of the ELF 39 phosphate reaction with alkaline phosphatase is a green-fluorescent precipitate. Because the product precipitates on the membrane, it can withstand washing and agitation. Agitation is avoided to prevent breaking the surface tension holding the small volume of substrate pipeted onto the surface of the blot. Washing can be performed to reduce background after the reaction is complete.
9. DDAO phosphate and Amplex Gold dye are unstable when stored at room temperature as an aqueous solution. Both dye stock solutions should be added to the buffer to make a staining solution immediately prior to use. Amplex Gold dye has a very bright signal but also produces a high background signal across the membrane. To minimize the background signal produced by Amplex Gold dye, do not add Amplex Gold dye to the staining solution until right before it is pipeted to the surface of the membrane.
10. The products of the reactions of DDAO phosphate and Amplex Gold dye with alkaline phosphatase and horseradish peroxidase, respectively, are both nonprecipitating. Because they do not adhere to the blotting membrane, they may diffuse from the site of the reaction if

washed or disturbed. Therefore, it is important to minimize the amount of solution in which the reaction takes place and to avoid washing the blot after the reaction is complete. It is also necessary to avoid agitation or disturbing the membrane during incubation for two other reasons: (1) agitation may cause greater diffusion of the product across the membrane; and (2) agitation can break the surface tension holding the small volume of substrate to the blot causing it to leave the membrane and reduce the amount of substrate available to the enzyme.

11. An increase in fluorescent signal concurrently with background is especially true for Amplex Gold dye. Amplex Gold dye generates a very bright signal and some background, which appears across the whole surface of the membrane. Amplex Gold dye is best visualized while it is wet because drying will cause the background signal to increase over the whole membrane, which may obscure the DDAO signal.

References

1. Berggren, K., Steinberg, T., Lauber, W., et al. (1999) A luminescent ruthenium complex for ultrasensitive detection of proteins immobilized on membrane supports. *Anal. Biochem.* **276,** 129–143.
2. Alba, F. and Daban, J. (1997) Nonenzymatic chemiluminescent detection and quantitation of total protein on Western and slot blots allowing subsequent immunodetection and sequencing. *Electrophoresis* **18,** 1960–1966.
3. Alba, F. and Daban, J. (1998) Rapid fluorescent monitoring of total protein patterns on sodium dodecyl sulfate-polyacrylamide gels and Western blots before immunodetection and sequencing. *Electrophoresis* **19,** 2407–2411.
4. Pretty On Top, K., Hattleburg, G., Berggren, K., et al. (2001) Green/red dual fluorescence detection of total protein and alkaline phosphate-conjugated probes on blotting membranes. *Electrophoresis* **22,** 896–905.
5. Martin, K., Kemper, C., Schulenberg, B., Jones, L., and Patton, W. (2002) Simultaneous red/green dual fluorescence detection on electroblots using BODIPY TR-X succinimidyl ester and ELF 39 phosphate. *Proteomics* **5,** 499–512.

8

Multiplexed Proteomics

Fluorescence Detection of Protein Differences by Two-Dimensional Gel Electrophoresis and Computer-Based Differential Display

Birte Schulenberg and Wayne F. Patton

1. Introduction

Two-dimensional gel electrophoresis (2-DGE) has become the central tool for proteome analysis and especially for the comparison of changes in protein expression levels arising from disease progression or therapeutic treatment modalities. Very often there are also changes involving post-translational modifications, such as phosphorylation or glycosylation. In order to study specific changes in protein post-translational modification, as well as overall changes in protein expression, the standard approach is to run two separate gels. One gel is used for the detection of a specific feature, like glycosylation (typically after Western blotting), and the second gel is used for total protein detection (usually employing silver stain or SYPRO® Ruby protein gel stain). This approach does, however, have some inherent problems and limitations associated with it. In order to map the glycoproteins, detected on an electroblot membrane, to the total protein profile from gels, the duplicate gels must run very reproducibly. Often, the overall dimensions of the gels change during fixation, staining, and transfer to the membrane, making it difficult to superimpose the gel profile directly with the blot profile. In addition, heavily glycosylated proteins usually are of high molecular weight and are difficult to transfer by electroblotting, thus leading to their loss from the analysis. Often, very hydrophilic glycoproteins bind poorly to transfer membranes as well, leading to their inefficient detection. Another major drawback to the procedure is that since numerous manipulations are involved in electroblotting, reliable quantitation becomes challenging.

Consequently, it has become necessary to devise a sequential staining method, referred to as *multiplexed proteomics*, that allows detection of two different features of a proteome in a single gel. The approach involves two fluorescent dyes that are fully compatible with mass spectrometric analysis methods. The development of cancer is often accompanied by increased glycosylation and changes to complex glycan branching structures. In order to demonstrate the feasibility of the multiplexed proteomics platform, liver tissue extracts from cancerous and normal tissues have been evaluated. (Both specimens were obtained as sodium dodecyl sulfate extracts from Geno-

From: *Handbook of Proteomic Methods*
Edited by: P. Michael Conn © Humana Press Inc., Totowa, NJ

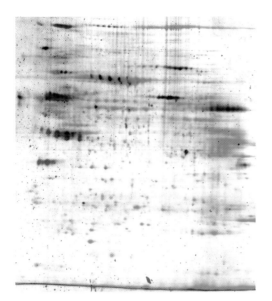 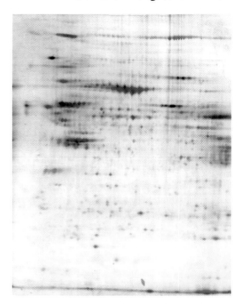

Fig. 1. Protein differential display maps produced using Z3 version 2 software (on the left) and Decodon's Delta2D version 3.0 software (on the right). The total protein profile is pseudo-colored green (Z3) or blue (Delta2D), whereas the glycoprotein pattern is pseudo-colored purple (Z3) or orange (Delta2D) in the images. Human liver tissue extract was used as a sample and run on pH 3–10 nonlinear immobilized pH gradient (IPG) strips as the first dimension and 12% sodium dodecyl sulfate-polyacrylamide gels as the second dimension.

Technology, St. Louis, MO.) The specimens were analyzed for changes in protein glycosylation, as well as overall protein expression levels. After separation of the proteins by 2-DGE the carbohydrate residues are detected by Pro-Q™ Emerald dye, with a detection sensitivity of 0.3–3 ng depending on the glycoprotein *(1)*. With the Pro-Q Emerald dye, the glycols present in glycoproteins are initially oxidized to aldehydes using periodic acid. The dye then reacts with the aldehydes to generate a highly fluorescent conjugate. After staining of the gels, glycoproteins are detected using a CCD camera-based imaging system and UV transillumination. After the images are obtained, the gels are poststained for total protein using SYPRO Ruby protein gel stain. The two different gel stains are imaged sequentially because the glycoprotein-specific stain is masked by the SYPRO Ruby dye.

The advantage of SYPRO Ruby dye over silver staining for the multiplexed proteomics application is its superior 3-log linear dynamic range, which allows collection of quantitative data by image analysis *(2–4)*. After all images are collected for both cancerous and noncancerous tissue extracts, image analysis software, such as Decodon's Delta2D (Greifswald, Germany) or Compugen's Z3 (Tel Aviv, Israel) package, is used to generate differential display maps, which permit ready visualization of the differences in protein expression and glycosylation profiles (*see* **Fig. 1** for an example of images obtained with each software package).

In order to characterize the glycosylation further, lectin blotting can be performed using two (or more) lectins with different carbohydrate epitope specificities. The lectins

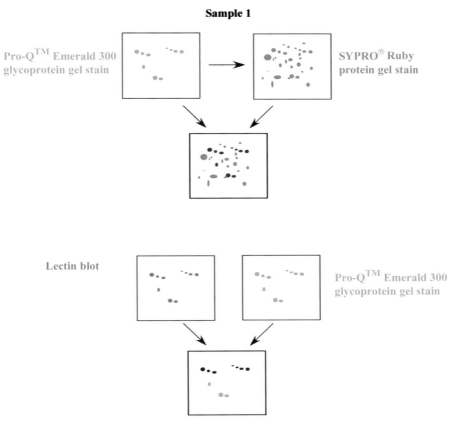

Fig. 2. Schematic diagram illustrating the use of multiplexed proteomics in glycoprotein analysis. The whole procedure can be performed on multiple samples, allowing for more comprehensive analysis than typically achieved with two-color dye pre-derivatization methods.

(we used ricin and concanavalin A in our study) are directly conjugated to alkaline phosphatase, thus obviating the need for a secondary antibody or streptavidin. Using DDAO phosphate as a substrate, the localized alkaline phosphatase activity leads to the production of a bright red-fluorescent signal that can be easily visualized using a laser scanner. These results too can be compared with the gels, providing more insight into the specific branching structures of the glycoproteins. **Figure 2** demonstrates the basic concept of multiplexed proteomics in the context of glycoprotein analysis.

The basic principles and processes of multiplexed proteomics as applied to glycoprotein analysis are described below. All experiments are described for large-format 2-D gels (1 mm × 20 cm × 20 cm) but can readily be performed on 1-D gels as well, using proportionally lower amounts of reagents.

2. Materials

All fluorescent stains were obtained from Molecular Probes (Eugene, OR). The volumes used for staining large-format 2-D gels are typically 500–600 mL/gel, unless otherwise noted. For smaller gels, a volume of stain that is roughly 10 times the volume of the gel should be employed (*see* **Notes 1** and **2**).

2.1. Glycoprotein Detection Using the Pro-Q Emerald 300 Glycoprotein Gel Stain

1. Pro-Q Emerald 300 glycoprotein gel stain kit (cat. no. P-21857, Molecular Probes). The kit contains the following components:
   a. Pro-Q Emerald 300 reagent (50×) in dimethylformamide (DMF), component A, 5 mL, enough for one large 2-D gel.
   b. Pro-Q Emerald 300 staining buffer (component B), 250 mL.
   c. Periodic acid, 2.5 g (component C).
2. Fix: 50% methanol, 10% acetic acid in dH_2O.
3. Wash: 3% acetic acid in dH_2O.
4. Oxidation: 1% periodic acid in 3% acetic acid. Add 250 mL 3% acetic acid to component C and mix until completely dissolved.
5. Staining: Dilute the Pro-Q Emerald 300 dye (component A) to 1× with Pro-Q Emerald 300 staining buffer (component B), just prior to staining. Each large-format 2-D gel is stained with 250 mL (absolute minimum; better to use 300 mL) of diluted Pro-Q Emerald 300 dye.

2.2 SYPRO Ruby Protein Gel Stain

1. SYPRO Ruby protein gel stain (cat. no. S-12000, Molecular Probes).
2. Wash: 10% methanol (or ethanol), 7% acetic acid.

2.3. Lectin Detection of Glycoproteins

1. Concanavalin A conjugated to alkaline phosphatase (Pro-Q Glycoprotein Blot Stain Kit no. 1 *with concanavalin A and DDAO phosphate (cat. no. P-21870, Molecular Probes). Make a stock solution of 2 mg/mL in dH_2O.
2. Ricin conjugated to alkaline phosphatase (cat. no. LA-200201, EY Labs, San Mateo, CA). Make a stock solution of 2 mg/mL in dH_2O. Care should be taken with the ricin conjugate as it is a toxin and should be handled with gloves at all times.
3. DDAO phosphate (cat. no. D-6487, Molecular Probes).
4. Wash buffer for blots: 50 mM Tris-HCl, 150 mM NaCl, pH 7.5.
5. Blocking buffer: 50 mM Tris-HCl, 150 mM NaCl, 0.2% Tween 20, 0.25% Mowiol® 4-88, pH 7.5. Mowiol 4-88 was obtained from Hoechst-Celanese (Charlotte, NC; cat. no. 50661910). It is easiest if a 10% Mowiol stock solution is prepared in 60°C water and then diluted into the buffer to the final concentration.
6. Incubation buffer: wash buffer plus 1 mM $CaCl_2$, 0.5 mM $MgCl_2$.
7. DDAO phosphate reaction buffer: 10 mm Tris-HCl, 1 mM $MgCl_2$, pH 9.5.
8. DDAO phosphate stock solution: 1.25 mg/mL in DMF; store at –20°C. This is stable for at least 6 mo. When the solution turns blue, the substrate has broken down and can no longer be used.

2.4. Imaging of Gels

1. CCD camera-based imaging system with a UV transilluminator and/or a dual excitation laser-based gel scanner with a 473–488-nm laser and 633–635-nm laser.

2.5. Image Analysis

1. Compugen's Z3 software or Decodon's Delta2D software.

3. Methods

3.1. 2-DGE

2-DGE was performed on a Genomic Solutions Investigator™ 2-D large-format system using the pHaser unit for isoelectric focusing according to standard protocols (as described in **ref. 5**). However, 2-DGE can be performed on any 2-DGE system available.

3.2. Staining of Glycoproteins

The Pro-Q Emerald 300 glycoprotein gel stain kit provides an easy method for detecting glycoproteins in the low nanogram range. The green-fluorescent Pro-Q Emerald 300 dye is excited at 280–300 nm and emits maximally at 530 nm. Alternatively, the Pro-Q Emerald 488 glycoprotein Gel Stain Kit may be utilized. The green-fluorescent dye in this kit is maximally excited at 510 nm and maximally emits at 520 nm. Although Pro-Q Emerald 488 dye is the glycoprotein stain of choice when imaging with a laser scanner with 473–488-nm laser, detection sensitivity is somewhat poorer than with the Pro-Q Emerald 300 dye on a UV transilluminator.

1. Fix the gel. Immerse the gel in fix solution (*see* **Subheading 2.1., step 2**) and incubate at room temperature with gentle agitation (e.g., on an orbital shaker at 50 rpm) overnight. This is to ensure that all sodium dodecyl sulfate is washed out of the gel (*see* **Note 3**). For large 2-D gels, use 500 mL of fix solution and incubate at room temperature overnight.
2. Wash the gel. Incubate the gel in 500 mL of wash solution (*see* **Subheading 2.1., step 3**) with gentle agitation for 15 min. Repeat this step twice.
3. Oxidize the carbohydrates. Incubate the gel in 500 mL of oxidation solution (*see* **Subheading 2.1., step 4**) with gentle agitation for 60 min.
4. Wash the gel. Incubate the gel in 500 mL of wash solution with gentle agitation for 10 min. Repeat this step three times more for large 2-D gels.
5. Prepare fresh Pro-Q Emerald 300 staining solution. Dilute the Pro-Q Emerald 300 reagent 50-fold into Pro-Q Emerald 300 staining buffer (provided in the kit). Large 2-D gels require 250 mL of staining solution.
6. Stain the gel. Incubate the gel in the dark in 250 mL of Pro-Q Emerald 300 staining solution (*see* **Subheading 2.1., step 5**) while gently agitating for 3 h. Staining overnight is not recommended.
7. Wash the gel. Incubate the gel in 700 mL of wash solution at room temperature for 15 min. Repeat this wash once for a total of two washes. Do not leave the gel in wash solution for more than 2 h, as the staining intensity will start to decrease.
8. Imaging. Image the gels on a CCD camera-based UV transillumination system using a 520-nm emission filter. Typically, optimal visualization of the signal requires up to 10-s exposures. *The gels must be imaged at this point*, before proceeding with any general protein stain (including SYPRO Ruby dye).

3.3. Staining for the Total Protein Profile

For detection of the total-protein pattern, any commercially available stain like silver stain, Coomassie Blue stain, or SYPRO Ruby protein gel stain may be used. However, for highest sensitivity and broadest linear dynamic range, SYPRO Ruby protein gel stain is recommended, since it allows one to obtain quantitative data over a 1000-fold range of protein concentrations, unlike silver stain and Coomassie Blue stain. SYPRO Ruby dye has two excitation maxima, which makes it suitable for UV transilluminator-based systems, as well as laser scanners. It generates a very bright orange fluorescence, which facilitates manual excision of protein spots on a UV transilluminator.

1. Staining. After the Pro-Q Emerald 300 glycoprotein gel stain is imaged, the gels are placed into 500–600 mL of SYPRO Ruby protein gel stain. The gels are stained overnight with gentle agitation.

2. Washing. In order to reduce the background arising from unbound dye trapped in the gel matrix, the gels are washed with 500–600 mL of 10% methanol (or ethanol; *see* **Notes 4–6**) and 7% acetic acid for 30 min before imaging.

3. Imaging. The gels can be imaged on either a UV transillumination system with a CCD camera or a laser scanner using a 488-nm excitation. A 580-nm emission filter is recommended.

If proteins are to be analyzed by mass spectrometry, the best point to excise the protein spots would be after the SYPRO Ruby protein gel stain, which makes the spots more readily visible.

3.4. Lectin Detection of Glycoproteins on Blots

1. Washing the blot. After electrotransfer of the proteins onto PVDF membrane, the membrane should be dried before proceeding. This reduces the background staining of the membrane support. Wash the blot with wash buffer (*see* **Subheading 2.3., step 4**) for 10 min and repeat this step two times more for a total of three washes.

2. Blocking nonspecific binding sites. Incubate the blot in blocking buffer (*see* **Subheading 2.3., step 5**) for 1–2 h or until the membrane is completely wet.

3. Reaction with concanavalin A-alkaline phosphatase conjugate. Dilute the concanavalin A-alkaline phosphatase conjugate (ConA AP) or ricin-alkaline phosphatase conjugate (Ricin AP) stock solution 2000-fold into incubation buffer (*see* **Subheading 2.3., step 6**) for a final concentration of 1 µg/mL. Remove the blocking buffer from the blot and incubate in the diluted ConA AP for 1 h at room temperature.

4. Washing the blot. Wash the blot in blocking buffer 2–3 times for 10 min each followed by another two washes in wash buffer for 2× 10 min. These last washes are to remove the Tween-20, which would inhibit the alkaline phosphatase reaction.

5. Reaction with DDAO phosphate. Dilute the DDAO phosphate stock (*see* **Subheading 2.3., step 8**) 1000-fold into the DDAO phosphate reaction buffer (*see* **Subheading 2.3., step 7**). Place the blot face down onto the DDAO phosphate solution that has been placed on a solid piece of plastic, such as a sheet protector, to produce a flat work surface. Do *not* use plastic wrap. Use 5–10 mL for a large size blot (20 × 20 cm). Incubate for 15–30 min. Air dry the membrane before imaging.

6. Imaging the blot. The DDAO phosphate can be imaged on a laser scanner equipped with a 633- or 635-nm laser, since it has an excitation maximum of 646 nm and an emission maximum of 659 nm.

3.4. 2-D Gel Analysis Using Z3

Image analysis with Z3 software works well with images in the TIFF format that do not exceed 4.5 Mb in file size. Version 2.0 of the software allows for multiple gels (more than two) to be analyzed, whereas the older version 1.5 only works with two gels at a time. The software serves as a useful visual tool for quick analysis of gels, as well as a program that allows quantitation of spots. In the case of two-gel experiments, each gel is assigned a color (purple for the comparative image and green for the reference image). When the gels are superimposed, all spots that are present in both gels appear black or gray. Differences appear purple or green. This allows for a quick visual analysis of 2-D gels and ready recognition of the appearance and the disappearance of proteins. If a more thorough analysis is needed, all spots may be quantitated automatically by the software and analyzed in a table format. However, it should be pointed out that the Z3 software expresses spot volumes in parts per million (ppm). This means the software takes the total area of the spots on a gel and then expresses a single spot relative to the total spot volume on a gel. If a project requires different protein loads on gels (e.g., gel 2 is loaded with twice the amount of material as gel 1), Z3 software should only be used for the detection of novel spots. The following protocol briefly outlines the workflow for the analysis of two gels. Prior manipulation of images in graphics programs should be avoided (even with respect to size and grayscale) unless absolutely necessary because this might result in a loss of data.

1. Open the reference image first (green) and then the comparative image (purple).
2. Detect the spots on both gels until enough features are indicated.
3. Edit the spots.
4. Overlay the gels. It is important that the gels have the same overall dimensions at this point. If the stains were imaged on different instruments, it might be necessary to adjust the image size in the Z3 program.
5. Add anchor points and register images until a satisfactory overlay is achieved. This can be done either manually or automatically. It generally works better if three to five anchors are placed manually before an automatic registration is attempted.
6. Match the spots (fully automatic).
7. Analyze the data.

If the registered image is to be used in a publication or for demonstration purposes, there are two ways to print that image. One way is to save the image in a GIF format directly from the Z3 software package. The image does lose some resolution but most of the time can be made into a publication-quality picture using a stand-alone graphics program. The other way is to print the screen and paste the image directly into a dedicated graphics program, which preserves the original image resolution. **Figure 1** shows a typical result for an overlay produced with Z3 version 1.5. In this example, human liver tissue lysate after Pro-Q Emerald 300 glycoprotein gel stain (purple) and SYPRO Ruby protein gel stain (green) were compared. If specific glycoproteins are to be readily recognized by eye from a profile of total protein, it seems to be easier to color the specific features (in this case the glycoproteins) in purple and the general stain (in this case the SYPRO Ruby dye) in green. The spot tables that contain all the data analysis results can be displayed in different manners, such as minimum spot areas, differential expression value, and so on. Afterward, these tables can be exported into spreadsheet programs (like Microsoft Excel) in a text file format.

3.5. Image Analysis Using Delta2D Version 3.0

Decodon's Delta2D software allows the same kind of analysis as Z3 software does, with the capability to generate virtual overlays of the gels and perform data analysis in spreadsheet format. However, there are some major differences between the programs worth noting. Pictures that have been taken with different imaging programs have to be set to the same resolution before opening them in Delta2D (e.g., 100 to 300 dpi). It is not possible to resize single images in the Delta2D software as it is in the Z3 package. Another minor difference is the way the programs deal with opening and closing images. Z3 software opens and closes images as most conventional graphics programs do. The Delta2D program only opens images and does not have a closing function. So, if one wants to open a different master image, just open a new picture and it will replace the older one. A major difference with Z3 software is that the spot finding and editing is performed after matching the gels and not before. The overall workflow in Delta2D software is as follows:

1. Open the master image.
2. Open the sample image.
3. Place anchor points (always place the anchor first on the master image and then on the sample image. The anchor point is displayed with the color of the image on which it has to be placed).
4. Perform either an automatic warp (rubber sheeting) or, if required, perform a global warp first. Add more anchor points if needed in specific areas of the gel and then warp the image exactly. It is always possible to remove single anchors too. Coloring of spots in the master and sample images can be changed very easy by going to <color scheme> and loading a different default from the scroll-down menu. This allows one to switch between black or white background, and so on.
5. After overlaying the images successfully, spot detection and editing are performed as well as all the data analysis.

Delta2D software allows the user to switch between a ppm-based spot quantitation or the spot volume, depending upon the project. If there is a large difference in the spot intensities between gels (e.g., one gel was loaded with twice the amount of material as the other) it is advisable to use spot volume, whereas if 70–90% of the spots should remain constant, the ppm value will be the more useful tool.

4. Notes

1. The recommended staining volume for a 0.75-mm-thick, 8 × 10-cm minigel is 50–100 mL.
2. Polypropylene dishes such as Rubbermaid® containers are best suited for large-format 2-D gels.
3. All gels should be agitated on an orbital shaker at 50 rpm. Faster speeds usually result in damage of the gels.
4. The SYPRO Ruby protein dye-stained gels can be washed with either ethanol or methanol-containing wash solution *(3)*. Ethanol has the advantage of being more environmentally friendly, but it might take a little bit more time (45 min instead of 30).
5. If nonspecific staining of nonglycosylated proteins is obtained, the fixation period in 50%, 7% acetic acid is not sufficient. In this case, it might be necessary to perform a 60-min fix in methanol/acetic acid first to remove the bulk of the sodium dodecyl sulfate and then immerse the gels overnight in the fixation buffer.
6. When spots (or bands) appear to have a nonfluorescent center and only the perimeter is brightly stained, this is an indication of overloading of the gel. Lower the load or, if very low-abundance proteins are to be visualized, simply do not use this spot for quantitation. However, in the case of heavily overloaded gels, the unspecific staining will increase in signal intensity too.

Acknowledgments

This work was supported by a grant from the National Cancer Institute (grant number IR33CA093292-01).

References

1. Steinberg, T. H., Pretty On Top, K., Berggren, K. N., et al. (2001) Rapid and simple single nanogram detection of glycoproteins in polyacrylamide gels and on electroblots. *Proteomics* **1,** 841–855.

2. Nishihara, J. C. and Champion, K. M. (2002) Quantitative evaluation of proteins in one- and two-dimensional polyacrylamide gels using a fluorescent stain. *Electrophoresis* **23,** 2203–2215.
3. Berggren, K. N., Schulenberg, B., Lopez, M. F., et al. (2002) An improved formulation of SYPRO Ruby protein gel stain: comparison with the original formulation and with a ruthenium II tris (bathophenanthroline disulfonate) formulation. *Proteomics* **2,** 486–498.
4. Berggren, K., Chernokalskaya, E., Steinberg, T. H., et al. (2000) Background-free, high sensitivity staining of proteins in one- and two-dimensional sodium dodecyl sulfate-polyacrylamide gels using a luminescent ruthenium complex. *Electrophoresis* **21,** 2509–2521.
5. Hanson, B. J., Schulenberg, B., Patton, W. F., and Capaldi, R. A. (2001) A novel subfractionation approach for mitochondrial proteins: a three-dimensional mitochondrial proteome map. *Electrophoresis* **22,** 950–959.

9

A Strategy for Characterizing Antibody/Antigen Interactions Using ProteinChip® Arrays

Alexandra Huhalov, Daniel I. R. Spencer, and Kerry A. Chester

1. Introduction

If antibody-based therapeutics are to be rationally designed to give optimal interactions with their target antigens, it is essential to understand these antibody/antigen interactions as fully as possible. This has been greatly advanced by the recent developments in mass spectrometry (MS), structural identification, and bioinformatics, which allow the relatively simple and rapid characterization of protein–protein interactions in a high-throughput manner. Previous methods of mapping an antibody-binding site (epitope) on an antigen have relied on peptide libraries *(1)*. Such methods are suitable for the identification of continuous amino acid sequences *(2)* but not for those that are conformationally dependent. This latter group forms a major category of antibody/antigen interactions.

In order to identify both continuous and conformational epitopes, an approach was introduced that involved the proteolytic digestion of preformed antibody/antigen complexes and the characterization of resulting peptide fragments. This was based on the findings that (1) antibodies are partially resistant to proteolytic degradation; (2) antigenic sites are protected from proteolytic enzymes in immune complexes; and (3) immune complexes do not dissociate upon proteolyis *(3–6)*. The approach had the advantage that the antibody/antigen interaction could be studied in a conformational state, but ambiguity problems arose from poor resolution of proteolytic peptide fragments when they were analyzed by polyacrylamide gel electrophoresis (PAGE) *(4)* and high-performance liquid chromatography (HPLC) *(7)*. These problems led to the application of matrix-assisted laser desorption/ionization mass spectrometry (MALDI MS) for the identification of the fragments *(8–10)*. However, MALDI MS can be laborious, reagent-intensive, and time-consuming *(9)*.

Because of these limitations, we have developed a method for the epitope mapping of antibody/antigen interactions using surface-enhanced laser desorption/ionization affinity mass spectrometry (SELDI AMS) *(11)*. In this process protein complexes are bound directly to specialized metal (ProteinChip® Array) chip surfaces and digested *in situ* with proteolytic enzymes. Nonreactive peptides are washed away, and the remaining bound peptides are identified by mass spectrometry. The SELDI AMS protocol is rapid, requiring very little starting material, and is amenable to high-throughput stud-

From: *Handbook of Proteomic Methods*
Edited by: P. Michael Conn © Humana Press Inc., Totowa, NJ

ies. The experimental time is much reduced in SELDI AMS as the removal of nonbinding antigen fragments is done by washing the chip, rather than by the lengthy separations used in MALDI MS. SELDI AMS also allows the study of proteins in crude supernatants or incompatible buffers because nonbinding proteins or salts are washed away. Very little starting material is required because the remaining antibody-associated peptides are laser desorped from the chip-bound antibody in the mass spectrometer, resulting in limited sample loss. Furthermore, many samples can be analyzed in parallel with their controls on a variety of chip surfaces, resulting in protocols that are versatile and readily optimized. Recently an interface has been developed to analyze samples on a ProteinChip® Array further using laser desorption ionization-tandem mass spectrometry, allowing the direct sequencing of peptides identified by SELDI AMS *(12)*.

This chapter outlines two examples of the method we have developed to characterize antibody/antigen interactions using SELDI AMS. First, to investigate the antigen-binding domain of a phage-derived single-chain Fv antibody fragment (scFv) *(13)* and second, to map the polyclonal response of patients to a recombinant scFv fusion protein using scFv libraries to determine immunogenic sites *(11)*.

In the first example we present the method used to partially characterize the binding domain of the phage-derived scFv, MFE-23 *(14)*, which is reactive with carcinoembryonic antigen (CEA) and has been used in two clinical trials *(15,16)*. CEA, a membrane-associated protein, consists of seven heavily glycosylated V-type, C2-type or I-type immunoglobulin fold domains (N-A1-B1-A2-B2-A3-B3) *(17,18)*. A structural prediction based on the CEA homology model *(19)* and the MFE-23 crystal structure *(20)* proposed that MFE-23 binding occurs at the N-terminal N-A1 domain pair of CEA *(20,21)*. In order to test these predictions, recombinant N-A1 domains were bacterially expressed in the presence of hexahistidine-tagged MFE-23, which was used to capture, purify, and couple the subsequent N-A1/MFE-23 complex to immobilized metal affinity chromatography (IMAC) ProteinChip® Arrays for SELDI AMS analysis. The complex was partially digested with trypsin, and the placement of the resulting antibody-bound peptides was studied in the homology model of CEA *(13)*.

In the second example, we modeled the polyclonal response in patients to a fusion protein consisting of MFE-23 genetically linked to carboxy-peptidase G2 (CPG2) *(22)*, used in antibody-directed enzyme prodrug therapy (ADEPT) *(23)* in order to neutralize its immunogenicity in patients. A murine scFv phage library was raised against CPG2, and a CPG2-reactive scFv was isolated from this library on the basis of its ability to inhibit CPG2 binding of sera from patients who had an immune response to this antigen *(11)*. This CPG2-reactive scFv was then covalently coupled to a preactivated ProteinChip® Array and used to capture CPG2. Their interaction was mapped by partial digestion with Glu-C, and the resulting peptides were modeled to two nonlinear sites in the crystal structure of CPG2 *(24)*, suggesting a conformational epitope. This was confirmed when subsequent alanine scanning mutagenesis of one of the predicted sites resulted in a reduction in immunogenicity of the mutant fusion protein when tested against patient sera *(11)*.

The methods to generate scFv libraries, screen against antigen, and produce recombinant scFv are not within the scope of this book and have been reviewed extensively elsewhere *(25,26)*. Here we describe our methods using SELDI AMS and structural analysis to characterize an antibody/antigen complex captured on (1) preactivated chips, and (2) IMAC chips, as well as (3) the use of simple bioinformatics for the prediction of epitopes in protein structures. These methods can be applied to any protein–protein interaction and are particularly suited to high-throughput applications.

2. Materials and Methods

There are several ProteinChip Array surfaces that are suitable for epitope mapping studies using SELDI AMS. Two of these are outlined below and include (1) preactivated chips with proteins covalently coupled to the chip surface, and (2) IMAC chips where proteins are coupled to the chip surface via polyhistidine tags, as well as methods for (3) the analysis of SELDI AMS data using protein software packages, and (4) the application of bioinformatics to fit the resulting information into structural models.

Calibration of the mass spectrometer is essential for converting time-of-flight data to accurate mass data. There are two types of calibration for ProteinChip arrays, external and internal. For external calibration, the masses of calibrants from one spot are applied to masses of proteins or peptides on another spot. External calibration results in mass accuracy of about 0.5% or better.[1] Internal calibration is more accurate and uses known masses in the sample for calibration. This type of calibration results in mass accuracy of about 0.05% or better.[1]

2.2. Epitope Mapping on Preactivated ProteinChip® Arrays

The PS10 and PS20 ProteinChip® Array surfaces are preactivated with carbonyl diimidazole chemistry that covalently bind to free primary amine groups.

1. Dilute test scFv, nonspecific scFv (if available), and antigen to 100–300 µg/mL in double-distilled water (ddH$_2$O). It is important to use ddH$_2$O when diluting protein for SELDI AMS as excess salt can disrupt the laser desorption of peptides.
2. Add 1–5 µL of diluted samples and phosphate-buffered saline (PBS; Dulbecco's PBS without calcium chloride; Sigma) to spots on a PS10 or PS20 chip (Ciphergen Biosystems) and incubate in a humidity chamber for 1 h at room temperature. To make a humidity chamber, fill the bottom half of an empty plastic pipet tip box half-way with ddH$_2$O and place the tip rack back on top to use as a support for the chips. When the lid is replaced firmly, it is sufficiently sealed to prevent drying of the chip for several hours. Do not allow the spots to dry.
3. Using a pipet, remove the solution from the first spot and immediately add 5 µL of 0.1% Triton X-100 (Sigma) in PBS (PBST), being sure not to allow the spots to dry. Repeat this procedure for all the spots used on the chip. Pass several streams of PBST over each spot using a pipet, making sure they do not cross over neighboring spots.
4. Submerge the chip into a tube filled with PBST and rotate for 5 min in a carousel shaker at high speed. Then repeat this step two more times with fresh PBST. Alternatively, washes can be carried out on a high-frequency shaker.
5. Rinse the chip twice in PBS as in **step 4**. Do not allow the spots to dry.
6. Block the spots by adding 5 µL of 1 *M* ethanolamine, pH 8.0 (BIAcore AB) to each spot and incubate in a humidity chamber for 15 min. For further blocking, add 5 µL of either 5% milk proteins/PBS, or 3% bovine serum albumin/PBS for 1 h in a humidity chamber.
7. Wash the chip as in **steps 3–5**. Do not allow the spots to dry.

---

[1]Taken from ProteinChip® software manual 3.0 Operation Manual.

8. Add 1–5 µL of antigen diluted in ddH$_2$O to 1 µL of PBST on spots and incubate in a humidity chamber at room temperature for 1 h. Leave one PBS-coated spot free of antigen for use as a control.

9. Wash the chip as in **steps 3–5**. Do not allow the spots to dry.

10. Add 2 µL of Endoproteinase Glu-C (Roche Molecular Biochemicals), modified sequencing-grade trypsin (Roche Molecular Biochemicals), or any sequencing-grade protease of choice to each spot diluted in PBST (including the spot containing only antigen for use as a control). Add enzyme at ratios of between 1:100 and 1:20 of the protein by weight. Digestion times and ratios are only given as guidelines and need to be optimized for individual antibody/antigen interactions. It is suggested a range of enzymes, enzyme ratios, digestion times, and temperatures be tested for optimal results.

11. Incubate at room temperature in the humidity chamber for 30–90 min. Include one PBS-coated spot containing no scFv or antigen as a control.

12. Wash spots as in **steps 3–5** with a final rinse in ddH$_2$O.

13. Add 0.5–1 µL Peptide All-In-One MW Standard (Ciphergen Biosystems) to a blank spot for external calibration. For internal calibration, add 0.5–1 µL MW standard to a sample spot after reading the chip, as calibrant peaks can swamp the sample peaks. After reading the sample spots, the intensity of the calibrants can be matched to the intensity of the sample, and the volume of MW standards can be added accordingly.

14. Apply 0.5–1 µL of the appropriate energy-absorbing matrix (EAM) to spots. EAM is an energy-absorbing compound capable of converting laser energy to thermal energy, thus facilitating the desorption/ionization of samples. It is important to use the appropriate EAM for analysis. For small molecules and peptides between 5 and 15 kDa, use a saturated solution of α-cyano-4-hydroxy cinnamic acid (CHCA; Ciphergen Biosystems) dissolved in 50% acetonitrile (Sigma) and 0.5% trifluoroacetic acid (Sigma) in ddH$_2$O. For molecules larger than 15 kDa, use sinapinic acid (SPA; Ciphergen Biosystems) in 50% acetonitrile and 500 µL of 0.5% trifluoroacetic acid in ddH$_2$O.

15. Dry the spots by placing a chip on top of the ProteinChip Reader. To dry spots more quickly, use a hair dryer set at cool. Hold the dryer about 10 cm away from the surface of the chip and for approx 1 min until the samples have dried.

16. Load the chip into the ProteinChip Reader.

17. Collect peptide mass data over the desired mass range. This will result in a set of mass spectra. For example, when analyzing the proteolytic fragments resulting from Glu-C digestion of CPG2 captured by a specific scFv covalently bound to a PS20 chip, four peptides are identified (**Fig. 1**).

2.3. Epitope Mapping on Immobilized Metal Affinity Chromatography (IMAC) ProteinChip® Arrays

The IMAC3 ProteinChip Arrays are coated with a nitrilotriacetic acid (NTA) functional group that binds transitional metals for capture of polyhistidine affinity-tagged proteins.

1. Draw an outline around each spot of an IMAC3 chip using a hydrophobic pen (ImmEdge™ Pen, Vector Laboratories).

2. Add 5 µL of 100% acetonitrile to each spot, incubate at room temperature for 1 min, and then wipe off with a tissue.

3. Add 1 µL of 50% acetonitrile diluted in ddH$_2$O to each spot; before it dries, add 5 µL of 50 m*M* nickel sulfate and incubate in a humidity chamber at room temperature for 30 min. IMAC ProteinChip® Arrays work best when coupled to nickel ions. Using Cu(II) and Fe(II) can affect the surface of the chip.

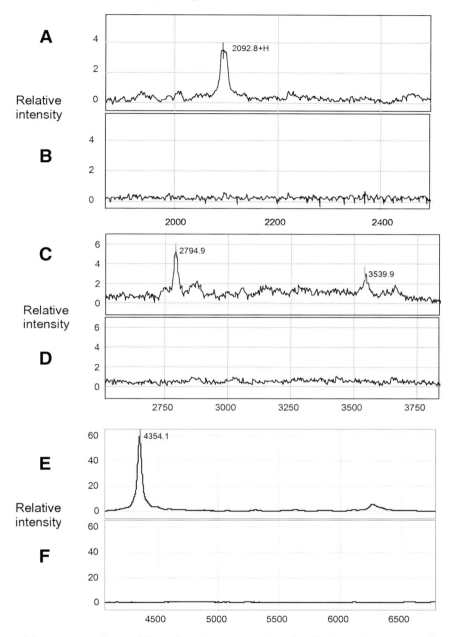

Fig. 1. Mass spectra obtained from the epitope mapping of antibody/antigen interactions using preactivated chips. This approach is illustrated by CPG2-reactive scFv, which are covalently bound to a preactivated ProteinChip Array and in turn, used to capture CPG2. The immune complex is digested with Glu-C, resulting in an array of peptides. Nonbinding peptides are removed by washing, and the remaining peptides are measured by their mass. Peaks are detected at **(A)** *m/z* 2092.8, **(C)** 2794.9, **(C)** 3539.9, and **(E)** 4354.1 Daltons. **(B,D,F)** No peaks are detected in the control spots containing a nonspecific scFv.

4. Reapply 5 µL of 50 m*M* nickel sulfate and incubate for 30 min in a humidity chamber at room temperature.

5. Submerge the chip in a tube filled with IMAC wash buffer (300 m*M* NaCl in PBS) and wash by rotating in a carousel shaker at high speed for 5 min. Binding and washing buffers benefit by the addition of salt (0.3–1 *M*). To increase selectivity of the polyhistidine-tagged protein, imidazole may be added to the binding buffer (5–10 m*M*) and/or wash buffer (10–100 m*M*).

6. Add 5 µL of polyhistidine-tagged scFv alone or in complex with antigen (containing no polyhistidine tag) diluted to 100–200 µg/mL in IMAC wash buffer. As a control add 5 µL of IMAC wash buffer to a spot and incubate in the humidity chamber for 30–60 min. Alternatively, incubations can be done on a high-frequency shaker for 5 min at room temperature providing the sample is not allowed to dry out.

7. Using a pipet, remove solution from spot and immediately add 5 µL of IMAC wash buffer being sure not to allow the spots to dry. Repeat this procedure for all the spots used on the chip.

8. Pass several streams of IMAC wash buffer over each spot using a pipet, making sure they do not cross over neighboring spots.

9. Submerge the chip in a tube filled with IMAC wash buffer and rotate in carousel shaker at high speed for 5 min. Then repeat this step two more times. Do not allow the spots to dry.

10. If scFv was added alone, add 5 µL of antigen (containing no polyhistidine tag) diluted to 100–200 µg/mL in IMAC wash buffer or add 5 µL of 40 m*M* imidazole in IMAC washing buffer and incubate in the humidity chamber for 1 h. Include one scFv-coated spot with no antigen as a control.

11. Wash the chip as in **steps 7–9**. Do not allow the spots to dry.

12. Add 2 µL of Endoproteinase Glu-C (Roche Molecular Biochemicals), modified sequencing grade trypsin (Roche Molecular Biochemicals), or other sequencing-grade protease of choice to each spot diluted in PBST added at ratios of between 1 : 100 and 1 : 20 of the protein by weight. Digestion times and ratios are only given as guidelines and need to optimized for individual antibody/antigen interactions. It is suggested that a range of enzymes, enzyme ratios, digestion times, and temperatures be tested for optimal results.

13. Incubate at room temperature in the humidity chamber for 30–60 min. Include one PBS-coated spot containing no scFv or antigen as a control. If scFv and antigen were captured onto the IMAC chip as a complex, include a control digestion done in solution. After the required digestion time, apply the control digestion to scFv and PBS-coated spots to control for nonspecific binding of proteolytic fragments to the nickel-saturated matrix.

14. Wash the chip as in **steps 7–9**.

15. Add 0.5–1 µL Peptide All-In-One MW Standard (Ciphergen Biosystems) to a blank spot for external calibration. For internal calibration, add 0.5–1 µL MW standard to a sample spot after reading the chip, as calibrant peaks can swamp the sample peaks. After reading of the sample spots, the intensity of the calibrants can be matched to the intensity of the sample and the volume of MW standards can be added accordingly.

16. Apply 0.5–1 µL of the appropriate EAM to spots.

17. Dry the spots.

18. Load the chip into the ProteinChip Reader.

19. Collect peptide mass data over the desired mass range.

2.4. Interpretation of Mass Spectra

There are several programs available that may be used to interpret data acquired by SELDI AMS. These programs enable the user to match the masses of identified peaks to peptide sequences resulting from digestion at every possible site in the molecule investigated. Here we describe methods for two such software packages freely distributed over the Internet.

2.4.1. PAWS

PAWS is a protein sequence analysis software package (Protein Information Retrieval On-line World Wide Web Lab website: http://prowl.rockefeller.edu/contents/resource.htm), which allows the user to paste in the amino acid sequence of a protein and modify it in a number of ways. It is particularly useful for determining proteolytic peptides arising from the cleavage of a protein.

1. Open the PAWS program, select "New" from the file menu and paste in the protein sequence of interest. Select the appropriate enzyme from the "Cleavage" menu. This will provide a list of all possible peptides resulting from the cleavage of the protein. For example, when analyzing the CPG2 sequence by Glu-C cleavage, 26 peptides are revealed. When acquiring cleavage peptides by Glu-C in PAWS, select the V8 protease for cleavage of both glutamic acid (E) and aspartic acid (D) as Glu-C is able to cleave at both these positions in phosphate buffer.
2. With this window open, select "A list of peptides" under the "Find" menu and type in any peptide masses identified by SELDI AMS. The program will provide a graphic representation of their placement in the sequence. Four proteolytic fragments were found for Glu-C digestion of CPG2 and were identified according to the CPG2 SWISS-PROT protein database reference (1CG2) **(Fig. 2)**. The observed peptide masses were, *m/z* 4354.1 assigned to CPG2 Ser 379-Lys 415 average mass of 4354.1 Daltons, *m/z* 2794.9 assigned to CPG2 Tyr391-Lys415 average mass of 2794.4 Daltons, *m/z* 3539.9 Daltons assigned to CPG2 Tyr 159-Glu189 average mass of 3539.8 Daltons, and *m/z* 2092.8 Daltons assigned to CPG2 Tyr159-Glu176 average mass of 2092.8 Daltons. The peptide fragments identified by SELDI AMS were located in two distinct and remote regions of the CPG2 sequence.

2.4.2. MS-Digest

MS-Digest is another protein sequence analysis software package (available on the Protein Prospector website: http://prospector.ucsf.edu/; *see* **ref. 27**), which provides a comprehensive list of proteolytic peptide fragments resulting from the cleavage of a user-specified protein.

1. Open the MS-Digest page and select "User protein" in the "Database" menu.
2. Select the appropriate enzyme in the "Digest" menu.
3. Set the maximum number of missed cleavages to between 5 and 10.
4. Set the program to recognize oxidized methionines and pyroglumatic acid containing peptides in the "Considered modifications."
5. Set the minimum and maximum fragment mass menus to the appropriate values. For example, when analyzing peptides obtained from the SELDI AMS of CPG2 using Glu-C, these parameters were set to 1000 and 6000 Daltons, respectively.
6. Paste the entire protein sequence into the "User protein sequence" menu.
7. Select "Perform digest."
8. The program will provide a list of possible peptides. Proteolytic fragment masses obtained from SELDI AMS can then be crossreferenced against the program-derived list of peptide fragments. All four of the proteolytic fragments derived from the Glu-C digestion of CPG2 were found in this list.

2.5. Mapping SELDI AMS-Identified Peptides in Structural Models

The application of structural models to map the placement of the proteolytic peptides identified by SELDI AMS provides invaluable information in the characterization

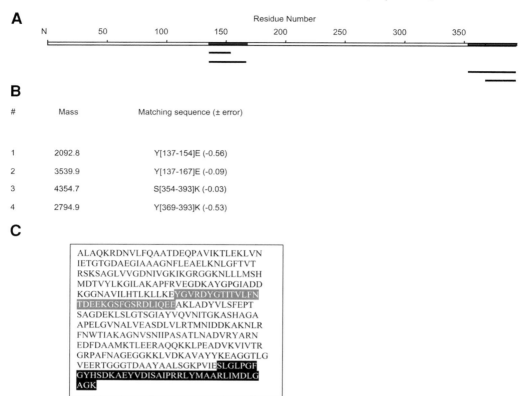

Fig. 2. The application of PAWS to analyze proteolytic peptide fragments identified by SELDI AMS. This technique is illustrated with data obtained from the SELDI AMS analysis of CPG2 captured on a preactivated ProteinChip® Array by a reactive scFv. The program determines all possible peptides resulting from cleavage of a protein sequence by a user-defined protease. Peptide masses obtained by SELDI AMS are inputted into the program and any matches are listed together with their placement in the protein sequence, which is graphically represented by lines. (**A**) Graphic representation of CPG2 amino acid sequence with the placement of scFv-binding peptides resulting from Glu-C digestion of CPG2 in complex with a specific scFv bound to preactivated chips represented by solid lines and (**B**) listed below with their exact placement in the sequence. (**C**) CPG2 amino acid sequence with SELDI AMS-identified peptides highlighted in gray (region 1) corresponding to mass peaks 2092.8 and 3539.9 Daltons, and in black (region 2) corresponding to mass peaks of 4354.1 and 2794.9 Daltons. The peptides correspond to two nonlinear regions along the protein sequence.

of protein-protein interactions. A crystal structure was available for CPG2, enabling the mapping of SELDI AMS identified peptides resulting from proteolytic digestions *(11)*.

Here we describe a method for mapping SELDI AMS-identified proteolytic fragments using Rasmol 2.7.1, a freely distributed molecular graphics visualization tool available over the Internet (Rasmol homepage: http://www.umass.edu/microbio/rasmol/; *see* **ref. 28**). This program allows the user to download Protein Data Bank (PDB; available on the Protein Databank website: http://www.pdb.org/; *see* **ref. 29**) files and study the structure interactively. This simple process enabled the mapping of the two regions

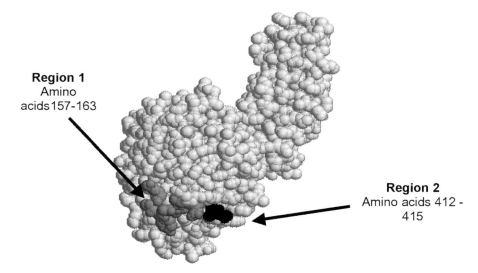

Region 1
Amino
acids157-163

Region 2
Amino acids 412 - 415

Fig. 3. Representation of the crystal structure of CPG2 showing the surface-exposed conformational epitope determined by SELDI AMS. Epitope mapping studies using a specific phage-derived scFv raised against CPG2 revealed reactive peptides corresponding to two non-linear sites along the CPG2 sequence concentrated at positions Y[159–189]E and S[376–415]K. These peptides were fit into the crystal structure of CPG2 and further characterized to obtain the surface-exposed regions of the epitope as determined by the define secondary structure of proteins (DSSP) solvent accessibility algorithm *(30)* revealing a conformational epitope at region 1 G[157–163]D (dark grey) and region 2 G[412–415]G (black).

identified by SELDI AMS in the structural model of CPG2 and shows that two sequentially remote regions are structurally adjacent, suggesting a conformational epitope.

1. Load the structural coordinates of the desired protein into the Rasmol program. For example, for the CPG2 tetrameric protein structure, load PDB reference file 1CG2.
2. Enter "Select *A" in the Rasmol command line to select the monomeric portion of the structure.
3. Enter "Save temp.pdb" to save the selection onto the desktop.
4. Enter "Zap" to clear the program.
5. Enter "Load temp.pdb" to load the selected monomer.
6. Select "Spacefill" under the "Display" menu in the main Rasmol window.
7. Select "Monochrome" under the "Colors" menu in the main Rasmol window.
8. Enter "Select 159–189" in the command window. This will select amino acids corresponding to their placement in the structure. These amino acids represent a region identified by SELDI AMS (sequences identified from mass data corresponding to peaks at 2092.8 and 3539.9 Daltons; *see* **Fig. 1A** and **C**).
9. Enter "Color purple." This will color the selected amino acids purple.
10. Enter "Select 376–415." These amino acids represent a region identified by SELDI AMS (sequences identified from mass data corresponding to peaks 2794.9 and 4354.7 Daltons; *see* **Fig. 1C** and **F**).
11. Enter "Color orange." This will color the selected amino acids orange.
12. Enter "Background white." This will color the background white.
13. Maneuver the structure using the mouse left and right buttons to the desired angle.
14. Save the final version by selecting an export file in the "Export menu" **(Fig. 3)**.

References

1. Benjamin, D. C., Berzofsky, J. A., East, I. J., et al. (1984) The antigenic structure of proteins: a reappraisal. *Annu. Rev. Immunol.* **2,** 67–101.
2. Berzofsky, J. A. (1985) Intrinsic and extrinsic factors in protein antigenic structure. *Science* **229,** 932–940.
3. Jemmerson, R. and Paterson, Y. (1986) Mapping epitopes on a protein antigen by the proteolysis of antigen-antibody complexes. *Science* **232,** 1001–1004.
4. Eisenberg, R. J., Long, D., Pereira, L., Hampar, B., Zweig, M., and Cohen, G. H. (1982) Effect of monoclonal antibodies on limited proteolysis of native glycoprotein gD of herpes simplex virus type 1. *J. Virol.* **41,** 478–488.
5. Moelling, K., Scott, A., Dittmar, K. E., and Owada, M. (1980) Effect of p15-associated protease from an avian RNA tumor virus on avian virus-specific polyprotein precursors. *J. Virol.* **33,** 680–688.
6. Schwyzer, M., Weil, R., Frank, G., and Zuber, H. (1980) Amino acid sequence analysis of fragments generated by partial proteolysis from large simian virus 40 tumor antigen. *J. Biol. Chem.* **255,** 5627–5634.
7. Sheshberadaran, H. and Payne, L. G. (1988) Protein antigen-monoclonal antibody contact sites investigated by limited proteolysis of monoclonal antibody-bound antigen: protein "footprinting." *Proc. Natl. Acad. Sci. USA* **85,** 1–5.
8. Suckau, D., Kohl, J., Karwath, G., et al. (1990) Molecular epitope identification by limited proteolysis of an immobilized antigen-antibody complex and mass spectrometric peptide mapping. *Proc. Natl. Acad. Sci. USA* **87,** 9848–9852.
9. Van de, W. J., Deininger, S. O., Macht, M., Przybylski, M., and Gershwin, M. E. (1997) Detection of molecular determinants and epitope mapping using MALDI-TOF mass spectrometry. *Clin. Immunol. Immunopathol.* **85,** 229–235.
10. Yi, J. and Skalka, A. M. (2000) Mapping epitopes of monoclonal antibodies against HIV-1 integrase with limited proteolysis and matrix-assisted laser desorption ionization time-of-flight mass spectrometry. *Biopolymers* **55,** 308–318.
11. Spencer, D. I., Robson, L., Purdy, D., et al. (2002) A strategy for mapping and neutralizing conformational immunogenic sites on protein therapeutics. *Proteomics* **2,** 271–279.
12. Reid, G., Gan, B. S., She, Y. M., Ens, W., Weinberger, S., and Howard, J. C. (2002) Rapid identification of probiotic *Lactobacillus* biosurfactant proteins by ProteinChip tandem mass spectrometry tryptic peptide sequencing. *Appl. Environ. Microbiol.* **68,** 977–980.
13. Huhalov, A. Spencer, D. I. R., Hawkins, R. E., Perkins, S. J., Begent, R. H. J., and Chester, K. A. (2002) Capture and analysis of interacting tagged-proteins: an application to map the carcinoembryonic antigen binding-site of single chain Fv molecule MFE-23. *J. Mol. Biol.,* in press.
14. Chester, K. A., Begent, R. H., Robson, L., et al. (1994) Phage libraries for generation of clinically useful antibodies. *Lancet* **343,** 455–456.
15. Begent, R. H., Verhaar, M. J., Chester, K. A., et al. (1996) Clinical evidence of efficient tumor targeting based on single-chain Fv antibody selected from a combinatorial library. *Nat. Med.* **2,** 979–984.
16. Mayer, A., Tsiompanou, E., O'Malley, D., et al. (2000) Radioimmunoguided surgery in colorectal cancer using a genetically engineered anti-CEA single-chain Fv antibody. *Clin. Cancer Res.* **6,** 1711–1719.
17. Oikawa, S., Imajo, S., Noguchi, T., Kosaki, G., and Nakazato, H. (1987) The carcinoembryonic antigen (CEA) contains multiple immunoglobulin-like domains. *Biochem. Biophys. Res. Commun.* **144,** 634–642.

18. Paxton, R. J., Mooser, G., Pande, H., Lee, T. D., and Shively, J. E. (1987) Sequence analysis of carcinoembryonic antigen: identification of glycosylation sites and homology with the immunoglobulin supergene family. *Proc. Natl. Acad. Sci. USA* **84,** 920–924.

19. Boehm, M. K., Mayans, M. O., Thornton, J. D., Begent, R. H., Keep, P. A., and Perkins, S. J. (1996) Extended glycoprotein structure of the seven domains in human carcinoembryonic antigen by X-ray and neutron solution scattering and an automated curve fitting procedure: implications for cellular adhesion. *J. Mol. Biol.* **259,** 718–736.

20. Boehm, M. K., Corper, A. L., Wan, T., et al. (2000) Crystal structure of the anti-(carcinoembryonic antigen) single-chain Fv antibody MFE-23 and a model for antigen binding based on intermolecular contacts. *Biochem. J.* **346,** 519–528.

21. Boehm, M. K. and Perkins, S. J. (2000) Structural models for carcinoembryonic antigen and its complex with the single-chain Fv antibody molecule MFE23. *FEBS Lett.* **475,** 11–16.

22. Michael, N. P., Chester, K. A., Melton, R. G., et al. (1996) In vitro and in vivo characterisation of a recombinant carboxypeptidase G2::anti-CEA scFv fusion protein. *Immunotechnology* **2,** 47–57.

23. Bagshawe, K. D. (1987) Antibody directed enzymes revive anti-cancer prodrugs concept. *Br. J. Cancer* **56,** 531–532.

24. Rowsell, S., Pauptit, R. A., Tucker, A. D., Melton, R. G., Blow, D. M., and Brick, P. (1997) Crystal structure of carboxypeptidase G2, a bacterial enzyme with applications in cancer therapy. *Structure* **5,** 337–347.

25. Hoogenboom, H. R. and Chames, P. (2000) Natural and designer binding sites made by phage display technology. *Immunol. Today* **21,** 371–378.

26. Verma, R., Boleti, E., and George, A. J. T. (2000) Antibody engineering: comparison of bacterial, yeast, insect and mammalian expression systems. *J. Immunol. Methods* **216,** 165–181.

27. Clauser, K. R., Baker, P., and Burlingame, A. L. (1999) Role of accurate mass measurement (+/– 10 ppm) in protein identification strategies employing MS or MS/MS and database searching. *Anal. Chem.* **71,** 2871–2882.

28. Sayle, R. A. and Milner-White, E. J. (1995) RASMOL: biomolecular graphics for all. *Trends Biochem. Sci.* **20,** 374.

29. Berman, H. M., Westbrook, J., Feng, Z., Gilliland, G., Bhat, T. N., Weissig, H., et al. (2000) The Protein Data Bank. *Nucleic Acids Res.* **28,** 235–242.

30. Kabsch, W. and Sander, C. (1983) Dictionary of protein secondary structure: pattern recognition of hydrogen-bonded and geometrical features. *Biopolymers* **22,** 2577–2637.

10

Stable Isotope Labeling with Amino Acids as an Aid to Protein Identification in Peptide Mass Fingerprinting

Robert J. Beynon

1. Introduction

Peptide mass fingerprinting is arguably the simplest one-dimensional method used in proteomics to identify proteins, whether recovered from bands or spots on one-dimensional (1-D) or two-dimensional (2-D) gels, as single proteins or mixtures isolated by chromatographic steps, or by affinity purification. In all instances, the protein to be identified is first isolated and then digested with a protease of defined specificity, such as trypsin or endopeptidase LysC. These peptides are then mass measured, usually by matrix-assisted laser desorption/ionization time-of-flight (MALDI-TOF) mass spectrometry (MS). Because the protein was isolated prior to proteolytic fragmentation, connectivity between the product peptides can reasonably be inferred and the resulting peptide mass fingerprint can be used to search protein databases *(1,2)*.

Although good-quality spectra can be matched with considerable confidence to the correct protein in the database, there are many factors that conspire to reduce the success of peptide mass fingerprinting. Poor-quality or low-intensity spectra may contain data on only a few peptides, and these may be inadequately matched for a high probability of identification. A MALDI-TOF mass spectrum originates from a stoichiometric mixture of tryptic peptides, but the intensities of the different ions are dramatically different, to the extent that some peptides are to all intents and purposes absent from the spectrum. It is well known that in a tryptic digest, Lys-terminated peptides are much less prevalent than Arg-terminated peptides in a MALDI-TOF mass spectrum. Finally, if the target protein is not present in the protein databases, then cross-species matching requires even more peptides to compensate for amino acid changes in homologous proteins *(3)*.

Under circumstances in which identification is problematic, there is some value in maximizing the information that is derived from a relatively simple MALDI-TOF mass spectrum before invoking more complex tandem in-space (orthogonal instruments such as a Q-TOF) or tandem in-time (e.g., ion trap) approaches to derive fragmentation data or partial sequence tags. Various strategies can be adopted. Chemical modification of the peptide can highlight the presence or absence of specific residues, thereby adding limited composition data to the search terms (e.g., **ref. *4***). For example, guanidination of lysine residues to homoarginine residues not only identifies lysine residues but also markedly enhances their representation in a MALDI-TOF mass spectrum *(5–10)*, presumably through the enhanced basicity of the homoarginine residue. However, such in

From: *Handbook of Proteomic Methods*
Edited by: P. Michael Conn © Humana Press Inc., Totowa, NJ

vitro approaches are beyond the scope of this chapter, although they might serve to enhance information gain from a peptide map.

An additional strategy to enhance the success of peptide mass fingerprinting derives from the incorporation of stable isotope labels through biosynthesis of the protein. The stable isotope is incorporated into a protein such that labeled peptides are readily identifiable during peptide mass fingerprinting, generating additional information about the peptide that can enhance identifiability. This chapter discusses the principles and practice of labelling in vivo, primarily to support protein identification by MALDI-TOF. For convenience, this overall process is abbreviated to SILAA (stable isotope labeling by amino acids; **Fig. 1**) based on the more specific term SILAC that has been used to describe a specific example of this approach *(11)*. SILAA would include the use of amino acids to label plants, animals, or microrganisms in vivo and thus includes SILAC as one component part.

In summary, SILAA can be used, in peptide mass fingerprinting, to provide additional composition data for the peptides in the mass spectrum. These additional composition data are an effective route to reduction of the search space during the protein identification process, and there are several good exemplars in the literature (**Table 1**; also *see* **refs. *11–21***).

2. Choice of Label for Incorporation of Stable Isotopes In Vivo
2.1. Metabolic Labeling Considerations

SILAAs can be used to label proteins at specific residues. However, administration and incorporation of labeled amino acids into intact organisms is problematic, and most studies to date have used cells grown in suspension culture or as monolayers (**Table 1**). This amino acid, supplied in growth media to cells growing in batch culture as a suspension or monolayer, is taken up and incorporated into proteins by protein synthesis. A variant that has thus far only been used in a single study is the use of continuous culture in a chemostat to maintain cells in a stable steady state *(21)*. By contrast, in batch culture the cells change state through the growth period, moving from lag phase to exponential growth and then shifting to stationary phase. These phase transitions are marked by changes in the profile of proteins that are being expressed and in the absolute and relative rates of turnover of individual proteins (*see* below).

The purpose of SILAA is to incorporate the precursor, ideally without modification or loss, into the proteins that are newly synthesized in the cells. The efficiency or extent of labeling will depend on several factors, including the atom percent excess (APE) of the label, the amount of unlabeled amino acid in the medium, dilution of the labeled amino acid in the intracellular pool by unlabeled amino acids either synthesized *de novo* or released by the process of intracellular protein degradation, or the deamination and oxidation of the labeled amino acid, which effectively removes the label from the precursor pool.

Fig. 1. *(see facing page)* Principles of SILAA. Stable isotope-labeled amino acids are incorporated into proteins by *de novo* synthesis, in a complex process that reflects overall protein turnover, dilution of label from unlabeled protein, and amino acid oxidation. The proteins are subsequently separated by an appropriate method, and the peptides are analyzed by MS. The incorporation of a stable isotope label into a peptide yields a clearly discriminable signal.

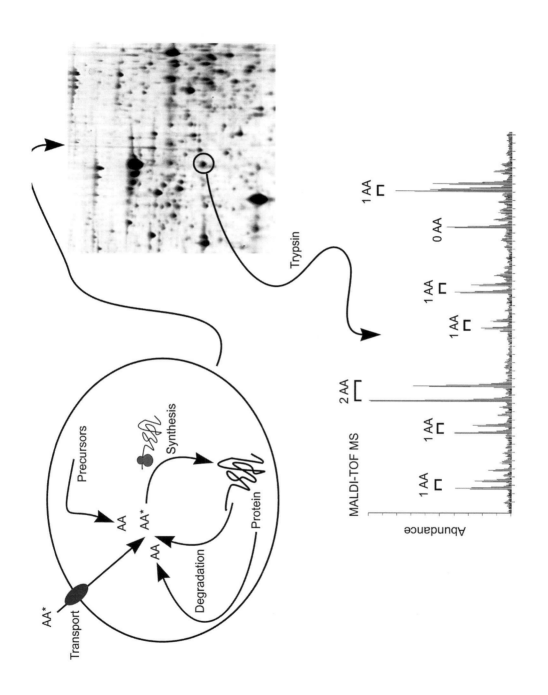

Precursors

AA

AA*

Synthesis

AA

Degradation

Protein

AA*

Transport

Trypsin

MALDI-TOF MS

Abundance

2 AA

1 AA

1 AA

1 AA

1 AA

0 AA

1 AA

Table 1
Examples of Studies Using Stable Isotope Labeled Amino Acids

Organism	Label	Growth conditions	Notes	MS	Ref.
Saccharomyces cerevisiae	^{15}N-enriched media (>96 APE)	Rich media, cells harvested mid-log phase	For relative quantification rather than identification	MALDI-TOF	*12*
Escherichia coli	L-[^{2}H$_3$] methionine (99.9 APE) [^{2}H$_2$] glycine (99.9 APE)	Minimal media or LB medium (control)	Mutants of strain used to enhance labeling of proteins.	MALDI-TOF MALDI-TOF PSD	*13*
Deinococcus radiodurans, B16 (mouse) cells	^{15}N enriched media (>98 APE)	Late log phase?		ESI-ion trap, ESI-FTICR	*14*
Saccharomyces cerevisiae	L-[^{2}H$_3$] methionine (99.9 APE) D,L-[^{2}H$_3$] serine (99.9 APE) L-[^{2}H$_2$] tyrosine (99.9 APE)	SC (synthetic complete) medium; cells harvested at mid-log phase.		MALDI-TOF	*15,16*
Saccharomyces cerevisiae	L-[^{2}H$_{10}$] leucine (99.4 APE)	Cells harvested at stationary phase		MALDI-TOF	*17*
Mouse C2C12 myoblast cells	L-[^{2}H$_3$] leucine (99.4 APE)	Eagle's minimal medium plus serum, six cell divisions	Dialysis of serum reduces input of unlabeled amino acid	MALDI-TOF, ESI-QTOF	*11*
Saccharomyces cerevisiae	[^{13}C$_6$] lysine (98 APE) [^{2}H$_3$] methionine [^{2}H$_2$] glycine [^{2}H$_2$] tyrosine [^{2}H$_4$] lysine [^{2}H$_3$] leucine	Synthetic medium, log phase	Lysine auxotroph	ESI-FTICR	*18*
Escherichia coli		Minimal media, supplemented as required with amino acids studies (APE not specified)	SILAA used to aid identification of pepsin-derived peptides in hydrogen exchange	ESI-ion trap	*19*
Escherichia coli HSF (human) cells	L-[^{2}H$_4$] lysine (96–98 APE)	Not stated	Analysis of recombinant proteins Endopeptidase Lys-C digestion to generate C-terminally labeled peptides	MALDI-TOF, ESI-ion trap	*20*
Saccharomyces cerevisiae	D,L-[^{2}H$_{10}$] leucine (98.5 APE)	Continuous culture at growth rate of 0.1 h^{-1}	Leucine auxotroph	MALDI-TOF	*21*

APE, atom percent excess; ESI, electroscopy ionization; FTICR, Lourier-transform ion cyclotron resonance; MALDI-TOF, matrix-assisted laser desorption/ionization time-of-flight; PSD, postsource decay.

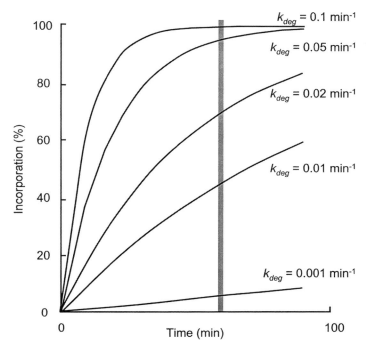

Fig. 2. The role of turnover in labeling of proteins. Proteins are replaced at different rates in the cell, and complete incorporation of a stable isotope label amino acid requires that sufficient half-lives be allowed for the pool of protein to have been substantially replenished by *de novo* synthesis. This can lead to a requirement for extensive labeling windows.

Whether the cells are growing or not, amino acids will be incorporated into proteins through the continuous process of intracellular protein turnover, whereby proteins are continually degraded and resynthesized *(22,23)*. The rate of turnover differs from protein to protein and will also vary in absolute terms according to the physiological state of the cell. Thus, the rate at which proteins incorporate label cannot be predicted for a short labeling window without some assumption about turnover rate. For example, with a labeling period of 60 min, a protein with a degradation rate of 0.001 min^{-1} will incorporate label extremely slowly and will only become approx 6% labeled during this time. By contrast, a protein with a turnover rate of 0.1 min^{-1} will be more than 99% labeled in the same period (**Fig. 2**).

If the cells are growing during this time, it follows that the rate of synthesis exceeds the rate of degradation (net accretion of protein), and thus, the efficiency of incorporation can be higher, because more of the labeled precursor is directed to new synthesis. Indeed, if the labeled variant is the only source of that amino acid, and if the cell is unable to synthesize the amino acid from other carbon sources (the cell is auxotrophic for that amino acid), then for every period over which the cell number doubles, the extent of labeling increases by 50%. This is exponential, with a rate constant equal to the first-order doubling rate. For a cell culture with a doubling rate of 0.1 h^{-1} (or a doubling time of 6.9 h), after 50 h of growth, more than 99% of the proteins would be labeled to the same APE as the precursor pool [calculated as $1 - e^{-(50 \times 0.1)}$. Thus,

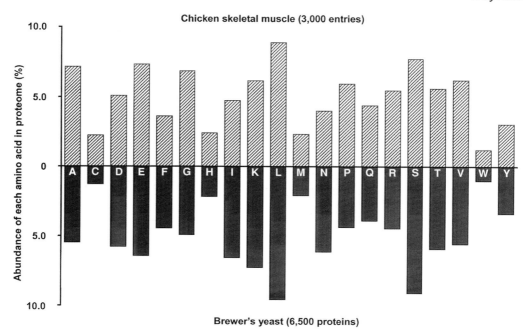

Fig. 3. The relative abundance of amino acids in proteomes. The choice of which amino acid to use in SILAA is dictated by many factors, but one of the most important may be the capability to label a very large number of peptides that derived from the fragmented proteome. The amino acid abundances are shown here for one complete proteome *(Saccharomyces cerevisiae)* and the proteins in chicken skeletal muscle.

even proteins with imperceptibly low rates of degradation will be virtually fully labeled as cell division and growth leads to the synthesis of new proteins.

2.2. Amino Acid Abundance in the Proteome

The distribution of amino acids in a proteome is very uneven, and some amino acids are far more abundant than others. The amino acid distributions from two proteomes, the entire *Saccharomyces cerevisiae* proteome and the chicken skeletal muscle proteome using currently available data, are given in **Fig. 3**. It is clear from these two rather disparate proteomes that the overall distribution of amino acids is very similar. For both, leucine is the most abundant amino acid, at approx 10% of all amino acids by frequency, and tryptophan is the least abundant, at around 1% by frequency. However, a more relevant consideration relates to the extent to which each amino acid is distributed through a set of peptides derived from a tryptic digest, for example. To illustrate, if one amino acid showed a marked tendency to occur in adjacent pairs, then each tryptic fragment would tend to contain this pair, and the useful representation of the amino acid is effectively halved. Of greater relevance, therefore, is the amino acid frequency in relation to its distribution in typical proteolytic digests, across the mass range of peptides (1000–3000 Daltons) that are most readily analyzed by MALDI-TOF (**Fig. 4**). The amino acids present in peptides that cover this mass range are, unsurprisingly, distrib-

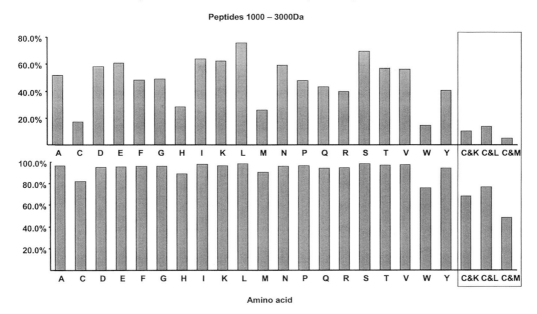

Peptides 1000 – 3000Da

Fig. 4. The utility of different amino acids in SILAA. This analysis is of a subset of tryptic peptides from the yeast proteome that range in mass between 1000 and 3000 Daltons. The data are expressed as the percentage of all peptides in that mass range that contain at least one of the appropriate amino acids. The lower panel is an analysis of the number of proteins that are represented by this selection criterion—having at least one peptide, in this mass range, that contains that amino acid. Finally, representative data are given for three amino acids used in combination with cysteine selection. Here the number of peptides falls dramatically, but the protein representation remains high.

uted similarly to the theoretical total proteome. For example, about half of the tryptic peptides between 1000 and 3000 Daltons in the yeast proteome contain at least one alanine residue; over three-fourths contain at least one leucine residue.

To pose the question differently: the alanine-containing peptides are derived from more than 96% of the entire theoretical proteome. Even tryptophan-containing peptides, of which there are only approx 15,000, ranging in mass from 1000 to 3000 Daltons, represent more than 75% of all yeast proteins. Of course, MALDI-TOF MS rarely generates a single peptide, and the bioinformatics tools would not give a confident assignment based on a single peptide, but a combination of SILAA, perhaps with one or more labeled amino acid precursors, supplemented by (for example) knowledge of the C-terminal amino (Lys or Arg) through guanidination might move some way toward a reliable identification. If one amino acid is used as a tagging set for simplification of the proteome, then the protein representation falls, but the reduction in complexity of the peptide mix declines very dramatically. This is illustrated in **Fig. 4** for three commonly used SILAA amino acids: leucine, lysine, and methionine, in combination with a cysteine-based trapping approach. Other affinity-based selection methods would yield similar results *(24)*. Slightly different arguments and ranges would apply to peptides analyzed by, for example, ion trap instruments, although the principles are the same.

2.3. Metabolic Lability

Implicitly, SILAA is based on extracellular administration of a labeled amino acid, after which it is transported across the cell membrane and exposed to the metabolic capabilities of that cell, including the ability to metabolize the label to carbon dioxide and water and in some circumstances to redistribute label from one amino acid to another (e.g., cysteine can by synthesized from methionine in many species). The most effective strategy seems to be the use of an essential amino acid (whether based on natural inabilities to synthesize that amino acid or through the use of auxotrophic mutants), to supply that amino acid predominantly in labeled form, through the use of synthetic media depleted of that amino acid, and to ensure that other nutrients are present in sufficiently large quantities to minimize the oxidation of the labeling amino acid with consequent loss of label.

The position of the label in the amino acid is also critical. For example, the $[^2H_{10}]$ leucine that we have used in our own investigations is supplied labeled at all positions other than at the amino and carboxyl groups (aqueous phase ionization of these two groups would rapidly remove the label). However, when peptides containing leucine residues are analyzed by MALDI-TOF, it became apparent that the mass separation of labeled and unlabeled peptides were in multiples of 9 Daltons, consistent with the loss of one deuterium atom (21). We believe that this is the α-deuterium atom, which would be readily lost by the process of transamination (**Fig. 5**), and we do not consider that the alternative explanation, loss of a single deuterium in the MALDI-TOF source (17), is a reasonable explanation. Although the transamination reaction is readily reversible, the probability of a deuterium atom being replaced at that position is very low. Ideally, selective labeling will use precursors that are labeled at atom centers that are metabolically stable. In this respect, ^{13}C-labeled amino acids might be less prone to problems as the carbon backbone of the amino acid remains intact until it is committed to oxidation. However, deuterium-labeled amino acids are less costly, and we have never experienced any problem with $[^2H_{10}]$-labeled leucine; the completeness of the transamination reaction means that no ambiguities arise.

2.4. Analytical Convenience

In most experimental systems, several amino acids will meet the criteria specified above, and one can be reasonably confident about obtaining specific labeling at high abundance. A further selection criterion that can be applied is that of analytical convenience. It has, for example, been reported that some stable isotope-labeled peptides or peptide derivatives can be resolved from their unlabeled counterparts by reversed phase chromatography (24,25). This particular consideration is less of a problem in analyses based only on MALDI-TOF.

Labeling with lysine residues has several advantages. First, the stable isotope shift allows clear discrimination between lysine and asparagine, which differ in monoisotopic residue mass by only 36 mDa, and which would require exceptional attention to calibration and mass accuracy in most routine MALDI-TOF instruments. If trypsin was used as the digestion enzyme, then peptides with a lysine labeling signature would be distinguished from arginine-terminated peptides, which could reduce the search space by a factor of 2 for each peptide. Against this must be factored the lower propen-

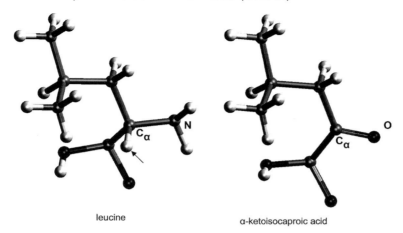

leucine   α-ketoisocaproic acid

Fig. 5. Metabolic loss of stable isotope labels. Leucine readily undergoes transamination to α-ketoisocaproate, with the consequent loss of the α-carbon hydrogen. Decadeuterated [^{2}H$_{10}$] leucine is incorporated into peptides yielding mass shifts that are multiples of 9 Daltons. This is almost certainly due to the loss of the α-carbon deuterium by transamination.

sity of lysine-terminated peptides to give strong MALDI-TOF signals, especially with α-cyano-4-hydroxy-cinnamic acid as the matrix, although there are solutions to enhancement of signal from lysine-terminated tryptic fragments *(5–10)*.

Leucine, the most abundant amino acid in the proteome, is commonly used as a label for SILAA. Most peptides in a peptide mass fingerprint will contain one or more leucine residues **(Table 1)** and thus maximize the composition information that can be recovered in a typical fingerprint. The use of leucine also permits discrimination from its isobaric partner, isoleucine, an information gain that is generally not sought because of the requirement for high-energy collision in a tandem experiment. Both leucine and isoleucine are abundant amino acids, and resolution of the two does provide an approximately twofold reduction in search space for each peptide for which this information is recovered. However, this can only be obtained in a tandem experiment, in which case, the acquisition of sequence tags obviates this additional composition information.

2.5. Backbone and Side-Chain Labeling with Nitrogen

A less commonly used approach uses ^{15}N labeling, usually derived from ammonium ions in the medium, such as ^{15}NH$_4$Cl or (^{15}NH$_4$)$_2$SO$_4$. In this labeling strategy, ^{15}N is incorporated into the amide nitrogen in the backbone and the side chains of lysine, arginine, histidine, glutamine, or asparagine residues. Arginine is readily labeled at the guanidino function through the reactions of the urea cycle. However, the side chains of lysine and histidine can only be labeled if the cell is competent in biosynthesis of these two amino acids. The side chains of asparagine and glutamine are readily labeled by all organisms. This approach is usually used with bacterial or yeast cells that are maintained in culture, and if it is possible to supply all of the nitrogen that the organism needs in the form of ^{15}N. Under these circumstances, the degree of labeling is substantial. However, each peptide contains different numbers of nitrogen atoms, and thus, the separation in *m/z* space between the labeled and unlabeled peptides is also variable and requires a high-resolution instrument to permit composition information, at the level of individual nitrogen atoms, to be recovered. Mixing the ^{14}N- and ^{15}N-labeled material in equal amounts means that the separation between the "light" and the "heavy" peaks can be determined as a relative mass separation, which therefore has more accuracy than comparison of two spectra, increasing the confidence with which the number of nitrogen atoms can be determined. There do not seem to be any commercial or publicly accessible software packages that can embody this additional information, however. Indeed, the differential ^{14}N : ^{15}N approach has predominantly been used to measure relative expression levels of proteins, not for identification *(15)*.

3. Analysis of Spectra after SILAA

Acquisition of SILAA-derived spectra may be straightforward, but subsequent data processing is not quite as straightforward. The requirement is for software that can either accommodate the stable isotope-labeled peptides alone, or in a mixture with their unlabeled counterparts. Of course, the mass of the labeled amino acid is greater than the mass that is coded in the identification software, and because it is not possible to predict which variant of the labeled amino acid is being used (e.g., [^{2}H$_7$] or [^{2}H$_{10}$] leucine), there is no simple approach to codifying all possible variants, even before additional loss of label through metabolism is brought into account. Many of the studies reported in **Table 1** have used local software solutions to identify the labeled amino acids.

For small numbers of samples, manual inspection of the spectra, whether the labeled and unlabeled forms are examined independently, or as a mixture of heavy and light forms, can indicate the mass of the unmodified peptide as well as the number of the labeled amino acid by virtue of the mass offset of the heavy variant. A typical example is given in **Fig. 6**. The 50 : 50 mixture of labeled and unlabeled peptides clearly shows the doublets corresponding to the presence of one or more labeled residues. However, we find that interpretation is aided by the spectra from a fully labeled and an unlabeled sample.

Manual inspection of the spectra generates a set of monoisotopic masses for the unlabeled peptides, and the additional information about the number of the target amino acid. There are relatively few software tools, however, that can make use of this information. Perhaps the simplest is the MASCOT search engine (http://www.matrixscience.co.uk; *see* **ref. 26**), which includes a search page that has the ability to supply composition data. Specifically, the list of peptide masses are supplement with additional text strings that indicate the number of amino acid residues, in the form **comp(n[AA])** where n is the number of the amino acid (AA) **(Fig. 7)**. In the example shown, a very high probability match was obtained with only four peptides. Inspection of the output of the MASCOT program indicates that the four peptides meet the criteria specified for the number of leucine residues.

An alternative approach is to treat the "heavy" amino acid as a pseudo-modification of the amino acid. Most peptide mass fingerprinting programs have an editable text file that allows new "modifications" to be introduced. It is then relatively straightforward to add new "modifications" that reflect the mass shifts commensurate with incorporation of labeled amino acid. Of course, this mass shift must be "attached" to the appro-

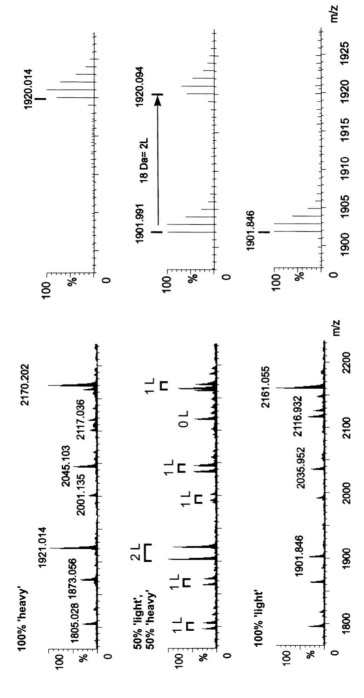

Fig. 6. Use of stable isotope labeling for protein identification. The spectra on the left are taken from a 2-D gel spot from yeast cells grown in the presence of [²H₁₀] leucine (top panel). An identical culture was grown in the presence of unlabeled leucine (bottom panel). Finally, cells of both cultures were mixed prior to another 2-D gel separation. The leucine-containing peptides are readily apparent from the doublets in the middle spectrum. The panel on the right zooms in on one of these peptides and indicates the 18-Dalton spacing that defines this peptides, of mass 1901.85, as having two leucine residues.

A

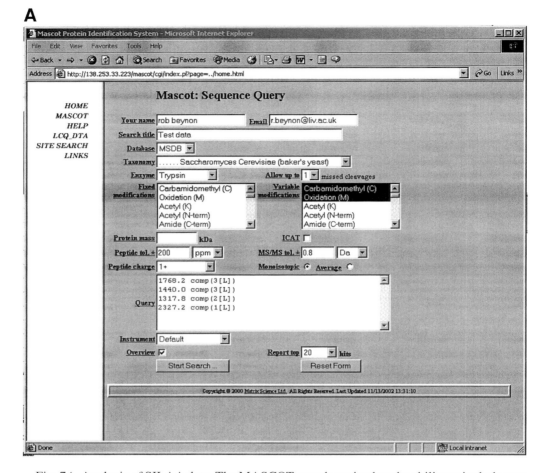

Fig. 7A. Analysis of SILAA data. The MASCOT search engine has the ability to include composition data in the peptide search terms.

priate amino acid. For the MASCOT search engine, the appropriate segment of the file for a 9-Dalton "heavy" leucine is:

```
Title:HeavyLeu (L)
Residues:L 122.13964 122.13964
*
Title:Carbamidomethyl (C)
Residues:C 160.03066 160.191
*
Title:Oxidation (M)
Residues:M 147.03541 147.191
```

4. Conclusions

SILAA is a valuable addition to the tools for comparative proteomics and for protein identification by mass spectrometry. It cannot be readily applied to all biological systems, and obtaining a high degree of labeling might be problematic with many mam-

B

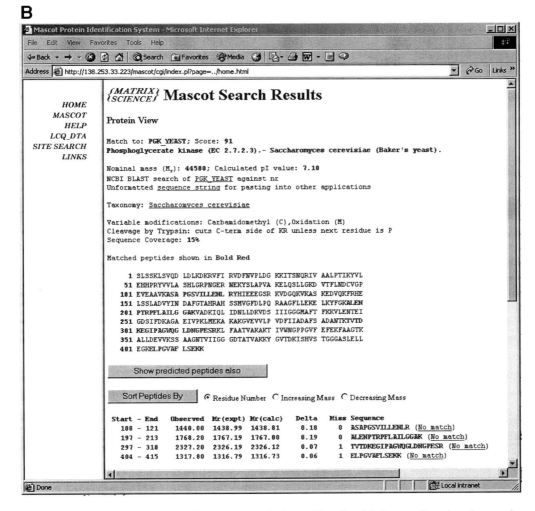

Fig. 7B. If identification is attained, the matched peptides should also confirm that the number of leucine residues ascribed to each peptide is correct.

malian species, for example. Nonetheless, it offers a substantial reduction of search space in peptide mass fingerprinting and is readily used in comparative studies. The approach can be further developed, for example, by multiplexed labeling protocols using different amino acids, each with a different heavy/light mass separation. Finally, although beyond the scope of this chapter, the use of SILAA to label intact proteins, coupled with the enhanced accuracy of mass measurement obtained with FTICR MS *(27,28)*, offers fascinating alternative strategies for protein and proteome characterization.

Acknowledgments

The work described in this chapter is supported by the Biotechnology and Biological Sciences Research Council. I am grateful to June Petty and Julie Pratt for the experimental data and to Jane Hurst for assistance with the analysis of peptide data.

References

1. Cottrell, J. S. (1994) Protein identification by peptide mass fingerprinting. *Pept. Res.* **7**, 115–124.
2. James, P., Quadroni, M., Carafoli, E., and Gonnet, G. (1994) Protein Identification in DNA Databases By Peptide Mass Fingerprinting. *Protein Sci.* **3**, 1347–1350.
3. Lester, P. J. and Hubbard, S. J. (2002) Comparative bioinformatic analysis of complete proteomes and protein parameters for cross-species identification in proteomics. *Proteomics* **2**, 1392–1405.
4. Brancia, F. L., Butt, A., Beynon, R. J., Hubbard, S. J., Gaskell, S. J., and Oliver, S. G. (2001) A combination of chemical derivatisation and improved bioinformatic tools optimises protein identification for proteomics. *Electrophoresis* **22**, 552–559.
5. Bonetto, V., Bergman, A. C., Jornvall, H., and Sillard, R. (1997) C-terminal sequence analysis of peptides and proteins using carboxypeptidases and mass spectrometry after derivatization of Lys and Cys residues. *Anal. Chem.* **69**, 1315–1319.
6. Keough, T., Lacey, M. P., and Youngquist, R. S. (2000) Derivatization procedures to facilitate de novo sequencing of lysine-terminated tryptic peptides using postsource decay matrix-assisted laser desorption/ionization mass spectrometry. *Rapid Commun. Mass Spectrom.* **14**, 2348–2356.
7. Brancia, F. L., Oliver, S. G., and Gaskell, S. J. (2000) Improved matrix-assisted laser desorption/ionization mass spectrometric analysis of tryptic hydrolysates of proteins following guanidination of lysine-containing peptides. *Rapid Commun. Mass Spectrom.* **14**, 2070–2073.
8. Beardsley, R. L. and Reilly, J. P. (2002) Optimization of guanidination procedures for MALDI mass mapping. *Anal. Chem.* **74**, 1884–1890.
9. Beardsley, R. L., Karty, J. A., and Reilly, J. P. (2000) Enhancing the intensities of lysine-terminated tryptic peptide ions in matrix-assisted laser desorption/ionization mass spectrometry. *Rapid Commun. Mass Spectrom.* **14**, 2147–2153.
10. Hale, J. E., Butler, J. P., Knierman, M. D., and Becker, G. W. (2000) Increased sensitivity of tryptic peptide detection by MALDI-TOF mass spectrometry is achieved by conversion of lysine to homoarginine. *Anal. Biochem.* **287**, 110–117.
11. Ong, S. E., Blagoev, B., Kratchmarova, I., Kristensen, D. B., Steen, H., Pandey, A., and Mann, M. (2002) Stable isotope labeling by amino acids in cell culture, SILAC, as a simple and accurate approach to expression proteomics. *Mol. Cell. Proteomics* **1**, 376–386.
12. Oda, Y., Huang, K., Cross, F. R., Cowburn, D., and Chait, B. T. (1999) Accurate quantitation of protein expression and site-specific phosphorylation. *Proc. Natl. Acad. Sci. USA* **96**, 6591–6596.
13. Chen, X., Smith, L. M., and Bradbury, E. M. (2000) Site-specific mass tagging with stable isotopes in proteins for accurate and efficient protein identification. *Anal. Chem.* **72**, 1134–1143.
14. Conrads, T. P., Alving, K., Veenstra, T. D., Belov, M. E., Anderson, G. A., Anderson, D. J., Lipton, M. S., Pasa-Tolic, L., Udseth, H. R., Chrisler, W. B., Thrall, B. D., and Smith, R. D. (2001) Quantitative analysis of bacterial and mammalian proteomes using a combination of cysteine affinity tags and ^{15}N-metabolic labeling. *Anal. Chem.* **73**, 2132–2139.
15. Hunter, T. C., Yang, L., Zhu, H., Majidi, V., Bradbury, E. M., and Chen, X. (2001) Peptide mass mapping constrained with stable isotope-tagged peptides for identification of protein mixtures. *Anal. Chem.* **73**, 4891–4902.
16. Zhu, H., Hunter, T. C., Pan, S., Yau, P. M., Bradbury, E. M., and Chen, X. (2002) Residue-specific mass signatures for the efficient detection of protein modifications by mass spectrometry. *Anal. Chem.* **74**, 1687–1694.

17. Jiang, H. and English, A. M. (2002) Quantitiative analysis of the yeast proteome by incorporation of isotopically labeled leucine. *J. Proteome Res.* **1,** 345–350.

18. Berger, S. J., Lee, S. W., Anderson, G. A., Pasa-Tolic, L., Tolic, N., Shen, Y., Zhao, R., and Smith, R. D. (2002) High-throughput global peptide proteomic analysis by combining stable isotope amino acid labeling and data-dependent multiplexed-MS/MS. *Anal. Chem.* **74,** 4994–5000.

19. Engen, J. R., Bradbury, E. M., and Chen, X. (2002) Using stable-isotope-labeled proteins for hydrogen exchange studies in complex mixtures. *Anal. Chem.* **74,** 1680–1686.

20. Gu, S., Pan, S. Bradbury, E. M., and Chen, X. (2002) Use of deuterium-labeled lysine for efficient protein identification and peptide *de novo* sequencing. *Anal. Chem.* **74,** 5774–5785.

21. Pratt, J. M., Robertson, D. H., Gaskell, S. J., Riba-Garcia, I., Hubbard, S. J., Sidhu, K., Oliver, S. G., Butler, P., Hayes, A., Petty, J., and Beynon, R. J. (2002) Stable isotope labeling in vivo as an aid to protein identification in peptide mass fingerprinting. *Proteomics* **2,** 157–163.

22. Pratt, J. M., Petty, J., Riba-Garcia, I., Robertson, D. H., Gaskell, S. J., Oliver, S. G., and Beynon, R. J. (2002) Dynamics of protein turnover, a missing dimension in proteomics. *Mol. Cell. Proteomics* **1,** 579–591.

23. Gerner, C., Vejda, S., Gelbmann, D., Bayer, E., Gotzmann, J., Schulte-Hermann, R., and Mikulits, W. (2002) Concomitant determination of absolute values of cellular protein amounts, synthesis rates, and turnover rates by quantitative proteome profiling. *Mol. Cell. Proteomics* **1,** 528–537.

24. Regnier, F. E., Riggs, L., Zhang, R., Xiong, L., Liu, P., Chakraborty, A., Seeley, E., Sioma, C., and Thompson, R. A. (2002) Comparative proteomics based on stable isotope labeling and affinity selection. *J. Mass Spectrom.* **37,** 133–145.

25. Zhang, R., Sioma, C. S., Wang, S., and Regnier, F. E. (2001) Fractionation of isotopically labeled peptides in quantitative proteomics. *Anal. Chem.* **73,** 5142–5149.

26. Perkins, D. N., Pappin, D. J., Creasy, D. M., and Cottrell, J. S. (1999) Probability-based protein identification by searching sequence databases using mass spectrometry data. *Electrophoresis* **20,** 3551–3567.

27. Martinovic, S., Veenstra, T. D., Anderson, G. A., Pasa-Tolic, L., and Smith, R. D. (2002) Selective incorporation of isotopically labeled amino acids for identification of intact proteins on a proteome-wide level. *J. Mass Spectrom.* **37,** 99–107.

28. Veenstra, T. D., Martinovic, S., Anderson, G. A., Pasa-Tolic, L., and Smith, R. D. (2000) Proteome analysis using selective incorporation of isotopically labeled amino acids. *J. Am. Soc. Mass Spectrom.* **11,** 78–82.

11

The Use of ^{18}O Labeling as a Tool
for Proteomic Applications

Ian I. Stewart, Ty Thomson, Daniel Figeys, and Henry S. Duewel

1. Introduction

In the past century, numerous significant advancements were made in our understanding of human disorders. The remaining uncured human diseases, such as cancer and AIDS, are often complex and involve multiple factors. Understanding these complex human conditions is the challenge of the 21st century. Often abnormal changes in cellular processes are the cause of disease and so, the ability to map these cellular changes will be key to understanding and treating them. Clearly, novel technologies are required to map cellular changes rapidly and globally.

Genomics was the first high-throughput technique able to monitor different change in gene expression levels simultaneously. One of the premises of genomics is that key elements in complex biological processes can be mapped by comparatively observing changes that occur at the gene expression levels. A whole industry has been built around gene expression analysis and has been successful in finding disease biomarkers and potential drug targets.

A similar premise was proposed for the field of proteomics for the differential expression of proteins. The importance of differential proteomics was clearly reinforced by studies that showed poor correlation between gene and protein expressions (1–4). Unfortunately, facile mapping of protein expression is far from being routine. In particular, one of the difficulties in proteomics has been the reliance on two-dimensional gel electrophoresis (2-DGE) as the method of choice for protein fraction and quantitation. 2-DGE is tedious to perform, difficult to automate, can have limited reproducibility, suffers from limited sensitivity when coupled to mass spectrometry (MS), and is unsatisfactory for membrane proteins (5,6).

Alternative methods to 2-DGE, based on MS, have been developed for the analysis of complex protein mixtures (7,8). Typically, these methods combine protein fractions with rapid on-line peptide separations and identification by MS. The elimination of 2-DGE provides improved sensitivity, detection of hydrophobic proteins, and ease of automation. However, it also eliminates the in-gel quantitation of proteins. Therefore, novel quantitation methods needed to be developed in order to quantify differential protein expression using MS. For example, a chemical derivitization approach, called isotope-coded affinity tagging (ICAT) (2,9), was developed for the differential protein expression analysis by MS. Briefly, the ICAT method consists of labeling the cysteine-

From: *Handbook of Proteomic Methods*
Edited by: P. Michael Conn © Humana Press Inc., Totowa, NJ

containing peptides in a sample with a light form of a chemical tag. A second sample can be labeled using the heavy form of a chemical tag (obtained by replacing a few hydrogens with deuteriums). The two samples are then mixed and analyzed by MS. Differential quantitation of individual peptides is established by the ratio of the heavy/light peptides recorded by the mass spectrometer.

We *(10)* and others *(11–14)* have developed an enzyme-based approach for the differential quantitation of proteins using MS. This approach consists of introducing a mass difference during the enzymatic digestion of proteins using ^{16}O water and ^{18}O water. The advantages of this approach are many fold: the sensitivity is as good as the enzymatic digestion, all the generated peptides are labeled, the heavy/light peptides coelute in reverse-phase chromatography, and the label does not affect the peptide detection or identification by tandem mass spectrometry (MS/MS). In this chapter, we review the biochemistry behind the enzymatic labeling using ^{16}O/^{18}O water, and we also discuss how MS signals are converted into quantitative information and what factors influence the quality of this data as well as applications of this method. The goal is to provide a clear explanation of the important considerations for users wishing to use this technique and perhaps offer new insights to current users.

2. ^{18}O and Labeling Biochemistry

2.1. Introduction of the Label: Mechanism of Protease-Catalyzed ^{18}O Incorporation

Isotopic labeling of peptides with ^{18}O is comparatively straightforward and can be readily integrated into any proteomics platform. The attraction of the method lies in the fact that the labeling step is a natural consequence of protease-catalyzed hydrolysis of proteins, a universal first step in protein analysis by MS. Furthermore, as demonstrated in the 1980s, ^{18}O can be incorporated into the C-terminal carboxyl group of peptides by simply using H_2^{18}O-enriched buffers in the course of protease digestion as a tool for MS-based peptide mapping *(15)*.

Although it is beyond the scope of this chapter to provide a comprehensive account of the classification (for an excellent reference source, *see* **ref. *16***) and mechanism of proteases (for suggested readings, *see* **refs. *17–22***), a fundamental knowledge is required to understand the essence of the labeling process. In nature, proteases are involved in a number of diverse physiological processes and are characterized by association to one of four broad families: (1) *the serine proteases*, typified by the mammalian chymotrypsin superfamily (e.g., trypsin, chymotrypsin, elastase, and thrombin) and the bacterial subtilisins; (2) *the cysteine proteases*, exemplified by the archetype papain; (3) *the aspartyl proteases*, including pepsin; and (4) *the metalloproteases*, of which bacterial thermolysin is the classic representative.

All proteases catalyze hydrolysis of peptide bonds in which a water molecule reacts with a peptide carbonyl group, and the bond to the amino group of the next amino acid is cleaved. The serine and cysteine proteases employ an active site Ser or Cys residue, respectively, to attack the peptide substrate, generating an acyl enzyme intermediate that is subsequently hydrolyzed to the products. Two half-reactions are involved, enzyme acylation and enzyme deacylation **(Fig. 1A)**. Serine proteases have independently evolved a similar catalytic apparatus, characterized by the so-called catalytic triad (Ser-195, His-57, and Asp-102 in the chymotrypsin superfamily). A similar catalytic triad

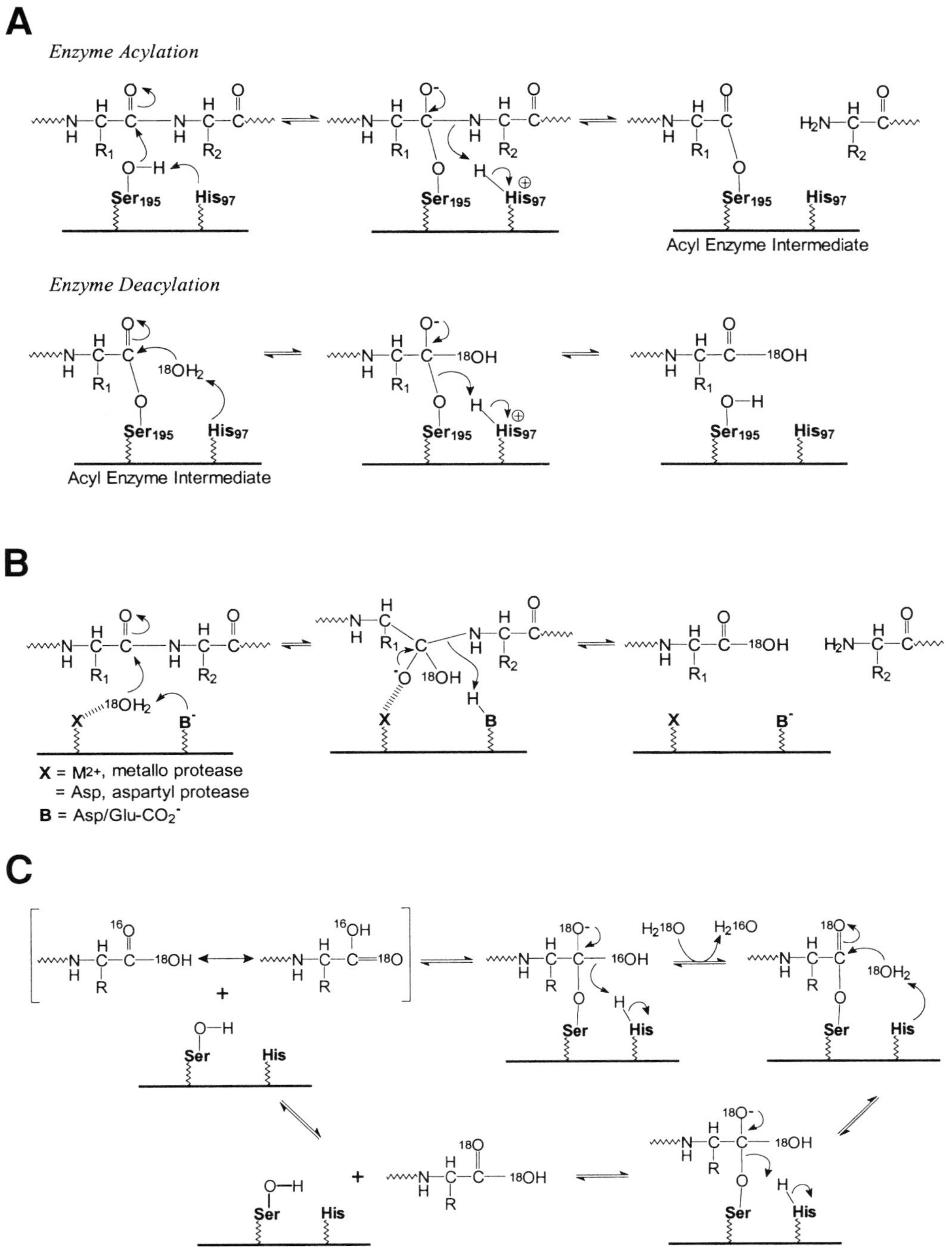

Fig. 1. Simplified representations of protease-catalyzed incorporation of ^{18}O into peptides. (**A**) General mechanism for serine proteases involving the acylation and deacylation half-reactions and the acyl enzyme intermediate. The overall mechanism is analogous for cysteine proteases. (**B**) General mechanism of aspartyl and metalloproteases via direct attack of a water molecule onto the peptide bond. In **A** and **B**, only relevant active-site residues of the protease are shown. See text for further details. (**C**) Proposed mechanism of enzyme-catalyzed back-exchange of the carboxyl oxygen atoms. Peptide fragments derived from the initial peptide bond cleavage event can serve as substrates for proteases, resulting in repeated exchange of the carboxyl oxygen atoms with the bulk solvent.

(or Cys/His dyad) is present in cysteine proteases. The simplified effect of the catalytic triad is the functioning of His-57 as a general base to promote nucleophilic attack of the Ser-195 hydroxyl on the peptide bond to form the acyl enzyme intermediate during the acylation reaction. The Asp-102 residue serves to orient the side chain of His-57 toward Ser-195 and to stabilize the developing positive charge on His-57. Once proto-nated, His-57 acts as a general acid for proton transfer to the amino group of the cleaved product peptide with the newly formed N-terminus. In the deacylation reaction, His-57 functions again as a general base abstracting a proton from water to promote its attack onto the enzyme-peptide ester in essentially the reverse of the acylation reaction. The aspartyl and metalloproteases function instead by activating the hydrolytic water mol-ecule as the direct attacking species on the peptide bond (**Fig. 1B**). A similar general acid/base catalytic regime takes place through involvement of an active site acidic residue.

By virtue of the mechanism for proteases described above, each proteolytic peptide generated (with exception of the C-terminal peptide of the protein) is expected to have at its carboxy terminus one oxygen atom originating from the original peptide carbonyl group and one oxygen atom incorporated from solvent water. Thus, when ^{18}O-enriched water is present in the solvent, and barring any kinetic isotope effects, the extent of single ^{18}O incorporation into each peptide will be a linear function of the relative $H_2^{18}O/H_2^{16}O$ ratio present in the digestion buffer (*see* **Subheading 2.2.**). However, as reported ini-tially *(14)* and as corroborated in subsequent studies *(10,12,23)*, several proteases in fact promote incorporation of two ^{18}O atoms into both positions of the carboxyl group. The proteases trypsin, Lys-C, and Glu-C (each are serine protease) will incorporate two oxygen atoms into their corresponding proteolytic fragments, whereas chymotrypsin and Asp-N (a metalloprotease) have been shown to catalyze only a single atom incor-poration *(14)*.

The incorporation of two oxygen atoms is rationalized by a mechanism of enzyme-catalyzed back-exchange of carboxyl oxygen atoms with bulk solvent. Here, proteolytic fragments initially generated from cleavage of the peptide bonds of the primary sub-strate are themselves substrates for the protease and enter a cycle of repeated binding and hydrolysis events (**Fig. 1C**). The back-exchange process will continue as long as protease activity persists. In the context of ^{18}O-labeling, since the likelihood for back exchange at each carboxyl oxygen position is the same for each repeated hydrolysis event, an equilibrium concentration of peptides labeled with 0, 1, or 2 ^{18}O atoms will ultimately be reached that is dependent on the enrichment percentage of $H_2^{18}O$ used in the digestion.

2.2. Factors Contributing to the Efficiency and Degree of Labeling

For discussion purposes, the *labeling efficiency* for a given peptide (labeled to non-labeled) is defined as the molar ratio of fragments containing an ^{18}O atom to fragments containing ^{16}O only. Thus, the greater the $^{18}O/^{16}O$ ratio, the greater is the labeling effi-ciency. In light of the incidence of protease-catalyzed back-exchange, both singly labeled ($^{18}O_1$) and doubly labeled ($^{18}O_2$) peptides are produced, and the labeling effi-ciency is more appropriately defined as the ratio of $^{18}O_{1+2}/^{16}O$. Accordingly, the *degree of labeling* is defined as the ratio of doubly labeled to singly labeled peptides, $^{18}O_2/^{18}O_1$, with a larger ratio indicating a higher proportion of peptides that have incorporated two ^{18}O atoms as a consequence of back-exchange with the bulk solvent.

Clearly, the labeling efficiency is ultimately dependent on the composition of the water during proteolysis. Experimentally, this is controlled by the digest buffer quantities of $H_2{}^{18}O$ and $H_2{}^{16}O$. The relative percentages of $H_2{}^{18}O$ and $H_2{}^{16}O$ present in a $H_2{}^{18}O/H_2{}^{16}O$ mixture can be represented by the values p and $(1 - p)$, respectively, where $[p + (1 - p)]$ is equal to 100%. Concurrent with the initial hydrolysis event accompanying peptide bond cleavage, the peptide fragment incorporates a single oxygen atom from the bulk solvent into the newly formed carboxyl group. In this manner, p can be regarded as the probability that an ^{18}O atom is incorporated from a molecule of $H_2{}^{18}O$, and $(1 - p)$ is the probability that the incorporated oxygen comes from a molecule of $H_2{}^{16}O$. Therefore, for proteases that do not promote back-exchange events or if the peptide itself does not participate in back-exchange (*see* below), the percentage of any given peptide fragment labeled with ^{18}O is equal to p, the percentage of $H_2{}^{18}O$ present in an $H_2{}^{18}O/H_2{}^{16}O$ mixture. In this case, the labeling efficiency is simply equal to $(p/1 - p)$ and is an exponential function of the percentage of $H_2{}^{18}O$ used during digestion. This relationship can be used to predict the theoretical labeling efficiency for any peptide based on a single hydrolysis event and the %$H_2{}^{18}O$ used during digestion.

When peptide fragments produced following the primary hydrolysis event are substrates for subsequent back-exchange events, the labeling efficiency and degree of ^{18}O incorporation will depend on several factors over and above a simple relationship to the %$H_2{}^{18}O$. Most important is the propensity of a particular peptide fragment to engage in repeated binding and hydrolysis events with the protease. If so, this implies that reoccurring back-exchange events are independent of each other and, furthermore, that either of the two carboxyl oxygens are equally likely to exchange with bulk solvent and become labeled with a probability of p by ^{18}O, or with a probability of $(1 - p)$ by ^{16}O. A schematic representation underlying this process is shown in **Fig. 2**. Repeated steps are multiplicative, and an equilibrium population of ^{16}O, $^{18}O_1$, and $^{18}O_2$ will eventually be reached. This effect is illustrated graphically in **Fig. 3** and illustrates that the equilibrium concentrations of labeled species for any given %$H_2{}^{18}O$ is essentially attained following 10–12 repeated hydrolysis events. Note that although a curve is presented for the situation resulting with a digest performed in 100% $H_2{}^{18}O$ for comparative purposes, this level of enrichment is ordinarily not feasible in routine experimentation because of practical considerations.

Ideally, if a digest could be performed in 100% $H_2{}^{18}O$, maximum labeling efficiency is reached concurrent with the initial hydrolysis event. In the case of more realistic conditions of $H_2{}^{18}O$ enrichment during digestion, the labeling efficiency will fall between two limits. The lower limit occurs following initial hydrolysis and the upper occurs after sufficient back-exchange events have occurred to reach equilibrium. For the examples shown in **Fig. 3**, the lower and upper limits of labeling efficiency ($^{18}O_{1+2}{:}{}^{16}O$) are: 90% $H_2{}^{18}O$, 9:1 and 99:1; 75% $H_2{}^{18}O$, 3:1 and 15:1; 50% $H_2{}^{18}O$, 1:1 and 3:1. Again, the experimentally observed labeling efficiency for a given peptide will lie within these limits and depend on the peptide's tendency to undergo back-exchange. Nonetheless, knowledge of the theoretical limits of the labeling efficiency may prove useful in the design of an experiment as an indicator of the level of $H_2{}^{18}O$ enrichment required to address a given situation.

What then controls if and to what extent a peptide enters a back-exchange cycle? It has been suggested that the size of an oligopeptide may serve as an indicator as to whether the peptide behaves as a good back-exchange substrate and becomes doubly

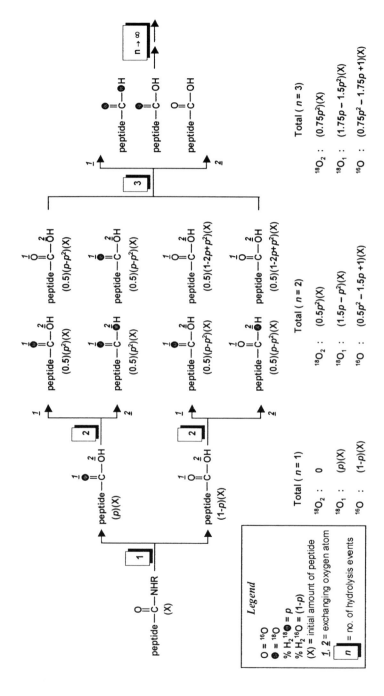

Fig. 2. Model illustrating the relative contribution of ^{16}O, $^{18}O_1$ and $^{18}O_2$ peptides as a consequence of protease-catalyzed back-exchange. The actual contribution from each species will depend on the percentage of $H_2^{18}O$ (p) present during digestion and the total number of hydrolysis events (n) that occur. The model is based on the principle of independent events and assumes an equal reactivity of $H_2^{18}O$ versus $H_2^{16}O$.

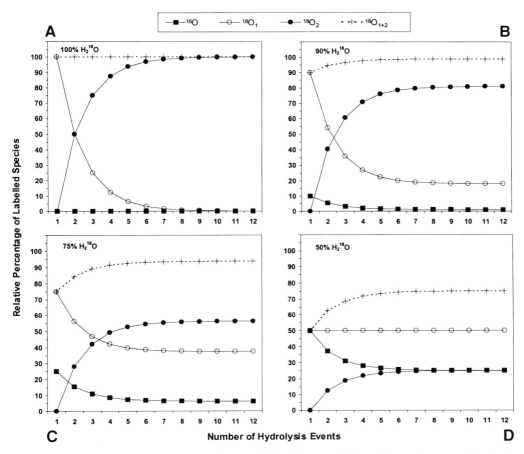

Fig. 3. **(A–D)** Graphical representations relating the cumulative effect of repeated hydrolysis events during back-exchange and the relative percentage of labeled species for a given peptide as a function of the % H$_2$^{18}O. Note that hydrolysis event 1 pertains to cleavage of the peptide bond, whereas events 2 and higher pertain to back-exchange events.

labeled *(14)*. We have observed that although the labeling efficiency of most peptides by trypsin is similar, variations in the degree of labeling do exist with no apparent correlation to the nature of the peptide *(10)*. To a first approximation, the extent of back-exchange will be governed by the kinetics of substrate binding and turnover. Two primary kinetic considerations will influence the susceptibility of a proteolytic fragment to undergo back-exchange. The first is the apparent binding affinity (K_m) of the protease for the peptide toward formation of a productive complex and the second is the overall rate constant for the process of solvent exchange with the carboxyl group by way of the catalytic mechanism. A peptide with a low K_m (high affinity) may readily form an enzyme-substrate complex but in a nonproductive mode. Productive binding can occur, but the complex may be slow to proceed through the exchange mechanism for a variety of mechanistic arguments. As the variation in the properties of peptide fragments will be extensive, it will be difficult to predict *a priori* the relative predisposition of a peptide to be a substrate for the protease and undergo back-exchange.

Experimental conditions are also expected to contribute to the degree of labeling. If we consider the case of a peptide that is a relatively poor substrate, then the time scale of the experiment may be insufficient to observe the back-exchange event. Under forcing conditions of increased digest time and protease concentration, the exchange process can be influenced to increase. From an analysis perspective, there are inherent advantages regarding the deconvolution of mass spectra for peptides that exhibit a greater level of ^{18}O labeling (*see* **Subheading 3.**) and the application of conditions to promote the degree of labeling may be desirable. Back-exchange may also depend on the complexity of the sample in that a high proportion of peptide substrates that are fast to exchange are expected to compete with poor peptide substrates for the protease. The trend is compounded if the concentration of the protein (e.g., a low-abundance protein) from which the slow-exchanging peptides are derived is considerably less than the average concentration of proteins in the mixture.

2.3. Experimental Design Considerations

Key to the success of any labeling strategy is a well-conceived experimental procedure. The design of the experiment will have an impact on the precision and accuracy of the MS data, which will influence the quality of the biological information and hence its relevance. An outline of a generic strategy as applied to cell lines (e.g., treated, untreated, disease, normal) is given in **Fig. 4**. A similar strategy can be extended to tissue samples. The first step involves a biological fractionation step by which the cell populations are converted into protein fractions either as whole cell lysates or as subcellular fractions (e.g., soluble, membrane, nuclear). Regardless of which method is used to generate the protein fraction, careful control of the sample processing for each fraction must be taken to minimize any potential losses. Differential losses between the two pools of samples prior to mixing and analysis will lead to inaccuracies in the comparative analysis of the relative protein levels in the two sample states. Differences that may occur in each sample set during processing may be assessed through the inclusion of unique protein standards that can be used to gauge overall recovery. The total protein content for each fraction should also be determined if possible.

Once the protein fraction has been generated, it will be digested and labeled with ^{18}O. With the exception of isotopically labeled amino acids that are incorporated into the protein in vivo, the postlysis labeling of proteins runs the risk of introducing experimental errors with every sample manipulation step required. This holds true for all chemical labels. However, because the process of labeling with ^{18}O is a natural consequence of protein digestion, it only requires that the digestion buffer contain as pure a concentration of $H_2^{18}O$ as possible and does not require exotic reagents or incorporate additional synthetic steps. Commercially available $H_2^{18}O$ can be obtained at more than 96% ^{18}O atom enrichment. For digestion buffers that are highly enriched with $H_2^{18}O$, the labeling reaction can be considered to be quantitative, an assumption that is not valid with chemical labeling. The caveat that applies to any labeling strategy is that the digestion efficiency in terms of its completeness must be the same among the labeled and nonlabeled samples. This in turn will be dependent on the trypsin activity and protein concentration in each sample. The key therefore is to ensure that the protein fractions produced using a given experimental design are easily partitioned into a digestion mixture that contains as pure a concentration of $H_2^{18}O$ as possible. A second consideration is that the fraction should not contain significant amounts of chemical agents that may

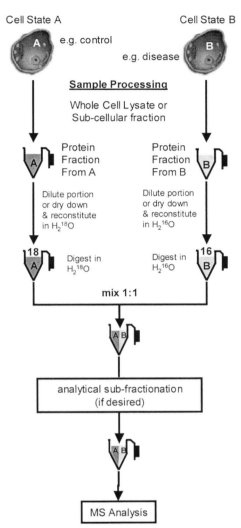

Fig. 4. A cartoon schematic illustrating the basic experimental approach for the comparative analysis of two different cell states using ^{18}O labeling and MS analysis. A critical step is ensuring that the $H_2^{16}O$ content in the fraction to be labeled can be exchanged with $H_2^{18}O$ quantitatively.

affect protease activity (e.g., sodium dodecyl sulfate [SDS]). Given the complex nature of most biological fractions and their reliance on detergents for solubilization, it is often quite difficult to achieve a compatible medium.

Preparation of the protein samples for digestion can be achieved by one of two general methods. These include dilution of the sample into $H_2^{18}O$-containing buffer or evaporation of the $H_2^{16}O$ in the original sample followed by reconstitution in $H_2^{18}O$-based buffers. With the dilution method, typically the protein sample should be diluted 20–100-fold into more than 95% $H_2^{18}O$ in order to maintain a high percentage of $H_2^{18}O$ post dilution. Although this can result in a dilution of analyte, it can also effectively reduce the relative concentration of deleterious additives used during protein isolation and solubilization (e.g., SDS), making the sample more compatible with trypsin digestion.

Although the evaporation strategy can effectively preconcentrate many sample components, the process must still be carefully thought out so that the residual $H_2^{16}O$ is kept to a minimum and matrix components present in the evaporated sample do not contribute to irreversible (or difficult) protein resolubilization or precipitation. Because of the number of possible whole-cell and subcellular fractionation techniques, it is important for the researcher to be aware of these requirements and plan for them before they start an experiment to avoid incompatibilities. An additional advantage of using ^{18}O labels is that when additional chromatographic subfractionation steps are used to reduce the complexity of the digested mixture containing labeled and unlabeled peptides, the two forms will coelute, preserving the analytical accuracy of the quantitation of the measurement *(24)*. With other isotopically labeled peptides that are $^2H/^1H$-based, significant separation of the peptides can occur that is especially pronounced when multiple label sites are present on a peptide.

2.4. Back-Exchange and ^{18}O-Label Stability

In any isotope labeling strategy, the success of the approach is dependent not only on the effectiveness of the forward labeling reaction to ensure near quantitative introduction at the desired target (e.g., Cys with ICAT; carboxy-termini with trypsin/^{18}O) but also on the stability of the introduced isotope label. Any loss of the isotope tag post labeling and post digestion must be rigorously avoided. Relevant to this discussion is the stability or retention of the incorporated ^{18}O labels under various experimental manipulations. In the case of amino acids, which can serve as models for ^{18}O-labeled peptides, the carboxyl oxygen atoms are prone to chemical exchange with aqueous solvent at rates that are pH- and temperature-dependent *(25)*. In general, only under extreme pH conditions and elevated temperatures is the extent of carboxyl oxygen exchange with solvent significant. The efficient chemical synthesis of ^{18}O-carboxyl-labeled amino acids requires prolonged treatment (2–3 d) of the amino acid in the presence of $H_2^{18}O$ under relatively strong reaction conditions (pH approx 0, 60–70°C). In contrast, the half-lives of ^{18}O-amino acids under progressively moderate conditions (e.g., pH 1.5, 20°C) in most cases, approach 1–2 mo, whereas under physiological conditions (neutral pH, 20–37°C), the extent of chemical back-exchange is essentially negligible *(25)*. Regardless, it is evident that strict measures should be followed to minimize the exposure of carboxyl-termini-labeled peptides to extreme acidic and alkaline conditions.

More important is the chemical stability of the ^{18}O label under conditions typically encountered for routine proteomics analysis. Of primary concern, for reasons just given, is the exposure of peptides to (extreme) acidic conditions in the course of sample manipulations, chromatographic peptide fractionation, and ionization techniques. Formic acid (FA) and trifluoroacetic acid (TFA) are commonly added (1–10% v/v) to quench peptide digests in order to inhibit protease activity by lowering the pH and to condition the sample for subsequent manipulations. For complex peptide samples, proteolytic digests, either previously quenched by acidification or not, are commonly separated into individual analyte fractions or separated in-line for liquid chromatography/electrospray ionization (LC/ESI) MS by C18-reverse-phase chromatography using water/acetonitrile gradients containing 0.01–0.1% FA or TFA. In the case of matrix-assisted laser desorption/ionization (MALDI) analysis, the sample is exposed to an acidic medium during the processes of drying in the presence of a large excess of acidic matrix (e.g., 2,5-dihyroxybenzoic acid, α-cyano-4-hydroxycinnamic acid) and desorption/ionization. The increasing interest and application of proteolytic ^{18}O labeling as a strategy for comparative quantification has encouraged groups to address these issues. Results from these investigations indicate that in general, the incorporated ^{18}O labels are stable following routine sample handling manipulations *(10)* and are retained during peptide fractionation and ionization methods *(12,14)*. However, these studies also report that the potential for exchange can occur when digests are quenched with 10% TFA *(14)* or when high-performance liquid chromatography (HPLC) fractions containing TFA are concentrated by evaporation *(12)*. Straightforward solutions to these problems, such as quenching by freezing and collecting HPLC eluents into buffered solutions, are suggested *(12,14)*.

In addition to acid catalyzed back-exchange, care must also be taken to ensure complete cessation of protease activity post mixing of the ^{16}O- and ^{18}O-labeled peptides. As discussed earlier, peptide fragments are themselves substrates for certain proteases (e.g., trypsin, Lyc-C, Glu-C) and can, depending on their particular kinetic properties, undergo repeated protease-mediated hydrolysis events. Although this phenomenon is advantageous during digestion in ^{18}O-enriched water to increase the population of doubly labeled ^{18}O peptides and the overall level of ^{18}O incorporation **(Fig. 3)**, it is equally deleterious to label stability upon dilution into a medium of lower H_2^{18}O content. The converse argument applies to peptides digested in 100% H_2^{16}O and then mixed with samples containing H_2^{18}O prior to analysis. Any quantification information for a given peptide pair labeled with ^{16}O or ^{18}O, respectively, is quickly rendered useless if protease-mediated back-exchange persists post mixing.

A somewhat inconsistent procedural element of protease digestions is the absolute concentration of protease added to the protein sample. Published protocols typically indicate addition of protease at a 0.01–0.1 (w/w) relative amount to the protein sample. In particular, model studies to probe back-exchange are generally designed in which concentrated test protein solutions are digested in enriched H_2^{18}O and then broadly diluted into the desired experimental "test" solution containing primarily ^{16}O buffer. As such, the carryover concentration of protease (and digest buffer components) can vary significantly in postdigest samples depending on the experimental procedures and postdigest manipulations.

We have reported on the effect of FA concentration as a quenching agent and its relationship to trypsin activity and back-exchange *(10)*. Under the experimental conditions examined, the results indicated that 2–5% FA (pH approx 3–4) was sufficient to prevent trypsin-catalyzed back-exchange without promoting acid-catalyzed back-exchange over the timecourse of a typical analysis. However, our observations had also indicated that halting digests with less than 1% FA did result in small but measurable back-exchange in certain peptides. Therefore, small perturbations in pH under quenching conditions appear to influence the extent of trypsin activity at low pHs. Indeed, elegant pH-dependent steady-state investigations have illustrated that the relative catalytic efficiency (k_{cat}/K_m) of trypsin at pHs between 3 and 4 is approx 0.05–0.1% that of the activity at optimum pH *(26)*. This level of trypsin activity may be considered to be inconsequential but can prove important in postdigest manipulated samples containing significant concentrations of carried over trypsin. In these cases, the degree of enzyme-catalyzed back-exchange will also be proportional with extended incubation periods.

To test these hypotheses, we have extended our analysis of trypsin-catalyzed back-exchange at low pH but under forcing conditions (unpublished results). To this end, tryptic digests of test proteins performed in 100% $H_2^{16}O$ were subjected to ultrafiltration to separate the product peptides from trypsin. The purified peptides were then diluted into 5% FA/80% $H_2^{18}O$ or diluted into 5% FA/80% $H_2^{18}O$ containing increasingly aggressive concentrations of trypsin, and samples were incubated at 4°C for various times prior to MS analysis. As expected, no back-exchange was observed in the samples diluted only into 5% FA. However, it quickly became apparent that a considerable portion of peptides underwent significant back-exchange in the presence of trypsin that was proportional to the trypsin concentration and incubation period. Therefore, although the relative activity of trypsin is greatly reduced at low pH *(26)*, precautions must be taken to characterize its effect on ^{18}O label stability under any specific set of experimental conditions. As discussed previously, 2–5% FA effectively quenches digests without compromising label stability under more conventional experimental conditions *(10)*. In any case, inclusion of lesser amounts of trypsin during digests and minimizing the time between pooling of the ^{16}O- and ^{18}O-digested samples and MS analysis will increase confidence regarding label stability.

Attention should also be paid to C18-separation of enzymatic digests. Craft et al. *(27)* reported the use of Poros-R2 in a column format to entrap proteins onto the resin followed by subsequent introduction of trypsin as a method of sample clean-up, concentration, and proteolytic digestion. Analogous conditions result when digests are subjected to C18-reverse-phase chromatography using water/acetonitrile gradients. As a consequence of sample loading, peptides and residual protease are concentrated many fold onto the head of the column relative to their solution concentrations. Although trypsin activity will be greatly reduced owing to the low pH of the mobile phases and because the active conformational integrity of trypsin may be compromised when binding to the solid phase, the potential for back-exchange between labeled peptides and the mobile phase *via* protease catalysis exists.

Ideally, any concerns regarding protease-catalyzed back-exchange can be dismissed if the protease is rendered completely inactive following digestion, with supporting experimental verification, by use of a protease inhibitor. Toluenesulfonyl lysine chloromethyl ketone (TLCK) is an active site-directed inhibitor of trypsin and exerts its inhibition through covalent alkylation of the catalytic histidine residue *(28)*. Once modified by TLCK, trypsin can no longer catalyze peptide bond cleavage or back-exchange since the active site is occupied by the inhibitor, preventing substrate binding, and the essential histidine residue can no longer serve as a general base/acid. In preliminary studies, we have found that treatment of tryptic digests with TLCK (100–500 μM) prior to and in concert with FA quenching is a highly effective method to prevent back-exchange (unpublished results). Other naturally occurring polypeptide inhibitors, such as soybean or bovine pancreas trypsin inhibitor, function by forming an essentially irreversible complex with trypsin by binding to the substrate-binding pocket *(29)*. The application of soybean trypsin inhibitor to stop digests in ^{18}O-labeling studies has been reported *(13)*. However, the efficacy of inhibition regarding back-exchange in this study was not explored in detail.

3. MS-Based Detection

3.1. The Effect of Labeling and Elemental Composition on Peptide Isotopic Abundance and Quantitation

The use of ^{18}O labeling as a quantitative or comparative tool in proteomics is based on the powerful method of isotope dilution mass spectrometry. Isotope dilution was first developed by Inghram and Hayden in 1954 *(30)* and has since had a significant impact on medical research in one form or another. In its simplest form, isotope dilution MS requires that the sample containing an element of interest be spiked with an isotopically modified form of the same element in a controlled manner. The spike is depleted in the major isotope that is normally used to quantitate the element and is enriched in a lesser abundance isotope. The mass spectrum recorded from the resultant mixture of sample and "spike" can then be used to determine the concentration of the target element in the original sample. The experiment is usually designed so that a single ratio (based on the intensity from the major isotope in the sample and the intensity of the major isotope from the spike that is different from that of the sample), can be used to provide the quantitative answer. For example, consider the isotopic abundance of elemental oxygen (^{16}O, 99.759%; ^{17}O, 0.037% and ^{18}O, 0.204%) where only a very small (negligible) amount of oxygen exists in the ^{18}O form. Producing and adding a known amount of a "spike" or standard that was enriched 100% in the ^{18}O form as part of an elemental analysis would allow quantitation of an unknown amount of ^{16}O-containing sample by MS.

This is illustrated in **Fig. 5A** including the appropriate equation. In this case, a spike containing twice as much isotopically pure (i.e., 100%) ^{18}O is added to the sample mixture. Here, the contribution to the spike intensity at *m/z* 18 owing to the natural abundance content of ^{18}O in the sample (0.2%) can be ignored, although it can easily be subtracted if the intensity of the most abundant isotope is known, and the relative isotope ratio (abundance) between ^{16}O and ^{18}O is known. The answer would be $50 \times 0.002 = 0.1$, and so at *m/z* 18, 99.1 units are due to the isotopically enriched ^{18}O spike and only 0.1 units are due to the natural abundance sample. Extending this example to natural abundance molecular oxygen and isotopically enriched molecular oxygen, the same process of calculation will hold; however, now that there are two ^{18}O atoms the shift in mass between the molecular ion and the spike is 4 amu **(Fig. 5B)**.

One of the major strengths of isotope dilution experiments is that for homogenous mixtures, the two elements, although isotopically different, will behave similarly during analysis. The ionization of sample and spike will be similar during the measurement and as such, the absolute ratio will not be affected by normal or drastic changes in the ionization process (i.e., fluctuations in the sample processing and analysis will affect sample and spike identically and therefore the ratio is by and large unaffected). In contrast to quantitative methods based on standards and calibration curves (and their associated inaccuracies), isotope dilution therefore provides a direct (one-step) method of quantitation. The challenge is to obtain good sources of isotopically enriched elements at affordable prices and to incorporate the "spikes" easily into the sample being analyzed. More in-depth discussions of isotope dilution MS can be found in the literature *(31)*.

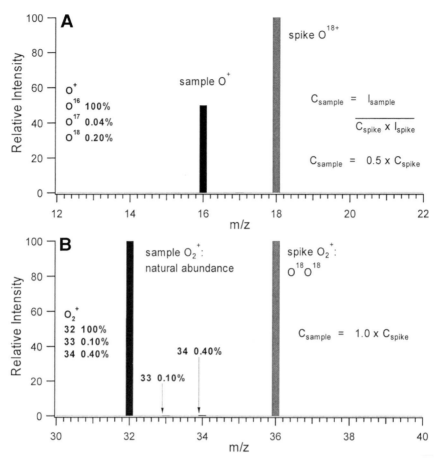

Fig. 5. Hypothetical isotope dilution quantitation experiments in which (**A**) elemental and (**B**) molecular oxygen are used. With little isotopic overlap, the quantitation is based on a straightforward equation given within the illustration (I = intensity, C = concentration).

The proteolytic labeling of peptides with ^{18}O can lead to significantly more complex isotopic patterns. As described above, either one or two ^{18}O labels can be incorporated into the C-terminus of a given peptide depending on the digestion conditions and the nature of the peptide. The incorporation of one or two ^{18}O labels will not significantly change the relative isotopic abundance of a given peptide; however, it will shift the mass of the monoisotopic ion by an amount equivalent to the number of ^{18}O labels incorporated. For example, when two labels are incorporated, this will shift the monoisotopic peptide mass by 4 amu; if only one label is incorporated, the monoisotopic peptide mass will be shifted by 2 amu. This is demonstrated in **Fig. 6**, in which the elemental composition and corresponding relative isotope abundances have been calculated from the peptide fragment YLYEIAR derived from the tryptic digestion of bovine serum albumin (BSA). Examination of the traces indicates that when labeled and unlabeled samples are mixed, there can be considerable overlap of the isotopic distributions. It is most severe for the singly labeled species and less severe for the doubly labeled species in

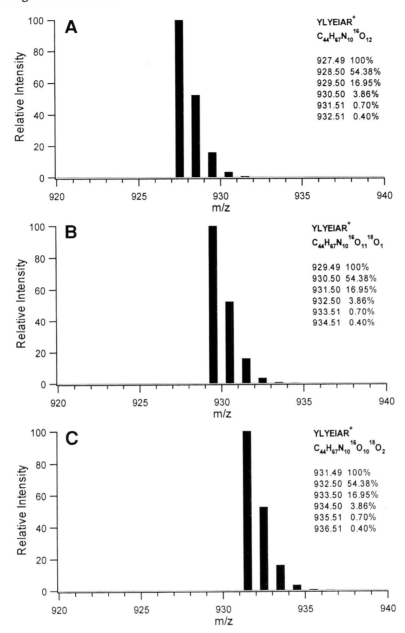

Fig. 6. The effect of the incorporation of ^{18}O labels on the elemental composition and relative isotopic distribution of the peptide YLYEIAR, derived from the tryptic digestion of bovine serum albumin. (**A**) Unlabeled. (**B**) Incorporation of one C-terminus ^{18}O. (**C**) Incorporation of two ^{18}O labels at the C-terminus.

this case. For larger peptides with more complex isotopic distributions, the overlap can be significant in both cases.

Fortunately the elemental composition for any given peptide can be used to calculate its isotopic pattern and calculate the relative contribution of each of the labeled states

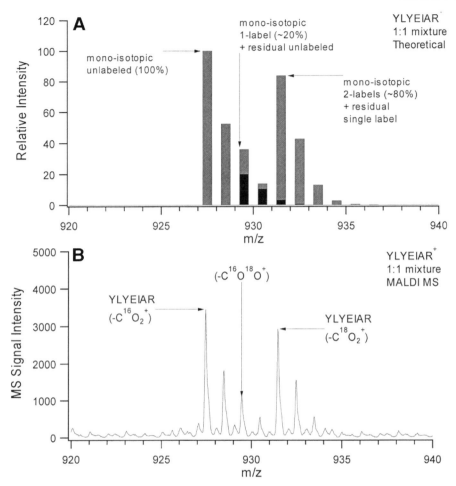

Fig. 7. In this example, equal amounts of unlabeled and ^{18}O-labeled YLYEIAR peptide (e.g., digested in approx 95% $H_2{}^{18}O$) are mixed. (**A**) Theoretical effect of incomplete labeling and isotopic overlap on the deconvolution and quantitation of mixtures of labeled and unlabeled peptides. (**B**) Experimental results from an actual MALDI MS experiment acquired on an AB/Sciex QStar Pulsar instrument. Using the isotope deconvolution scheme presented in **Equations 1a–c**, the ratio is calculated as 1.01.

in overlapping patterns. For example, consider a hypothetical mixture of labeled and unlabeled YLYEIAR peptide, as illustrated in **Fig. 7**. From the discussion above, it can be seen that it is often the case that the degree of labeling can vary even under conditions of high $H_2{}^{18}O$ enrichment owing to residual ^{16}O content and the potential for back-exchange upon mixing. It is therefore common to have between 10 and 50% of the labeled peptides containing only one ^{18}O label at the C-terminus and as such have multiple species present in a given isotopic structure or "peak pair." The deconvolution of these isotopic patterns is straightforward providing that the identity of the peptide (i.e., sequence) is known so that the elemental composition can be used to derive the relative isotopic abundance of the peptide and there is no significant variation in the MS measurement of the relative isotopic peaks (i.e., the relative intensities of the isotopes are truly reflective of the theoretical relative abundances). Many mathematical equations can be found in the literature describing this process for ^{18}O labels *(12,13)*. If it is assumed that there are no more than three labeled states (0, 1, and 2 labels) and the unlabeled monoisotopic *m/z* value can be identified, the deconvolution is straightforward. The relative intensities of the monoisotopic peaks from each of the labeled states can be calculated from the following equations:

$$I_{0L} = I_0 \tag{1a}$$

$$I_{1L} = I_2 - [I_{0L} \times R_{0L}(0/2)] \tag{1b}$$

$$I_{2L} = I_4 - [I_{0L} \times R_{0L}(0/4) + I_{1L} \times R_{1L}(0/2)] \tag{1c}$$

In the above, I_{0L} refers to the relative intensity of the unlabeled (0) monoisotopic peak, I_{1L} refers to the relative intensity of the singly labeled (^{18}O$_1$) monoisotopic peak, and I_{2L} refers to the relative intensity of the double-labeled (^{18}O$_2$) monoisotopic peak. I_0, I_2, and I_4 refer to the raw intensities of the first peak in the envelope (unlabeled monoisotopic) and of the peaks at 2 and 4 amu higher, respectively. The ratios $R_{0L}(0/2)$ and $R_{0L}(0/4)$ refer to the theoretical isotopic relative abundance ratios between the monoisotopic (0) and the peaks 2 and 4 amu higher, respectively. Once the relative contributions have been calculated, the ratio of unlabeled over total labeled peptides can be determined. The simplest form of this is given in the following equation:

$$[\text{Peptide }^{16}\text{O}]/[\text{Peptide }^{18}\text{O}] = I_{0L}/(I_{1L} + I_{2L}) \tag{2}$$

Returning to the example given in **Fig. 7A**, the relative contributions can be calculated as follows: $I_{0L} = 100$, $I_{1L} = I_2 - [I_{0L} \times R_{0L}(0/2)] = 36 - 100 \times 0.17 = 19$ and $I_{2L} = I_4 - [I_{0L} \times R_{0L}(0/4) + I_{1L} \times R_{1L}(0/2)] = 84 - (100 \times 0.007) - (19 \times 0.17) = 80$. The relative ratio of unlabeled to labeled is calculated as $100/(19 + 80) = 100/99$ or $1:1$. The results from an actual MALDI MS experiment are shown in **Fig. 7B**. Using the equations given above, the ratio of unlabeled to labeled peptide was calculated to be 1.01.

From the above example it should be understood that, unlike other labeling schemes in which the light and heavy labeled peptides are often fully resolved on the *m/z* axis, ^{18}O labeling often does not generate fully resolved peak pairs. For low-molecular-weight peptides that exhibit simple isotope patterns and a high degree of labeling (i.e., ^{18}O$_2 \gg {}^{18}$O$_1$) the mass spectrum will often appear as a true peak pair. When higher molecular weight peptides (e.g., $M_r > 2000$) that exhibit more complex isotopic patterns are mixed, the labeling will appear more like an "extended" envelope and is more difficult to discern visually. An example of this given in **Fig. 8**, in which the tryptic peptide GLVLIAFSQYLQQCPFDEHVK derived from BSA is used for illustration. Here, the mass spectrum is of the doubly charged form of the peptide. Although the isotopic patterns can become quite complex and conceptually more difficult to quantitate than simple well-resolved structures, using the general formula provided in **Equations 1a–c**, quantitation can be straightforward providing that good signal statistics are present across the isotopic patterns during measurement.

It should be noted that if the peptides' identity and elemental composition are known, the relative isotopic abundances can be calculated manually through classical means. This is laborious and can be avoided by using software or many of the free web-based calculators such as those found on the UCSF Protein Prospector web site *(32)*. Using these tools, the elemental composition and isotopic abundances can be calculated read-

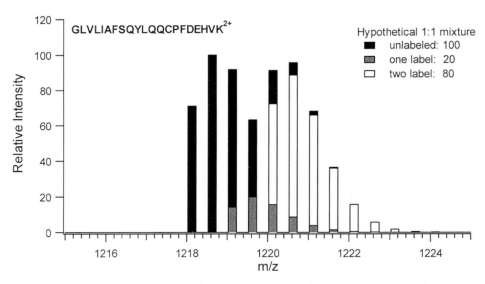

Fig. 8. The effect of increased peptide mass on the complexity of the isotope distribution for a hypothetical 1:1 unlabeled versus labeled mixture. In this example double labeling is only 80% efficient. At higher mass, isotopic peak pairs become less easily discernable. The peptide is derived from the tryptic digestion of BSA, and the unlabeled singly charged mono-isotopic mass is 2435.24 based on an elemental composition of $C_{13}H_{172}N_{27}O_{31}S_1$.

ily if the amino acid peptide sequence is known. In addition, the "averagine"-based relative isotopic abundance can also be derived for peptides for which only the mass of the peptide is known and not its amino acid sequence. The averagine-based isotopic composition is calculated from the average elemental composition of peptides at a given mass *(32)*. In most cases it closely resembles the actual or natural isotopic abundance, differing only when exotic elements are present or for peptides with irregular sequences (e.g., high cysteine content). For example, a rapid estimate of the quantitation of the peptide given in **Fig. 8** can be derived by using an averagine mass of 2435 and knowing the raw MS intensities of the first peak in the series (I_0), the third peak in the series (I_{0+2}), and the fifth peak in the series (I_{0+4}). In this example, $I_0 = 71$, $I_{0+2} = 92$, and $I_{0+4} = 91$, resulting in a ratio of approx 0.96, in close agreement with theory.

3.2. MS Performance Characteristics Required for Accurate Peptide Quantitation Using ^{18}O Labeling

The purpose of this section is to provide a basic understanding of the important features of an MS instrument for delivering high quality isotope ratio information from ^{18}O-labeled samples. It is not the intention to review commercially available instruments and their suitability for quantitative proteomics using ^{18}O labels.

The description of how labeled and unlabeled peak pairs are "deconvoluted" and quantified provides a good basis for understanding what instrument features are required to generate accurate high-quality MS information using ^{18}O labels. In considering the potential complexity of the isotopic patterns involved, it can be argued that an instrument should provide near baseline resolution of complex peptide isotope distributions. This is necessary to provide an accurate measure of the unobscured intensity of the individual isotopes in a given isotopic pattern so that they can be used for quantitation. For instruments using MALDI ion sources this will typically require isotopic resolution for singly charged peptide ions from approx 800 to 3000 amu. For nanoES (electrospray) and LC applications (nanoLC), the MS should be capable of resolving multiply charged peptide ions. Ultimately, the ability to baseline resolve 4+ ions up to *m/z* 2000 would be desirable, but given the complexity of the isotopic patterns at those masses and that the mass shift is only 4 amu, it would be sufficient to resolve 2+ and 3+ ions up to *m/z* 1500. It can be difficult to generate good-quality MS/MS spectrum necessary for identification past these masses on some commercial instruments.

A second important feature is the instrument detection limit, sensitivity, and linear dynamic range (LDR). The instrument sensitivity is usually defined in terms of the minimum detectable quantity whose signal *(S)* is statistically different from that of the noise *(N)* at a given mass (e.g., $3 \times S/N$). The ultimate sensitivity of a given MS system sets the lower limit of the instrument's LDR, or the range in which the detected response is linear with peptide concentration. In the simplest sense, the LDR defines the instrument's capability to measure the intensity of a peptide accurately at low concentration compared with the intensity of a similar (labeled) peptide present at high concentration in a mixture. The larger the dynamic range of an instrument, the greater the ability to quantitate small and large differences accurately in peptide pairs present at low and at high concentrations. Given the fact that there can be significant diversity in the expression of proteins in a given biological system of interest *(1)*, the greater the MS instrument's LDR the more effective it will be at accurately measuring the resultant $^{18}O/^{16}O$ ratios expressed in these systems and generating better quality comparative data from two sample states. Three to four orders of magnitude seems to represent an upper level in terms of instrument performance, whereas the majority of instrumentation used in proteomics applications usually exhibit only 2–3 orders of magnitude with respect to LDR. This is in the context of short acquisition times for measurement over extended *m/z* ranges (e.g., full scan mode). It is possible to achieve much larger LDRs in different acquisition contexts, in which, for example, instruments specifically designed for elemental isotope ratio measurements can normally achieve 8–12 orders LDR.

A very simple illustration of the concept of LDR as applied to a peak pair with a 10-fold excess of ^{18}O labeled peptide is given in **Fig. 9**. The point here is that significant differences (e.g., 10-fold) can be measured near the baseline corresponding to lower concentration species, in the midrange and even for elevated concentrations. In addition, the figure also illustrates some of the quantitation errors that can be introduced as the upper and lower LDR limits are approached. At the upper level, the isotopic peaks of the higher concentration species will be attenuated, resulting in a smaller true ratio being calculated. At the lower level, the pair may appear as a single or lone peptide peak, indicating a larger than expected ratio. Further discussion of the practical implications of dynamic range as it applies to ^{18}O labeling is given below (*see* **Subheading 3.3.**).

The acquisition of MS data containing relative isotope distributions that reflect the expected natural abundance-based ratios of a given peptide in a consistent manner is a third extremely important quality. The instrument should be capable of acquiring good statistics of peak intensities and relative peak intensities, especially considering that measuring accurate relative intensities is the foundation for these quantitative measurements. A number of factors can contribute to fluctuations in the peak signal intensity across an isotopic distribution during acquisition; the most significant are related

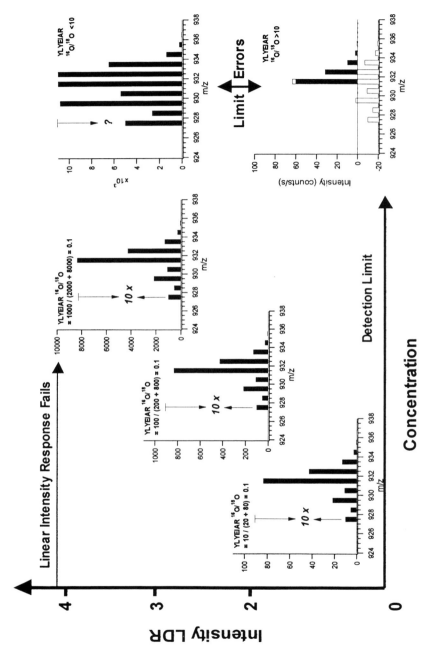

Fig. 9. An illustration depicting the concept of quantitation up to 4 orders of magnitude of signal intensity (LDR). On the left, there are two illustrations designed to depict the concept of quantitation errors that can occur at the upper and lower detection limits. At the upper limit of the LDR the intensity response will often become attenuated, resulting in a smaller true quantitation measure. At the lower limit, if one of the pairs falls below the detection limit, the more abundant state will often appear as a single unpaired peak and thus the ratio will be skewed more than it might truly be in solution.

to the nature of the source used and the requirements of the measurement. With most scanning instruments, multiple measurements can be acquired in a cumulative or averaged manner until the desired relative peak intensity statistics are achieved. This implies that there are no time constraints on the measurement and that the user is not sample-limited. Such measurements can be acquired using MALDI sources and nanoES sources when focused analyses are required. Averaging or accumulating over longer periods of time to improve signal quality loses its effectiveness when the intensity of an analyte peak is not significantly larger than the chemical background. In such cases, the background signal will be accumulated at the same rate as the analyte peak and there is no improvement in the overall *S/N*. With MALDI sources, the nature of the matrix can have a significant impact on the background signal, often limiting the ultimate utility at low levels; a similar effect can be observed with nanoES.

Depending on the nature of the experiment, a certain degree of error can easily be accepted. For instance, if the experiment is designed to quantitate a particular peptide accurately in a sample mixture, then deviation in the relative isotope ratio measurement for peaks in a distribution should be as small as possible, certainly less than 5%. For many experiments the goal is to provide a comparative survey of peptides in a given sample with the focus on determining which ones are "differentially expressed." Often the researcher is only interested in differences greater than 5–10-fold, and therefore larger errors in the measured isotopic ratios owing to counting statistics may be acceptable. This is especially true considering that significant error may have already propagated through the matching and sample processing steps. The researcher must determine what errors are acceptable when choosing a method or instrument.

A further (and perhaps the most) critical consideration is the quality of signal that can be acquired over a wide *m/z* range (e.g., 400–2000 amu) in short time periods. This is necessary for LC MS acquisitions or LC MS/MS acquisitions when the entire spectrum must be surveyed accurately on the time scale of 1 s or less. The ability to produce reasonable signal statistics on these time scales across a wide dynamic range may be difficult with traditional scanning instruments like triple-quadrupole or ion-trap instruments. However, hybrid quadrupole time-of-flight instruments (Qq-TOF or Q-TOF) *(33–35)* and Fourier-transform ion cyclotron resonance (FTICR) instruments *(8,36,37)* are quite capable of acquiring high-quality spectrum in short acquisition periods.

The fourth important consideration is related to the quality of the MS/MS fragment ion mass spectrum that an instrument produces. This will, in part, be related to the sensitivity, mass accuracy *(38)*, and mass range of the fragment ions that an instrument can measure. In general, lower energy fragmentation spectrum that are rich in (continuous) y and b ion series are easy to sequence manually and also often score well with most commercial database search engines. As is illustrated in **Fig. 10**, the major effect of *18O* labeling is that any fragment ion series that contains the C-terminus of the peptide will have a mass different by an amount equal to the mass of the substituted *18O* atoms. The example given in **Fig. 10** illustrates a representative b and y ion series as it would appear with 0, 1, and 2 *18O* atoms substituted at the C-terminus. The b ion series masses are unaffected; however, the y ion series will be shifted. Most commercially available database search engines [e.g., Mascot *(39)*] are able to take these modifications into account, and the user can add one or two *18O* labels as variable or fixed modifications as part of

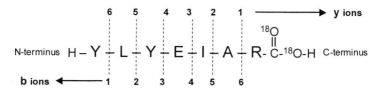

series	b	y¹⁶O¹⁶O	y¹⁶O¹⁸O	y¹⁸O¹⁸O
			(+2)	(+4)
1		175.12	177.12	179.12
2	277.16	246.16	248.16	250.16
3	440.22	359.24	361.24	363.24
4	569.26	488.28	490.28	492.28
5	682.35	651.35	653.35	655.35
6	753.38	764.43	766.43	768.43

Fig. 10. The effect of the substitution of C-terminus ^{16}O atoms for ^{18}O atoms on the fragment ion masses generated. Only the major b- and y-ion series are shown for simplicity. The major effect is that the mass of any fragment ion containing the C-terminus can be changed depending on how many ^{18}O atoms the C-terminus contains.

the search routine. In the authors' experience, a comparison of the scoring between labeled and unlabeled peptides generally yields similar results, with the scoring for the labeled peptides being 20% lower on average than the unlabeled peptides. Nonetheless, the overall quality of the MS/MS spectra for these experiments is similar, suggesting that improvements to the scoring algorithms may reduce this scoring difference.

Considering that true biological samples will often contain hundreds to thousands of proteins in a given fraction, yielding tens of thousands of peptides after digestion, it is often not enough for an instrument to have the capability to generate MS/MS fragment ion spectra manually. The use of automated acquisition techniques such as information-dependent acquisition (IDA) or data-dependent acquisition (DDE) routines are rapidly becoming industry standards. In particular these acquisition routines are critical for automated acquisition during an LC MS/MS run. For example, in an IDA experiment like those used on AB/Sciex QStar instruments, the instrument performs a 1-s survey of a given mass range (e.g., 400–1400 amu). From this survey scan, precursor ions are automatically chosen for fragmentation based on specified features such as the intensity of the monoisotopic mass and its charge state. Each survey scan might be accompanied by three MS/MS product ion scans that pick the three most intense ions having either 2⁺ or 3⁺ charge states. When IDA routines are used to acquire data from isotopically labeled peptide mixtures (i.e., $^{18}O/^{16}O$), each IDA cycle will contain quantitative information on the precursor ions in the survey scan and identification information in each of the accompanying MS/MS experiments. Using the list of peptides identified as a guide, each accompanying survey scan will contain the corresponding isotopic ratios and quantitation information. Linking the two datasets and manually calculating the quantitation values is straightforward. For large datasets, this will be an extremely time-consuming effort and will only provide information on the peptides identified. Depending on the complexity of the sample and the objectives of the experiment, a single-pass or "shotgun" approach may only contain information on the most abundant proteins. Additional experiments or subfractionation may provide a deeper level of information.

In terms of nanoLC MS or nanoLC MS/MS detection, one of the fundamental advantages of choosing ^{18}O labeling over other isotopic labeling techniques is that the labeled and unlabeled forms of a given peptide present in a mixture will coelute during the analysis. With other isotopic labeling techniques (e.g., ^2H-based labeling), the apex of the light and heavy labels will exhibit different elution times during a chromatographic run *(24)*. There are two consequences of this. The first is that the data must be manually screened to determine the chromatographic apex for the light and heavy species, and the ratio for the two species must be calculated at this point. As a manual operation this is time-consuming, and software-based alignment tools would be a prerequisite when working with large datasets. Perhaps more importantly, if there is a significant difference in the elution time, it is possible that one of the components may elute in the presence of an unrelated higher intensity ion, effectively suppressing the electrospray signal of the ion of interest and introducing a quantitation error *(40)*. In essence the separation and detection of non-coeluting peak pairs is counterproductive to the advantage of isotope dilution-based techniques. Zhang et al. *(24)* highlighted the shortcomings of this effect and its impact on data quality. With ^{18}O labels, the unlabeled and singly and doubly labeled peptides will all coelute.

An example of this is illustrated as part of the LC MS experiment given in **Fig. 11.** The intensities for the corresponding extracted ion currents for a selected peptide are given in **Fig. 11B**, illustrating the chromatographic overlap of the ^{16}O and ^{18}O peptides. Quantitation can be performed at any point across the chromatographic peak, providing there is useful signal intensity, and is therefore not limited to identifying the apex. In this experiment the ^{16}O- and ^{18}O-digested sample mixture contained a number of proteins present in different ratios. The overall quantitation result for phosphorylase b based on the identification and quantitation of 20 unique peptides is consistent with the theoretical value. For the more abundant proteins or peptides in this sample, the ability to quantitate on the rising or falling edge of a chromatographic peak means that quantitation errors associated with strong signal intensities that may be present at the apex can be avoided. By using the chromatographic peak shape response, the best signal intensity for the quantitation calculation can be carefully selected, allowing accurate result for any concentration range. Using coeluting peaks, there should effectively be no upper limit quantitation error.

3.3. ^{18}O Labeling: Practical Considerations

As stated above, the dynamic range of an MS instrument is a highly important consideration because it provides a limit as to the ultimate differences that can be measured. Using ^{18}O labels there are, however, a number of practical limitations that are specific to the label. The first involves the enrichment percentage of $H_2^{18}O$ that can be achieved during protease digestion. Although it is possible to purchase suitable quantities of highly enriched (e.g., 95–98%) $H_2^{18}O$, the working concentration will probably be

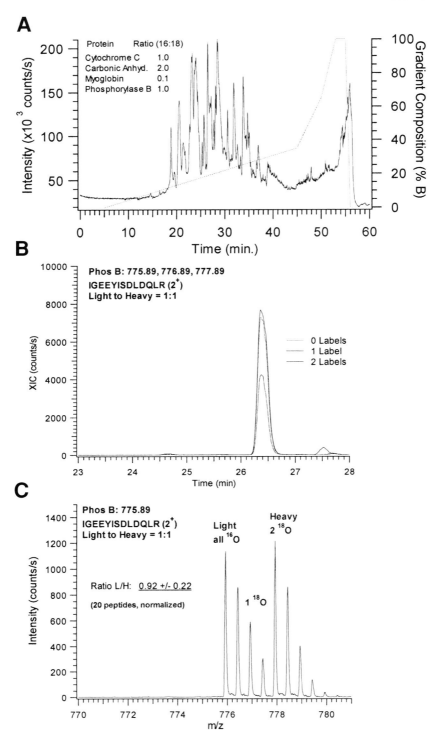

Fig. 11. (**A–C**) LC MS data illustrating the coelution of labeled and unlabeled peptides as well as an indication of the quantitation precision obtained from 20 phosphorylase B peptides. The nanoLC MS data were acquired using an Agilent 1100 HPLC-AB/Sciex QStar pulsar.

reduced from sample manipulations. The end result is that there will always be a limited % $H_2{}^{16}O$ (v/v) present in the digestion mixture that will directly influence the ultimate measurable difference between the labeled and unlabeled mixtures. For example, consider the case in which the labeled peptide was present in excess of an unlabeled peptide in a mixture derived from samples digested in $H_2{}^{18}O$ and $H_2{}^{16}O$, respectively. If the residual $H_2{}^{16}O$ content in the $H_2{}^{18}O$ digest results in a contribution of ^{16}O peptide that is 5% of the total peak intensity for the ^{18}O labeled peak, then the maximum measurable ratio that can be calculated in the mixed sample will be 20.

More importantly, if the unlabeled peptide (from $H_2{}^{16}O$) is present at a concentration that is less than the residual ^{16}O peptide naturally present in the ^{18}O-labeled sample, then it will not be detected. More specifically, the calculated ratio will be that of the residual and not of the sample and hence will be false. Therefore looking for differences greater than 20-fold in many samples may be of limited utility. Because of these sorts of errors, it is always best to run the labeled sample by itself to evaluate the labeling efficiency and degree of labeling prior to mixing, and thus ensure a greater level of confidence in the data. It is often the case, however, that an arbitrarily defined difference is chosen as "significant." If it is considered that differences of 10 or greater are significant or interesting between proteins in two samples, then small percentages of residual ^{16}O labeling and associated quantitation errors become less of an issue. In such cases the analysis would be more like a survey of pairs that are different rather than a systematic quantitation of all peptides. Considering the amount of data that can be generated from one LC MS run from a sample containing hundreds or thousands of proteins, this approach seems to be quite reasonable.

A second limitation is related to the calculation of large ratio differences when the labeled or heavy peptide is the less abundant peptide. This problem is illustrated by considering the situation in which a lesser abundant labeled peptide is present at values approaching 5% of the intensity of an unlabeled peptide. Referring to the discussions in **Subheadings 3.1.** and **3.2.**, if the absolute intensity of the labeled peak is less than the precision of the isotope ratio measurement, then the ratio cannot be calculated accurately. A solution to this problem, referred to as "inverse labeling" by Wang et al. *(11)*, has been described in the literature. A cartoon depicting the basic concept of inverse labeling is given in **Fig. 12**. Protein fractions A and B are each split, and then one is digested using $H_2{}^{16}O$ and the second is digested using $H_2{}^{18}O$. Of the four samples, there exist two complimentary but reverse labeled pairs (inverse labeling). Both sample mixtures are analyzed, and for peptides that do not have quantitation errors the measured ratios should be the inverse of each other (i.e., $1:1 = 1:1$ and $1:10 = 10:1$). Thus, two levels of sample validation can be obtained. For those samples that suffer this type of quantitation error, the sample containing the lower intensity unlabeled peptide should provide the more accurate data.

4. Applications of ^{18}O Labeling

4.1. Application of ^{18}O Labeling for Quantitative Proteomics

The recent interest in applying isotopic labeling techniques for the comparative analysis of protein expression in two different sample states *(2,9,41–50)* has led to considerable interest in ^{18}O labels. Although much initial research has been focused on validating the use of ^{18}O-labeled peptides to quantitate protein expression *(10,13,14,23,51)*, the chal-

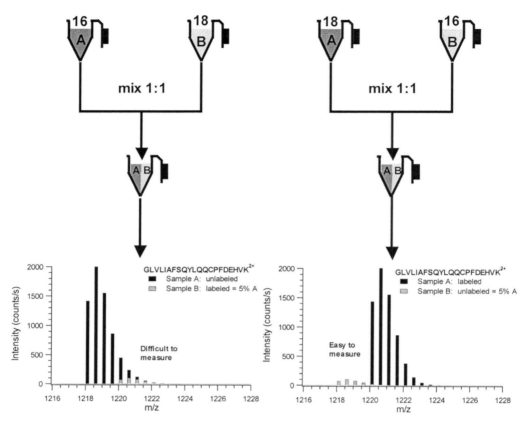

Fig. 12. A schematic diagram illustrating the inverse labeling strategy as a means of overcoming quantitation errors. This approach will also provide a second level of validation for peptides that are not affected by quantitation errors because the two results are expected to agree.

lenge lies in applying these to real experimental systems *(11,12)*. An outline of a generic strategy as applied to cell lines (treated, untreated, diseased, normal) is given in **Fig. 4**.

The report by Yao et al. *(12)* represents one of the first proteomic studies that compares protein expression between two "real samples." In their work they compared virion proteins from two serotypes (Ad5 and Ad2) of adenovirus. Virus particles were isolated using standard methodology and dried to remove excess $H_2^{16}O$. Proteins from the Ad5 and Ad2 serotypes pellets were then subjected to sequential Lys-C and trypsin digestion in ^{16}O and ^{18}O buffers, respectively. The quantitative aspect of the analysis was achieved by subjecting peptide fractions to MALDI MS coupled to an IonSpec FTICR for high-resolution identification of peak pairs. Proteins were identified manually through their fragment ion spectra acquired using nanoES on an AB/Sciex hybrid Qq-TOF instrument. The adenovirus system has been well characterized and represents a limited number of quantifiable proteins and peptides. They were able to use this tandem approach to compare the experimentally determined ratios for a number of peptides with expected values and conclude that ^{18}O labeling "enables a quantitative shotgun approach for proteomic studies."

To date, however, reports in which an ^{18}O labeling strategy has been applied to highly complex mixtures have been elusive. This is probably because of the enormous challenge that the corresponding dataset would represent in terms of the time required to calculate ratios for the thousands of peptides that might be present. Although many MS instrument manufacturers and third-party companies have produced software for the quantification of labeled samples both "on the fly" and post MS acquisition, there currently does not exist a suitable solution for ^{18}O-labeled data. The ability to convert large sets of MS data rapidly into meaningful quantitative information represents the next big challenge in making ^{18}O labeling a more applicable tool for proteomics. Nevertheless, Washburn et al. *(35)* have included ^{18}O labeling as part of an encompassing quantitative proteomic scheme for complex proteomic mixtures.

4.2. Application of ^{18}O Labeling for Unique Sequencing Applications

Peptide sequencing or partial sequencing has rapidly become a powerful means to identify peptides and ultimately proteins using MS/MS. The peptide sequence or partial peptide sequence deduced can be searched against catalogued or previously uncatalogued proteins in a database. In MS/MS, an intact peptide is isolated in the first stage of MS and then fragmented in the second stage, producing and measuring a number of daughter ions characteristic of the peptide. These ions can be from different sized C-terminal portions of the peptide (termed the y-ion series), different sized N-terminal portions of the peptide (termed the b-ion series), or various internal fragments. This concept is illustrated in **Figure 10**, for the b- and y-ion series of the peptide YLYEIAR. ^{18}O labeling will clearly identify which ions belong to the y-ion series through a corresponding mass shift or unique pairing, seriously decreasing the complexity and ambiguity of information in the MS/MS spectrum.

When a protein is digested in a mixture of $H_2^{16}O$ and $H_2^{18}O$, the resultant peptides will contain zero, one, or two ^{18}O atoms. When performing MS/MS on one of these peptides, the mass range that is allowed through for fragmentation can be set to include more than one of the labeled states of the peptide (termed low or open resolution). Since the C-terminus of the peptide is contained in each of the y-ion series, and in no other series, the pairs or triplets of ions spaced 2 amu apart can easily be identified as y ions. When the y-ion series exists as a unique set or grouping of peaks, they are readily identified, making the MS/MS spectrum easier to interpret.

This technique can be used for both protein identification and *de novo* sequencing of a protein. For protein identification, the MS/MS spectrum is compared against a protein database that uses a routine (e.g., variable modification) that takes into account the mass shift in the series containing the C-terminus. Normally two variable modification settings are required: one that accounts for one ^{18}O label and a second that accounts for two ^{18}O labels. When an MS/MS spectrum is searched against a database, all the peptides in the database are matched up against the spectrum and scored based on how well their theoretical fragmentation pattern matches that observed in the spectrum. No sequence is actually determined directly from the spectrum. The ^{18}O labeling allows more certainty in potential matches, as the shifted y ions in the spectrum must be matched up against the y ions in the fragmentation pattern along with a shift in the labeled monoisotopic parent ion mass.

For *de novo* sequencing, the ability to identify the y ions rapidly is critical. For this task, the MS/MS spectrum cannot simply be compared with the fragmentation patterns of a number of peptides. It also cannot be compared with the fragmentation patterns of all possible peptides in a certain mass range, as this represents a considerable undertaking for larger peptides. Instead, a sequence is constructed based on the fragmentation pattern observed in the MS/MS spectrum. The y-ion series can be used to construct a sequence for the peptide, based on the mass differences between the ions. If not all the y ions are present, a partial sequence can be generated and used to identify some ions in the b-ion series. The remainder of the ions can be used to fill in the gaps in the sequence. Without the ^{18}O label to identify the y ions, the sequence is constructed based on the mass differences between the ions, but there are many more ions to consider, as the y- and b-ion series cannot be readily differentiated. This results in more false sequences and more incomplete sequences. In addition to increasing the ease of *de novo* sequencing, ^{18}O labeling lends itself to the automation of *de novo* sequencing *(33)*, which is normally done manually by an expert owing to the complexity of the spectrum. Qin et al. *(52)* have used the *de novo* technique to sequence peptides, with 7–16 amino acid residues acquired during an LC MS/MS run on an ion-trap mass spectrometer.

It is important to note that the C-terminal peptide of a protein generally cannot be labeled in the same manner. The peptides are labeled because their C-terminal amino acid allows them to be cleaved from the protein and also to reassociate with the protease to allow continued oxygen exchange between the carboxylic acid at the C-terminus and bulk solvent. The protein's C-terminus will not necessarily be an amino acid that allows it to associate with the protease, and thus oxygen exchange will not occur with its carboxylic acid. Although ^{18}O labeling cannot be used for the sequencing of these peptides, it can be used to differentiate or identify the protein C-terminal peptide because it will be the only one that is not labeled or contain a unique labeling pattern. This concept has been discussed in the literature *(14,33,53,54)*. Back et al. *(55)* have also used ^{18}O labeling in a unique application for identifying crosslinked peptides for protein interaction studies. If a protein has been modified in some way by a proteolytic enzyme, then it is possible that it may have a new C-terminus. If the sequence is just cut in the middle, or a segment is removed and the pieces not ligated, then it will have multiple C-termini. The detection of these modifications depends on the enzyme that is used to digest the protein having a different specificity from the enzyme that cleaved the protein *in vivo*. Unfortunately, this only allows for the detection of new C-termini and no other proteolytic modifications.

In a recent paper by Liu and Regnier *(51)*, a unique heavy isotope labeling strategy is described for the analysis of all tryptic peptides including the protein C-terminus and N-terminally blocked peptides. The dual labeling approach attaches different stable isotope ($^{2}H_3$ at the primary amine sites and ^{18}O at the C-terminus) labels to either end of a peptide. This produces unique "coding" combinations that, when interpreted correctly, yield information to distinguish C-terminal peptides, recognize N-terminal peptides from proteins in which the amino terminus is acylated, and identify primary structure variations between proteins from different sources.

4.3. Post-Translational Modification Identification Applications

^{18}O labeling can also be used to identify a variety of post-translational modifications including *N*-glycosylation *(23,56)*. Glycosylation is an important post-translational modification that often occurs on the asparagine residue in the consensus sequence NXS/T (where X is any amino acid other than proline). These *N*-linked carbohydrates can be removed with a highly specific enzyme, such as peptide *N*-glycosidase F

(PNGase F). *N*-glycosylation represents perhaps the earliest modification that ^{18}O labeling was used to detect. In 1992, Gonzalez et al. *(58)* described a method of incorporating ^{18}O atoms into *N*-glycosylation sites during enzymatic deglycosylation. Later, Küster and Mann *(56)* expanded this method to detect *N*-glycosylation sites, as well as many occurrences of the consensus sequence that accompany *N*-glycosylation. During the deglycosylation, the β-aspartylglycosylamine bond is hydrolyzed, cleaving the carbohydrate from the peptide and converting the asparagine to an aspartic acid residue. If the reaction is carried out in a mixture of ^{16}O and ^{18}O water, either an ^{18}O or a ^{16}O is incorporated into the aspartic acid side chain. In this manner, the sites of attachment of carbohydrates are marked with a $[M+1]^+/[M+3]^+$ pattern that is readily identified in a mass spectrum. Because the PNGase F enzyme does not reassociate with aspartic acid residues, there is no back-exchange of oxygen, and it is only possible for one ^{18}O atom to be incorporated per glycosylation site. Because only one ^{18}O atom can be added, the ^{16}O/^{18}O ratio in each of these sites should be the same as the ^{16}O/^{18}O ratio of the water in the deglycosylation medium *(10)*. The ^{16}O/^{18}O pattern generated in these experiments is useful on several levels. It can be used to confirm which consensus sequences in a protein were linked to carbohydrates. Second, basic quantitation of the extent of *N*-glycosylation can be performed using the ^{16}O/^{18}O pattern, and last, it also increases the ease of detection of an unknown protein from a complex mixture.

4.3. Other Applications of ^{18}O Labeling

Although much initial research has been focused on validating the use of ^{18}O-labeled peptides to quantitate protein expression, there has been relatively little reported on their use as diagnostic tools. The use of isotope labeling for diagnostic studies for the purpose of quantifying sample loss during various stages of sample handling in proteomic processes seems a natural extension and logical step toward optimizing these steps. **Figure 13** represents the basic strategy employed. The first step is to generate labeled and unlabeled peptide standards from a common stock solution. A concentrated protein stock solution is first created, and then small volumes of this stock solution are diluted in larger volumes of digestion buffers made with either normal natural abundance water ($H_2{}^{16}O$) or isotopically enriched water ($H_2{}^{18}O$). By diluting by 100-fold, for example, only ~approx 1% v/v of $H_2{}^{16}O$ water will be introduced to the $H_2{}^{18}O$-containing buffer, minimizing any quantitation errors. These standards can then be diluted to any desired level necessary at which one (e.g., unlabeled) will be used as the test sample and the other (e.g., labeled) will serve as the control. For best results, the control should be subjected to the same dilution steps using the same solvents as the sample and storage conditions. In this manner, at the end of the experiment, the sample and control can be mixed easily in a 1 : 1 ratio, and any differences can be attributed to sample loss associated with the process under investigation.

To date there have been very few reports in the literature describing this approach. We have recently used this strategy to evaluate a number of common sample processing steps used in proteomics *(10)*. In particular, we have found a decrease in sample recovery efficiency at lower total peptide concentrations when Ziptip purification is used. Similarly, we followed peptide losses during speed vac dry-down and recovery and found that, on average, these losses could be as large as 20–30%, consistent with reports from the literature *(60)*. Finally, we examined the effect of sample concentration on the in-solution digestion efficiency for standard proteins under controlled con-

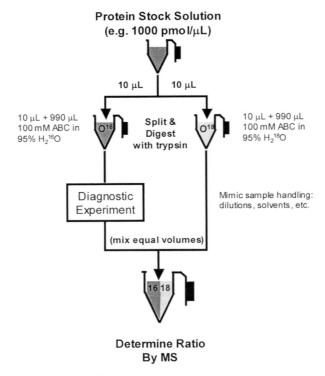

Fig. 13. Strategy employed for diagnostic evaluation of target steps or processes. Test (^{16}O) and control (^{18}O) samples are treated in exactly the same manner aside from the actual diagnostic step so that when they are recombined and measured, any differences in a 1:1 mixture will be owing to losses during the diagnostic application under investigation.

ditions. The data are consistent with previous findings *(61)* demonstrating a decrease in protein digestion efficiency with concentration and confirm the important relationship between digestion efficiency and protein concentration for fixed conditions.

5. Concluding Remarks

The labeling of peptides with ^{18}O can be achieved readily in concert with the protease digestion of proteins based on well-understood biochemistry. It is a simple and straightforward procedure that depends on the nature of the peptides formed, the digestion conditions, and the relative ratio of ^{16}O/^{18}O in the digestion buffer mixture. As a result, its use as a tool for proteomics has found many useful applications specifically in terms of quantitation and sequence information including post-translational modifications. There remain however, a number of issues that must be addressed before ^{18}O labeling can truly be considered a mature proteomics technique.

Considering the possible scale and intent of many proteomics efforts, the number of experiments required can be truly daunting. Thus very careful experimental design and the development of new technology will be important to meet these challenges. In particular, the price and availability of reasonable quality $H_2{}^{18}$O is not always guaranteed *(62)*. Rising or variable costs may mean that it is impractical to perform these experiments on a large scale or in a sustained manner until efficient methods or techniques are developed to minimize the absolute volume requirements per sample.

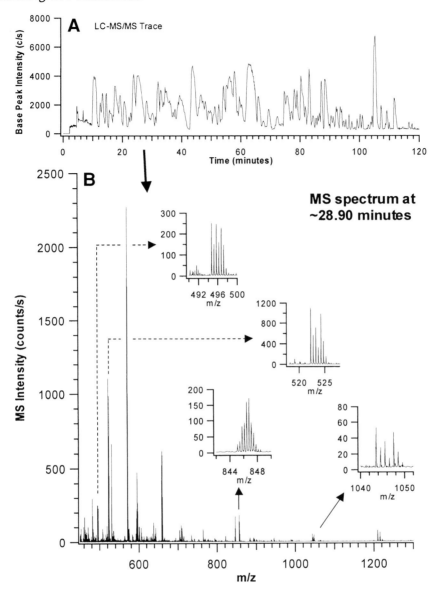

Fig. 14. An LC MS/MS trace acquired from a mixture of labeled and unlabeled whole-cell lysate sample digests. A base peak total ion chromatogram for the mixture is given in (**A**) and a single MS trace is given in (**B**). In the MS scan there are numerous peptide peak pairs that must be identified (sequenced) and quantified. The data were acquired using an Agilent 1100 HPLC-AB/Sciex QStar pulsar system.

It is possible to identify several hundred proteins from an LC MS/MS run of a complex mixture. It is likely, however, that the number or peptide pairs present will be significantly greater because the majority of the peptides pairs present in a complex sample may not be selected for MS/MS owing to limited IDA or DDE duty cycle (i.e., only so many MS/MS per unit time are possible). An example of this is given in **Fig. 14**, in which a single 1-s MS acquisition from an LC MS/MS trace acquired from a complex sample derived from a cell lysate mixture is illustrated. In the MS spectrum, there are numerous labeled peak pairs at various relative intensities, all of which must be quantified and identified (sequenced) by MS/MS. In this example, the spectrum contains more than 20 labeled peak pairs, and zoomed images of only a few representative peak pairs are shown. Each of these peaks will have overlapping yet finite chromatographic duration in which they can be quantified and sequenced. The amount of raw data to be processed and interpreted from this 120-min LC MS/MS run of a complex mixture therefore represents an enormous challenge (e.g., approx 10,000–15,000 pairs). This is especially true considering the tedious nature of manually calculating the relative peptide quantitation as described above.

In order for labeling to be useful on larger scales in the manual context, highly focused strategies must be employed so that specific data are targeted rather than trying to determine what is interesting from a larger dataset. The sorting of larger datasets of labeled peaks becomes realistic with the aid of software processing and sorting tools. To this end, a differential analysis software package that can deconvolute, process, and sort $^{16}O/^{18}O$ labeled peak pairs by their relative intensity ratios has recently been described *(63,64)*. The peak lists can then be analyzed and relevant peak pairs identified for further analysis and identification by MS/MS and database searching. In this way the entire data set is sorted *en masse*. Although many companies provide commercially available differential software for a variety of labeling strategies, few if any are presently capable of the deconvolution required for the rapid and accurate calculation of ratios necessary for ^{18}O labeling. It is likely that the most significant advances in the rapidly growing field of quantitative proteomics based on MS detection will be driven by powerful software packages and acquisition strategies that enable the researcher to access any or all of the rich peptide information present in a sample on competitive time scales.

References

1. Corthals, G. L., Wasinger, V. C., Hochstrasser, D. F., and Sanchez, J. C. (2000) The dynamic range of protein expression: a challenge for proteomic research. *Electrophoresis* **21,** 1104–1115.
2. Gygi, S. P., Rist, B., Gerber, S. A., Turecek, F., Gelb, M. H., and Aebersold, R. (1999) Quantitative analysis of complex protein mixtures using isotope-coded affinity tags. *Nat. Biotechnol.* **17,** 994–999.
3. Anderson, L. and Seilhamer, J. (1997) A comparison of selected mRNA and protein abundances in human liver. *Electrophoresis* **18,** 533–537.
4. Futcher, B., Latter, G. I., Monardo, P., McLaughlin, C. S., and Garrels, J. I. (1999) A sampling of the yeast proteome. *Mol. Cell. Biol.* **19,** 7357–7368.
5. Fey, S. J. and Larsen, P. M. (2001) 2D or not 2D. Two-dimensional gel electrophoresis. *Curr. Opin. Chem. Biol.* **5,** 26–33.
6. Gygi, S. P., Corthals, G. L., Zhang, Y., Rochon, Y., and Aebersold, R. (2000) Evaluation of two-dimensional gel electrophoresis-based proteome analysis technology. *Proc. Natl. Acad. Sci. USA* **97,** 9390–9395.
7. Figeys, D. (2002) Proteomics approaches in drug discovery. *Anal. Chem.* **74,** 412A–419A.
8. Martin, S. E., Shabanowitz, J., Hunt, D. F., and Marto, J. A. (2000) Subfemtomole MS and MS/MS peptide sequence analysis using nano-HPLC micro-ESI fourier transform ion cyclotron resonance mass spectrometry. *Anal. Chem.* **72,** 4266–4274.

9. Aebersold, R., Rist, B., and Gygi, S. P. (2000) Quantitative proteome analysis: methods and applications. *Ann. NY Acad. Sci.* **919,** 33–47.

10. Stewart, I. I., Thomson, T., and Figeys, D. (2001) 18O labelling: a tool for proteomics. *Rapid Commun. Mass Spectrom.* **15,** 2456–2465.

11. Wang, Y. K., Ma, Z., Quinn, D. F., and Fu, E. W. (2001) Inverse 18O labeling mass spectrometry for the rapid identification of marker/target proteins. *Anal. Chem.* **73,** 3742–3750.

12. Yao, X., Freas, A., Ramirez, J., Demirev, P. A., and Fenselau, C. (2001) Proteolytic 18O labeling for comparative proteomics: model studies with two serotypes of adenovirus. *Anal. Chem.* **73,** 2836–2842.

13. Mirgorodskaya, O. A., Kozmin, Y., Titov, M. I., Korner, R., Sonksen, C. P., and Roepstorff, P. (2000) Quantitation of peptides and proteins by matrix assisted laser desorption/ionization mass spectrometry using 18O-labeled internal standards. *Rapid Commun. Mass Spectrom.* **14,** 1226–1232.

14. Schnolzer, M., Jedrezewski, P., and Lehmann, W. D. (1996) Protease-catalyzed incorporation of 18O into peptide fragments and its application for protein sequencing by electrospray and matrix-assisted laser desorption ionization mass spectrometry. *Electrophoresis* **17,** 945–953.

15. Rose, K., Simona, M. G., and Offord, R. E. (1983) Amino acid sequence determination by g.l.c.-mass spectrometry of permethylated peptides. Optimization of the formation of chemical derivatives at the 2–10 nmol level. *Biochem. J.* **215,** 261–272.

16. Barrett, A. J., Rawlings, N. D., and Woessner, J. F. (eds.) (1998) *Handbook of Proteolytic Enzymes.* Academic Press, San Diego.

17. Neurath, H. (1984) Evolution of proteolytic enzymes. *Science* **224,** 350–357.

18. Steitz, T. A. and Shulman, R. G. (1982) Crystallographic and NMR studies of the serine proteases. *Annu. Rev. Biophys. Bioeng.* **11,** 419–444.

19. Kraut, J. (1977) Serine proteases: structure and mechanism of catalysis. *Annu. Rev. Biochem.* **46,** 331–358.

20. Dodsona, G. and Wlodawerb, A. (1998) Catalytic triads and their relatives. *Trends Biochem. Sci.* **23,** 347–352.

21. Walsh, C. (eds.) (1979) *Enzymatic Reaction Mechanisms.* W. H. Freeman, New York.

22. Ash, E. L., Sudmeier, J. L., Day, R. M., et al. (2000) Unusual 1H NMR chemical shifts support (His) Cepsilon 1 . . . O=C H-bond: proposal for reaction-driven ring flip mechanism in serine protease catalysis. *Proc. Natl. Acad. Sci. USA* **97,** 10,371–10,376.

23. Reynolds, K. J., Yao, X., and Fenselau, C. (2002) Proteolytic 18O labeling for comparative proteomics: evaluation of endoprotease Glu-C as the catalytic agent. *J. Proteome Res.* **1,** 27–33.

24. Zhang, R., Sioma, C. S., Wang, S., and Regnier, F. E. (2001) Fractionation of isotopically labeled peptides in quantitative proteomics. *Anal. Chem.* **73,** 5142–5149.

25. Murphy, R. C. and Clay, K. L. (1979) Synthesis and back exchange of 18O labeled amino acids for use as internal standards with mass spectrometry. *Biomed. Mass Spectrom.* **6,** 309–314.

26. Antonini, E. and Ascenzi, P. (1981) The mechanism of trypsin catalysis at low pH. Proposal for a structural model. *J. Biol. Chem.* **256,** 12,449–12,455.

27. Craft, D., Doucette, A., and Li, L. (2002) Microcolumn capture and digestion of proteins combined with mass spectrometry for protein identification. *J. Proteome Res.* **1,** 537–547.

28. Shaw, E., Mares-Guia, M., and Cohen, W. (1965) Evidence for an active-center histidine in trypsin through use of a specific reagent, 1-chloro-3-tosylamido-7-amino-2-heptanone, the chloromethyl ketone derived from N-alpha-tosyl-L-lysine. *Biochemistry* **4,** 2219–2224.

29. Bode, W. and Huber, R. (2000) Structural basis of the endoproteinase-protein inhibitor interaction. *Biochim. Biophys. Acta* **1477,** 241–252.

30. Inghram, M. G. and Hayden, R. J. (1954) *Handbook on Mass Spectroscopy, Nuclear Series.* Report no. 14., NRC-USA.

31. de Bievre, P. (1994) Isotope dilution mass spectrometry (IDMS), in *Trace Element Analysis in Biological Specimens* (Herber, R. F. M. and Stoeppler, M., eds.) Elsevier, New York, pp. 169–184.

32. UCSF, Protein Prospector (http://prospector.ucsf.edu).

33. Shevchenko, A., Chernushevich, I., Ens, W., Standing, K. G., Thomson, B., Wilm, M., and Mann, M. (1997) Rapid 'de novo' peptide sequencing by a combination of nanoelectrospray, isotopic labeling and a quadrupole/time-of-flight mass spectrometer. *Rapid Commun. Mass Spectrom.* **11,** 1015–1024.

34. Shevchenko, A., Loboda, A., Ens, W., and Standing, K. G. (2000) MALDI quadrupole time-of-flight mass spectrometry: a powerful tool for proteomic research. *Anal. Chem.* **72,** 2132–2141.

35. Washburn, M. P., Ulaszek, R., Deciu, C., Scheiltz, D. M., and Yates, J. R. 3rd (2002) Analysis of quantitative proteomic data generated via multidimensional protein identification technology. *Anal. Chem.* **74,** 1650–1657.

36. Quenzer, T. L., Emmett, M. R., Hendrickson, C. L., Kelly, P. H., and Marshall, A. G. (2001) High sensitivity fourier transform ion cyclotron resonance mass spectrometry for biological analysis with nano-LC and microelectrospray ionization. *Anal. Chem.* **73,** 1721–1725.

37. Smith, R. D. (2002) Advanced mass spectrometry methods for the rapid and quantitative characterization of proteomes. *Comp. Funct. Genomics* **3,** 143–150.

38. Clauser, K. R., Baker, P., and Burlingame, A. L. (1999) Role of accurate mass measurement (+/– 10 ppm) in protein identification strategies employing MS or MS/MS and database searching. *Anal. Chem.* **71,** 2871–2882.

39. Mascot Database Search Engine, Matrix Science (http://www.matrixscience.com).

40. Kebarle, P. and Tang, L. (1993) From ions in solution to ions in the gas phase—the mechanism of electrospray mass spectrometry. *Anal. Chem.* **65,** 972A–986A.

41. Chen, X., Smith, L. M., and Bradbury, E. M. (2000) Site-specific mass tagging with stable isotopes in proteins for accurate and efficient protein identification. *Anal. Chem.* **72,** 1134–1143.

42. Steen, H. and Pandey, A. (2002) Proteomics goes quantitative: measuring protein abundance. *TRENDS Biotech.* **20,** 361–364.

43. Conrads, T. P., Alving, K., Veenstra, T. D., et al. (2001) Quantitative analysis of bacterial and mammalian proteomes using a combination of cysteine affinity tags and ^{15}N-metabolic labeling. *Anal. Chem.* **73,** 2132–2139.

44. Goshe, M. B., Conrads, T. P., Panisko, E. A., Angell, N. H., Veenstra, T. D., and Smith, R. D. (2001) Phosphoprotein isotope-coded affinity tag approach for isolating and quantitating phosphopeptides in proteome-wide analyses. *Anal. Chem.* **73,** 2578–2586.

45. Griffin, T. J., Gygi, S. P., Rist, B., et al. (2001) Quantitative proteomic analysis using a MALDI quadrupole time-of-flight mass spectrometer. *Anal. Chem.* **73,** 978–986.

46. Kelleher, N. L., Nicewonger, R. B., Begley, T. P., and McLafferty, F. W. (1997) Identification of modification sites in large biomolecules by stable isotope labeling and tandem high resolution mass spectrometry. The active site nucleophile of thiaminase I. *J. Biol. Chem.* **272,** 32,215–32,220.

47. Gygi, S. P., Rist, B., and Aebersold, R. (2000) Measuring gene expression by quantitative proteome analysis. *Curr. Opin. Biotechnol.* **11,** 396–401.

48. Zhou, H., Ransish, J. A., Watts, J. D., and Aebersold, R. (2002) Quantitative proteome analysis by solid-phase isotope tagging and mass spectrometry. *Nature Biotech.* **19,** 512–515.

49. Washburn, M. P., Wolters, D., and Yates, J. R. 3rd (2001) Large-scale analysis of the yeast proteome by multidimensional protein identification technology. *Nature Biotech.* **19,** 242–247.

50. Regnier, F. E., Riggs, L., Zhang, R., et al. (2002) Comparative proteomics based on stable isotope labeling and affinity selection. *J. Mass Spectrom.* **37,** 133–145.

51. Liu, P. and Regnier, F. E. (2002) An isotope coding strategy for proteomics involving both amine and carboxyl group labeling. *J. Proteome Res.* **1,** 443–450.

52. Qin, J., Herring, C. J., and Zhang, X. (1998) De novo peptide sequencing in an ion trap mass spectrometer with ¹⁸O labeling. *Rapid Commun. Mass Spectrom.* **12,** 209–216.

53. Rose, K., Savoy, L. A., Simona, M. G., Offord, R. E., and Wingfield, P. (1988) C-terminal peptide identification by fast atom bombardment mass spectrometry. *Biochem. J.* **250,** 253–259.

54. Kosaka, T., Takazawa, T., and Nakamura, T. (2000) Identification and C-terminal characterization of proteins from two-dimensional polyacrylamide gels by a combination of isotopic labeling and nanoelectrospray Fourier transform ion cyclotron resonance mass spectrometry. *Anal. Chem.* **72,** 1179–1185.

55. Back, J. W., Notenboom, V., de Koning, L. J., et al. (2002) Identification of cross-linked peptides for protein interaction studies using mass spectrometry and ¹⁸O labeling. *Anal. Chem.* **74,** ASAP.

56. Kuster, B. and Mann, M. (1999) ¹⁸O-labeling of N-glycosylation sites to improve the identification of gel-separated glycoproteins using peptide mass mapping and database searching. *Anal. Chem.* **71,** 1431–1440.

57. Adamczyk, M., Gebler, J. C., and Wu, J. (2002) Identification of phosphopeptides by chemical modification with an isotopic tag and ion trap mass spectrometry. *Rapid Commun. Mass Spectrom.* **16,** 999–1001.

58. Gonzales, J., Takao, T., Hori, H., Besada, V., and Rodriquez, R. (1992) *Anal. Biochem.* **205,** 151–158.

59. Ficarro, S. B., McCleland, M. L., Stukenberg, P. T., et al. (2002) Phosphoproteome analysis by mass spectrometry and its application to saccharomyces cerevisiae. *Nat. Biotech.* **20,** 301–305.

60. Speicher, K. D., Kolbas, O., Harper, S., and Speicher, D. W. (2000) Systematic analysis of peptide recoveries from in-gel digestions for protein identifications in proteome studies. *J. Biomolec. Tech.* **11,** 74–86.

61. Locke, S. and Figeys, D. (2000) Techniques for the optimization of proteomic strategies based on head column stacking capillary electrophoresis. *Anal. Chem.* **72,** 2684–2689.

62. Ad Hoc Committee of the North American Society for the Study of Obesity (1999) Report on the supply and demand of ¹⁸O enriched water (http://www.naaso.org/newsflash/oxygen.htm).

63. Caldwell, J. A., Orsi, C. A., White, F. M., et al. (2002) *Comprehensive comparative proteome analysis*, in Proceedings of the 50th Annual Conference on Mass Spectrometry and Allied Topics, June 2–6, Orlando, FL.

64. MDS Proteomics Develops Software for Automating Differential Protein Analysis (6/24/02) ProteoMonitor, www.proteomonitor.com.

12

Automated Nanoflow Liquid Chromatography/ Tandem Mass Spectrometric Identification of Liver Mitochondrial Proteins

Bart Devreese, Frank Vanrobaeys, Elke Lecocq, Joël Smet, Rudy Van Coster, and Jozef Van Beeumen

1. Introduction

Since the introduction of the term *proteomics* (*1*), different methodologies have been developed that allow global analyses of gene expression at the protein level. The original papers mainly described the analysis of differential gene expression using two-dimensional polyacrylamide gel electrophoresis (2-D-PAGE). Nowadays, beside 2-D-PAGE, approaches involving differential isotopic labeling (*2,3*) or multidimensional chromatography (*4*) have been introduced. Although these new methods certainly have significant advantages, particularly in terms of quantitation, they should rather be regarded as being additional and, so far, they do not completely replace differential 2-D-PAGE analysis, as stated in reviews by several authors (*5,6*). Also, in other research areas related to proteomics, gel electrophoretic methods are being used. A striking example is the study of the *interactome*. A particular approach toward identification of the proteins that interact with a specific target protein involves copurification using affinity techniques, followed by separation of the copurified proteins by 1-D or 2-D gel electrophoresis (*7*).

After their gel electrophoretic separation, proteins of interest have to be identified. Mass spectrometry (MS) is absolutely the method of choice to accomplish this objective. Proteins, therefore, are usually digested *in situ* with trypsin. The resulting peptides are extracted from the gel slices and then analyzed by MS.

Several strategies have been developed to use the mass spectral data from the protein digestion mixture for database searching and identification of the corresponding protein. The simplest method is certainly the peptide mass fingerprinting (PMF) strategy using matrix-assisted laser desorption/ionization time-of-flight (MALDI-TOF) instrumentation (*8*). In this method, one simply determines the masses of the tryptic peptides. The values obtained are a kind of "fingerprint" of the protein and are used in a database search. The proteins deposited in this database are digested *in silico*, and the resulting theoretical digestion patterns are compared with the experimental fingerprint. Identification is achieved if the best fitting protein from the database has a considerable number of theoretical tryptic peptide masses that correspond to the experimental

From: *Handbook of Proteomic Methods*
Edited by: P. Michael Conn © Humana Press Inc., Totowa, NJ

pattern. Several PMF algorithms exist, of which some are freely accessible through the Internet. Generally, at least five peptide masses should fit to allow a reliable identification, and a good score depends greatly on the mass accuracy of the measurement.

PMF is certainly a simple, easily automated, and fast method, but it suffers from several drawbacks. It handles rather poorly with mixtures of proteins (gel bands or spots may contain multiple proteins), post-translational modifications are generally ignored, and certain groups of proteins do not display enough tryptic fragments within the optimal range of the MALDI mass spectrometer (800–3000 Daltons) to achieve a reliable score [e.g., small proteins (<10 kDa), membrane proteins]. Very often, PMF is used as a preliminary identification tool, and the proteins that remain ambiguously identified are forwarded to further analysis.

More stringent protein identification involves (partial) sequence determination of the tryptic peptides. This requires fragmentation of the peptide and analysis of the fragments, aimed at unraveling the sequence of the peptide. Although efforts have been made to involve postsource decay fragmentation on a MALDI-TOF mass spectrometer, most of the work so far has been done using tandem mass spectrometers (MS/MS), equipped with electrospray ionization sources. More specifically, the introduction of nano-electrospray sources have allowed us to increase the sensitivity of the method to the low femtomole range, which is absolutely essential to meet the needs for analysis of most biological samples.

1.2. Nano-Electrospray in Proteomics: Off-Line or On-Line?

The original concept of nano-electrospray, as introduced by Wilm and Mann *(9)* was based on theoretical considerations demonstrating maximal efficiency of the electrospray process by miniaturization of the spray emitter. Their setup involves the use of Au/Pd-coated borosilicate needles in which the sample is loaded using a syringe. The needle is placed inside a specifically designed source. The spray is initiated when a voltage is applied (0.8–1.2 kV) that is supported by a small backpressure. The flow rates achieved are in the 10 nL/min range. The concept allows one to spray a volume of a few microliters for at least 30 min, which allows the analyst to perform several experiments, including tandem mass spectral analyses of several peptides from a digestion mixture. This requires, however, several manual operations for precursor selection and optimization of fragmentation settings and is therefore not suited for the high-throughput demands in proteomics. Moreover, protein digestion is performed in buffers that are poorly compatible with electrospray ionization (ESI), and desalting may be required, which is often not as evident with the small volumes that are to be used.

A major advantage of ESI is that it is well suited for direct interfacing with liquid chromatography (LC) instrumentation, particularly of the reversed-phase high-performance liquid chromatography (HPLC) type in which highly compatible solvent systems are used. Hyphenation of HPLC with ESI MS, moreover, allows the use of automation tools that are well developed within the chromatography field, such as autosamplers and valve-switching systems. The need for sensitivity in proteomics has driven the development of more sensitive separation systems, mainly by miniaturization of columns and detectors. In their original paper describing the development of a commercial nanoLC system, Chervet et al. *(10)* noted that upon scaling down the column internal diameters from those conventionally used (4.6 mm to 75 µm), the detection limit for several proteins was reduced to the low fmol range as required for, e.g., protein identification from 2-D-PAGE gels *(10)*. The flow rate used in such a system is 150–180 nL/min, which is still higher than the theoretical "ideal" flow rate of 10–20 nL/min as calculated by Wilm and Mann for nano-electrospray *(9)*, but is well handled by commercial nano-electrospray sources.

Some 5 years ago, we implemented this nanoLC concept, coupled to a quadrupole/TOF-type mass spectrometer, mainly for the characterization of proteins from microorganisms, i.e., *Shewanella oneidensis* separated on 2-D-PAGE *(11)*. Several other groups have used this method to separate tryptic digests from gel-separated proteins for automated LC-MS/MS purposes. Interesting applications cover the analysis of cerebrospinal fluid *(12)* or peroxisomal membrane proteins *(13)*. Others used it in a 2-D-chromatographic setup combined with immobilized metal affinity chromatography (IMAC) for specific analysis of phosphorylated peptides *(14)*. Both hybrid quadrupole/TOF (designated as Q-TOF or Qq-TOF, depending on the manufacturer) and ion trap instruments may be used to obtain sequence information of the eluted peptides. Also, the more powerful technique of Fourier-transform ion cyclotron resonance (FTICR) has been used to analyze nanoLC-separated peptides *(15)*. The current software delivered with these instruments allows for automated MS to MS/MS switching: any peptide eluting from the column of which a molecular ion is generated in the ESI source and that produces a signal above a certain threshold is automatically recognized and selected for fragmentation. In particular, the Q-TOF-type instruments perform analysis sufficiently fast to allow multiple peptides from a single peak to be selected and fragmented in the case coelution of peptides occurs. We describe below the details of the experimental setup for the automated LC-MS/MS system we are currently using.

2. Materials and Methods

2.1. Experimental Setup

We use a commercial nanoLC system, i.e., an Ultimate Micro LC system combined with a FAMOS autosampler (LC-Packings/Dionex, Amsterdam, The Netherlands). No pumps are yet available to provide stable gradients below a flow rate of 0.5 μL/min. Therefore, in this system a classical reciprocal pump is built that pumps at 150 μL/min together with a flow-splitting device that reduces the flow rate to 150 nL/min. Essentials of the setup are outlined in **Fig 1**.

1. The samples, generally 4 μL of a peptide digest mixture, are loaded onto the column (PEPMAP, 75 μm ID, 15 cm, LC-Packings) via an on-line preconcentration step on a microprecolumn (800-μm ID, 2-mm) cartridge. This has been proved to be an essential step for both desalting and reducing sample loading times. It should, indeed, be realized that loading of a 4-μL sample at a flow rate of 100 nL/min already takes some 20 min, which would increase analysis time.

2. The washing step is performed using 0.1% formic acid in water, delivered at 10 μL/min by a separate 130A syringe pump (Applied Biosystems, Foster City, CA), a procedure that increases the lifetime of the (expensive) analytical column. Note that we use formic acid rather than trifluoroacetic acid, which is generally used in reversed-phase chromatography, because the latter causes a significant decrease in ionization efficiency for peptides. The columns were specifically manufactured to allow the use of formic acid with limited effects on the separation.

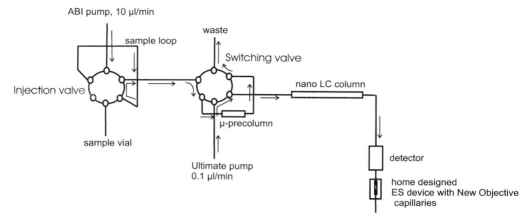

Fig. 1. Schematic drawing of the nanoLC setup. The sample is loaded via the injection valve and transferred at a flow rate of 10 µL/min toward the micro-precolumn. After some period, the switching valve position is changed to allow connection between the two columns and the gradient is started.

3. After 10 min, valve A is switched to connect the precolumn to the separating column, and the gradient is started. We particularly apply a linear gradient, from 5% acetonitrile/0.1% formic acid in water to 80% acetonitrile/0.1% formic acid in water, over a period of 50 min.
4. The column is connected to the UV detector equipped with a U-shaped cell and linked directly, via a 25-µm ID fused silica capillary, to an ESI device developed at home, using Teflon sleeves. This device holds a New Objective PicoTip needle (Woburn, MA), which is a Au/Pd-coated nano electrospray needle having a 10-µm tip.
5. The mass spectrometer is a Micromass Q-TOF instrument (Manchester, UK). The instrument is set to perform an MS survey scan of 1 s over an *m/z* range of 400–1500. Any peak with a threshold above 100 counts/s is automatically detected and the corresponding molecular ion is selected with the quadrupole for fragmentation.
6. The collision gas is Ar (1 bar), and the collision energy is kept at 32 V.
7. The scan time is 1 s, with *m/z* from 50 to 2000.
8. The instrument is used in the positive ion mode and calibrated prior to analysis, using horse myoglobin.

2.2. Nanoliquid LC-MS/MS Identification of Mitochondrial Liver Proteins

Our experimental setup has proved useful in different research projects. We have recently been focusing on the analysis of mitochondrial proteins, mainly for the purpose of studying deficiencies in the pathway of oxidative phosphorylation. Several diseases, some of them lethal at a very young age, are known to originate from mutations in the enzymes from this pathway, or from defects in the assembly genes that are necessary to construct the large multienzyme complexes correctly *(16)*. The technique of blue native polyacrylamide gel electrophoresis (BN-PAGE), combined in a 2-D approach with SDS-PAGE, has been shown to be very useful for mapping the oxidative phosphorylation system, because it is able to separate the five OXPHOS multienzyme complexes in the first dimension and the individual subunits in the second *(17)*. The fingerprints displayed by this technique might become a very useful tool for analysis of the origin of disease, as it rapidly shows the lack of individual subunits, or even complete shifts in the patterns. We recently published a BN-2-D-PAGE map of human heart mitochondrial proteins in which we identified (using a mass spectrometric approach) all the proteins visualized in lanes corresponding to OXPHOS complexes *(18)*. This map is of considerable use for further analysis of *postmortem* material or bioptic samples.

For some subunits, tissue specific isoforms exists *(19)*. Also, the effects of genetic deficiencies in the OXPHOS and related genes are often tissue-dependent. Therefore, we also attempted to analyze the liver OXPHOS proteome. Liver samples, however, contain relatively more enzymes of other metabolic pathways, particularly from the β-oxidation of fatty acids, which we expect to appear in the gel pattern. Although the main pattern is similar, the BN-2-D-PAGE of liver mitochondria has some differences compared with that of heart, and therefore it was necessary to make a global analysis of the liver BN-2-D-PAGE map. In the article on the heart mitochondria *(18)*, we have discussed the need for LC-MS/MS identification of these samples for the reasons already mentioned: the relatively poor resolution of the BN-PAGE technique generates spots that very often contain complex mixtures of multiple proteins, and some of the proteins are very small or membranar and have a limited number of tryptic peptides. In the current work, we also focused on those proteins whose bands were outside the OXPHOS lane, in order to gain more information on the presence of other, potentially interesting, mitochondrial protein complexes.

2.3. The Liver BN-PAGE Map

The mitochondria from 110 mg of liver sample were prepared as described previously, and also sample preparation and BN-2-D-PAGE were similar to that for the heart mitochondrial analysis *(18)*.

1. Some 900 µg of protein extract is loaded. The pattern obtained, shown in **Fig. 2**, compares the liver mitochondrial with the heart mitochondrial proteome map.
2. All the labeled protein spots from the liver map are cut from the gel, and digested with trypsin; the resulting digestion mixture is analyzed by LC-MS/MS using the setup described above.
3. An example of such an identification of a spot, spot 1 on the figure, is shown in **Fig. 3**. **Figure 3A** shows the total ion current chromatogram, which is comparable to a UV chromatogram in classical chromatography.
4. From each peak, an MS spectrum, followed by an MS/MS spectrum is taken, the latter under the condition that the MS signal intensity reached a particular value.
5. The MS spectrum of the labeled peak is shown in **Fig. 3B**. In this particular case, the peptide was doubly charged, which can easily be seen from the isotopic distribution. Any biological molecule, apart from the monoisotopic peak, shows smaller satellite peaks that correspond to species containing naturally abundant isotopes, e.g., ^{13}C atoms, which yield a mass difference of 1 Dalton for each molecular isotope. Since mass spectrometry detects m/z values, this mass difference reduces to $1/n$, with n being the charge of the peptide. A doubly charged peptide can therefore be recognized owing to the presence of a 0.5-Dalton mass difference between the isotopes.
6. The MS/MS spectrum is given in **Fig. 3C**. The very good signal-to-noise ratio in the spectra, and the observation that many spectra are strongly dominated by y'' ion series, makes manual interpretation feasible. This type of fragmentation occurs at the peptide bond, as shown in **Fig. 4**, with charge retention at the C-terminal fragment.

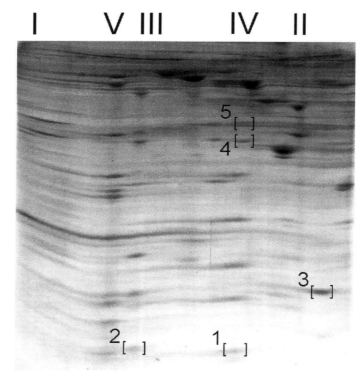

Fig. 2. Liver BN-PAGE gel. The labeled proteins are identified in this project.

The good quality of the spectra very often allows us to deduce a peptide sequence of significant length that can be used in similarity search programs such as FASTA or BLAST. This has the major advantage that it allows identification of proteins from organisms for which little database information is available by looking for homology with proteins of better characterized related organisms, especially those from which the genome sequences have been sequenced. An alternative is the excellent Peptide Sequence Tag algorithm, which allows protein identification based on very short sequence stretches *(20)*. The method uses the sequence data of a set of three amino acid residues, including the masses of the smallest and the largest fragment from the peak series from which these residues were derived. Again, the method is limited to proteins that are already available in the database.

Fig. 3. *(see facing page)* Identification of a protein by nanoLC-MS/MS (spot 1 on the liver BN-2-D-PAGE gel). (**A**) The total ion current chromatogram, displaying the intensities of the mass spectral signal in the function of time. The peak labels show the masses of the precursor ions selected for fragmentation. (**B**) The peak labeled 749.9 corresponds to the signal of the fragment ions from the doubly charged ion with a mass-to-charge ratio of 749.90. (**C**) The corresponding MS/MS spectrum from which the sequence L/INMDAL/I is easily deduced. (Note that this type of fragmentation does not distinguish between Leu and Ile.) The protein is identified as propionyl-CoA carboxylase (α-chain).

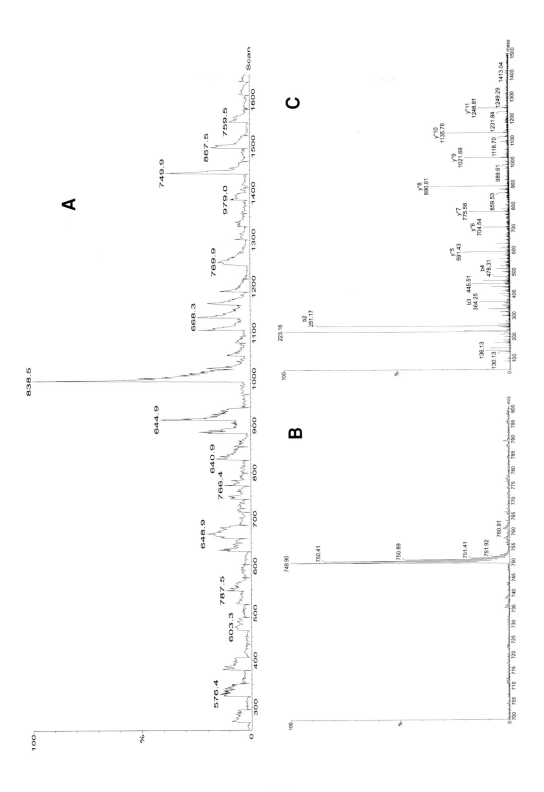

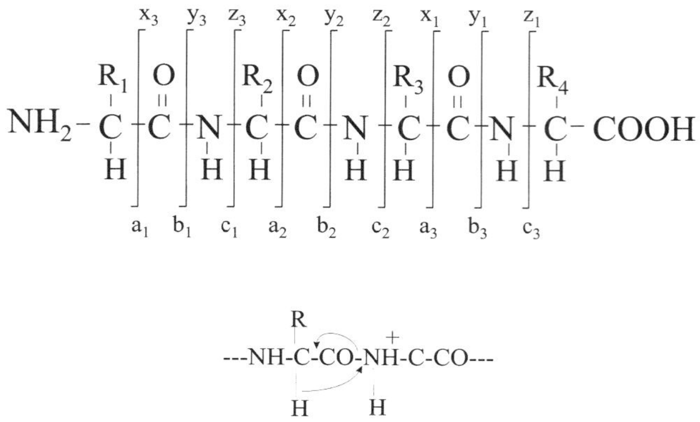

$$x_3 \quad y_3 \quad z_3 \qquad x_2 \quad y_2 \quad z_2 \qquad x_1 \quad y_1 \qquad z_1$$

$$NH_2 - \overset{R_1}{\underset{H}{C}} \overset{O}{\overset{\|}{-C}} + N - \overset{R_2}{\underset{H}{C}} \overset{O}{\overset{\|}{-C}} + N - \overset{R_3}{\underset{H}{C}} \overset{O}{\overset{\|}{-C}} + N - \overset{R_4}{\underset{H}{C}} - COOH$$

$$a_1 \quad b_1 \quad c_1 \qquad a_2 \quad b_2 \quad c_2 \qquad a_3 \quad b_3 \quad c_3$$

$$\overset{R}{\underset{H}{|}} \qquad \overset{+}{\underset{H}{|}}$$
$$---NH-C-CO-NH-C-CO---$$

Fig. 4. Nomenclature of the MS/MS fragmentation of peptides. In low-energy collisions y″ ions are typically generated. In the bottom part of the figure the formation of these ions is explained. A mobile proton can move along the peptide chain in the gas phase, weakening the amide bond between two amino acids.

Recently, several manufacturers have developed software packages that are helpful in the automated sequence determination from these spectra. Many of them use algorithms that allow an input of the complete LC-MS/MS data, without prior manual processing, and provide identifications based on both the precursor ion mass and the uninterpreted MS/MS data. One of the most widely used algorithms, MASCOT, is publicly available (www.matrixscience.com; *see* **ref. 21**).

In this study, we focused mainly on proteins that were different from the earlier published heart BN-PAGE. A striking example of the power of LC/MS-MS analysis was the ability to distinguish between the tissue-specific isoforms of subunit VIIa of the cytochrome C oxidase complex. On the heart proteome map, both the liver/heart and heart isoform were detected, whereas in the liver map (spot 2) the first isoform was exclusively detected. **Figure 5** shows the MS/MS spectra that distinguish these forms. Compared with the heart proteome map, we could identify one extra subunit (spot 3), the MLRQ subunit of complex I (**Fig. 6**). This spot is far out of the lane corresponding to the entire complex, and it may be that it is dissociating from the complex, unlike the other subunits. Apart from the OXPHOS proteins, we detected several subunits from other mitochondrial complexes such as the trifunctional enzyme (spot 4, β-subunit and 5, α-subunit). We are currently investigating the reproducibility of the presence and location of these proteins in the 2-D-PAGE patterns to see whether the method is suitable for investigating a wider range of mitochondrial proteins for the study of non-OXPHOS-related disease.

3. Conclusions

NanoLC-MS/MS allows the identification of individual proteins separated by any gel electrophoretic technique. The scale-down of the separation process provides the

---

Fig. 5. *(see facing page)* Distinguishing between two isoforms of complex IV, subunit VIIa. **(A)** The spectrum of a peptide of the liver/heart-type isoform (sequence LFQEDDEIPLYLK). **(B)** The heart isoform (sequence LFQEDNDIPLYLK), differing by only two amino acids.

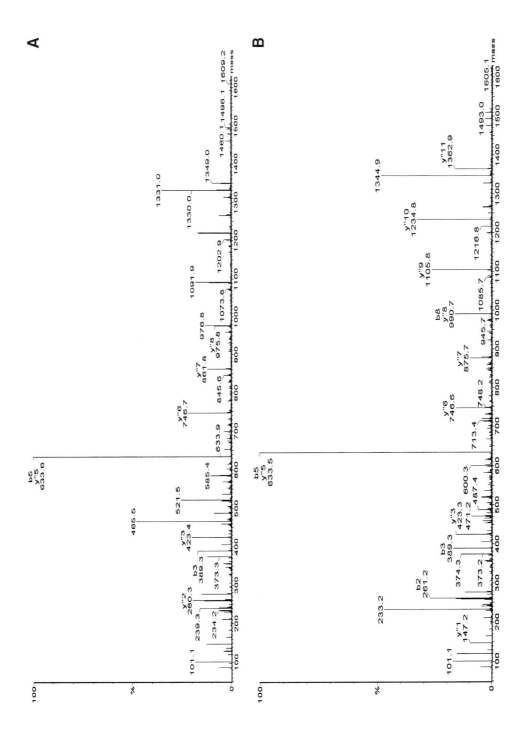

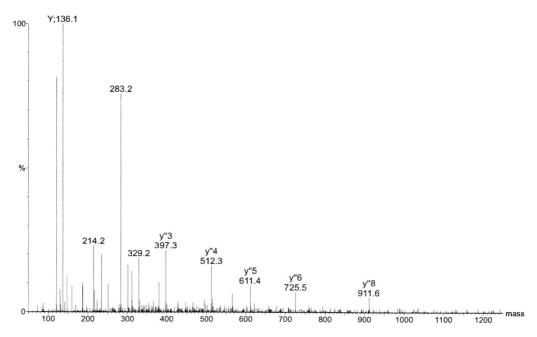

Fig. 6. The MS/MS spectrum that allowed us to identify the complex I MLRQ subunit by BN-2-D-PAGE of the liver.

necessary sensitivity and the interfacing with nano-electrospray sources, and the process is easily automated at the levels of both the chromatography and the automated data acquisition by the mass spectrometer. Moreover, the consequent process of database searching is now also completely automated.

References

1. Wilkins, M. R., Pasquali, C., Appel, R. D., et al. (1996) From proteins to proteomes: large scale protein identification by two-dimensional electrophoresis and amino acid analysis. *Biotechnology* **14,** 61–65.
2. Gygi, S. P., Rist, B., Gerber, S. A., Turecek, F., Gelb, M. H., and Aebersold, R. (1999) Quantitative analysis of complex protein mixtures using isotope-coded affinity tags. *Nat. Biotechnol.* **17,** 994–999.
3. Ong, S.-E., Blagoev, B., Kratchmarova, I., Kristensen, D. B., Steen, H., Pandey, A., and Mann, M. (2002) Stable isotope labelling by amino acids in cell culture, SILAC, as a simple and accurate approach to expression proteomics. *Mol. Cell. Proteomics* **1,** 376–386.
4. Opiteck, G. J., Ramirez, S. M., Jorgenson, J. W., and Moseley, M. A. (1998) Comprehensive two-dimensional high-performance liquid chromatography for the isolation of overexpressed proteins and proteome mapping. *Anal. Biochem.* **258,** 349–361.
5. Rabilloud, T. (2002) Two-dimensional gel electrophoresis in proteomics: old, old fashioned, but it still climbs up the mountains. *Proteomics* **2,** 3–10.
6. Fey, S. and Larsen, P. M. (2001) 2D or not 2D. *Curr. Opin. Chem. Biol.* **5,** 26–33.
7. Neubauer, G., King, A., Rappsilber, J., et al. (1998) Mass spectrometry and EST-database searching allows characterization of the multi-protein spliceosome complex. *Nat. Genet.* **20,** 46–50.

8. Pappin, D. J. C., Höjrup, P., and Bleasby, A. J. (1993) Rapid identification of proteins using peptide-mass fingerprining. *Curr. Biol.* **3,** 327–332.

9. Wilm, M. and Mann, M. (1996) Analytical properties of the nanoelectrospray ion source. *Anal. Chem.* **68,** 1–8.

10. Chervet, J. P., Ursem, M., and Salzmann, J. P. (1996) Instrumental requirements for nanoscale liquid chromatography. *Anal. Chem.* **68,** 1507–1512.

11. Devreese, B., Vanrobaeys, F., and Van Beeumen, J. (2000) Automated nanoflow liquid chromatography/tandem mass spectrometric identification of proteins from *Shewanella putrefaciens* separated by two-dimensional polyacrylamide gel electrophoresis. *Rapid Commun. Mass Spectrom.* **15,** 50–56.

12. Raymackers, J., Daniels, A., De Brabandere, V., et al. (2000) Identification of two-dimensionally separated human cerebrospinal fluid proteins by N-terminal sequencing, matrix-assisted laser desorption/ionization-mass spectrometry, nanoliquid chromatography-electrospray ionization-time-of-flight mass spectrometry, and tandem mass spectrometry. *Electrophoresis* **21,** 2266–2283.

13. Schafer, H., Nau, K., Sickmann, A., Erdmann, R., and Meyer, H. E. (2001) Identification of peroxisomal membrane proteins of *Saccharomyces cerevisiae* by mass spectrometry. *Electrophoresis* **22,** 2955–2968.

14. Ficarro, S. B., McCleland, M. L., Stukenberg, P. T., et al. (2002) Phosphoproteome analysis by mass spectrometry and its application to *Saccharomyces cerivisiae*. *Nat. Biotechnol.* **20,** 301–305.

15. Martin, S. E., Shabanowitz, J., Hunt, D. F., and Marto, J. A. (2000) Subfemtomole MS and MS/MS peptide sequence analysis using nano-HPLC micro ESI-Fourier transform ion cyclotron resonance mass spectrometry. *Anal. Chem.* **72,** 4266–4274.

16. Schoffner, J. M. (2001) An introduction: oxidative phosphorylation diseases. *Semin. Neurol.* **21,** 237–250.

17. Schägger, H. and von Jagow, G. (1991) Blue native electrophoresis for isolation of membrane protein complexes in enzymatically active form. *Anal. Biochem.* **199,** 223–331.

18. Devreese, B., Vanrobaeys, F., Smet, J., Van Beeumen, J., and Van Coster, R. (2002) Mass spectrometric identification of mitochondrial OXPHOS-subunits separated by two-dimensional blue-native polyacrylamide gel electrophoresis. *Electrophoresis* **23,** 2525–2533.

19. Van Kuilenburg, A. B. P., Van Beeumen, J. J., Van der Meer, N. M., and Muijsers, A. O. (1992) Subunits VIIa,b,c of human cytochrome c oxidase. Identification of both 'heart-type' and 'liver type' isoforms of subunit VIIa in human heart. *Eur. J. Biochem.* **203,** 193–199.

20. Mörtz, E., O'Connor, P. B., Roepstorff, P., et al. (1996) Sequence tag identification of intact proteins by matching tandem mass spectral data against sequence data bases. *Proc. Natl. Acad. Sci. USA* **93,** 8264–8267.

21. Perkins, D. N., Pappin, D. J., Creasy, D. M., and Cottrell, J. S. (1999) Probability-based protein identification by searching sequence databases using mass spectrometry data. *Electrophoresis* **20,** 3551–3567.

13

In Silico Proteomics

Predicting Interactions from Sequence

Joel R. Bock and David A. Gough

1. Introduction and Overview

This *Handbook of Proteomic Methods* largely comprises current experimental technologies to identify, quantify, and characterize expressed proteins and their interactions within cells, tissues, and body fluids. These techniques have evolved rapidly with an impetus from the industrial biotechnology sector. Nevertheless, experimental elucidation of all proteomic constituents within an organism and the documentation of their interactions remain formidable tasks. This is further complicated by the broad diversity in protein expression guaranteed by alternative splicing of pre-mRNA or post-translational modifications. In one dramatic example, more than 38,000 different isoforms of Down syndrome cell adhesion molecule (DSCAM) were observed in *Drosophila melanogaster (1)*. Obviously, the combinatorics required for comprehensively explicating all protein–protein interactions, especially for higher eukaryotes, are prohibitive, even with the use of advanced high-throughput approaches.

These facts motivate the investigation of *in silico* methods to complement experimental proteomics technologies. In this chapter, we present a machine learning method to generate hypotheses about protein–protein interactions. In classical biological modeling, one starts from conservation laws and thermodynamics, develops system equations and appropriate boundary conditions, and focuses on mechanisms behind experimental observations *(2)*. However, often one of the following cases is encountered: (1) the system equations are known, but meaningful computation is unfeasible (e.g., protein folding simulation); (2) existing models don't explain all experimental observations[1]; or (3) the system equations are unknown, but many data examples are available. The third case applies to the prediction of protein–protein interaction networks or protein–ligand interactions.

In this chapter, we summarize our research on the automatic prediction of biomolecular interactions using machine learning. The design ethos motivating these investi-

---

[1]A compelling example was the recent demonstration that coexpression of genes in *Drosophila* occurs along blocks of chromosomally proximal genes, accounting for 20% of the fly genome. These genes were not functionally linked; therefore the idea that coexpression implies functional association must be revisited *(3)*.

From: *Handbook of Proteomic Methods*
Edited by: P. Michael Conn © Humana Press Inc., Totowa, NJ

gations is to model these interactions using simple features to represent characteristics that are hypothesized to contribute to binding. Two types of biomolecular interactions are studied: protein–protein and small molecule–protein. Predicting protein–protein interactions has important implications for assembling networks of interactions within living cells, which is a step toward understanding biological processes as integrated systems. Protein–small molecule prediction may some day provide the means to target pharmaceuticals to inhibit the activity of key proteins within signaling networks associated with disease states.

For these investigations, we use support vector machine (SVM) learning *(4,5)* to build discrimination functions that separate input features into classes, resulting in a hypothesis as to whether or not (or how strongly) the corresponding biomolecules will interact. These discrimination functions are based on training datasets of known interactions.

The main contribution of this research is the demonstration that predictions of protein interactions are possible *without information on three-dimensional protein structure.* In particular, amino acid sequence-based descriptors are used to represent proteins. This implies the possibility of proceeding *directly* from genomic sequence to inference of the encoded protein interaction pairs.

This chapter presents highlights from four different studies. We begin with numerical experiments on a large, heterogenous database of experimentally derived protein interaction pairs. Next, the methodology is applied to prediction of all the protein–protein interactions within a single organism. This is followed by discussion of a framework for generalizing the prediction technique across species. Finally, we propose a new method to estimate the binding free energy between a ligand and receptor.

2. Interactions in a Broad Database

A previous investigation *(6)* was intended to demonstrate the feasibility of predicting protein–protein binding based solely on primary structure and associated physicochemical properties. This publication was apparently the first to frame the prediction of protein–protein interactions in terms of a classification problem, in other words, training a computer to learn to discriminate between the classes *interaction* and *noninteraction* on the basis of correlated patterns of amino acid sequence and substructure in the interacting pairs.

2.1. Experimental Methods

The Database of Interacting Proteins (DIP) contains experimentally determined interactions between proteins observed in a wide variety organisms *(7).* We sampled this database to create positive examples for machine learning training. Negative examples were generated by randomizing amino acid sequences sampled from DIP, while preserving both (1) amino acid composition and (2) di- and tripeptide k-let frequencies *(8).*

Data were partitioned into balanced training and testing sets, at approximately a 1 : 1 ratio. Training data were used to construct the decision function. Next, testing data were separated to evaluate the performance of the trained system. An ensemble of datasets was created by randomly sampling the DIP interacting proteins and newly created "shuffled" amino acid sequences; these data were then subjected to repeated statistical trials. A different SVM was trained for each k-let correlation frequency and experimental trial. The results of these trials were averaged to eliminate potential biases owing to chance sampling of the dataset.

Table 1
System Generalization Accuracy Summary[a]

k-let freq.	No. of Examples (Train, Test)	Inductive accuracy (%)
1	(2190, 2189)	80.96 ± 1.42
2	(2192, 2192)	80.19 ± 0.86
3	(2203, 2195)	80.13 ± 0.89

[a]Inductive accuracy is the percentage of correct protein interaction predictions on test data not previously seen by the system. N = 10 trials.

The performance of each SVM was evaluated using the inductive accuracy of predictions made on previously unseen test examples as the performance metric. "Inductive accuracy" is defined here as the percentage of correct protein interaction predictions on the test set, consisting of nearly equal numbers of positive and negative interaction examples.

2.2. Feature Representation

The problem of feature selection is to find a set of salient attributes to represent the concept that is to be learned. Here, attributes representing residue charge, hydrophobicity, and surface tension were selected to represent proteins in complex. For each amino acid sequence of a protein–protein complex, feature vectors were assembled from encoded representations of tabulated residue properties including charge, hydrophobicity, and surface tension for each residue in sequence. As an index of amino acid hydrophobicity, we chose to use the consensus normalized hydrophobicity scale of Eisenberg *(9)*. The surface tension scale used for residue features in this research is described in Bull and Breese *(10)*. The charge for each amino acid was taken from the set $\{-1, 0, +1\}$.

2.3. Results and Discussion

The main results of the protein–protein interaction predictions are summarized in **Table 1**. Each row in the table corresponds to a constant k-let frequency used to generate the negative training and testing examples. Data in column 2 indicate the average total number of each type of examples for each case. These data have been averaged over an ensemble of N = 10 trials, a sufficient sample, as indicated by the low variance shown in column 3.

The inductive accuracy of the learning machines, as summarized in **Table 1**, is encouraging, given the depth of the DIP database. For each statistical background comprising k-let orders 1–3, about *four of five* potential protein interactions are correctly estimated by the system. It bears reiteration here that only primary structure data have been used to train the SVM. We submit that some implicit information regarding structural, chemical, and biological affinity has been represented and learned by virtue of the affirmative labeling of protein interaction pairs. The implications of the results shown in **Table 1** for future proteomics research are intriguing.

However, these results must be interpreted with caution. An important objective is to ascertain the extent to which the present machine learning approach may provide utility to the proteomics community. Therefore, several important issues must be brought to light.

2.3.1. Problems with Accuracy Statistics

The purpose of this study was to demonstrate the feasibility of predicting protein–protein binding, by posing interactions in terms of a *classification* problem. To apply this prediction methodology in more realistic proteomics experiments, additional statistics of classification performance should be computed.

It has been noted that the use of *accuracy* alone may misrepresent the generalization potential of a classifier, in particular when comparing different classification architectures against one another *(11)*. Accuracy as a statistic assumes (1) equal misclassification costs (for false positives and false negatives), and (2) a known class distribution in the "target environment."

If one of the classes in a two-class problem is "rare" relative to the other one, it is trivial to produce a highly accurate classifier by simply predicting the majority class. To illustrate, imagine a bacterial organism, *Bacillus X*, with a proteome of size $N = 1000$ containing potentially 499,500 unique pairwise interactions. Let us assume that there are five interactions per protein. Then we would expect to find 5000 genuine positive interactions in Nature and 494,500 negatives for this organism. In this case, a classifier always predicting the negative class (–) has 99% accuracy!

Therefore, presentation of results only in terms of the prediction accuracy may not provide sufficient information to evaluate one classifier critically over another on a given dataset. In subsequent sections of this chapter, additional statistics are used to explore the prediction performance of SVM classifiers as applied to different datasets.

2.3.2. Problems with Balanced Data

An important complication concerns the distribution of *actual* positive and negative examples in Nature. Remember that our training and testing datasets were *balanced*—comprising nearly equal numbers of positive and negative examples. The true state of Nature is probably highly *imbalanced* with respect to the distribution of classes. It is expected that the number of "noninteractions" would greatly outweigh the number of actual interactions when considering all pairwise combinations of proteins for a given proteome. When data classes are imbalanced, replicating the naturally occurring proportions of each class within the training set produces classifiers that perform poorly on minority-class examples *(12)*.

As done here, one can present both positive and negative examples in a nearly balanced proportion, so that new data points are recognized as members of the correct class in generalization *(13)*. This should be done without regard to the prior probabilities associated with positive and negative data points in the biological sample. However, in designing a balanced sampling within the artificial (machine learning) environment, a variance with respect to the distribution of data classes in the real, biological world is generated.

Our ultimate objective is to make useful predictions in biology. Therefore, we are forced to deal with this dilemma, which is sometimes referred to as a "needle in a haystack" problem in data mining. When making predictions on all possible pairwise combinations in a different organism, if the classifier is characterized by the same false positive rate estimated from the training data set, the number of false positives would increase significantly. The precision associated with these predictions would be seriously degraded relative to the training data. The sensitivity, or true positive rate, would be expected to remain the same. Predictions made for the "minority class" (interacting protein pairs) would tend to have a much higher error rate than those of the majority class.

The "cost" of making a false positive decision is generally different from the cost of a false negative one. Because of the relative scarcity of positive interactions, it might be argued that false negatives are inherently more costly than false positives. A commercial drug discovery entity might take the opposite position, as false positive leads used to initiate expensive lead validation wet chemistry experiments carry potentially significant economic costs. Therefore, it may be appropriate to take prediction *probabilities* into account because we envision that the predictions should be subjected to some degree of manual curation by human experts. SVM classifiers output a binary decision, so doing this would require transforming the outputs to produce continuous-valued, *a posteriori* probabilities (e.g., *see 14,15*).

2.3.3. Other Issues

One way to account for cost sensitivity in learning is to analyze a group of classifiers using lift charts, recall-precision charts or receiver operating characteristic (ROC) curves *(16)*. ROC analysis makes no assumptions about relative distributions of the data classes, or about misclassification costs *(11)*. This type of analysis is carried out in **Subheading 3.4.** of this chapter.

When data classes are expected to be largely imbalanced, another possibility is to equalize the cost basis for the associated disproportionate misclassification costs. Practical strategies to accomplish this are presented by Elkan *(17)*, who proposes (1) changing the proportion of examples of the majority class (here, the noninteracting protein–protein pairs) and then retraining the classifier using estimated costs; or (2) applying a classification rule involving the cost-weighted posterior probabilities of class membership for each training pattern. In the second case, the (artificial) even class distribution is maintained.

Notwithstanding these concerns, in **Subheading 4.** we present results of an investigation in which we use experimentally derived protein–protein interactions within one organism (*Helicobacter pylori*, 1039 data points) as a training dataset, and make predictions of a complete network of interactions in a different organism (*Campylobacter jejuni*, 5367 predicted interactions). Strictly following the "needle in the haystack" line of thought here, finding only 5367 interactions in *C. jejuni* is a surprising result, as one might expect a much larger number of interactions ($O[100,000]$!) to occur by extrapolation from raw numbers calculated from the precision and sensitivity cross-validation estimates. This observation prompts the question: does the SVM learn to "detect" an interaction on a constant percentage of data points, or only when a data point representing a true protein–protein interaction is present? The predicted interactions are valid hypotheses awaiting empirical confirmation or falsification.

3. Interactions in One Species

Having demonstrated that this approach is feasible on a broad database of protein interactions, in this section our objective is to predict all the protein–protein interactions within a single organism. The investigation summarized here is reported in greater detail in **ref. *18***. The model organism subject to investigation is the yeast *Saccharomyces cerevisiae*, chosen for this study because its genome was the first to be completely sequenced *(19)*, it shares a core set of conserved proteins for metabolic processes, protein folding, trafficking, and degradation with many higher eukaryotes *(20)*, and it is an excellent experimental platform for genetic engineering *(21)* and targeted drug discovery *(22)*.

Tradeoffs that arise between the precision and sensitivity of the protein–protein interaction predictions are explored, based on results obtained from cross-validation experiments on a nonredundant database. We also studied the importance of eliminating redundant protein–protein interactions from the training dataset.

3.1. Experimental Methods

The yeast–protein interaction dataset was collected by Ito and co-workers *(23)*. These investigators used a large-scale variation of the two-hybrid technique, known as interaction mating, which exploits the fact that haploid yeast cells of opposite mating type (MATa, or MATα) fuse to form diploids when brought into contact with each other. "Bait" protein A (fused to the DNA-binding domain) and "prey" protein B (fused to the activation domain) are expressed in different haploid strains, each of opposite mating type. The combinatorial mating of these strains and consequent fusion of haploids indicates interaction of the corresponding fused proteins *(24)*.

Ito et al. *(23)* assert that approx 95% of the open reading frames (ORFs) in the yeast genome are represented in all possible combinations of DNA-binding and activation domain proteins.[2] The interaction mating procedure resulted in a set of 4549 independent interactions among 3278 distinct protein interactions after redundancy minimization. The amino acid sequences of these 4549 positive interacting pairs were compiled from online sequence databases using their respective ORF designations.

To preclude bias from redundant examples in numerical experiments, redundant or similar interacting pairs in the dataset were identified using the Smith–Waterman dynamic programming algorithm *(25)* and removed prior to the study. We used an affine insertion/deletion model *(26)*, with common penalty values for gap opening and extension (12, 2, respectively) in the alignment process. Pointwise, amino acid mutation probabilities were modeled using the BLOSUM62 substitution matrix *(27)*. Sequences identified as redundant at the 99% significance level ($p < 0.01$) in an all-against-all pairwise similarity analysis were removed. Note that our objectives here differ fundamentally from sequence database searches in which the highest scoring similarities are desired. Such searches typically use *p*-values many orders of magnitude smaller than those used here.

After comprehensive forward redundancy elimination, the dataset comprised a positive example set of 3011 protein interactions, roughly two-thirds the size of the orig-

---

[2]The yeast interaction dataset is available from Ito and co-workers at http://genome.c. kanazawau.ac.jp/Y2H/.

inal dataset. Negative examples were derived from the balance of the proteome of *S. cerevisiae*. From 6408 ORFs, we found a set of 6360 protein sequences in on-line databases, which was sampled randomly to construct noninteracting pairs, and designated as "noninteracting" by virtue of not belonging to the positive set derived from the all-against-all Y2H screen Ito *(23)*. Positive and negative amino acid interaction sequence sets were validated to ensure mutual disjointness and assembled into separate database files, keyed by corresponding ORF pairs. The final sample contained 3011 positive and 6360 negative protein interaction pairs.

3.2. Feature Representation

A parsimonious set of features characterizing the design sample was developed, using only amino acid sequence descriptors as described in **Subheading 2.2.**

Different combinations of features associated with the amino acids were studied. We repeatedly trained and evaluated a collection of predictive systems, using different sets of features $\mathcal{F}i$ comprising all combinations of one, two, and three physicochemical attributes in concert:

$$\mathcal{F}1 \subset \{\{C\}; \{H\}; \{T\}\}$$

$$\mathcal{F}2 \subset \{\{C, H\}; \{C, T\}; \{H, T\}\}$$

$$\mathcal{F}3 = \{C, H, T\}$$

where C, H, and T stand for charge, hydrophobicity, and surface tension, respectively, and $\{*\}$ denotes a particular set of features.

The results reported here are based on numerical predictions obtained using features from the seven distinct sets $\mathcal{F}1$, $\mathcal{F}2$, and $\mathcal{F}3$ described above. Objective evaluation of the output of the trained machine was performed to determine the rates of correct and incorrect predictions and to estimate the generalization error rate upon application of the trained system to inference on other species.

3.3. Prediction Objectives and Metrics

Several practical objectives for computational inference of protein–protein interactions are conceivable. Three cases are noteworthy, each corresponding to different goals, and each suggesting a distinct means by which prediction success should be quantified.

1. If the objective is to detect all of the possible protein–protein interactions in a given proteome, minimizing the occurrence of false negative predictions (misses) would be important.
2. If maximizing the correct positive prediction rate (hits) is important (for instance, when scrutinizing a large set of potential drug targets to enhance the efficiency of drug discovery), confidence in the affirmative decision that a protein–protein interaction has been detected takes precedence.
3. In general, it may be preferred to use an overall estimate of the accuracy of the system classification rate, including both positive and negative predictions.

To address each of these cases, we calculated the machine learning performance statistics known as "sensitivity," "precision," "specificity," and "accuracy" *(28)*. For case 1, the relevant metric is the *sensitivity*, which measures how many actual protein interactions present in the data are found by the system. Sensitivity is calculated as

$S = TP/(TP + FN)$, where TP = number of true positive interaction decisions and FN = number of false negative decisions ("misses"). Case 2 calls for the use of the *precision*, which describes the rate at which a positive interaction decision is correct. Precision is computed as $P = TP/(TP + FP)$, where FP is the number of false positives declared by the system.

For the general case 3, the prediction *accuracy* may be useful. Accuracy expresses the general error of the system, computed as $A = (TP + TN)/(TP + TN + FP + FN)$. Here, TN represents the number of true negative classifications. Recall the difficulties with the use of accuracy listed in **Subheading 2.3.**, which restrict the robustness with which different classifiers may be compared in practice. An alternative in case 3 is to construct a ROC curve, a locus of points expressing tradeoffs between the sensitivity and the (1 minus) specificity, as a function of variation in a detection parameter *(29)*. *Specificity* conveys the rate at which negative examples in the data are correctly classified and is equal to $1 - FP/(FP + TN)$.

Using the interacting protein-encoded features (*see* **Subheading 2.2.**) as input, a series of polynomial kernel support vector machines *(5)* were trained to differentiate between pairs of proteins that did (and did not) interact within a biological context.

We divided a random subsample (2729 of 3011 positive, 3560 of 6360 negative) of the encoded protein interactions into 10 distinct subsets, allocating positive and negative examples to each subset in approximate proportion to their frequency in the aggregate experimental sample ($2729 : 3560$, or $\approx 1 : 1.3$). This sample was used to serially train and test different SVM classifiers and to estimate their expected generalization error rates, precision, and sensitivity by the statistical technique of 10-fold cross-validation *(30)*.

3.4. Receiver Operating Characteristic (ROC) Analysis

As mentioned in **Subheading 3.3.**, if greater priority is given to uncovering all the protein interactions present in an organism (case 1), different performance statistics should be applied when designing the predictive system. Here, detection is at a premium, as opposed to confidence in the correctness of a positive system decision. Sensitivity indicates the rate of correct positive predictions on all the available data, or the "true positive" rate *(TPR)*. ROC curves, introduced by Peterson and Birdsall *(29)*, elucidate the relationship between true positive and false positive rates, offering insight into the cost (in terms of false alarm probability) of a given rate of sensitivity. A group of different classifiers may be compared in ROC space, where each classifier's ROC curve is typically parameterized by a constant value of the detection threshold. The threshold is selected to reject the noise background of varying degrees, and its effect is to regulate the trade between precision and sensitivity. Higher detection thresholds decrease the rate of false alarms but cause an increase in the rate of false negatives.

We explored a "quasi-ROC" space containing the SVM classifiers of this investigation. This space is related to the ROC curves of signal detection theory in the sense that it provides an analysis of groups of classifiers in (*FPR*, *TPR*) coordinates. However, quasi-ROC curves here are actually loci of discrete points, each corresponding to a given SVM polynomial kernel order d. Whereas each separate curve in classical ROC space represents a different classifier, with the ratio *FPR/TPR* varying continuously, progressing in quantum steps along the present "curves" corresponds to changing SVM architectures. In further contrast to conventional ROC analysis, the "curves" connecting the d-values denote different feature sets used to represent proteins.

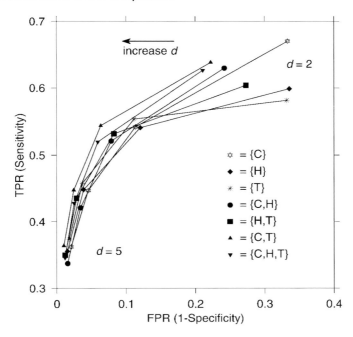

Fig. 1. Quasi-ROC curves for predictions of protein–protein interactions in *S. cerevisiae*. Symbols represent different attribute sets used to encode the amino acid sequences of the constituent proteins comprising a biological interaction. Each data point was obtained from 10-fold cross-validation estimates of SVM performance. The arrow indicates the direction of increasing SVM polynomial kernel order $d \in \{2, 3, 4, 5\}$, which corresponds to increasing precision (decreasing sensitivity).

There are issues with ROC analysis worthy of mention. ROC curves for protein–protein interactions are not likely to be as widely applicable as in the fields of radar *(31)* or sonar signal processing *(32)*, for which the signal and noise backgrounds have been thoroughly characterized. The protein interaction ROC space will appear differently, in terms of curve shape and absolute magnitudes, under different relative distributions of data used in their creation. Moreover, this effect might be encountered for each different organism under investigation.

3.5. Results and Discussion

The results of this analysis are presented in **Fig. 1**. Each curve plotted in the figure corresponds to a different feature set, denoted by a different symbol. These features are indicated in the figure legend. Points along the curves are mapped by varying the SVM polynomial order d and recording the *(FPR, TPR)* values from 10-fold cross-validation.[3] The area under a particular ROC curve at a given value of the abscissa is an indicator of the accuracy of the associated classifier, now decoupled from *a priori* assumptions about relative distributions of the positive–negative interaction classes or misclassification costs *(11)*.

---

[3]Polynomial order assumes integer values, resulting in piecewise-linear segments for these ROC "curves."

Table 2
Prediction Performance Statistics for Two Different SVM Classifiers[a]

		$d = 2$	$d = 5$
TNR	$\dfrac{TN}{TN+FP}$	0.778	0.990
S(TPR)	$\dfrac{TP}{TP+FN}$	0.638	0.363
FNR	$\dfrac{FN}{FN+TP}$	0.362	0.636
FPR	$\dfrac{FP}{FP+TN}$	0.222	0.010
P	$\dfrac{TP}{TP+FP}$	0.687	0.965

[a]Data again correspond to the feature set $\{C,T\}$ dominating the ROC space of **Fig. 1**. *TNR*, true negative rate; *FNR*, false negative rate. Other statistics are defined in **Subheadings 3.3.** and **3.4.** Total example count: $N^+ = 2729$, $N^- = 3560$.

Various performance statistics computed for the two extreme SVM polynomial orders studied are summarized in **Table 2**. These data correspond to the feature set $\{C, T\}$ found to dominate the ROC space as visualized in **Fig. 1**. *TNR* is the true negative rate, and *FNR* is the false negative rate.

We find it interesting that protein–protein interactions in *S. cerevisiae* may be reliably inferred using only a minimal description of their sequential amino acid characteristics. This technique is exciting in its simplicity and compelling predictive performance and may prove useful in programs of targeted pharmaceutical discovery, for example, when small effector molecules are to be directed at protein interaction networks *(19)*, or in therapeutic disruption of disease-related signal transduction cascades *(33)*.

There are, however, several important caveats associated with this computational screening approach. The experimental results indicate that a high precision, in excess of 90%, is obtained at the cost of a low sensitivity (approx 36%). This means that although confidence in positive predictions is high, many actual protein–protein interactions are not detected by the system. This may not be acceptable if the objective is to uncover all interactions in a proteome under investigation.

An important question is how to specify properly the "negative" (noninteracting) examples used to train the system. One strategy is to assume that an experimental dataset is comprehensive in the sense that all protein interactions detectable by the experimental system used in its derivation are represented; all protein pairs not contained within this experimental set are declared as negatives. This naïve approach, taken here, is on average a reasonable approximation. The 6408 ORFs in *S. cerevisiae* translate into more than 20.5 million possible distinct protein–protein interactions, assuming (conservatively) that only one protein is produced per ORF [for a proteome of size N proteins, there are at least $(N^2 + N)/2$ distinct interaction pairs].[4]

Our approach to minimizing redundant protein interaction pairs was carried out to eliminate bias in the predictor owing to similar training and testing examples. To study the effect on predictive acuity without any such filtering, cross-validation experiments were conducted on the complete design sample from **ref. 23**, intentionally using all available interactions in cross-validation experiments. Here, each of the feature sets $\mathcal{F}_i$ were scrutinized. As anticipated, prediction success as quantified by the objective measures discussed in **Subheading 3.3.** was enhanced relative to the nonredundant dataset results summarized in **Fig. 1**. Whereas both precision and accuracy were characterized by modest improvements (3.9 and 5.2%, respectively), the observed sensitivity rate increased by 20.6%, averaged over all feature sets. This result suggests that an apparent sensitivity rate, if obtained by indiscriminate predictions without redundancy elimination processing, may be significantly overstated.

4. Interactions across Species

In this section, we summarize recent research that proposes a methodology to extrapolate across species using the machine learning approach *(35)*. We introduce an algorithm (the *phylogenetic bootstrap*), which suggests traversal of a phenogram, interleaving rounds of computation and experiment, to develop a knowledge base of protein interactions in genetically similar organisms on a proteome-wide scale.

The idea is to use an analogy between the proteomes of two closely related organisms to predict protein–protein interactions. A "template" or design organism provides a network of experimentally derived interactions. These data are used to train the system and to estimate the standard error of its generalization capability. The trained system is used to infer the structure of an interaction network in a related organism. Given a list of experimental interactions, all that is required to infer the proteome-wide interaction map are the amino acid sequences of the target organism. We refer to this approach as "interaction mining," in association with the concept of data mining, which concentrates on the application of specific algorithms for extracting structure from data *(36)*.

4.1. Experimental Methods

We trained a learning system to recognize correlated patterns of primary structure within protein interaction pairs taken from the human gastric bacterium *Helicobacter pylori*. A high-quality *H. pylori* interaction dataset was collected by Rain and colleagues *(37)* and is publicly available on-line. This *H. pylori* experimental dataset comprised the template organism. A close phylogenetic neighbor, *Campylobacter jejuni*, was selected for the predictions. Both *H. pylori* and *C. jejuni* are microaerophilic, gram-negative, flagellate, spiral bacteria. Analysis of their major constituent protein domains shows a high degree of similarity.

---

[4]Owing to alternative RNA splicing, the actual number of proteins produced by a gene is likely to be much higher. Genes with dozen or more transcripts are commonly observed *(34)*.

The SVM *(5,38)* can be trained to classify labeled empirical data points by constructing an optimal high-dimensional decision surface that simultaneously maximizes the separation between data classes and minimizes the "structural risk":

$$R(\alpha) = \int_Z Q(z,\alpha)dF(z), \ \alpha \,|\, \Lambda \qquad (1)$$

with respect to parameters α using an independent, identically distributed (i.i.d.) sample $Z = \{z_1, z_2, ..., z_l\}$ generated by an (unknown) underlying probability distribution F, where Q is an indicator function, and Λ is a set of parameters. The sample points $z_i = (x_i, y_i)$ comprise protein features $x_i \in R^n$ and their classifications $y_i \in \{-1, +1\}$. In practice, the learning task converges rapidly as a constrained quadratic programming is solved. The resultant decision function h represents an hypothesis generator for inference on novel data points, mapping them onto the discrete set y, or $h{:}x{\rightarrow}y$. This is a binary decision ($+1{\Rightarrow}interaction$, $-1{\Rightarrow}nointeraction$).

The assumption in **Eq. 1** of a fixed generative probability distribution $F(Z)$ is a key issue in the design of this data mining application. A direct consequence of this assumption is that a decision function h, developed from a training sample Z_a taken from species S_a, may be used to predict protein–protein interactions on a sample Z_b *from another species S_b*, provided that features of their respective proteomes are not too dissimilar in some sense, or:

$$\rho\,[F(Z_a), F(Z_b)] \leq \delta \qquad (2)$$

where ρ is a measure of distance between its arguments, and δ is a constant. The statistic ρ is general and may be taken to signify cross-species similarity based on genome-level "edit distance" *(39)*, whole-proteomic content *(40)*, or proximity within phylogenies constructed from multidomain orthologous protein sequences *(41)*, to cite only three of many possibilities.

We introduce here the phylogenetic bootstrap algorithm. Bootstrap methods in applied statistical inference are numerical techniques for estimating the standard error of arbitrary test statistics *(42)*. The phylogenetic bootstrap for protein–protein interaction mining does not compute a statistic *per se* but suggests a method for incrementally "walking" laterally across a phenogram, interleaving rounds of computation and experiment, to develop a knowledge base of protein–protein interactions in genetically related organisms. Using the hypothesis $h{:}x{\rightarrow}y$ [based on an assumed common probability distribution $F(Z)$], we infer the interactions within a sample taken from a distinct, evolutionarily similar proteome. These predictions are a function of the generalization confidence level derived from 10-fold cross-validation error estimation *(30)*. The probability of correctness of a novel prediction may be estimated by

$$Pr\{\hat{y} = y \,|\, h\} = g(\delta)(1 - \varepsilon_{cv}) \qquad (3)$$

where $\hat{y}$ is the predicted interaction for a putative interacting protein pair, y is the true state of nature, ε_{cv} is the cross-validation error rate, and $g(\delta)$ is a decreasing function of the interproteomic distance (**Eq. 2**). A simple plausible (and conservative) form for the function g is an exponential

$$g(\delta) = e^{-\lambda\delta} \qquad (4)$$

where λ is the rate of decay. Substituting this function in **Eq. 3**, the prediction confidence becomes

$$Pr\{\hat{y} = y \,|\, h\} = e^{\lambda\delta}(1 - \varepsilon_{cv}), \ \lambda > 0; \ \delta \in [0, \infty) \qquad (5)$$

Note that this representation is schematic. The value of the decay parameter λ and calibration of the distance in **Eq. 2** can only be determined after experimental validation of the numerical predictions.

Upon completion of this process, predicted protein–protein interactions in the novel organism may be used to design successive genetic or biochemical experiments. The results of these selected experiments are fed back to refine the current model, and flesh out empirical protein interactions within the new proteome. This iterative process may continue as long as certain criteria on acceptable estimated prediction error rate and proteome similarity remain satisfied. The steps comprising the phylogenetic bootstrap as proposed in this investigation may be distilled into an algorithm, as described next.

4.2. Algorithm

The phylogenetic bootstrap algorithm is summarized in this section.

1. *Input.* First, it is necessary to specify the species S_a, S_b subject to investigation. In general, some existing protein interaction data may be at hand for each proteome, although their relative cardinality may be quite skewed. Our line of thought assumes that no interaction data are available for S_b; we have only a set of labels $\{Y_a\}$ corresponding to experimentally verified interactions sampled from the proteome of species S_a. These labels, along with the amino acid sequence sets $\{s_a\}$ and $\{s_b\}$ comprising the species' respective proteomes, are inputs to the algorithm. Other inputs required are the interproteome distance δ **(Eq. 2)** and the maximum acceptable rate of generalization error, ε_{cv}^{max}, where $0 < \varepsilon_{cv}^{max} < 0.5$.

2. *Construct features from training sample*, based on attributes of the primary structure sequences s_a from the training data set. Encoded attributes X_a for entire proteomes may be derived from tabulated residue properties including charge, hydrophobicity, and surface tension as described previously *(6)*. At this stage, data preprocessing including normalization and filtering should be performed to produce a useful sampled attribute set $\{x \mid x \in R^n, x \subset X\}$. A total of l data points z are constructed by adding labels y to the accepted feature vectors $\{x\}$, or $z_i = (x_i, y_i)$, $i = 1,...l$. The union of positively and negatively labeled examples constitutes the training sample $\{Z_a\}$.

3. *Compute decision rule.* Design an optimal support vector machine to classify data points in the sample $\{Z_a\}$. After learning, the system builds a decision rule h that maps input data vectors x_i onto the classification space $y_i \in [+1, -1]$. The numerical sign of y_i is interpreted as the likelihood that the two proteins represented by x_i will interact.

4. *Estimate CV error.* Perform k-fold cross-validation experiments on the training set. Segregate the observations $\{z^k\}$ within each data fold k, and train a different SVM using data $\{z^m\}$ from each of the $k - 1$ disjoint data folds $\{z^m \mid z^m \in Z_a, m \neq k\}$. Predict the class membership of the omitted points $\{z^k\}$. Accumulate the total number of misclassifications observed in this process. Take the final k-fold average cross-validation error as the estimated expectation of generalization error rate ε_{cv} of the learner h. The magnitude of this error estimate in practice will be extended by some function of interproteomic distance, say $g(\delta)$.

5. *Construct features from novel sample.* Construct features $\{X_b\}$ from sequences $\{s_b\}$ for the unlabeled proteome S_b. All-vs-all pairwise interactions may be represented in the prediction set. The same data preparation process should be applied as carried out in **step 1**.

6. *Predict novel interaction network.* Predict a new network of protein–protein interactions $\{\hat{Y}_b\}$ via the trained system $h(\alpha): x_b \rightarrow Y_b$, where α are parameters of the model. To the extent that the assumption of proteomic similarity $\rho [F(Z_a), F(Z_b)] < \delta$ is satisfied, each point estimate is expected to be accurate with a probability $g(\delta)(1 - \varepsilon_{cv})$, or $Pr\{\hat{y} = y \mid h\} = g(\delta)(1 - \varepsilon_{cv})$.

7. *Validate sample experimentally.* Take a random sample from the protein interaction prediction set $Z_b = \{(x, \hat{y}) \mid x \subset X_b, \hat{y} \subset Y_b\}$ and verify the predicted protein interactions (both positive and negative) using experimental proteomics techniques. Compare the experimentally observed and calculated estimated prediction error rates. Assert that the following statement holds true: $\varepsilon_{cv}^{\ v} \leq \varepsilon_{cv} < \varepsilon_{cv}^{\ max}$, where the superscript v denotes validation by biological experiment.

8. *Input.* Select sequences $\{s_c\}$ from a new, related organism $\{S_c\}$. The similarity assumption $\rho[F(Z_a), F(Z_b)] < \delta$ must still be maintained.

9. *Update training sample.* Add sequences from the validated prediction set to the training set, and consider this expanded set as the training set for the next iteration: $\{s_a\} = \{s_a\} + \{s_b\}$. Update the class labels by adding the prediction label set $\{Y_a\} = \{Y_a\} + \{\hat{Y}_b\}$. Protein interactions for organism $\{S_c\}$ can now be computed.

10. *Iterate.* Return to **step 1** and repeat the process. The stopping condition for this iteration is violation at any time of the assertions regarding the generalization error rate, i.e., when the error rate from cross-validation, ε_{cv}, exceeds the specified limit $\varepsilon_{cv}^{\ max}$, or when the experimental observations contain more frequent errors than the calculated rate, or $\varepsilon_{cv}^{\ v} \geq \varepsilon_{cv}$.

4.3. Results and Discussion

For the design organism *Helicobacter pylori* strain 26695, a total of 1039 protein interactions were selected for analysis. Interactions were identified from the database provided online at http://pim.hybrigenics.com. From the nominal *H. pylori* proteomic complement of $N = 1555$ sequences, a sample of 1039 noninteracting sequences was selected. This created a balanced representation of each data class to train the learning system, the total sample length being $l = 2078$ observations. Each sample point $z_i = (x_i, y_i)$, $i = 1,...,l$ was constructed from primary structure features $x_i \in R^n$ and their interaction class labels $y_i \in \{-1, +1\}$.

The learning machine generates an interaction hypothesis $\hat{y}$ for each data point x via the computed decision surface $h: x \rightarrow y$. Define the null H_0 and alternative hypotheses H_A as:

$$H_0: y \mid x = -1, \quad (nointeraction),$$

$$H_A: y \mid x = +1, \quad (interaction\ present).$$

There are two types of statistical errors that may occur on each decision $\hat{y}$.

1. If H_0 is true and is rejected ($\hat{y} = +1$, $y = -1$), the machine commits a type I error, or "false positive" decision.

2. If H_0 is false (interaction present) and is not rejected ($\hat{y} = -1$, $y = +1$), a type II, or "false negative" error, is made.

The 10-fold cross-validation prediction error estimates obtained on the design sample are presented in **Table 3**. Results are shown for *sensitivity*, *precision*, and *accuracy* (defined in **Subheading 3.3.**). Recalling **Eqs. 3–5**, the expected precision and sensitivity of the classifier's performance in the novel organism will be less than the estimates shown in the table. The actual performance decrement cannot be evaluated until biological experiments validate or invalidate the testable hypotheses comprising the network of interactions. At present, we can only estimate upper bounds on performance for this set of generated hypotheses.

Table 3
Ten-fold Cross-Validation Performance Estimate
Derived from Classifiers Trained on Examples
from the Design Organism *H. pylori*[a]

Precision	Sensitivity	Accuracy
80.2	68.6	75.8

[a]High precision indicates the suppression of type I (false positive) errors. High sensitivity means that type II errors are suppressed by the decision function (i.e., low false negative rate). Numbers are expressed as percentages. Data sample size $N = 2078$.

Fig. 2. Predicted whole-proteome interaction map for *Campylobacter jejuni*. In this diagram, individual proteins are represented as vertices, and the interactions between pairs of proteins are indicated by edges connecting nodes. Proteins with a large number of partners (>15; 1% of all predictions) are colored black; light gray colored nodes signify that relatively few proteins (≤5; 61% of predictions) are expected to interact with that node. Dark gray nodes represent proteins with 6–14 interaction partners.

4.3.1. C. jejuni *Interaction Hypotheses*

The level of estimated generalization obtained from leave-one-out analysis of the *H. pylori* proteome supports confidence in the prediction of protein-protein interactions in *Campylobacter jejuni*. *C. jejuni* and *H. pylori* are close phylogenetic relatives (*see*, e.g., **Fig. 1** in **ref. 43**), displaying highly similar constituent protein domains and genomic content (**Fig. 2** in **ref. 40**). The *C. jejuni* proteome contains 1613 proteins, of

Table 4
Comparison of Proteome-Wide Interaction Map Connectivities for Different Organisms Found in the Literature[a]

Refs.[b]	Organism	Method	Proteomic coverage	Average connectivity
1	*S. cerevisiae*	Experiment	0.55	1.388
2	*S. cerevisiae*	Experiment	0.26	1.523
3	*E. coli*	Prediction	0.10	2.14
4	*C. jejuni*	Prediction	1.00	**3.33**
5	*H. pylori*	Experiment	0.47	3.36
6, 7	*S. cerevisiae*	Experiment	0.17	3.2, 4.5–5.8
8	*C. elegans*	Experiment	NR	5.4

[a]Proteome coverage is the estimated number of distinct proteins involved in interactions as a fraction of either the total proteomic complement or assay depth for a given organism. Average connectivity refers to the average number of interaction partners per protein comprising the map.

[b]References: 1 *(23)*; 2 *(72)*; 3 *(73)*; 4 (present investigation); 5 *(37)*; 6 *(74)*; 7 *(75)*; 8 *(76)*. Note: in **ref. 75**, a retrospective reanalysis of data originally reported in **ref. 74** resulted in an updated estimated average connectivity of 4.5–5.8 for *S. cerevisiae*.

NR, not reported.

which all possible unique pairwise protein–protein interactions (1,300,078 pairs) were encoded as features and added to the sample X_b for interaction mining. Using one of the 10 classifiers $h(\alpha, x)$ developed during cross-validation analysis on the design organism, an interaction hypothesis was generated for each data point in this sample. A total of 5367 distinct protein–protein interactions were declared by the decision function. Each protein comprising the *C. jejuni* interaction map was predicted to have, on average, biological connections with 3.33 other proteins.

4.3.2. Scaling Properties of Predicted Interaction Map

Networks of interactions in a number of natural and man-made systems display conserved motifs of substructural connections, suggesting universal design patterns that correlate with successful information processing or evolutionary fitness *(44)*. We observed here that the inferred *C. jejuni* protein–protein interaction map shares a key topological scaling property in common with previous proteome-wide investigations: the average connectivity of the interaction network.

Table 4 lists data collected from several different proteome-scale investigations on different organisms. It can be seen that on average, 3.33 proteins are linked to each protein in the *C. jejuni* interaction map. This level of connectivity compares favorably with the other investigations cited in the table, especially to the experimental data from **ref. 37**, which provided the design sample for training the learning system in the present investigation.

4.3.3. Map Visualization

We present a visualization of a complete hypothesized protein interaction map for *C. jejuni* in **Fig. 2**. In the figure, individual proteins are represented as vertices, and the interactions between pairs of proteins are indicated by edges connecting nodes. Proteins

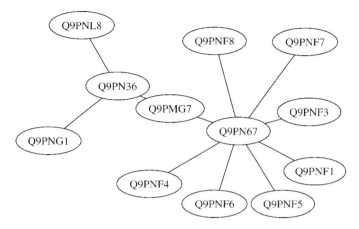

Fig. 3. Principal components of an hypothesized two-component thermoregulation signaling pathway in *C. jejuni*. Shown is a subnetwork of interactions comprising the primary interaction partners of the sensor (Q9PN36) and regulator (Q9PN67) proteins. Each protein node is labeled by its corresponding ORF designation. The previously uncharacterized protein Q9PMG7 may play a role in transferral of the message from sensor to regulator in the thermoregulation signaling pathway.

with a large number of partners (>15; 1% of all predictions) are colored black, whereas light gray nodes signify that relatively few proteins (≤5; 61% of predictions) are expected to interact with that node. Dark gray nodes represent proteins with 6–14 interaction partners.

4.3.4. Selected Biological Examples

In this section, we present specific biological examples of protein–protein interactions predicted for *C. jejuni*, exemplifying the type of information that may be extracted from the application of this approach. This represents only a sampling of the subnetworks automatically generated by the interaction mining procedure.

4.3.4.1. THERMOREGULATION

Two-component signal transduction systems are essential in the regulation of many bacterial functions, including chemotaxis, metabolism, and the response to environmental stress. The two-component mechanism constitutes a membrane environmental sensor and a cytoplasmic regulator. This mechanism typically involves autophosphorylation of histidine residues on the sensor protein, which then acts as a kinase for the regulator, the phosphorylation of which induces transcriptional activation appropriate to the chemical or thermal stimulus *(45)*.

Elements of a hypothesized two-component thermoregulation signaling pathway in *C. jejuni* are presented in **Fig. 3** and **Table 5**. The figure displays only a subnetwork of interactions comprising the primary interaction partners of the sensor and regulator proteins. Each protein node is labeled by its corresponding ORF designation. The two-component sensor (Q9PN36) is functionally linked to the putative heat-shock regulator (Q9PN67) via an intermediary protein Q9PMG7. Heat-shock proteins are known to solubilize misfolded or denatured proteins in case of extreme thermal insult to the cell *(46)*.

Table 5
Principal Components of a Hypothesized Two-Component Thermoregulation
Signaling Pathway in *C. jejuni*[a]

ORF	Status	Annotation	Partners
Q9PN36	A	Two-component sensor	Q9PNL8,Q9PNG1,Q9PMG7
Q9PN67	P	Heat shock regulator	Q9PMG7,Q9PNF8,Q9PNF7, Q9PNF3,Q9PNF1,Q9PNF5, Q9PNF6,Q9PNF4
Q9PMG7	H	Protein Cj1495c	Q9PN36,Q9PN67

[a]Status refers to the functional annotation status of the ORF, with H=hypothetical, P=putative, and A=annotated.

The intermediate protein Q9PMG7 is designated as "hypothetical," meaning it has sequential similarity to other proteins of unknown function. This 180-residue protein contains two possible sites for phosphorylation (casein kinase II and tyrosine) as detected by PROSITE search *(47)*. It is a feasible hypothesis that this previously uncharacterized protein may play a role in transferral of the message from sensor to regulator in the *C. jejuni* thermoregulation signaling pathway.

If elements of this inferred pathway are validated in wet-biological studies, we suggest the possibility of its manipulation or obstruction using antibiotic agents. As recently noted, targeted inhibition of histidine kinase signal transduction pathways in bacteria may have beneficial effects for host mammals, in which cellular signal transduction proceeds according to a different mechanism *(48)*.

4.3.4.2. FERRIC UPTAKE AND REGULATION

The storage and regulation of iron levels is a fundamental aspect of cellular survival for Gram-negative bacteria. Iron is a nonabundant essential nutrient that is toxic in excessive concentrations, necessitating its regulation within the cell. In *C. jejuni*, ferritins (iron-storage proteins) are also involved in oxidative stress resistance *(49)*.

A subnetwork of putative protein interactions integral to ferric uptake and regulation processes is shown in **Fig. 4**. This interaction group comprises proteins linking the extracellular signal (Q9PJA5, putative integral membrane protein) to the regulatory (P48796, ferric uptake regulation) and transcriptional machinery (Q9PNK3, leucyl-tRNA transferase; Q9PN44, polyribonucleotide nucleotidyltransferase) within the cell. Such a connection is required to respond to dynamically changing requirements for iron storage or removal. Q9PNK3 is predicted to interact with Q9PMS3, a putative ferredoxin that may play a role in the intracellular redox system.

Another key protein in this figure is Q9PMD5 (possible bacterioferritin) that may be instrumental in redox stress resistance, by storing iron in a soluble and nontoxic form. Q9PMD5 is linked to a 30S ribosomal protein (Q9PI17), which may suggest that this system is also involved in protection of the ribosomal machinery from iron toxicity. It is of interest to note that the hypothetical protein Q9PMG7 appears again in this inferred scenario of iron regulation. Although a functional role has not been assigned for this protein, it is possible that it participates in many pathways within the cell. In **ref. *50***, it

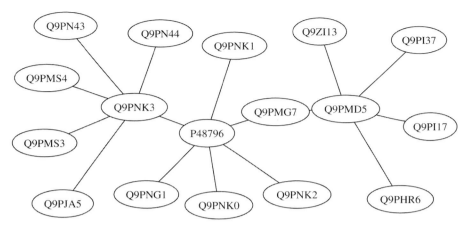

Fig. 4. Principal components of an hypothesized ferric uptake regulation pathway in *C. jejuni*. Each protein node is labeled by its corresponding ORF designation. The figure shows a subnetwork of predicted protein interactions linking the extracellular signal (Q9PJA5, putative integral membrane protein) to the regulatory (P48796, ferric uptake regulation) and transcriptional machinery (Q9PNK3, leucyl-tRNA transferase; Q9PN44, polyribonucleotide nucleotidyltransferase). Such connection is required to respond to changing requirements for iron storage or removal. Protein Q9PMD5 (possible bacterioferritin) may participate in redox stress resistance, by storing iron in a soluble, nontoxic form. Q9PMD5 is linked to a 30S ribosomal protein (Q9PI17), suggesting that this system may be involved in protection of the ribosomal machinery from iron toxicity.

was argued that the most highly connected proteins in protein interaction networks are most crucial to a cell's viability. Perhaps this protein carries such significance within *C. jejuni*. This question awaits further proteomic study and validation.

The protein components central to the hypothesized ferric uptake interaction cluster are summarized in **Table 6**.

5. Protein–Small Molecule Interactions

Genetics, genomics, and proteomics technologies have generated huge numbers of drug targets. Accordingly, lead identification and optimization steps have assumed critical importance. High-throughput experimental screening assays (*51*) have been complemented recently by computational ("virtual screening") approaches to identify and filter potential ligands, given the characteristics of the target receptor structure of interest (*52,53*). In virtual screening, databases of compound libraries are searched, and scoring or discrimination functions are used to select the "best" candidate compounds for biological activity analysis (*54*).

Scoring ligands in virtual screening is often associated with computational docking simulations that mate receptor and cognate small-molecule ligand in three-dimensional space. To provide broad generalization in "chemical diversity" space, computing this score requires the accurate prediction of binding affinities of many structurally distinct ligands (*55*).

Regression-based scoring functions, as exemplified by the work of Böhm (*56*), are fast but require a three-dimensional structure of the receptor. This prohibits their use

Table 6
Principal Components of a Hypothesized Ferric Uptake Regulation
Pathway in *C. jejuni*[a]

ORF	Status	Annotation	Partners
P48796	A	Ferric uptake regulation protein	Q9PNK3,Q9PNK2,Q9PNK1, Q9PNG1,Q9PMG7
Q9PNK3	A	Leucyl-tRNA synthetase	Q9PMS3,Q9PN43,Q9PMS4, Q9PN44,Q9PJA5
Q9PMD5	A	Possible bacterioferritin	Q9PI17,Q9PHR6,Q0ZI13, Q9PI37,Q9PMG7

[a]Status refers to the functional annotation status of the ORF, with H = hypothetical, P = putative, and A = annotated.

in cases in which the structure is difficult to obtain, such as with transmembrane proteins. The accuracy of such methods has also been called into question. A recent investigation concluded that "no significant correlation" existed between Böhm-type scores and experimentally determined binding affinities for a group of 15 complexes (*57*).

This section discusses our research on a new method to estimate the binding free energy between a ligand and receptor (*58*). Using support vector regression (SVR) (*59*), we trained a system to learn the functional mapping between a set of ligand-receptor features and the total free binding energy of the complex, a continuously valued function. As in the preceding sections of this chapter, we use simple descriptors to represent proteins and ligands; no three-dimensional information about either is required. This method provides the capability for large-scale "virtual screening" of receptors against a library of ligands.

5.1. Experimental Methods

The ligand-receptor dataset used in this investigation was aggregated automatically using information located in a number of disparate online resources, coupled with local computations. An object database was constructed from these data and subsequently sampled to generate examples for training and testing the performance of the regression estimation system. The experimental database consisted of 2956 objects, each having attributes as summarized in this section.

5.1.1. Ligand-Receptor Complex

Ligand-receptor data were extracted from the Computed Ligand Binding Energy (CLiBE)[5] database, a compendium of information on complexed receptors and ligands. Each record in CLiBE contains computed values for the total ligand-receptor potential energy field ΔG^0, given by

$$\Delta G^0 = \Delta G_v + \Delta G_h + \Delta G_e + \Delta G_s \tag{6}$$

[5]CLiBE circa August 2002 had 14,731 records, with 2803 distinct ligands and 2256 distinct receptors. *See* http://xin.cz3.nus.edu.sg/group/clibe/clibe.asp.

where the right-hand-side partitioning represents energy contributions owing to non-bonded van der Waals interactions, hydrogen bonds, electrostatic forces, and ligand desolvation energies, respectively *(60)*. Methods underlying the computation of binding energies comprising the database subject to this investigation are described in **ref. *61***.

The complexes within this resource are themselves based on "heterogen" records found in the Protein Data Bank (PDB) *(62)*[6] for which a chemical identity has been assigned to the ligand. PDB is a public domain repository of experimentally determined structures of biological macromolecules.

5.1.2. Ligand Structures and Chemical Names

Data files with entries representing ligand structures and their associated chemical names were obtained from the National Cancer Institute (NCI) Open Database of Compounds.[7] The data entries were represented as SMILES strings, where SMILES (Simplified Molecular Input Line Entry System) is a specification and nomenclature for describing molecules as a compact, one-dimensional string of characters, including atoms, bonds, aromatic rings, and branches *(63)*.

5.1.3. Molecular Connectivity

The SMILES representation for each ligand molecule was converted to a two-dimensional connectivity matrix using a computational chemistry package [JOElib *(64)*].[8] The rows and columns of this matrix reflect the cardinality of constituent atoms established by the SMILES representation. At row i and column j, a unit-valued entry is made if the corresponding atoms in the molecule are covalently connected; otherwise the value of that matrix element is zero. Diagonal elements of this matrix store the appropriate atomic number, as suggested previously *(65)*.

5.1.4. Molecular Synonyms

To maximize the chemical diversity of objects potentially available for numerical experiments, a list of common chemical synonyms corresponding to each ligand were obtained using the on-line ChemFinder service.[9] Each ligand synonym within its list was used in a lexical similarity search of the NCI compound files to obtain SMILES representations when different chemical names were used for identical ligands across databases.

5.2. Feature Representation

Each ligand-receptor complex was transformed into a vector of numerical features presumed to be salient for learning the target concept. Receptor and ligand feature vectors constructed as outlined in this section are concatenated and labeled with the value of their total free binding energy. These vectors are subjected to SVM regression training and cross-validation testing to evaluate how keenly the system learned the concept as posed.

---

[6]PDB contains 18,294 structures as of July 23, 2002. *See* http://www.rcsb.org/pdb/.

[7]Available at http://cactvs.cit.nih.gov/ncidb2/download.html, this resource currently contains over 250,000 compounds.

[8]Open source, available at http://joelib.sourceforge.net/.

[9]*See* http://chemfinder.cambridgesoft.com/result.asp.

5.2.1. Receptor

Receptor protein features were generated as described previously *(6)*, considering tabulated physicochemical properties (charge, hydrophobicity, and surface tension) of the amino acid sequence to be prototypical of binding characteristics of the receptor. Each residue in sequence was replaced by floating point numbers with values corresponding to these physical properties. This vector of numbers was then mapped onto a fixed-length interval, to provide a basis for comparison between receptor proteins of varying sequence length.

5.2.2. Ligand

Exemplars for the ligand component of each molecular complex required a novel approach. The design ethos followed here dictates beginning with a minimal, elemental group of features, in order to develop intuition regarding the feature space.

In accordance with this approach, the two-dimensional molecular connection matrix was supplemented by additional arrays, each of which contained numerical values for fundamental, measurable chemical properties characterizing the atoms comprising the molecule. These properties included the atomic *ionization potential energy*, which represents the energy necessary to remove the outermost electron from the ground state of a neutral atom, and the *electron affinity*, which is a measure of energy change upon adding an electron to a neutral atom *(66)*. Ionization energies are always positively valued, whereas electron affinities may assume either positive or negative numerical values.

For each small-molecule ligand, three two-dimensional arrays representing molecular topology, electronic structure, and chemical behavior of the component elements, were concatenated into a single, wide matrix. The resulting aggregate data matrix was then factorized using the singular value decomposition *(67)*. The singular values computed in this factorization are extracted, representing a projection onto one-dimensional space of the essential characteristics of molecular bond topology and, it is hypothesized, the spatial distribution of molecular properties important for binding with a receptor.

Burden *(65)* introduced the idea of computing the eigenvalues of a hydrogen-suppressed molecular bond graph with atomic number on the diagonal and numbers indicating bond presence and type at off-diagonal positions. This matrix was used as a means to group substructures for chemical similarity search. In that work, it was maintained that the smallest eigenvalue embodied information on *all* molecules, and therefore was sufficient as a topological descriptor. Here, all singular values are retained, regardless of their relative magnitudes, as discarding the entire set is not justifiable. This vector is finally stretched (or compressed) onto a fixed length interval, as was performed for the receptor features.

5.2.3. Learning Concept

The concept to be learned is the function $y = f(x)$ that maps ligand–protein feature vectors x to the corresponding free energy of binding y. How well the SVR machine learns this concept will be quantified using the statistics described below, collected from the averaged results of 10 different 10-fold cross-validation experiments, representing 100 different training/testing procedures.

5.3. Prediction Metrics

One measure of effectiveness for regression estimation is the normalized mean squared error, given by

$$nmse = \frac{1}{\sigma^2} \frac{1}{N} \sum_{k=1}^{N} (y_k - \hat{y}_k)^2 \tag{7}$$

where N is the number of target points predicted, σ^2 is the actual sample variance, and y_k and $\hat{y}_k$ are the actual and estimated target values of the k-th data point, respectively *(68)*. Because *nmse* is normalized by the sample variance, it may be used to compare different regression studies on a more equitable basis than would be possible using the conventional rms error; intuitively, a given prediction experiment is less challenging when the variance in the data is small. Note that if we replace the prediction terms $\hat{y}_k$ with the arithmetic mean $\hat{y}$ in **Eq. 7**, the value of the statistic is 1. This trivial case results when the predictor simply outputs the mean value of the data. Low values of *nmse* indicate good overall predictive acuity.

Pointwise predictions of ligand binding may be evaluated using the normalized mean absolute error, defined by

$$nmae = \frac{1}{\sigma} \frac{1}{N} \sum_{k=1}^{N} |y_k - \hat{y}_k| \tag{8}$$

This statistic is normalized by the sample variance for the same reasons as were cited for *nmse* above. Furthermore, its value may be interpreted as the number of standard deviations, on average, that predictions differ from the target values across the test set. The lower the value of *nmae*, the better the system pointwise predictive ability.

In some ligand screening situations [such as "virtual screening" *(52)*], predicting the relative ranking of binding strengths among a set of ligand-receptor pairs may be desired. The output of such an analysis would be a list of predicted binding energies, sorted according to predicted magnitudes $\Delta\hat{G}^0$. In such cases a measurement of non-parametric or rank correlation, such as represented by Kendall's τ coefficient *(69)*, is informative. In cross-validation, given an ordered array of N "(actual, predicted)" values $(y_1, \hat{y}_1),...,(y_N, \hat{y}_N)$, we systematically compare the numerical signs of individual bivariate pairs $X = (y_i, \hat{y}_i)$ and $Y = (y_j, \hat{y}_j)$ for $i = 1,..., N, j = (i+1),..., N$.

If either (a) $y_i > y_j$ and $\hat{y}_i > \hat{y}_j$, or (b) $y_i < y_j$ and $\hat{y}_i < \hat{y}_j$ is observed, X and Y are said to be "concordant." Otherwise, the points are "discordant." Kendall's τ expresses the tendency of two ordered lists y and $\hat{y}$ to increase or decrease coordinately and is computed as

$$\tau = \frac{N_C - N_D}{\sqrt{N_C + N_D + T_X} \ \sqrt{N_C + N_D + T_Y}} , \ -1 \le \tau \le +1. \tag{9}$$

where N_C is the total number of concordant pairs, N_D is the number of discordant pairs, and T_X, T_Y are counts of the "ties" found in X and Y pairs, respectively. A large positive (or negative) value of τ indicates that the rank ordered values y and $\hat{y}$ are positively (or negatively) correlated.

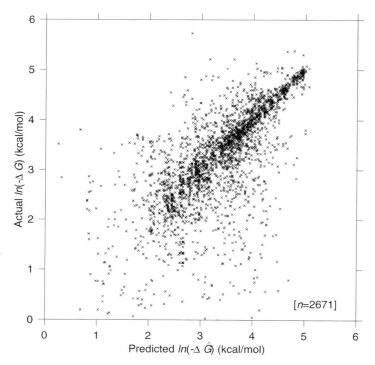

Fig. 5. Actual vs predicted binding free energy. Shown are typical results from one complete 10-fold cross-validation experiment on the ligand-receptor database. Sample size $n = 2671$.

5.4. Results and Discussion

The total sample used in these experiments comprised 2671 distinct ligand–receptor complexes. The output of the trained system is a predicted level of binding free energy y (in kcal/mol) given a set of features abstracted from a given input complex x. A qualitative glimpse of typical results from one complete 10-fold cross-validation test is offered in **Fig. 5**, which shows a scatter plot of actual versus predicted binding energy. The figure shows that some degree of correlation between prediction and truth exists.

The principal results obtained in this investigation are summarized in **Table 7** and **Fig. 5**. The table compares the 10 10-fold cross-validation error estimates with a number of studies reported in the literature. In contrast to the present results (shown in boldface), all the competing methodologies shown in the table are derived from scoring functions or simulations predicated on knowledge of the three-dimensional structure of receptor and ligand complex. The columns in **Table 7** comprise test sample size n; the mean target binding energy $\hat{y}$ and standard deviation (σ_y), in kcal/mol; normalized mean square error (*nmse*, **Eq. 7**); normalized mean absolute error (*nmae*, **Eq. 8**); and Kendall's tau (τ, **Eq. 9**).

The records in the table are listed in order of increasing *nmse*. This statistic is proposed as the primary objective indicator of accuracy for direct prediction of binding free energy.

Of particular note on consideration of **Table 7** are the sample size and mean free binding energies characterizing the ligand-receptor data used here, when contrasted to the

Table 7

Comparison of Predictions of Ligand-Receptor Binding Free Energies in the Present Investigation (Boldface Font) and Various Studies Reported in the Literature[a]

Ref.[b]	n	$\hat{y}$ (kcal/mol)	$\sigma\hat{y}$ (kcal/mol)	*nmse*	*nmae*	τ
1	14	−4.09	1.179	0.198	0.344	0.753
2	12	−0.98	0.332	0.271	0.401	0.667
3	11	−4.25	0.711	0.377	0.466	0.455
4	**2671**	**−37.76**	**35.106**	**0.419**	**0.377**	**0.552**
5	13	−3.93	0.796	0.440	0.497	0.632
6	30	−8.897	2.591	0.720	0.661	0.418
7	17	−8.17	3.785	0.789	0.621	0.358
8	63	−1.45	0.560	1.342	0.836	0.307
9	13	−10.27	6.683	1.466	0.511	0.533

[a]Test data statistics are sample size *(n)*, target value mean *(ŷ)*, and standard deviation $(\sigma\hat{y})$. Results are shown for normalized mean square error (*nmse*, **Eq. 7**), normalized mean absolute error (*nmae*, **Eq. 8**), and Kendall's tau (τ, **Eq. 9**).

[b]References: 1, Table 3 *(70)*; 2, Table 3 *(56)*; 3, Table 4 *(71)*; 4, (present investigation); 5, Table 4 *(70)*; 6, Table 4 *(77)*; 7, Table 1 *(78)*; 8, Table 1 *(79)*; 9, Table 5 *(80)*. *Note:* results for the present investigation are average values from 10 10-fold cross-validation experiments.

other investigations. The current sample size ($n = 2671$) is a factor of 42 times larger than the next largest dataset. The mean free binding energy is seen to be −38 kcal/mol, significantly stronger than the other data summarized in the table. Moreover, it can be seen that the present dataset is highly variable, as the standard deviation (35 kcal/mol) is on the same order as the mean.

Recall from the previous discussion that *nmse* values on the order of 1 are tantamount to trivial prediction of the mean value of a test dataset. Lower values of *nmse* are associated with genuine learning of underlying patterns in the data and with effective generalization. On this basis, the highest predictive accuracy (#1; *nmse* = 0.198) observed in this comparative study was realized by Head and co-workers *(70)*, who present a hybrid approach combining ligand-receptor three-dimensional structural information and parameters derived from molecular mechanics. The test set comprised 14 ligand-receptor complexes.

The second best *nmse* in this group was achieved by Böhm *(56)* using a regression-based empirical scoring function based on hydrogen bonds, electrostatics, complementary surface areas, and other characteristics of receptor-ligand pairs for which the three-dimensional structure has been previously determined.

Next on our list of prediction results is the investigation reported in Wang et al. *(71)*. Their approach uses another empirical scoring function for binding free energy that explicitly accounts for contributions owing to Van der Waals interactions, metal-ligand bonding, hydrogen bonds, desolvation energies, and different kinematic effects. A regression equation is developed using these terms derived from known receptor-ligand complexes. All 11 data points in the test sample were based on endothiapepsin receptor complexes.

The current method, based on support vector regression, obtained the fourth-best prediction error (*nmse* = 0.419) averaged over 10 different 10-fold cross-validation tests. We suggest that this error rate represents a significant step, for the following reasons:

1. The error rate and rank correlation value are surprisingly competitive with other investigations, in light of the relatively large variance and extremely large sample size of the underlying dataset. Note that the fifth lowest *nmse* value in **Table 7** was also obtained by Head et al. *(70)*, for a different dataset than they used in entry 1. Group 5 comprised 13 HIV-1 protease/HIV protease inhibitor complexes and showed a value of *nmse* = 0.440. So the same methodology by the same research group, applied on a different dataset, realized quite different predictive results. This demonstrates the variability in results that are possible when using small sample sizes, while providing confidence in the robustness of our current method and results, which were based on a sample size $n = 2671$.
2. The features used to represent the ligand–protein complexes in the support vector regression do not require any information about three-dimensional structure. All that is required as input data are the amino acid sequence of the receptor, a connection table representing the ligand structure in two dimensions, and the atom characteristics at the nodes of this connection table.
3. There is no limitation on the protein family membership of the putative receptor(s) or on the type (organic, synthetic) or size of ligand used.
4. The results obtained in this study suggest that it may be possible to infer binding energies for complexes involving newly sequenced or difficult-to-crystallize proteins, or for ligands that only exist in computer memory, awaiting synthesis upon successful *in silico* screening.

5.4.1. Rank Correlation

We draw the reader's attention to the trend in Kendall's rank correlation statistic τ in **Table 7**. It is apparent that there is a general inverse correlation between the magnitude of binding energy prediction errors (*nmse*, *nmae*) and the value of τ. That is, low values of prediction error are associated with high values of the correlation coefficient. τ measures the tendency of two ordinal random variables (here, actual and predicted binding energy rank) to increase or decrease coordinately. If direct prediction of the physical binding energy is reasonably accurate, we would expect to see a positive and nontrivial correlation between the corresponding rank-ordered variables.

The ability to rank a set of ligand-receptor complexes reliably during lead optimization (vs directly computing binding energy) remains important in the area of drug discovery. Such a procedure may add value, for example, as a decision aid when downselecting a set of ligands for chemical synthesis. In connection with the current methodology, we recognize that training the SVR requires example data representing estimated or measured values of binding free energy. The output of a computational technique cannot exceed the accuracy of its input; this is especially true with systems that learn from examples. Therefore, at present the qualitative analysis or ranking of potential ligands may be the main utility of the SVR technique.

6. Conclusions

The investigations and results presented in this chapter have demonstrated that explicit information about three-dimensional biomolecular structure is not necessary to make predictions of protein–protein and protein–ligand interactions. Using simple descriptors of physicochemical characteristics of amino acid sequences (for proteins) and molecular connection tables (for small-molecule ligands), the machine learning techniques developed here have been shown to predict these interactions successfully at rates greatly exceeding chance.

Even though machine learning strives to generate hypothesized protein interactions automatically within a proteome, it is important to realize that techniques such as this complement—but cannot supplant—experiments. Machine learning approaches generate empirically falsifiable hypotheses to be subsequently supported (or not supported) experimentally. An iterative coupling between successive rounds of computer prediction and experimental validation must be accomplished. Only in so doing can the regions of applicability and limitations of the present approach be discovered. We anticipate that further development along these lines may produce a robust computational screening technique that may be useful to reduce the set of putative candidate protein interactions in an organism, tissue, or physiological state of interest.

References

1. Schmucker, D., Clemens, J. C., Shu, H., et al. (2000) *Drosophila* DSCAM is an axon guidance receptor exhibiting extraordinary molecular diversity. *Cell* **101,** 671–684.
2. Fung, Y. C. (1993) *Biomechanics: Mechanical Properties of Living Tissues*, 2nd ed. Springer-Verlag, New York.
3. Spellman, P. T. and Rubin, G. M. (2002) Evidence for large domains of similarly expressed genes in the *Drosophila* genome. *J. Biol.* **1,** 5.1–5.8.
4. Boser, B. E., Guyon, I. M., and Vapnik, V. N. (1992) A training algorithm for optimal margin classifiers, in *Proceedings of the Fifth Annual ACM Workshop on Computational Learning Theory* (Haussler, D., ed.), ACM Press, Pittsburgh, PA, pp. 144–152.
5. Vapnik, V. N. (1995) *The Nature of Statistical Learning Theory.* Springer-Verlag, Heidelberg, Germany.
6. Bock, J. R. and Gough, D. A. (2001) Predicting protein-protein interactions from primary structure. *Bioinformatics* **17,** 455–460.
7. Xenarios, I., Rice, D. W., Salwinski, L., Baron, M. K., Marcotte, E. M., and Eisenberg, D. (2000) DIP: The database of interacting proteins. *Nucleic Acids Res.* **28,** 289–291.
8. Kandel, D., Mathias, Y., Unger, R., and Winkler, P. (1996) Shuffling biological sequences. *Discrete Appl. Math.* **71,** 171–185.
9. Eisenberg, D. (1984) Three-dimensional structure of membrane and surface proteins. *Ann. Rev. Biochem.* **53,** 595–623.
10. Bull, H. B. and Breese, K. (1974) Surface tension of amino acid solutions: a hydrophobicity scale of the amino acid residues. *Arch. Biochem. Biophys.* **161,** 665–670.
11. Provost, F., Fawcett, T., and Kohavi, R. (1998) The case against accuracy estimation for comparing induction algorithms, in *Proceedings of the Fifteenth International Conference on Machine Learning (IMLC-98)*, Morgan Kaufmann, San Francisco, CA, pp. 445–453.
12. Weiss, G. M. and Provost, F. (2001) The effect of class distribution on classifier learning: an empirical study. Technical Report ML-TR-44, Department of Computer Science, Rutgers University.
13. Swingler, K. (1996) *Applying Neural Networks: A Practical Guide.* Academic, London, UK.
14. Kwok, J. T. (1999) Moderating the outputs of support vector machine classifiers. *IEEE Trans. Neural Net.* **10,** 1018–1031.
15. Platt, J. C. (1999) Fast training of support vector machines using sequential minimal optimization, in *Advances in Kernel Methods: Support Vector Learning*, MIT Press, Cambridge, MA, pp. 185–208.
16. Witten, I. H. and Frank, E. (1999) *Data Mining: Practical Machine Learning Tools and Techniques with Java Implementations.* Morgan Kaufmann, San Francisco, CA.

17. Elkan, C. (2001) The foundations of cost-sensitive learning, in *Proceedings of the Seventeenth International Joint Conference on Artificial Intelligence (IJCAI)*, Seattle, WA, pp. 973–978.

18. Bock, J. R. and Gough, D. A. (2003) Machine learning inference of protein-protein binding in *Saccharomyces cerevisiae*, in review.

19. Goffeau, A., Barrell, B. G., Bussey, H., et al. (1996) Life with 6000 genes. *Science* **274,** 563–567.

20. Chervitz, S. A., Aravind, L., Sherlock, G., Ball, C. A., Koonin, E. V., and Dwight, S. S. (1998) Comparison of the complete protein sets of worm and yeast: orthology and divergence. *Science* **282,** 2022–2028.

21. Mumberg, D., Muller, R., and Funk, M. (1995) Yeast vectors for the controlled expression of heterologous proteins in different genetic backgrounds. *Gene* **156,** 119–122.

22. Munder, T. and Hinnen, A. (1999) Yeast cells as tools for target-oriented screening. *Appl. Microbiol. Biotechnol.* **52,** 311–320.

23. Ito, T., Chiba, T., Ozawa, R., Yoshida, M., Hattori, M., and Sakaki, Y. (2001) A comprehensive two-hydrid analysis to explore the yeast protein interactome. *Proc. Natl. Acad. Sci. USA* **98,** 4569–4574.

24. Bartel, P., Chien, C. T., Sternglanz, R., and Fields, S. (1993) Elimination of false positives that arise in using the two-hybrid system. *Biotechniques* **14,** 920–924.

25. Smith, T. F. and Waterman, W. S. (1981) Identification of common molecular subsequences. *J. Mol. Biol.* **147,** 195–197.

26. Altschul, S. F. and Gish, W. (1996) Local alignment statistics. *Methods Enzymol.* **266,** 460–480.

27. Henikoff, S. and Henikoff, J. G. (1992) Amino acid substitution matrices from protein blocks. *Proc. Natl. Acad. Sci. USA* **89,** 10,915–10,519.

28. Kohavi, R. and Provost, F. (1998) Glossary of terms. *Machine Learning* **30,** 271–274.

29. Peterson, W. W. and Birdsall, T. G. (1953) The theory of signal detectability. Technical Report TR-13, Communications and Signal Processing Laboratory, University of Michigan, Ann Arbor, MI.

30. Stone, M. (1974) Cross-validatory choices and assessment of statistical predictions. *J. Roy. Stat. Soc.* **36,** 111–147.

31. Skolnik, M. I. (1980) *Introduction to Radar Systems*, 2nd ed. McGraw-Hill, New York.

32. Urick, R. J. (1983) *Principles of Underwater Sound*, 3rd ed. McGraw-Hill, New York.

33. Druker, B. J., Talpaz, M. T., Resta, D. J., et al. (2001) Efficacy and safety of a specific inhibitor of the BCR-ABL tyrosine kinase in chronic myeloid leukemia and acute lymphoblastic leukemia. *N. Engl. J. Med.* **344,** 1031–1037.

34. Black, D. L. (2000) Protein diversity from alternative splicing: a challenge for bioinformatics and post-genome biology. *Cell* **103,** 367–370.

35. Bock, J. R. and Gough, D. A. (2003) Whole-proteome interaction mining. *Bioinformatics* **19,** 125–135.

36. Bradley, P. S., Fayyad, U. M., and Mangasarian, O. L. (1998) Mathematical programming for data mining: formulations and challenges. Technical Report MSR-98-01, University of Wisconsin Data Mining Institute, Madison, WI.

37. Rain, J. C., Selig, L., De Reuse, H., et al. (2001) The protein-protein interaction map of *Helicobacter pylori*. *Nature* **409,** 211–215.

38. Burges, C. (1998) A tutorial on support vector machines for pattern recognition. *Data Mining Knowledge Discovery* **2,** 121–167.

39. Sankoff, D., Leduc, G., Paquin, B., Lang, B. F., and Cedergren, R. (1992) Gene order comparisons of phylogenetic inference: evolution of the mitochondrial genome. *Proc. Natl. Acad. Sci. USA* **89,** 6575–6579.

40. Tekaia, F., Lazcano, A., and Dujon, B. (1999) The genomic tree as revealed from whole proteome comparisons. *Genome Res.* **9,** 550–557.
41. Brown, J. R., Douady, C. J., Italia, M. J., Marshall, W. E., and Stanhope, M. H. (2001) Universal trees based on large combined protein sequence data sets. *Nat. Genet.* **28,** 281–285.
42. Efron, B. and Gong, G. (1983) A leisurely look at the bootstrap, the jackknife, and cross-validation. *Am. Stat.* **37,** 36–48.
43. Eisen, J. A. (2000) Assessing evolutionary relationships among microbes from whole-genome analysis. *Curr. Opin. Microbiol.* **3,** 475–480.
44. Milo, R., Shen-Orr, S., Itzkovitz, S., Kashtan, N., Chklovskii, D., and Alon, U. (2002) Network motifs: simple building blocks of complex networks. *Science* **298,** 824–827.
45. Klumpp, S. and Krieglstein, J. (2002) Phosphorylation and dephosphorylation of histidine residues in proteins. *Eur. J. Biochem.* **269,** 1067–1071.
46. Alberts, B., Bray, D., Lewis, J., Raff, M., Roberts, K., and Watson, J. D. (1989) *Molecular Biology of the Cell*, 2nd ed. New York.
47. Bairoch, A., Bucher, P., and Hofmann, K. (1997) The PROSITE database, its status in 1997. *Nucleic Acids Res.* **25,** 217–221.
48. Matsushita, M. and Janda, K. D. (2002) Histidine kinases as targets for new antimicrobial agents. *Bioorg. Med. Chem.* **10,** 855–867.
49. Andrews, S. C. (1998) Iron storage in bacteria. *Adv. Microb. Physiol.* **40,** 281–351.
50. Jeong, H., Mason, S. P., Barabási, A.-L., and Oltvai, Z. N. (2001) Lethality and centrality in protein networks. *Nature* **411,** 41–42.
51. Cunningham, M. J. (2000) Genomics and proteomics: the new millennium of drug discovery and development. *J. Pharmacol. Toxicol. Methods* **44,** 291–300.
52. Bissantz, C., Folkers, G., and Rognan, D. (2000) Protein-based virtual screening of chemical databases. 1. Evaluation of different docking/scoring combinations. *J. Med. Chem.* **43,** 4759–4767.
53. Waszkowycz, B. (2002) Structure-based approaches to drug design and virtual screening. *Curr. Opin. Drug Discovery Dev.* **5,** 407–413.
54. Langer, T. and Hoffmann, R. D. (2001) Virtual screening: an effective tool for lead structure discovery? *Curr. Pharma. Design* **7,** 509–527.
55. Gohlke, H. and Klebe, G. (2001) Statistical potentials and scoring functions applied to protein-ligand binding. *Curr. Opin. Struct. Biol.* **11,** 231–235.
56. Böhm, H. J. (1998) Prediction of binding constants of protein ligands: a fast method for the prioritization of hits obtained from *de novo* design or 3D database search programs. *J. Comput. Aided Mol. Design* **12,** 309–323.
57. Moret, E. E., van Wijk, M. C., Kostense, A. S., and Gillies, M. B. (1999) Scoring peptide(mimetic)-protein interactions. *Med. Chem. Res.* **9,** 604–620.
58. Bock, J. R. and Gough, D. A. (2002) A new method to estimate ligand-receptor energetics. *Mol. Cell. Proteomics* **1,** 904–910.
59. Smola, A. J. and Schölkopf, B. (1998) A tutorial on support vector regression. Technical Report NC-TR-98-030, Royal Holloway College, University of London, London.
60. Ortiz, A. R., Pisabarro, M. T., Gago, F., and Wade, R. C. (1995) Prediction of drug binding affinities by comparative binding energy analysis. *J. Med. Chem.* **38,** 2681–2691.
61. Chen, Y. Z. and Zhi, D. G. (2001) Ligand-protein inverse docking and its potential use in the computer search of protein targets of a small molecule. *Proteins* **43,** 217–226.
62. Berman, H. M., Westbrook, J., Feng, Z., et al. (2000) The Protein Data Bank. *Nucleic Acids Res.* **28,** 235–242.
63. Weininger, D. (1988) SMILES, a chemical language and information system. 1. Introduction to methodology and encoding rules. *J. Chem. Inform. Comput. Sci.* **28,** 31–36.

64. Wegner, J. and Zell, A. (2002) JOELib: a Java based computational chemistry package, in *6th Darmstädter Molecular-Modelling Workshop*, Technische Universität, Darmstadt, Germany.

65. Burden, F. R. (1989) Molecular identification number for substructure searches. *J. Chem. Inform. Comput. Sci.* **29,** 225–227.

66. Boikess, R. S. and Edelson, E. (1981) *Chemical Principles*, 2nd ed. Harper & Row, New York.

67. Golub, G. H. and van Loan, C. F. (1989) *Matrix Computations*, 2nd ed. Johns Hopkins University Press, Baltimore, MD.

68. Gershenfeld, N. A. and Weigend, A. S. (1993) *The Future of Time Series: Learning and Understanding*, vol. XV of *Sante Fe Institute Studies in the Sciences of Complexity.* Addison-Wesley, Reading, MA, pp. 1–70.

69. Kendall, M. G. (1938) A new measure of rank correlation. *Biometrika* **30,** 81–93.

70. Head, R. D., Smythe, M. L., Oprea, T. I., Waller, C. L., Green, S. M., and Marshall, G. R. (1996) VALIDATE: a new method for the receptor-based prediction of binding affinities of novel ligands. *J. Amer. Chem. Soc.* **118,** 3959–3969.

71. Wang, R., Liu, L., Lai, L., and Tang, Y. (1998) SCORE: a new empirical method for estimating the binding affinity of a protein-ligand complex. *J. Mol. Modeling* **4,** 379–394.

72. Schwikowski, B., Uetz, P., and Fields, S. (2000) A network of protein-protein interactions in yeast. *Nat. Biotechnol.* **18,** 1257–1261.

73. Wojcik, J. and Schächter, V. (2001) Protein-protein interaction map inference using interacting domain profile pairs. *Bioinformatics* **17(suppl. 1),** S296–S305.

74. Uetz, P., Goit, L., Cagney, G., et al. (2000) A comprehensive analysis of protein-protein interactions in *Saccharomyces cerevisiae. Nature* **403,** 623–627.

75. Tucker, C. L., Gera, J. F., and Uetz, P. (2001) Towards an understanding of complex protein networks. *Trends Cell Biol.* **11,** 102–106.

76. Walhout, A., Boulton, S., and Vidal, M. (2000) Yeast two-hybrid systems and protein interaction mapping projects for yeast and worm. *Yeast* **17,** 88–94.

77. Wang, R., Lai, L., and Wang, S. (2002) Further development and validation of empirical scoring functions for structure-based binding affinity prediction. *J. Comput. Aided Mol. Design* **16,** 11–26.

78. Rarey, M., Kramer, B., Bernd, C., and Lengauer, T. (1996) Time-efficient docking of similar flexible ligands, in Biocomputing: *Proceedings of the 1996 Pacific Symposium*, Hunter, L. and Klein, T., eds., January 3–6, World Scientific Publishing, Singapore.

79. Zhang, T. and Koshland, D. E. (1996) Computational method for relative binding energies of enzyme-substrate complexes. *Protein Sci.* **5,** 348–356.

80. Schapira, M., Totrov, M., and Abagyan, R. (1999) Prediction of the binding energy for small molecules, peptides and proteins. *J. Mol. Recog.* **12,** 177–190.

II

POST-TRANSLATIONAL MODIFICATIONS, VARIANTS, AND ISOFORMS

14

Predicting Glycan Composition from Experimental Mass Using GlycoMod

Catherine A. Cooper, Elisabeth Gasteiger, and Nicolle H. Packer

1. Introduction

GlycoMod (http://www.expasy.org/tools/glycomod/) *(1)* is a computational tool that finds all possible compositions of a glycan structure from its experimentally determined mass. It may be used to calculate the possible compositions of free or derivatized glycan structures, or compositions of glycans attached to glycoproteins and glycopeptides.

Standard proteomics methods identify the proteins in a sample by mass spectrometric analysis of protease-derived peptides. The experimental masses are matched to those generated by computational digestion of a database of protein sequences. What is commonly observed is that many of the masses in the spectrum do not match the predicted peptides. These unmatched masses are often caused by post-translational modifications such as glycosylation. Protein glycosylation alters in such conditions as cancer, rheumatoid arthritis, inflammation, infection, and the immune response. In addition, recombinant proteins differ in their glycosylation depending on the expression system used.

GlycoMod is closely integrated with the protein identification and characterization tools on the ExPASy server (http://www.expasy.org/tools/) *(2)*. These tools, in particular PeptIdent, FindMod *(3)*, and FindPept *(4)*, are closely integrated and hyperlinked with SWISS-PROT and TrEMBL *(5)* entries on ExPASy, and among each other. Navigation among database entries, data submission forms, and program results is made easy (in both directions), and the number of copy/paste and mouse click operations is minimized. This means that in addition to direct submission to GlycoMod, this program can be accessed seamlessly from other tools that identify a potential candidate glycoprotein from peptide-mass fingerprinting.

Both *N*- and *O*-linked glycan compositions can be calculated using GlycoMod. Compositions can be calculated from the mass of free glycans, which have been released from a glycoprotein or glycopeptide by enzymatic or chemical methods. Masses of derivatized glycans can also be used, such as peracetylated, permethylated, or any derivative produced by reductive amination of the reducing terminus. GlycoMod can suggest glycan compositions from the mass of a glycopeptide when either the sequence of the peptide or the protein from which it originated is known. GlycoMod can theoretically digest proteins and peptides, with common modifications to cysteine and

From: *Handbook of Proteomic Methods*
Edited by: P. Michael Conn © Humana Press Inc., Totowa, NJ

methionine, using a variety of peptidases including trypsin, Lys C, Asp N and Glu C, or with the chemical cyanogen bromide. The program calculates possible glycan compositions by subtracting the generated peptide masses from the experimentally obtained masses.

2. Materials

GlycoMod can be accessed at http://www.expasy.org/tools/glycomod/. An online user manual is available at http://www.expasy.org/tools/glycomod/glycomod-doc.html.

3. Methods

3.1. Entering a Query

1. Open the GlycoMod input page, as shown in **Fig. 1**, found at http://www.expasy. org/tools/glycomod/.
2. Enter the experimental mass of either the oligosaccharide (free or derivatized) or a glycopeptide. If multiple masses are to be analyzed, separate by spaces or new lines, or use masses listed in a text file (each mass must be listed on a new line).
3. Select mass type—monoisotopic or average.
4. Enter a mass tolerance in either Daltons or ppm.
5. Select an ion mode and adduct—[M], [M+H]⁺, [M+Na]⁺, [M+K]⁺, [M+H]⁻, [M+CH₃COO]⁻ (acetate), [M+TFA]⁻ (trifluoroacetic acid) or enter your own adduct and its mass.
6. Select the glycan type—*O*-linked, via the hydroxyl group of serine or threonine (*see* **Note 1**); or *N*-linked, via the amide nitrogen of an asparagine residue.
7. Select the glycan form from the associated drop-down list.

     For *N*-linked: Free/PNGase released oligosaccharides (*see* **Note 2**)
          ENDO H or ENDO F released oligosaccharides (*see* **Note 3**)
          Reduced oligosaccharides
          Derivatized oligosaccharides
          Glycopeptides (motif N-X-S/T/C (X not P) will be used) (*see* **Note 4**).
     For *O*-linked: Free oligosaccharides (*see* **Note 5**)
          Reduced oligosaccharides (*see* **Note 6**)
          Derivatized oligosaccharides
          Glycopeptides (only those containing S or T will be used) (*see* **Note 4**)
8. If a free or reduced oligosaccharide, go to **step 11**.
9. If a glycopeptide:
     a. Enter a peptide/protein sequence, a SWISS-PROT/TrEMBL *(5)* ID or AC (*see* **Note 7**), or a set of unmodified peptide masses ([M], where the masses are average or monoisotopic as specified in **step 2**).
     b. Select a protease and/or digestion condition from the drop-down list (*see* **Notes 8–10**). Select the number of missed cleavage sites: 0, 1, 2, or 3.
     c. Choose how any cysteines in the protein may be modified (*see* **Note 11**): reduced and alkylated using (i) iodoacetamide, (ii) iodoacetic acid, or (iii) vinyl pyridine; or acrylamide adducts (*see* **Note 12**).
     d. Select if methionines may be oxidized (*see* **Note 11**).
10. If a derivatized oligosaccharide:

     For reductive derivatization, i.e., where the oligosaccharide has been derivatized at the reducing terminus by a process of reductive amination, insert the derivative name and mass [M] where specified. The mass required is the monoisotopic or average mass (as specified in **step 2**) of the non-reacted derivative, e.g., 2-aminopyridine has monoisotopic mass = 94.053 (*see* **Note 13**).

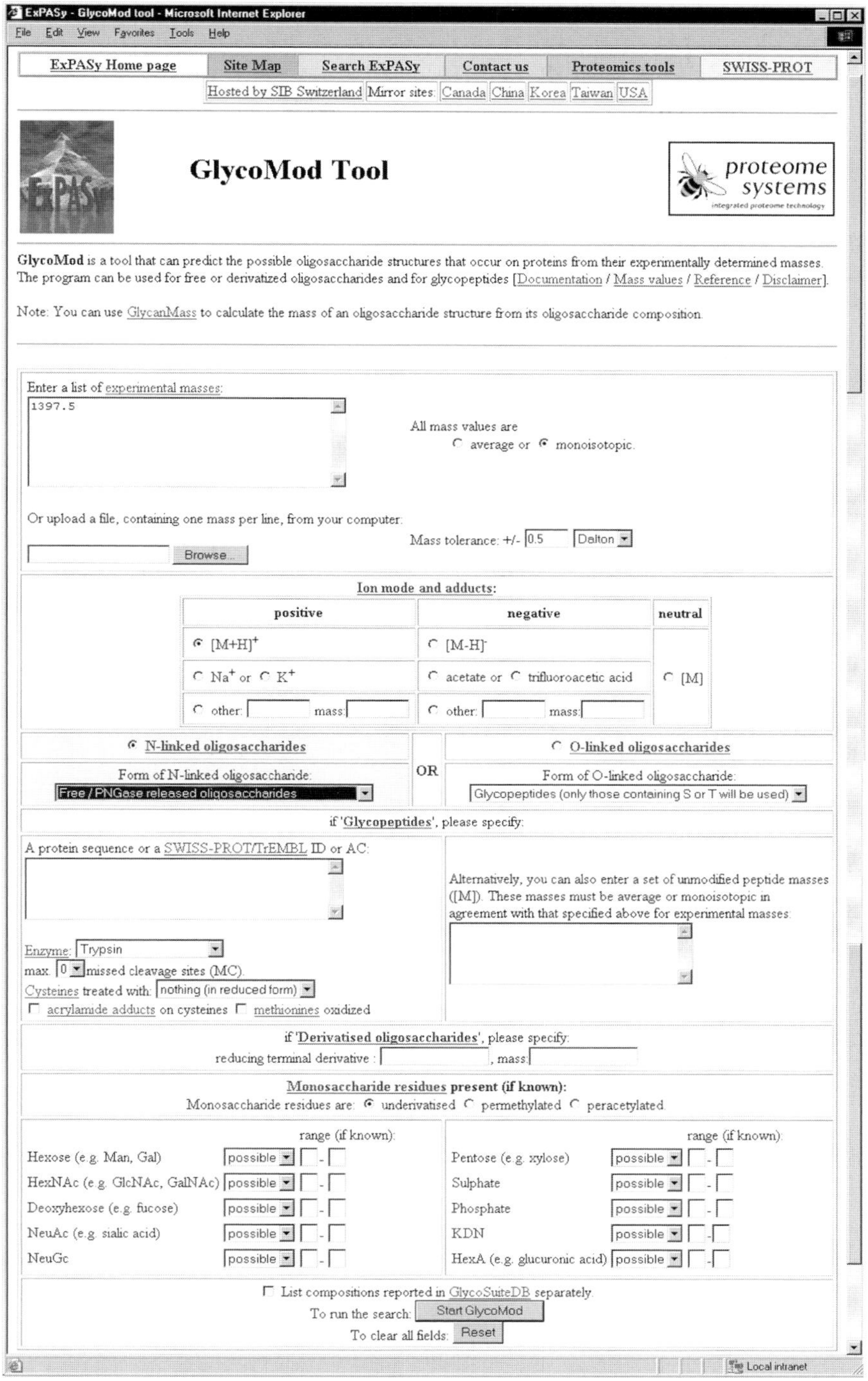

Fig. 1. Input page of GlycoMod (http://www.expasy.org/tools/glycomod/).

Table 1
**Total Numbers of Individual Monosaccharides Allowed
for *N*-Linked and *O*-Linked Glycan Compositions**

Monosaccharide residue	*N*-linked oligosaccharides	*O*-linked oligosaccharides
Hexose	0–20	0–14
HexNAc	0–20	0–14
Deoxyhexose	0–6	0–6
NeuAc	0–5	0–7
NeuGc	0–5	0–7
Pentose	0–3	0–3
Sulfate	0–3	0–6
Phosphate	0–2	0–6
KDN	0–0	0–2
HexA	0–0	0–2

For permethylation or peracetylation select the appropriate button listed under "Monosaccharide residues" (*see* **step 11**).

11. Select the monosaccharide residue type: underivatized, permethylated, or peracetylated.
12. Stipulate which monosaccharides are/are not/ or may possibly be present in the glycan (*see* **Notes 14–17** and **Table 1**).
13. If desired you may select to list separately all oligosaccharide compositions listed in GlycoSuiteDB (http://www.glycosuite.com) (*see* **ref. 6**, Chap. 15 of this book, and **Note 18**).
14. Start GlycoMod.

3.2. Interpreting the Results

1. The output from GlycoMod (an example of which is shown in **Fig. 2**) is divided into two main sections—a header and one or two tables depending on whether the box to "List compositions reported in GlycoSuiteDB separately" was checked.
2. The header section lists the monosaccharide compositional information that was specified in the query (*see* **Subheading 3.1., step 12**).
3. If the glycopeptide option was selected, the header section also lists the protein name as annotated in SWISS-PROT/TrEMBL (if applicable), the number of miscleavages selected, the number and type of cysteine modifications, and any methionine modifications specified. Also listed are the potential glycopeptides and their corresponding masses calculated from the protein sequence (*see* **Subheading 3.1., step 9** and **Note 19**).
4. The output tables report the glycoform mass (*see* **Note 20**), delta mass (*see* **Note 21**) in Daltons or ppm (depending on units entered by the user on the input form), and the monosaccharide compositions whose theoretical masses match the experimental masses (*see* **Subheading 3.1., step 2**) after any stated derivatives or peptide masses are subtracted.
5. If the glycan type is *N*-linked, the predicted glycan core type is given (*see* **Notes 22–24**).
6. If a composition has been reported in the GlycoSuiteDB database, a hyperlink to the corresponding GlycoSuiteDB entry is provided.
7. For a glycopeptide output, the table/s contain additional information on the peptide mass ([M]), peptide sequence, theoretical glycopeptide mass, and the possible modifications used in the calculation.

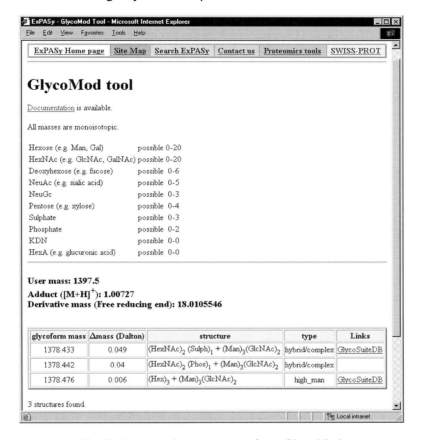

Fig. 2. An example output page from GlycoMod.

8. For users who wish to export the GlycoMod results into an external application (e.g., a spreadsheet program), a raw text version of the results is provided, without any html formatting. This text file can be downloaded from the end of the page displaying the results.

4. Notes

1. Oligosaccharides may also be *O*-linked via the hydroxyl group of hydroxylysine, hydroxyproline, or hydroxytryosine, but these linkages are less common and are not considered in the current version of GlycoMod.
2. *N*-linked glycans released from a glycoprotein using PNGase F, PNGase A, or anhydrous hydrazine, followed by regeneration of the reducing terminus, are considered to be "free oligosaccharides."
3. *N*-linked glycans released from a glycoprotein using ENDO F or ENDO H cleaved at the GlcNAc(β1-4)GlcNAc core linkage to asparagine. The released glycan therefore has one less GlcNAc moiety. The GlycoMod output takes this into account and predicts the composition of the complete *N*-linked glycan.
4. When a protein sequence or a SWISS-PROT/TrEMBL ID or AC is entered, GlycoMod only considers those peptides with the sequence NX[S/T/C] where X≠P for *N*-linked glycans, and peptides that contain S and/or T for *O*-linked glycans.
5. *O*-linked glycans released using *O*-glycanase or by mild hydrazinolysis, followed by regeneration of the reducing terminus or by nonreductive β-elimination are considered to be "free oligosaccharides."

6. Traditional β-elimination, also known as reductive β-elimination, releases the oligosaccharides with alkali, and then reduces them to alditols (i.e., reduced oligosaccharides) to prevent base degradation or "peeling" of the oligosaccharides.

7. Where there are multiple sites within a sequence, GlycoMod calculates the glycan composition as if there was only one site. Therefore, the resulting glycan composition may actually consist of more than one glycan structure.

8. Possible enzymes are trypsin, Lys C, Arg C, Asp N, Asp N + *N*-terminal Glu, Glu C in a bicarbonate buffer, Glu C in a phosphate buffer, and chymotrypsin.

9. Cleavage rules for all the enzymes may be found at http://www.expasy.org/tools/peptide-mass-doc.html#table1.

10. The digestion may also be performed using CNBr.

11. When modifications to cysteine and/or methionine are selected, GlycoMod considers peptides with both unmodified and modified amino acids. When more than one potentially modifiable amino acid occurs in a peptide, the masses of all possible combinations of modified and unmodified residues are calculated. For example, if cysteine modification has been selected and a peptide contains three cysteine residues, then GlycoMod considers the masses for that peptide containing 0, 1, 2, and 3 modified residues.

12. This is common when the protein is prepared using polyacrylamide gel electrophoresis.

13. When calculating the addition of a derivative, GlycoMod automatically adds the mass of two hydrogen atoms. This is necessary to calculate the resultant glycan derivative from the monosaccharide residue mass plus derivative mass after reductive amination (*1*).

14. The total number of individual monosaccharides in any glycan composition has been limited to those shown in **Table 1**. These limits were set after careful examination of the literature.

15. Five rules have been added to GlycoMod to reduce the number of possible *N*-linked glycan compositions from a given mass. Again, these rules were implemented after careful investigation of the known *N*-linked glycan structures.

   a. A composition may not contain both sulfate and phosphate.

   b. The sum of the number of hexose plus HexNAc residues must be greater than or equal to the number of sulfate or phosphate residues.

   c. The sum of the number of hexose plus HexNAc residues cannot be zero.

   d. The number of fucose residues plus 1 must be less than or equal to the sum of the number of hexose plus HexNAc residues.

   e. If the number of HexNAc residues is less than or equal to 2 and the number of hexose residues is greater than 2, then the number of NeuAc and NeuGc residues must be zero.

16. There are no rules to reduce the number of possible *O*-linked glycan compositions from a given mass, owing to their known diversity.

17. An upper limit on the total mass of the glycoform has been set. This limit is:

   5000 Daltons for underivatised *O*-linked glycans.

   8000 Daltons for underivatised *N*-linked glycans.

   7000 Daltons for permethylated *O*-linked glycans.

   10,000 Daltons for permethylated *N*-linked glycans.

   9500 Daltons for peracetylated *O*-linked glycans.

   13,000 Daltons for peracetylated *N*-linked glycans.

18. GlycoSuiteDB is a curated database of glycan structures. It contains information from the scientific literature on glycoprotein-derived glycan structures, their two-dimensional structures, biological sources, the methods used by the researchers to determine each glycan structure, and the literature references used to obtain the information.

19. If a SWISS-PROT AC was entered, GlycoMod considers any post-translational modifications to the amino acids in the peptides as annotated in the SWISS-PROT entry.

20. The glycoform mass is the oligosaccharide residue mass with no derivatization or ion adducts.

21. The delta mass is the difference between the theoretical and experimental glycoform masses.

22. The structure of *N*-linked glycans is generally well conserved, with a core region consisting of two *N*-acetylglucosamine (HexNAc) residues and three mannose (hexose) residues. To aid the user, when the composition contains at least two HexNAc residues and three hexose residues, these are removed from the overall composition and written separately, e.g., $(Hex)_1(HexNAc)_2(Deoxyhexose)_2(NeuAc)_1 + (Man)_3(GlcNAc)_2$.

23. If the number of HexNAc residues equals 2 and the number of hexose residues is greater than or equal to 5, then the glycan is of the type "high mannose."

24. If the number of HexNAc residues is greater than or equal to 3 and the number of hexose residues is also greater than or equal to 3, then the *N*-linked glycan is of the type "hybrid/complex."

References

1. Cooper, C. A., Gasteiger, E., and Packer, N. H. (2001) GlycoMod—a software tool for determining glycosylation compositions from mass spectrometric data. *Proteomics* **1,** 340–349.

2. Bairoch, A., Gasteiger, E., Gattiker, A., Hoogland, C., Lachaize, C., Mostaguir, K., Ivanyi, I., and Appel, R. D. (2002) The ExPASy proteome WWW server in 2002. http://www.expasy.org/doc/expasy.pdf.

3. Wilkins, M. R., Gasteiger, E., Gooley, A., Herbert, B., Molloy, M. P., Binz, P. A., Ou, K., Sanchez, J.-C., Bairoch, A., Williams, K. L, and Hochstrasser, D. F. (1999) High-throughput mass spectrometric discovery of protein post-translational modifications. *J. Mol. Biol.* **289,** 645–657.

4. Gattiker, A., Bienvenut, W. V., Bairoch, A., and Gasteiger, E. (2002) FindPept, a tool to identify unmatched masses in peptide mass fingerprinting protein identification. *Proteomics* **2,** 1435–1444.

5. O'Donovan, C., Martin, M. J., Gattiker, A., Gasteiger, E., Bairoch, A., and Apweiler, R. (2002) High-quality protein knowledge resource: SWISS-PROT and TrEMBL. *Briefings Bioinformatics* **3,** 275–284.

6. Cooper, C. A., Harrison, M. J., Wilkins, M. R., and Packer, N. H. (2001) GlycoSuiteDB: a new curated relational database of glycoprotein glycan structures and their biological sources. *Nucleic Acids Res.* **29,** 332–335.

15

Querying GlycoSuiteDB

Catherine A. Cooper, Hiren J. Joshi, Mathew J. Harrison, Marc R. Wilkins, and Nicolle H. Packer

1. Introduction

GlycoSuiteDB (http://www.glycosuite.com) *(1)* is a curated and annotated relational database of glycoprotein glycan structures. The database contains more than 7500 entries of validated *N*- and *O*-linked structures reported in the literature. Entries **(Fig. 1)** contain information, when available, on the (1) glycan structure (including glycan type, linkage, mass, and composition); (2) biological source (e.g., native and recombinant species, tissue and/or cell type, strain, blood group, and disease); (3) protein (name, known sites of glycosylation, and the attachment sites of the individual glycan structure where known); (4) literature references; and (5) analytical methods used to determine the structure. Each entry also has a two-dimensional image of the glycan structure.

2. Materials

GlycoSuiteDB can be accessed at http://www.glycosuite.com. Usage requires a license agreement.

3. Methods

GlycoSuiteDB has a powerful interface for querying the database with one, or a combination, of many fields. These fields include those of glycan composition, mass, taxonomy, tissue or cell type, attached protein, disease, and reference. More recently, we have introduced a query by structure approach.

The following sections first describe each of the simple query methods, then the advanced query method, and finally reindexing and subselecting the results. All the query types, including the advanced query, are accessed directly from the GlycoSuiteDB home page (http://www.glycosuite.com).

3.1. Query by Structure

This query method allows the user to search for a particular structure or for a particular epitope within glycan structures.

From: *Handbook of Proteomic Methods*
Edited by: P. Michael Conn © Humana Press Inc., Totowa, NJ

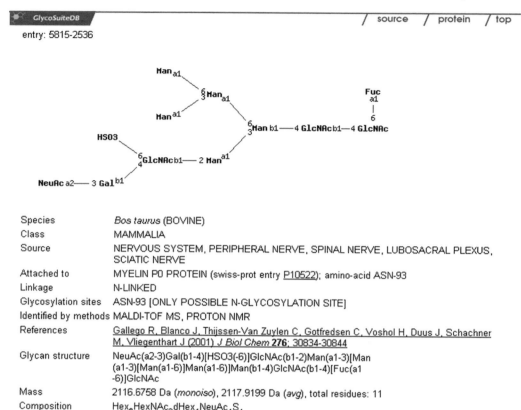

Fig. 1. An example entry from GlycoSuiteDB.

1. First you need to construct a glycan structure or partial structure. We have created a tool that allows the user to construct a glycan structure visually on-screen. To draw a new structure, select "New Structure" from the drop-down list and click "Edit." A new window will appear.
2. Choose either: a. or b., as follows:
    a. Choose a monosaccharide residue from the drop-down list to appear furthest toward (or at) the reducing terminus of your query structure. Tick the checkbox if this residue must be located at the reducing terminus (i.e., is the monosaccharide residue directly linked to the protein backbone). If this checkbox is not ticked, the query structure may be located at the reducing terminus, or internally within the resulting glycan structure. The anomeric configuration and the hydroxyl group on the adjacent residue to which this monosaccharide is attached may be entered (not applicable if the residue must be located at the reducing terminus). Click "Build."
    b. Select a predefined structure.
3. To extend the structure, click on the monosaccharide residue on which to build. It will turn red. Select the hydroxyl group to which you want to link. Choose a monosaccharide or substituent in the drop-down list and select its anomeric configuration where appropriate. Tick the checkbox for phosphodiester linkage if required. Click "Build." Repeat as necessary.

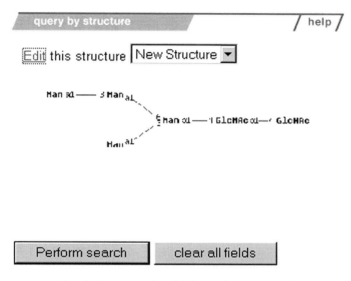

Fig. 2. An example of "Query by structure."

4. "Undo" will delete the last addition. "Prune" will delete the selected residue and anything attached to the nonreducing terminus (left-hand side) of it. "Reset Structure" will clear the structure and return the user to the beginning.
5. When your query structure is complete, click "Update Query." The structure should then be visible in the main GlycoSuiteDB query page (**Fig. 2**), and the query can be executed by clicking on "Perform Search."
6. To edit a structure that you have previously drawn, select it from the drop-down list and click "Edit." A new window will appear. Your structure may be extended or pruned as above.
7. The results will be returned as a summarized list (**Fig. 3**), indexed by taxonomy. This list can then be manipulated and the GlycoSuiteDB entries shown (*see* **Subheading 3.10.**).

3.2. Query by Composition

This query method allows the user to search for a glycan structure based on the type and number of monosaccharides it contains.

1. Enter a number or a number range, e.g., 0, 1, 3–5, <4, >5 into the boxes corresponding to the residue/s by which you wish to query. By default, fields left blank imply zero to infinity of this monosaccharide or group. Selecting the "Exact composition" checkbox will change this (*see* **Notes 1** and **2**).
2. It is also possible to query the database by the total number of monosaccharides in a glycan structure. Enter either a number or a range of numbers into this field (*see* **Note 3**).
3. The results will be returned as a summarized list (**Fig. 3**), indexed by composition. This list can then be manipulated and the GlycoSuiteDB entries shown (*see* **Subheading 3.10.**).

3.3. Query by Mass

This query method allows the user to search for a glycan based on results from mass spectrometry (MS), specifically, using parent masses of glycans generated from matrix-assisted laser desorption/ionization time-of-flight (MALDI-TOF) or liquid chromatography (LC) mass spectrometry (LC-MS) (*see* **Note 4**).

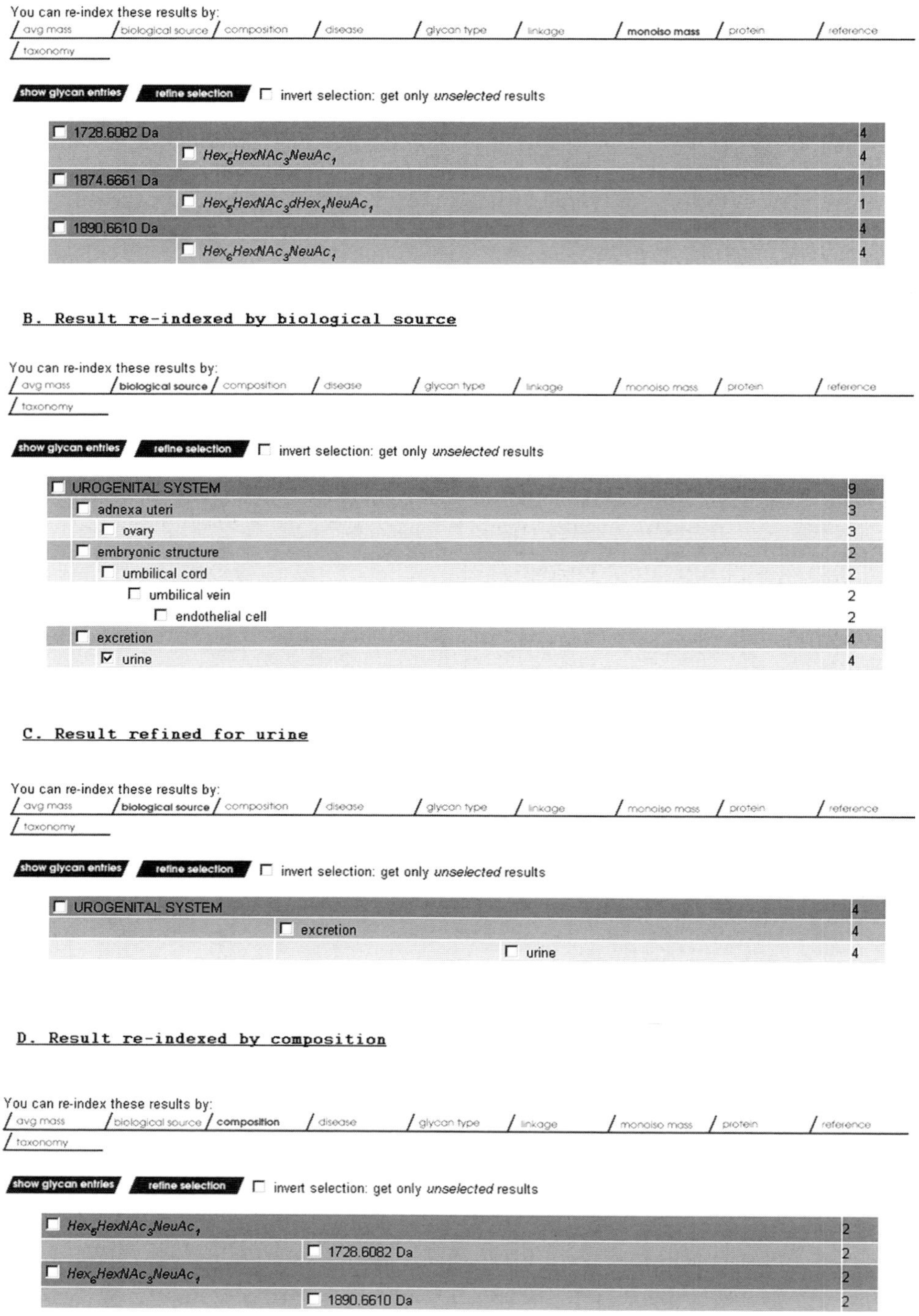

You can re-index these results by:
/ avg mass / biological source / composition / disease / glycan type / linkage / **monoiso mass** / protein / reference
/ taxonomy

[show glycan entries] [refine selection] ☐ invert selection: get only *unselected* results

☐ 1728.6082 Da		4
	☐ $Hex_5HexNAc_3NeuAc_1$	4
☐ 1874.6661 Da		1
	☐ $Hex_5HexNAc_3dHex_1NeuAc_1$	1
☐ 1890.6610 Da		4
	☐ $Hex_6HexNAc_3NeuAc_1$	4

B. Result re-indexed by biological source

You can re-index these results by:
/ avg mass / **biological source** / composition / disease / glycan type / linkage / monoiso mass / protein / reference
/ taxonomy

[show glycan entries] [refine selection] ☐ invert selection: get only *unselected* results

☐ UROGENITAL SYSTEM		9
☐ adnexa uteri		3
☐ ovary		3
☐ embryonic structure		2
☐ umbilical cord		2
☐ umbilical vein		2
☐ endothelial cell		2
☐ excretion		4
☑ urine		4

C. Result refined for urine

You can re-index these results by:
/ avg mass / **biological source** / composition / disease / glycan type / linkage / monoiso mass / protein / reference
/ taxonomy

[show glycan entries] [refine selection] ☐ invert selection: get only *unselected* results

☐ UROGENITAL SYSTEM		4
	☐ excretion	4
		☐ urine 4

D. Result re-indexed by composition

You can re-index these results by:
/ avg mass / biological source / **composition** / disease / glycan type / linkage / monoiso mass / protein / reference
/ taxonomy

[show glycan entries] [refine selection] ☐ invert selection: get only *unselected* results

☐ $Hex_5HexNAc_3NeuAc_1$		2
	☐ 1728.6082 Da	2
☐ $Hex_6HexNAc_3NeuAc_1$		2
	☐ 1890.6610 Da	2

Fig. 3. Indexing, reindexing, and refining GlycoSuiteDB results.

1. Enter a single value, e.g.: "1234.56"; a numerical range, e.g.: "1234-1345.5"; or a list of single masses and/or ranges, e.g.: 1234.56, 2345-2350 (*see* **Note 5**).
2. A mass tolerance can be entered where single masses, or a list of single masses, have been specified (*see* **Note 6**).
3. The results will be returned as a summarized list **(Fig. 3)**, indexed by monoisotopic mass. This list can then be manipulated and the GlycoSuiteDB entries shown (*see* **Subheading 3.10.**).

3.4. Query by Taxonomy

This query method allows the user to search for all glycans originating from a particular species or taxonomic class (*see* **Note 7**). In the case of recombinant proteins, both the origins of the protein and the expression system are searched against.

1. Choose one of the species or class fields from the drop-down list (*see* **Note 8**).
2. The keyword field may be used to query by any species name, common name, class, or part thereof (*see* **Notes 9** and **10**).
3. The results will be returned as a summarized list **(Fig. 3)**, indexed by taxonomy. This list can then be manipulated and the GlycoSuiteDB entries shown (*see* **Subheading 3.10.**).

3.5. Query by Biological Source

This query method allows the user to select glycan structures found in a particular system, tissue, or cell type (*see* **Note 11**). Using this information it is possible to note which glycan structures are expressed in a particular tissue or cell type, thereby reflecting the availability of precursors and the activity of glycosyltransferases.

1. Choose one of the system, tissue, or cell type fields from the drop-down list. Selecting a system will query for all tissues and/or cell types in that system. The default, ANY, will query for all systems, tissues, and cell types.
2. Or, enter any system, tissue, or cell type in the keyword field (*see* **Notes 9** and **10**).
3. The results will be returned as a summarized list (*see* **Fig. 3**), indexed by biological source. This list can then be manipulated and the GlycoSuiteDB entries shown (*see* **Subheading 3.10.**).

3.6. Query by Attached Protein

This query method allows the user to search for all glycan structures that have been found attached to a particular protein.

1. Insert any protein name or part of a protein name (*see* **Notes 12–14**) into the keyword field box (*see* **Notes 9** and **10**).
2. Or, enter one or multiple SWISS-PROT accession numbers separated by any combination of space, commas, or semicolons into the SWISS-PROT field box. If multiple proteins are to be searched for, OR will join them. For example, P00750 P01215 will be searched as P00750 OR P01215.
3. The results will be returned as a summarized list **(Fig. 3)**, indexed by protein name. This list can then be manipulated and the GlycoSuiteDB entries shown (*see* **Subheading 3.10.**).

3.7. Query by Disease

This query method allows the user to search for all glycan structures that have been found in tissue or cells isolated from a species with a particular disease.

1. Choose one of the disease states from the drop-down list. The default, ANY, will query for all disease states, including no disease. If NONE is selected, the query will return only glycans found where no disease was present. If ALL is selected the database will query for glycans where disease was present.
2. Or, enter a disease in the keyword field (*see* **Notes 9** and **10**).
3. The results will be returned as a summarized list (**Fig. 3**), indexed by disease. This list can then be manipulated and the GlycoSuiteDB entries shown (*see* **Subheading 3.10.**).

3.8. Query by Reference

This query method allows the user to search for database entries entered from a particular journal article or from a series of articles (e.g., by the same author or in the same year or published in the same journal).

1. Enter an author's name or part of an author's name. A single initial can be added after the author's name, separated by a space (*see* **Notes 9** and **10**).
2. Any publication year or part of the year may be given as keyword(s). Multiple publication years must be separated by any combination of space, commas, or semicolons. By default, multiple publication years will be joined by OR. Ranges of years can also be given as starting year-finishing year. For example, 1995; 1995, 1998, 2001; 1995, 1998–2001.

3.9. Advanced Query

This query method is a combination of all the query methods explained above and allows the user to perform complex queries. For example, it is possible to query GlycoSuiteDB for glycan structures that have been isolated from tissue or cells orginating from the human urogenital system, have a monoisotopic mass below 2000 Daltons, contain at least 1 *N*-acetylneuraminic acid residue, are *N*-linked, and contain the substructure Man(a1-3)Man(a1-6)[Man(a1-3)]Man(b1-4)GlcNAc(b1-4)GlcNAc (**Fig. 2**) located at the reducing terminus. You can select the result to be indexed by any of the indexes listed in the drop-down box (*see* **Note 15**). This list can then be manipulated and the GlycoSuiteDB entries shown (*see* **Subheading 3.10.**).

3.10. Reindexing and Subselecting the Results

A feature of GlycoSuiteDB is the ability to reindex (*see* **Note 15**) and subselect results. For example, the above advanced query results in the output shown in **Fig. 3A** when indexed by monoisotopic mass. These results can be reindexed (**Fig. 3B**) by clicking on an index type shown in the menu listed above the results. From any index it is possible to refine the results by clicking one or more checkboxes in the summarized list and clicking "Refine selection" (**Fig. 3B** and **C**). The process of reindexing and refining the results can be repeated until the desired entries are selected. These entries can be displayed by clicking "Show glycan entries" (*see* **Note 16**).

4. Notes

1. The P (phosphate) field queries for attached phosphate as well as phosphodiester linkages. The S (sulfate) field queries for attached sulfates.
2. The methyl and acetyl fields refer to additional methyl and acetyl groups, respectively. For example, a search for Neu5,9Ac$_2$ would require the user to input 1 NeuAc and 1 Ac.
3. Additional substituents such as Me, Ac, phosphate, and sulfate groups are not included in the total number of monosaccharide residues.

4. For a more comprehensive program, see GlycoMod (http://www.expasy.org/tools/glycomod) (*see* Chap. 14).

5. If multiple masses are to be analyzed, separate by a comma.

6. Mass tolerance does not apply to mass ranges.

7. Species names and taxonomic classes are standardized to the NCBI taxonomy database (http://www.ncbi.nlm.nih.gov/Taxonomy).

8. Selecting a class will query for all species in that class. The default, ANY, will query for all species.

9. Multiple keywords may be given, separated by any combination of spaces, commas or semi-colons; these will be joined by AND. The wildcards "*" (0 or more of any character), and "?" (exactly one character) may also be used. For example, human, hum*, hum?n.

10. By default the wildcard character "*" is used both before and after any letters given. For example, "john" would search and return all words that contained John, including Johnson and McJohn. Where given, if you select the box "Select exactly this keyword" then only the exact keyword as typed will be returned. This feature can be used in conjunction with wildcards. For example, "carcinoma*" with "Select exactly this keyword" will return Carcinoma, Hepatocellular but not Adenocarcinoma.

11. GlycoSuiteDB uses an anatomy classification system adapted from the anatomy categories of the National Library of Medicine's medical subject headings (MeSH) for describing the tissue or cell type source of glycans. Each entry in GlycoSuiteDB has the tissue or cell type described in a maximum of five columns: system, division1, division2, division3, and division4, where each is a nested subset of the previous column.

12. The protein name used in GlycoSuiteDB is the prefered name given by the SWISS-PROT/TrEMBL protein databases *(2)*. We plan to include the ability to search for a protein by one of its synonyms in the next release (due early 2003).

13. If a protein is not found in SWISS-PROT or TrEMBL, the protein name given is that recorded by the authors of the relevant literature article.

14. There are many (approx 2000) entries in GlycoSuiteDB that do not list a protein name. This is because in many instances researchers look at the total glycans present in a mixture, rather than analyze glycans from purified proteins (because of limited amounts).

15. Results may be indexed or reindexed by: average mass, biological source, composition, disease, glycan type, linkage, monoisotopic mass, protein, reference, taxonomy.

16. GlycoSuiteDB shows a maximum of 50 entries per page. The summarized list at the top of each page details only the results shown on that page. Subsequent pages may be reached by clicking on the individual page numbers listed under the summarized list.

References

1. Cooper, C. A., Harrison, M. J., Wilkins, M. R., and Packer, N. H. (2001) GlycoSuiteDB: a new curated relational database of glycoprotein glycan structures and their biological sources. *Nucleic Acids Res.* **29**, 332–335.

2. O'Donovan, C., Martin, M. J., Gattiker, A., Gasteiger, E., Bairoch, A., and Apweiler, R. (2002) High-quality protein knowledge resources: SWISS-PROT and TrEMBL. *Briefings Bioinformatics* **3**, 275–284.

16

New Tools for Quantitative Phosphoproteome Analysis

Li-Rong Yu, Van M. Hoang, and Timothy D. Veenstra

1. Introduction

The success of the various genome sequencing projects *(1)* and the ability to measure differences in gene expression routinely at the transcription level *(2)*, has shifted a considerable amount of focus towards proteomics—the characterization of gene expression at the protein level *(3)*. Much of this focus has concentrated on methods to identify and measure effectively the relative abundance of a large number of proteins in a rapid fashion *(4)*. Major developments in the areas of protein and peptide fractionation and instrument technology have been required to accomplish these goals. In the vast majority of global proteomic projects most proteins are identified by mass spectrometry (MS)-based methods. The proteins are identified either by matching a set of peptide masses measured by MS to a list of predicted peptide masses or by generating sequence-related information of a single peptide by tandem MS and searching this MS/MS spectrum against an appropriate database. This identification strategy usually provides the identity of the analyzed proteins and sequences of their degraded peptides.

Although this level of characterization may be sufficient to identify a protein in a complex mixture, it does not describe the true nature of a protein adequately. Proteome characterization requires knowledge of protein expression levels, location within the cell, interactions with each other and with other biomolecules, functions, and so forth. One of the key descriptors of a protein, which is amenable to proteomics technology, is the delineation of post-translational modifications (PTMs). The delineation of a protein's function solely from a change in its abundance provides a limited view since numerous vital activities of proteins are modulated by PTMs that may not be reflected by changes in protein abundance.

Arguably, the most important PTM that is used to modulate protein activity and propagate signals within cellular pathways and networks is phosphorylation *(5)*. Cellular processes ranging from cell cycle progression, differentiation, development, peptide hormone response, and adaptation are all regulated by protein phosphorylation. It has been estimated that as many as one-third of the proteins within a given eukaryotic proteome undergoes reversible phosphorylation. A change in a protein's phosphorylation state does not necessarily coincide with a change in its expression level. For example, the treatment of tumor-promotion-sensitive murine JB6 epidermal cells with either 12-*O*-tetradecanoylphorbol 13-acetate (TPA) or epidermal growth factor (EGF) results in no

From: *Handbook of Proteomic Methods*
Edited by: P. Michael Conn © Humana Press Inc., Totowa, NJ

change in the expression of the mitogen-activated protein kinases (MAPK) Erk1 and
Erk2; however, both Erk1 and Erk2 show an increase in their phosphorylation state *(6)*.
In addition, a similar result has been shown for the activation of Akt and MEK1 by
growth hormone in murine 32D leukemic cells *(7)*. These results show that the protein
activation, and hence cellular processes, are often controlled by an alteration in their
phosphorylation state and not their abundance within the cell.

There are many classical techniques designed to determine whether a protein is
phosphorylated. The predominant method has been the use of affinity reagents such as
anti-phosphoamino acid-specific monoclonal antibodies (MAbs). Although affinity-
based detection methods can determine whether a protein is modified they cannot identify the specific site of modification without prior knowledge of the identity of the
phosphorylated site. Knowledge of the phosphorylated site(s) is important since a modification at a different site within the same protein most often has a different effect on
the protein's activity. In addition, several different kinases may modify a single protein, each providing an indication into which cell pathway may be active. For example,
protein kinases A and C, as well as cyclic adenosine monophosphate (cAMP)-
dependent protein kinase II all catalyze the phosphorylation of the α-amino-3-hydroxy-
5-methyl-4-isoxazolepropionate (AMPA) receptor *(8)*. Each of these kinases phosphorylates a different site within the AMPA receptor, resulting in a different function
being imparted onto the receptor. Although site-specific MAbs can be produced, they
are only useful in hypothesis-driven studies to validate the phosphorylation site within
a protein but are impractical in discovery-driven studies designed to identify novel sites
of phosphorylation.

MS provides the best available instrumental technology for the site-specific identification of PTMs. The present attributes of MS of obtaining mass measurements as well
as sequence information of peptides are directly applicable to the site-specific phosphorylation identification. Recently, there has been an increase in the development of
methods, primarily centered on the use of stable isotope labeling, to use MS to quantify the relative phosphorylation state of proteins. In this chapter, we review the current progress in the identification and quantitation of the phosphorylation state of
proteins. Although most of the development has been done using single proteins or
simple mixtures of proteins, they provide the potential for the identification and quantitation of phosphoproteins in complex proteome samples.

2. Classical Protein Phosphorylation Detection

Although two-dimensional polyacrylamide gel electrophoresis (2-D-PAGE) has been
used primarily for the separation and quantitation of the relative abundance of proteins
from different systems, methods are also being developed to characterize the phosphoproteome of cells using these methods. The primary difference in detecting phosphorylated proteins is the staining method that is selected. The optimum labeling or
staining procedures will be specific to phosphorylated proteins and not interact with
nonphosphorylated species. Because many phosphorylation-regulated proteins are present in low amounts within the cell, the staining or labeling method must also be sensitive. Both of these criteria are met using both ^{32}P-labeling (when possible) and
immunostaining. Although the use of ^{32}P-labeling is only applicable to cultured cells,
it does represent a sensitive method of phosphoprotein detection. After in vivo label-

ing of the proteins within the cell with ^{32}P, they are fractionated using procedures such as 2-D-PAGE or high-performance liquid chromatography (HPLC). To determine the amino acid types that are modified, the phosphoproteins are completely hydrolyzed, and the phosphoamino acid content is determined. The specific site(s) of phosphorylation can be determined by proteolytic digestion of the radiolabeled protein, separation, and detection of phosphorylated peptides (e.g., by 2-D peptide mapping), followed by Edman sequencing. To measure differences in relative abundances of phosphoproteins, ^{32}P-labeled proteomes are separated by 2-D-PAGE, and the gels are visualized by autoradiography. Since only the radioactively labeled proteins produce an image, the phosphorylation state of proteins can be compared by measuring their relative spot intensities *(5)*. Unfortunately, the use of ^{32}P$_i$ to label proteins does not lend itself to high-throughput proteome-wide analysis because of issues with handling radioactive compounds and the associated contamination of analytical instrumentation.

Immunostaining is more universally applicable and is still highly sensitive. Furthermore, antibodies can be used to discriminate among seryl, threonyl, and tyrosyl phosphorylation. The general low abundance of phosphoproteins often requires a two-pronged approach for detection and subsequent identification of these proteins. As shown in **Fig. 1**, two distinct 2-D-PAGE gels of each sample, an analytical and preparative gel, can be run *(9)*. The analytical gel is used to detect the phosphoproteins specifically. In immunostaining, the proteins from the analytical gel are transferred to a polyvinylidene fluoride (PVDF) membrane, which is then immunoblotted using a phosphospecific antibody of choice. The visualized phosphoproteins can then be traced back to their spots on the preparative gel in which more total protein has been separated and they have been stained using a conventional staining method such as Coomassie blue or silver stain. To measure changes in protein phosphorylation state, two distinct proteome samples can be run on analytical gels that are then separately transferred to PVDF membranes and visualized by immunoblotting. The intensity of the immunoblotted phosphoprotein spots can be compared between the two gels. Since gel-to-gel reproducibility is always an issue in 2-D-PAGE, the immunostained membrane can also be stained to provide landmarks to pinpoint the exact spot of the phosphoprotein on the preparative gel. Some of the stains used for this include India ink, and colloidal gold stain *(10)*. Colloidal gold stain has the advantage of being more sensitive, however, it can result in high background on nitrocellulose membranes. However, the inherent limitations of gel-based separation constrain the overall utility of this methodology as a tool to characterize phosphoproteome comprehensively.

Fortunately, MS-based methods have been developed that provide more effective methods to identify, and potentially quantify, specific sites of phosphorylation. In its simplest form, MS can be used to provide an accurate mass measurement of an intact phosphorylated protein. Comparing this mass with the calculated mass of the unmodified protein or the mass of the protein after phosphatase treatment allows the number of bound phosphate groups to be calculated *(11)*. Unfortunately, analysis of the intact protein by this method does not provide any information related to the specific site of phosphorylation—a key piece of information that can directly affect the protein's function. To identify the specific phosphorylated residues requires analysis of the protein at the peptide level. Peptides are generated by enzymatic or chemical digestion of the intact protein which are then analyzed by either MS or tandem MS *(12)*. Although MS

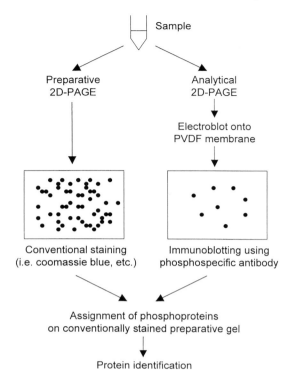

Fig. 1. The identification of phosphoproteins using 2-D-PAGE. A preparative and analytical gel of the proteome sample is run because of the generally low abundance of phosphoproteins within proteome samples. The analytical gel is immunoblotted using a phosphospecific antibody to localize the phosphoproteins within the gel. The visualized spots are aligned with their corresponding spots on the preparative gel that has been stained using a conventional protein staining method, such as colloidal Coomassie. The phosphoprotein is identified using standard MS-based methods such as tandem MS or peptide mapping.

measurements can confirm the presence of a phosphate group on a peptide, tandem MS is still necessary to establish the specific site of phosphorylation when two or more phosphorylatable residues are present. Because a majority of the identification of phosphorylation sites is made at the peptide level, this chapter focuses on MS analysis of phosphopeptides.

3. MS-Based Phosphoprotein Characterization

3.1. Enrichment of Phosphopeptides

The relatively low abundance of phosphopeptides compared with unmodified peptides within a complex mixture is one of the major difficulties in confidently identifying true phosphopeptides *(13)*. Methods that can selectively capture phosphopeptides, thereby generating a mixture enriched for these species, not only increase the ability to detect this class of peptides but also aid in the downstream analysis since the identification of a greater percentage of phosphopeptides can be anticipated. Phosphopeptide enrichment combined with high sensitive MS technologies provides great potential for phosphoproteome characterization. Several strategies, both classical and developmental, have been (or are being) developed to enrich samples for phosphorylated peptides or phosphoproteins prior to MS analysis.

3.2. Immunoaffinity Chromatography

One of the earliest means to enrich a sample for phosphopeptides prior to MS analysis is the use of antibodies *(14)*. Antibodies can be used to immunoprecipitate a phosphorylated protein(s) from a mixture or select phosphopeptides from a mixture containing all types of modified and unmodified peptides. The selection of the antibody to use depends on many factors including prior knowledge of the sample. If it is known that a protein of interest is phosphorylated and the question is at which site(s), an antibody directed toward a specific epitope within the protein may be used. If there is no prior knowledge about the sample and the goal is to identify unknown phosphorylation sites, a phospho-specific antibody is the best choice. A number of phosphoseryl (pSer)-, phosphothreonyl (pThr)-, and phosphotyrosyl (pTyr)-specific antibodies have been developed against specific phosphoproteins *(15)*. The potential specificity of such antibodies makes site-specific phosphorylation analysis feasible, providing the relevant antibodies can be prepared. In the cases of phospho-specific antibodies, the phospho-amino acid is part of the recognition epitope but is often not the major recognition factor between the antibody and phosphoprotein. The contribution of the neighboring residues allows high-affinity antibodies to be prepared, so that the phosphorylation status of single proteins within complex mixtures can be monitored.

A common procedure is the use of antibodies whose affinity is more dependent on the phosphorylation state of a specific residue type (i.e., Ser, Thr, or Tyr) and not the neighboring primary sequence *(16)*. These antibodies are generally used to generate mixtures of proteins or peptides that are phosphorylated at a specific residue type or to indicate the presence of phosphorylated proteins within a complex mixture that has been fractionated, for example, using 2-D-PAGE. The detected phosphoproteins are then identified and characterized using MS technologies. Yanagida et al. *(17)* used this strategy to study changes in the phosphorylation states of proteins extracted from murine fibroblast L929 cells treated with tumor necrosis factor-α (TNF-α). In this study, proteins were extracted from L929 cells at various time points after TNF-α treatment. The protein extracts were separated by 2-D-PAGE and either electroblotted onto a PVDF membrane or silver-stained. The blot was immunostained using an anti-pTyr monoclonal antibody to enable the identification of phosphoproteins as well as quantify any changes in the phosphorylation state of the proteins over time. The protein spots that immunostained with the anti-pTyr antibody were correlated with their position on the silver-stained gel. These gel spots were then excised from the 2D-PAGE gel, digested with trypsin and analyzed by matrix-assisted laser desorption/ionization time-of-flight (MALDI-TOF) MS to identify the phosphoprotein. Twenty-one different phosphoproteins were identified within the L929 cell lysates, including eight that showed a time-dependent change in their phosphorylation state based on the immunostaining of the PVDF membrane.

Although the above describes the use of phosphoamino acid-specific antibodies to isolate or detect intact phosphoproteins prior to digestion and MS analysis, these antibodies may also be used post-digestion to enrich for phosphopeptides. This strategy was used to identify the phosphorylated residues within EphB2, a receptor tyrosine kinase involved in neuronal axon guidance, neural crest cell migration, the formation of blood vessels, and the development of facial structures and the inner ear *(18)*. In this study, EphB2 was first isolated using an anti-EphB2 antibody followed by tryptic digestion of the intact protein. An anti-pTyr antibody was then used to isolate the phosphopeptides from this mixture. Eight major peaks observed using MALDI MS represented phosphopeptides containing nine different pTyr residues.

3.3. Immobilized Metal Affinity Chromatography

One of the most popular affinity-based methods to extract phosphoproteins or phosphopeptides from complex mixtures is immobilized metal affinity chromatography (IMAC) *(19)*. In IMAC, trivalent cations such as Fe^{3+} or Ga^{3+} are bound to a solid support such as iminodiacetic or nitrilotriacetic acid. Passing a complex mixture over an IMAC column results in enrichment for the phosphorylated species owing to the affinity of the phosphate moiety for the metal ion. After washing the column, the remaining bound species are eluted using high pH or phosphate buffer. Recently, Cao and Stults *(20)* developed a 2-D separation procedure incorporating IMAC followed by capillary electrophoresis (CE) coupled directly on-line with electrospray ionization (ESI)-MS/MS to allow the preconcentration, separation, and identification of phosphopeptides. The system was initially demonstrated using phosphorylated β- and α-casein and shows considerable promise for the analysis of complex mixtures of phosphoproteins. However, nonspecific binding of nonphosphopeptides to the IMAC column is a common issue for the IMAC technique. This nonspecific binding mostly could be contributed by the carboxylate groups of peptides such as those from C-terminus, glutamic, and aspartic acids. By converting carboxylic acid groups to methyl esters, Ficarro et al. *(21)* demonstrated that the nonspecific binding of peptides to the IMAC column was reduced by at least two orders of magnitude. In their work, the phosphoproteome of yeast was enriched using this strategy, and the resulting phosphopeptides were analyzed by nanoflow HPLC/ESI-MS. A total of 216 phosphopeptides were identified, among them 156 peptides that were multiply phosphorylated. The quantitation of phosphopeptides was not fully demonstrated in that work, but this methodology can easily be modified to allow quantitation of phosphoproteins by converting the peptides to the corresponding methyl esters using either normal isotopic or deuterated methanol prior to IMAC enrichment of phosphopeptides. In addition, custom-made, nanoscale IMAC columns have been used to enrich phosphorylated peptides from proteins that had been previously separated by sodium dodecyl sulfate (SDS)-PAGE *(22)*. These miniaturized IMAC columns have been combined with MALDI MS to characterize the in vivo phosphorylated peptides from the human p47/phox phosphoprotein.

3.4. Metabolic Labeling

Brian Chait's group presented one of the first examples utilizing stable isotope labeling to quantify changes in the phosphorylation state of a protein *(23)*. This group cultured *Saccharomyces cerevisiae* in both natural isotopic abundance (i.e., normal) and ^{15}N-enriched medium. Whereas all the proteins expressed in the cells grown in normal medium would be made up of atoms reflecting the natural isotope distribution of N (i.e., 99.635% ^{14}N and 0.365% ^{15}N), proteins expressed in the cells grown in ^{15}N-enriched medium would be made up primarily of ^{15}N atoms. The net result is two samples in which the proteins expressed in each medium can be distinguished by MS since the same

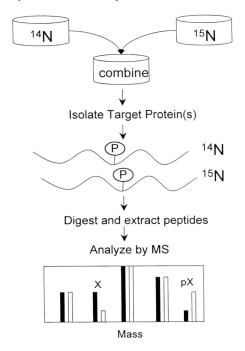

Fig. 2. Schematic of site-specific quantitation of changes in the level of protein phosphory-lation using metabolic isotope labeling and phosphoprotein isolation. Peptides that remain unchanged in the two cell pools are assumed to be present in equal abundance, and the level of phosphoporylation of peptide A is assumed to change from 30% (pool 1) to 70% (pool 2), leading to a decrease in the measured intensity ratio of unphosphorylated peptide X and an increase for phosphorylated peptide pX.

peptide in each sample will have a distinctive mass difference based on the number of nitrogen atoms within each peptide. Although immuno- and IMAC columns are effective at enriching mixtures of phosphopeptides, they do not provide any direct method to quantify the degree of phosphorylation. It would be useful to measure the relative phosphorylation state of a peptide between two different samples, for instance, when one is treated with a growth factor, drug, or other substance. Chait's laboratory was the first to show the ability to measure the relative abundance of a phosphopeptide from two distinct samples *(23)*. The general strategy is shown in **Fig. 2**. In this initial demonstration, a wild-type and G1 cyclin Cln2-deficient yeast strain were grown in either natural isotopic abundance and ^{15}N-enriched media, respectively. After combining the cells and extracting the proteins, the Cln2- dependent protein STE20 was isolated and analyzed by ESI-MS. Mass spectral measurement of the intensity ratios of the isotopically labeled (*cln2*$^-$) vs unlabeled (*CLN2*$^+$) phosphopeptides showed that at least four sites exhibit large increases in phosphorylation in the *CLN2*$^+$ cell pool. These Cln2-dependent phosphorylation sites appear to be consensus cyclin-dependent pSer/pThr, consistent with direct phosphorylation of Ste20 by Cln2-Cdc28 *(24)*. The two disadvantages of this approach are that because the protein itself is isotopically labeled using heavy isotope-enriched medium, the method is only amenable to organisms that can be metabolically labeled, and it does not provide any means to enrich a mixture specifically for phosphopeptides beyond the use of IMAC or the antibodies described above.

3.5. Chemical Modification and Isotopic Labeling of Phosphopeptides

Fortunately, two new methods have been developed that provide for the specific enrichment and quantification of phosphopeptides. Both methods incorporate stable isotopes to label the samples to be compared differentially and employ subsequent MS analysis for the identification and quantitation of the enriched phosphopeptides.

The first strategy to isolate and quantify phosphopeptides was developed concurrently, and independently, by two groups *(25,26)*. Although there are subtle differences in the specific procedures, the overall approaches are similar. The reaction scheme illustrating the labeling of the pSer and pThr residues of phosphoproteins is outlined in **Fig. 3**. Unfortunately the chemistry involved in this procedure does not make it amenable to pTyr residues. The first step involves blocking reactive sulfhydryls such as those from cysteinyl residues via reductive alkylation or performic acid oxidation. In the following step, phosphate groups are removed via hydroxide ion-mediated β-elimination from the pSer and pThr residues, resulting in their conversion to dehydroalanyl and β-methyl-dehydroalanyl residues, respectively. These newly formed α,β-unsaturated double bonds render the β-carbon in each sensitive to nucleophilic attack; hence, the next modification involves a Michael addition of the bifunctional reagent 1,2-ethanedithiol (EDT). The addition of EDT to either the dehydroalanyl or the β-methyl-dehydroalanyl residue results in the residues with a free sulfhydryl, which now serves as a reactive site that can be covalently modified with iodoacetyl-PEO-biotin, a thiol-reactive molecule containing a free biotin group. The end result is the covalent modification of phosphoryl residues with a biotinylated tag in place of the former phosphate moiety. Relative quantification of phosphopeptides is achieved using stable isotopic labeling with commercially available sources of either the light ($HSCH_2CH_2SH$ or EDT-D_0) or the heavy ($HSCD_2CD_2SH$ or EDT-D_4) isotopic version of EDT. Once the samples have been modified they are combined and digested with trypsin (or another proteolytic enzyme of choice), and the modified peptides are specifically isolated using immobilized avidin chromatography, taking advantage of the high affinity of the biotin/avidin interaction. These isolated peptides are analyzed using reversed-phase liquid chromatography (LC) coupled directly on-line with MS.

The identification and quantitation of the phosphorylated peptides is usually achieved using two MS operation modes. In the MS mode, the masses of the intact peptides are measured, and the signal intensity of the individual species provides a direct measure of phosphorylation status of the peptides. The signals originating from the modified versions of the phosphopeptides are usually recognized as pairs separated by a mass difference between the EDT-D_0 and EDT-D_4 labels (i.e., 4.0 Daltons or its multiples). The mass spectrum of the phosphorylated peptide FQS^{P}EEQQQTEDELQDK from β-casein in which the pSer residue has been modified with EDT (D_0 or D_4) and iodoacetyl-PEO-biotin is shown in **Fig. 4A** *(25)*. The 2.01 *m/z* difference between the [M+2H]$^{2+}$ ions at *m/z* 1236.96 and 1238.97 corresponds to the expected 4.02-Dalton mass difference.

In MS/MS mode, the labeled phosphopeptide ion collides with an inert gas (usually nitrogen or helium), which causes it to fragment into smaller ions. These fragment, or daughter, ions provide partial sequence information that is used in conjunction with com-

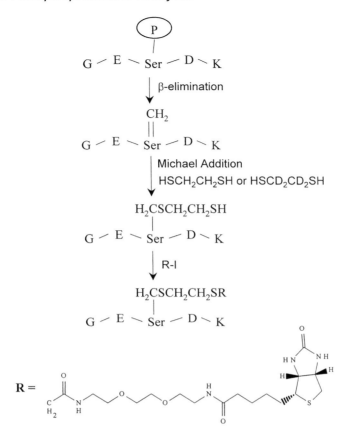

Fig. 3. Phosphopeptide isotope affinity tag (PhIAT) strategy for isolating and quantifying phosphopeptides. Proteins containing phosphothreonyl or phosphoseryl residues are modified with reagents containing an isotypically labeled linker and biotin group. The modified peptides are isolated by using immobilized avidin affinity chromatography. The quantitation of the relative phosphorylation state of phosphopeptides extracted from two different sources is achieved through the use of a light (EDT-D_0) and heavy isotopic version (EDT-D_4) of 1,2-ethanedithiol (EDT).

mercially available computer algorithms to identify the peptide. The key attribute of this labeling strategy to identify phosphopeptides is that the modification remains attached to the Ser or Thr residue during MS/MS fragmentation of the peptide. In a typical experiment studying unmodified phosphopeptides, the phosphate group can readily dissociate from the peptide during MS/MS (indeed, even in MS analyses as well), preventing the site-specific assignment of the phosphate modification. The modification strategy described above, however, allows the exact phosphorylation site to be determined by MS/MS, as shown in **Fig. 4B** *(25)*. In this example the MS/MS spectra for the EDT-D_0 and D_4 modified versions of a peptide are shown. The mass difference between the y_{13}^{2+} and y_{14}^{2+} daughter ions is equal to the mass of a seryl residue modified as described above. Although there is at least one other possible site of phosphorylation within this peptide, the MS/MS spectrum clearly identifies the seryl residue as the site of phosphorylation.

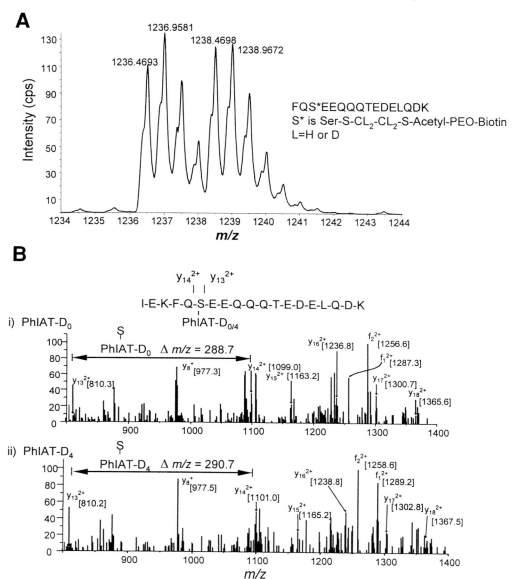

Fig. 4. (**A**) Mass spectra of EDT-D_0/D_4 labeled β-casein peptide. The [M+2H]²⁺ ion pair corresponds to the mass of the derivatized β-casein phosphopeptide FQS*EEQQQTEDELQDK where S* is covalently modified with -SCL₂CL₂S-acetyl-PEO-biotin and L is either H (EDT-D_0 label) or D (EDT-D_4 label). (**B**) Tandem MS identification of β-casein phosphopeptides. The tandem MS/MS spectra of a phosphopeptide modified and affinity isolated using the (**i**) light and (**ii**) heavy isotopic versions of EDT and iodoacetyl-PEO-biotin are shown. Both labeled versions of the phosphopeptide were identified in a single LC data-dependent MS/MS analysis of the enriched mixture.

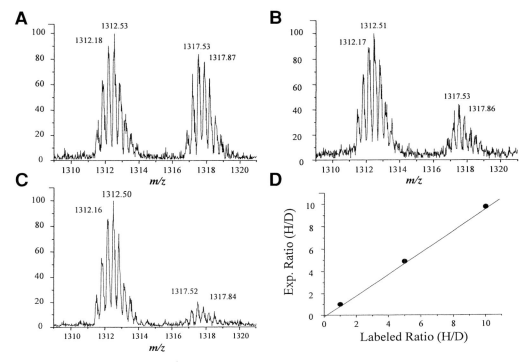

Fig. 5. Stoichiometric isotopic labeling of β-casein phosphopeptides. Samples containing sto-ichiometric ratios of β-casein of (A) 1:1, (B) 5:1, and (C) 10:1 were labeled with EDT-D_0:EDT-D_4 and analyzed by LC-MS. The relative quantity of each species was calculated by integrating their reconstructed ion chromatograms. (D) Excellent agreement between the expected and experimental ratios were observed when they were plotted against one another.

Another novel attribute of the labeling strategy described above is its ability to quantify the relative phosphorylation of a peptide from two different samples. As an illustration, several stoichiometric amounts of β-casein were processed as described in **Fig. 3**. As shown for a modified phosphopeptide in **Fig. 5**, the ratios integrated for each mass spectrum correlate well with the stoichiometric concentration ratios of the 1:1, 5:1, and 10:1 used in the labeling experiment. The two protein samples are pooled after the isotopic labeling step, eliminating any variations associated with subsequent sample handling.

A similar labeling scheme was developed concurrently by Chait's group *(26)*. Their procedure for modifying the phosphopeptides was essentially the same as that described above except that (+)-biotinyl, 3-maleimidopropionamidyl-3,6-dioxoctanediamine was used in place of iodoacetyl-PEO-biotin. The result of this substitution is that two versions of each phosphopeptide separated by 18 Daltons are observed within the mass spectrum. This doublet results from partial opening of the maleimide ring through hydrolysis. Beyond testing their procedure using experiments similar to those described by Goshe et al. *(25)*, this group also tested the ability of their approach to the identification and characterization of phosphoproteins by directly isolating the biotinylated peptides from a trypsin digest of an unfractionated protein mixture. They doped a whole yeast cell extract with a high concentration of chicken ovalbumin, an exogenous phos-

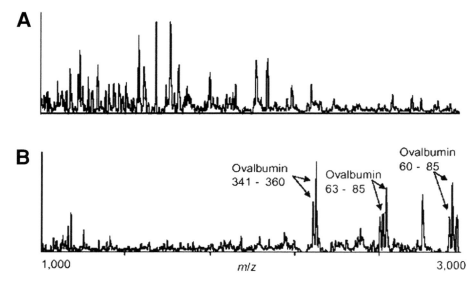

Fig. 6. Enrichment and identification of modified phosphopeptides. (**A**) The MALDI MS spectrum of a yeast cell lysate to which 2% (w/w) ovalbumin that has been modified as described in **ref. *26*** has been added. (**B**) The MALDI MS spectrum of the same sample that has been passed over a monomeric avidin column to extract the modified phosphopeptides. Signals corresponding to modified phosphopeptides from ovalbumin were readily observed in this spectrum.

phorylated protein. They then recorded the MALDI MS spectra of the complex peptide mixture before and after avidin purification (**Fig. 6**). Although the initial spectrum is quite complex, three intense pairs of peaks dominate the spectrum obtained after avidin purification. These peaks not only corresponded to the molecular masses of known phosphorylated peptides in ovalbumin (i.e., residues 341–360, 63–85, and 60–85), they were also identified by ESI-LC-MS/MS. Consistent with experiments described by Goshe et al. (*25*), the MS/MS spectra were able to define the sites of biotinylation and hence the site of phosphorylation. This experiment effectively demonstrated the value in being able to enrich a mixture for phosphopeptides prior to MS analysis. Although this group did not incorporate a stable isotope labeling step to quantitate differences in the relative amounts of the phosphopeptides, this step can be easily incorporated into their scheme, as demonstrated by Goshe et al. (*25*)

Like many other methodologies for phosphopeptide detection and identification, the strategies developed by the above two groups have their disadvantages. The hydroxide-mediated β-elimination not only reacts to pSer and pThr, it also occurs to the *O*-glycosylated Ser and Thr residues in spite of relatively lower reactivity, which may result in ambiguity in distinguishing between phosphorylation and glycosylation if a sample has both types of protein modifications. In addition, the diastereomers formed from the same peptide upon Michael addition of EDT can be slightly resolved by high-resolution reversed-phase HPLC, which complicates the overall separation of peptides if complex peptide mixtures are analyzed and subsequently reduces the ability to detect and identify low-abundance species based on LC-MS/MS method.

The second labeling method to isolate and quantify phosphopeptides has been developed in the laboratory of Reudi Aebersold *(27)*. Although the chemistry involved is more complex, this method potentially provides greater enrichment since the phosphopeptides are covalently linked to a solid support during the enrichment step. The sequence of chemical reactions for selectively isolating phosphopeptides from a peptide mixture consists of six steps, as shown in **Fig. 7**. To eliminate potential intra- and intermolecular condensation, the peptide amino groups are protected using *t*-butyl-dicarbonate (tBoc) chemistry *(28)*. Following this the carboxylate and phosphate groups are modified via a carbodiimide-catalyzed condensation reaction to form amide and phosphoramidate bonds. The phosphoramidate bonds are then hydrolyzed via an acid-mediated reaction to deprotect the phosphate group and cystamine is attached to the regenerated phosphate group via another carbodiimide-catalyzed condensation reaction. A free sulfhydryl group is generated at each phosphate group by reduction of the internal disulfide of cystamine, which allows the phosphopeptides to be captured by reacting the sulfhydryl groups with iodoacetyl groups immobilized on glass beads. The covalent attachment of the phosphopeptides to the solid phase allows stringent washing conditions to be used, thereby reducing the number of nonspecifically bound components being recovered with the phosphopeptides of interest. The phosphopeptides are recovered by cleavage of phosphoramidate bonds using trifluoroacetic acid at a concentration that also removes the tBoc protection group, thus regenerating peptides with free amino and phosphate groups. The carboxylate groups, however, remain blocked from step 2.

Although the labeling strategies described above were developed using model systems, they have also been assessed with cell lysates since the greatest utility will be in the application to proteome-wide identification and quantitation of phosphopeptides. In the case of the strategy proposed by Aebersold et al. *(27)*, phosphopeptides were isolated from a *S. cerevisiae* cell lysate and analyzed by LC-MS/MS, with collision-induced dissociation (CID) spectra recorded and searched against the sequence database. Greater than 80% of the useful MS/MS spectra were identified as phosphopeptides. This method yielded mixtures highly enriched in phosphopeptides with minimal contamination from other peptides. Because this strategy does not require the removal of the phosphate group, it is equally applicable to pSer-, pThr-, and pTyr- containing peptides. The MS/MS spectra of the modified phosphopeptides were of high enough quality to allow the peptides to be identified using sequence database searching. The MS/MS spectra could discriminate between pSer/pThr and pTyr- containing peptides since pSer and pThr readily lose the H_3PO_4 group during CID, allowing these residues to be identified via the fragment ions corresponding to the loss of 98 Daltons *(27,29)*. However, definitive identification of the exact phosphorylation sites in multiply phosphorylated peptides is difficult in many cases owing to the neutral loss of phosphate. Phosphotyrosyl residues are more stable and do not lose their phosphate group during fragmentation. Although this strategy does not provide a direct method to quantify changes in phosphorylation state between peptides from two different samples, the blocking of the carboxylates using either normal isotopic abundance or deuterated ethanolamine (i.e., ethanolamine-d_4) would allow for incorporation of stable isotope tags that later can be differentiated and quantified by MS.

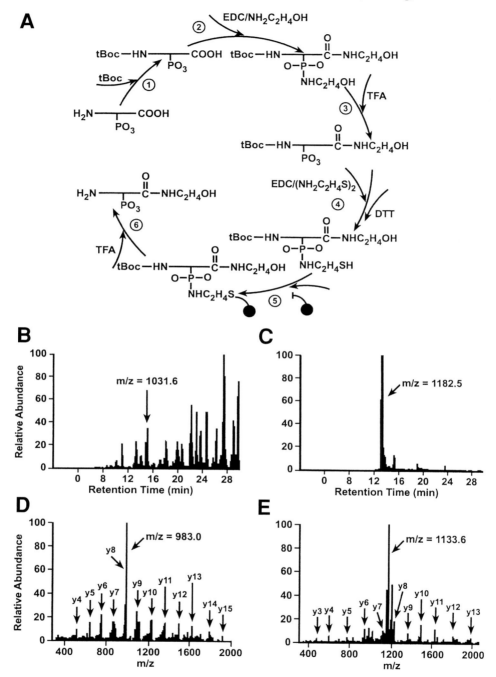

Fig. 7. Phosphopeptide isolation strategy and validation with phosphoprotein β-casein. (**A**) Schematic illustration of the chemistry involved in selective phosphopeptide isolation. A tryptic digest of β-casein (10 pmol) was analyzed by LC-MS/MS both before (**B, D**) and after (**C, E**) phosphopeptide isolation according to the procedure in **A**. (**B**) Ion chromatogram of 1 pmol of β-casein digest prior to phosphopeptide isolation. The doubly charged form of the expected tryptic phosphopeptide from β-casein is represented by the signal at $m/z = 1031.6$. (**C**) Ion chromatogram of the isolated phosphopeptides of β-casein digest. The peak at $m/z = 1182.5$

4. Conclusions

Although the information obtained from sequencing a relative table genome is not enough to understand a biological system, proteome analysis should provide much information on how proteins work and execute their functions within a cell. Quantitative measurement of proteome alterations at the protein expression level is basically important to understand how a cell changes its functional effectors in response to perturbations, however, such a measurement is only one kind of inspection of many changes within a cell. Reversible phosphorylation as a key signal transduction mechanism within a cell plays essential roles in regulating a cell's function in the context of molecular networks. Investigation of phosphoproteins at a proteome level or phosphoproteome, therefore, helps us to understand better how a cell is regulated. This kind of investigation involves phosphoprotein detection, site localization of phosphorylation, qualitative determination of phosphorylation and dephosphorylation, and quantitative measurement of phosphoprotein changes.

For many years, classical phosphoprotein detection methods have been used in many biological studies to examine phosphoproteins. The recent MS technology provides a highly sensitive tool to determine the phosphorylation state of a protein or proteome, taking advantage of its high mass measurement accuracy. The phosphopeptide enrichment and chemical isotopic labeling strategies described above enhance our ability to detect these low-level phosphoproteins and even quantify the subtle changes of phosphorylation states. Although these methods have their own limitations and need to be further exploited and evaluated in the global phosphoproteome analysis for cells, they are undoubtedly the valuable tools that make such investigation practical and closer to the realm of quantitative phosphoproteome. By incorporating multidimensional chromatographic separation, these methodologies should enable detect extremely low-level phosphopeptides and further expand the overall dynamic range of the analysis. The continued development of both sample preparation for enriching and labeling phosphoproteins and instrumentation for sensitive detecting those isolated phosphoproteins or phosphopeptides will ultimately reach the goals of phosphoproteome analysis.

Acknowledgments

This project has been funded with federal funds from the National Cancer Institute, National Institutes of Health, under contract no. NO1-CO-12400.

By acceptance of this article, the publisher or recipient acknowledges the right of the U.S. government to retain a nonexclusive, royalty-free license and to any copyright covering the article. The content of this publication does not necessarily reflect the views or policies of the Department of Health and Human Services, nor does

---

Fig. 7 *(continued)* represents the doubly charged form of the same tryptic phosphopeptide from β-casein indicated in **B**, modified by ethanolamine at its seven carboxylate groups. **(D)** CID spectrum of β-casein digest in **B**. The peak at $m/z = 983.0$ represents the doubly charged form of the selected parent ion ($m/z = 1031.6$) minus the H_3PO_4 group. **(E)** CID spectrum of isolated phosphopeptides of β-casein digest in **C**. Again, the peak at $m/z = 1133.6$ represents the doubly charged form of the selected parent ion ($m/z = 1182.5$) minus H_3PO_4, and the y-ion series used for peptide identification are indicated.

mention of trade names, commercial products, or organizations imply endorsement by the U.S. government.

References

1. Broder, S. and Venter, J. C. (2000) Sequencing the entire genomes of free-living organisms: the foundation of pharmacology in the new millennium. *Annu. Rev. Pharmacol. Toxicol.* **40,** 97–132.
2. Brown, P. O. and Botstein, D. (1999) Exploring the new world of the genome with DNA microarrays. *Nat. Genet.* **22,** 33–37.
3. Dongre, A. R., Opiteck, G., Cosand, W. L., and Hefta, S. A. (2001) Proteomics in the post-genome age. *Biopolymers* **60,** 206–211.
4. Aebersold, R. and Goodlett, D. R. (2001) Mass spectrometry in proteomics. *Chem. Rev.* **101,** 269–295.
5. Cohen, P. (2000) The regulation of protein function by multisite phosphorylation—a 25 year update. *Trends Biochem. Sci.* **25,** 596–601.
6. Huang, C., Ma, W. Y., Young, M. R., Colburn, N., and Dong, Z. (1998) Shortage of mitogen-activated protein kinase is responsible for resistance to AP-1 transactivation and transformation in mouse JB6 cells. *Proc. Natl. Acad. Sci. USA* **95,** 156–161.
7. Liang, L., Jiang, J., and Frank, S. J. (2000) Insulin receptor substrate-1-mediated enhancement of growth hormone-induced mitogen-activated protein kinase activation. *Endocrinology* **141,** 3328–3336.
8. McDonald, B. J., Chung, H. J., and Huganir, R. L. (2001) Identification of protein kinase C phosphorylation sites within the AMPA receptor GluR2 subunit. *Neuropharmacology* **41,** 672–679.
9. Kaufmann, H., Bailey, J. E., and Fussenegger, M. (2001) Use of antibodies for detection of phosphorylated proteins separated by two-dimensional gel electrophoresis. *Proteomics* **1,** 194–199.
10. Dunn, M. J. (1999) Detection of total proteins on western blots of 2-D polyacrylamide gels. *Methods Mol. Biol.* **112,** 319–329.
11. Han, J. M., Kim, J. H., Lee, B. D., Lee, S. D., Kim, Y., Jung, Y. W., Lee, S., Cho, W., Ohba, M., Kuroki, T., Suh, P. G., and Ryu, S. H. (2002) Phosphorylation-dependent regulation of phospholipase D2 by protein kinase Cdelta in rat pheochromocytoma PC12 cells. *J. Biol. Chem.* **277,** 8290–8297.
12. Molloy, M. P. and Andrews, P. C. (2001) Phosphopeptide derivatization signatures to identify serine and threonine phosphorylated peptides by mass spectrometry. *Anal. Chem.* **73,** 5387–5394.
13. Annan, R. S., Huddleston, M. J., Verma, R., Deshaies, R. J., and Carr, S. A. (2001) A multidimensional electrospray MS-based approach to phosphopeptide mapping. *Anal. Chem.* **73,** 393–404.
14. Sun, T., Campbell, M., Gordon, W., and Arlinghaus, R. B. (2001) Preparation and application of antibodies to phosphoamino acid sequences. *Biopolymers* **60,** 61–75.
15. Tsai, E. M., Wang, S. C., Lee, J. N., and Hung, M. C. (2001) Akt activation by estrogen in estrogen receptor-negative breast cancer cells. *Cancer Res.* **61,** 8390–8392.
16. Marcus, K., Immler, D., Sternberger, J., and Meyer, H. E. (2000) Identification of platelet proteins separated by two-dimensional gel electrophoresis and analyzed by matrix assisted laser desorption/ionization-time of flight-mass spectrometry and detection of tyrosine-phosphorylated proteins. *Electrophoresis* **21,** 2622–2636.
17. Yanagida, M., Miura, Y., Yagasaki, K., Taoka, M., Isobe, T., and Takahashi, N. (2000) Matrix assisted laser desorption/ionization-time of flight-mass spectrometry analysis of pro-

teins detected by anti-phosphotyrosine antibody on two-dimensional-gels of fibrolast cell lysates after tumor necrosis factor-alpha stimulation. *Electrophoresis* **21,** 1890–1898.

18. Kalo, M. S., Yu, H. H., and Pasquale, E. B. (2001) In vivo tyrosine phosphorylation sites of activated ephrin-B1 and ephB2 from neural tissue. *J. Biol. Chem.* **276,** 38940–38948.

19. Gaberc-Porekar, V. and Menart, V. (2001) Perspectives of immobilized-metal affinity chromatography. *J. Biochem. Biophys. Methods* **49,** 335–360.

20. Cao, P. and Stults, J. T. (1999) Phosphopeptide analysis by on-line immobilized metal-ion affinity chromatography-capillary electrophoresis-electrospray ionization mass spectrometry. *J. Chromatogr. A* **853,** 225–235.

21. Ficarro, S. B., McCleland, M. L., Stukenberg, P. T., Burke, D. J., Ross, M. M., Shabanowitz, J., Hunt, D. F., and White, F. M. (2002) Phosphoproteome analysis by mass spectrometry and its application to *Saccharomyces cerevisiae*. *Nat. Biotechnol.* **20,** 301–305.

22. Stensballe, A., Andersen, S., and Jensen, O. N. (2001) Characterization of phosphoproteins from electrophoretic gels by nanoscale Fe(III) affinity chromatography with off-line mass spectrometry analysis. *Proteomics* **1,** 207–222.

23. Oda, Y., Huang, K., Cross, F. R., Cowburn, D., and Chait, B. T. (1999) Accurate quantitation of protein expression and site-specific phosphorylation. *Proc. Natl. Acad. Sci. USA* **96,** 6591–6596.

24. Wu, C., Whiteway, M., Thomas, D. Y., and Leberer, E. (1995) Molecular characterization of Ste20p, a potential mitogen-activated protein or extracellular signal-regulated kinase kinase (MEK) kinase kinase from *Saccharomyces cerevisiae*. *J. Biol. Chem.* **270,** 15984–15992.

25. Goshe, M. B., Conrads, T. P., Panisko, E. A., Angell, N. H., Veenstra, T. D., and Smith, R. D. (2001) Phosphoprotein isotope-coded affinity tag approach for isolating and quantitating phosphopeptides in proteome-wide analyses. *Anal. Chem.* **73,** 2578–2586.

26. Oda, Y., Nagasu, T., and Chait, B. T. (2001) Enrichment analysis of phosphorylated proteins as a tool for probing the phosphoproteome. *Nat. Biotechnol.* **19,** 379–382.

27. Zhou, H., Watts, J. D., and Aebersold, R. (2001) A systematic approach to the analysis of protein phosphorylation. *Nat. Biotechnol.* **19,** 375–378.

28. Bodanszky, A. B. M. (ed.). (1984) *The Practice of Peptide Synthesis*, vol. 21. Springer-Verlag, New York.

29. Bennett, K. L., Stensballe, A., Podtelejnikov, A. V., Moniatte, M., and Jensen, O. N. (2002) Phosphopeptide detection and sequencing by matrix-assisted laser desorption/ionization quadrupole time-of-flight tandem mass spectrometry. *J. Mass Spectrom.* **37,** 179–190.

17

Computer-Aided Strategies for Characterizing Protein Isoforms

Frédéric Nikitin and Frédérique Lisacek

1. Introduction

A large number of proteins of a given sample can be separated on a gel or through a column. In proteomics, separation techniques are coupled with mass spectrometers *(1,2)*. Besides, proteins usually undergo an enzymatic digestion prior or posterior to the separation step. Resulting peptide mass data are analyzed and matched with masses of theoretically digested proteins found in selected databases. The selection of a reference protein database is therefore crucial to guarantee the success and the precision of identification. The common situation of such a protein identification process, known as peptide mass fingerprinting (PMF), is illustrated in **Fig. 1**. Further analysis of the sequences that result from this first identification step is usually called *characterization*, which relies on careful examination of the cross-references commonly found in sequence databases **(Fig. 1)**.

In many instances, mass data confirm the presence of various forms of a single protein. Indeed, a given sample is likely to contain more than one gene product. Depending on the organism and the tissue under investigation, from 2 to more than 10 protein isoforms can be identified in a given sample [see the index to two-dimensional polyacrylamide gel electrophoresis (2-D PAGE) databases at www.expasy.org/ch2d/2d-index.html].

The present chapter describes and illustrates how sequence databases can be used for protein identification purposes and how they can be mined to track potential protein isoforms, that is, for characterization purposes. Both topics are approached through the particular use of expressed sequence tags (ESTs).

2. Reference Data Resources for Protein Identification in Higher Eucaryotes

Two criteria are essential to select an appropriate reference protein database: exhaustiveness and nonredundancy. This section mainly describes the general purpose protein databases.

From: *Handbook of Proteomic Methods*
Edited by: P. Michael Conn © Humana Press Inc., Totowa, NJ

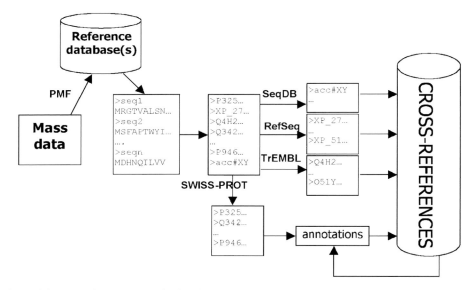

Fig. 1. Given a reference protein database, the outcome of peptide mass fingerprinting (PMF) is a list of protein candidates. SWISS-PROT is the only protein sequence database with high-level annotation and numerous crosslinks with other databases. All other web-based resources are minimally cross-referenced to gene/genome sequence and literature databases. The proteome annotation process is fed through these interacting components.

2.1. SWISS-PROT and TrEMBL

Three unique features characterize the SWISS-PROT database *(3)*:

1. Extended manual annotation of each entry.
2. Minimal redundancy (same protein sequences are merged into one single entry).
3. Cross-reference with other databases (currently approx 60 databases covering information relative to the literature, structure, biochemistry, protein families, etc.).

In SWISS-PROT, annotation encompasses a variety of topics:

1. Function(s) of the protein.
2. PTM(s).
3. Domains and sites.
4. Secondary structure.
5. Quaternary structure.
6. Similarities to other proteins.
7. Disease(s) associated with deficiencies in the protein.
8. Sequence conflicts, variants, and so on.

Most of the annotation is found in the comment lines (CC), in the feature table (FT), and in the keyword lines (KW) of an entry (*see* www.expasy.org).

TrEMBL is a computer-annotated supplement to SWISS-PROT that contains the translations of EMBL nucleotide sequence entries not yet integrated into SWISS-PROT *(3)*.

SWISS-PROT matches the nonredundancy criterion but covers various species unevenly. TrEMBL provides more sequence data but is less detailed, to the detriment of nonredundancy.

2.2. RefSeq and LocusLink

RefSeq and LocusLink aim at providing a comprehensive directory for genes and reference sequences for genomes of model organisms *(4)*. RefSeq focuses on reference sequence standards for genomes, transcripts, and proteins. In LocusLink, information on genes, gene families, variation, gene expression, and annotation is organized at each genome locus. Together, RefSeq and LocusLink offer a nonredundant and exhaustive view of model genomes (www.ncbi.nlm.nih.gov/LocusLink/). However, the automated procedures implemented for genome annotation are error-prone, and consistency checks are imperative, as illustrated later in this chapter.

2.3. Others

Various specialized databases, such as genome databases, are available to serve as reference. Obviously, a translated genome database offers both qualities of being exhaustive and nonredundant. The choice of the reference is thus highly dependent on the nature of the proteomic study, and all options cannot be described in this chapter. Furthermore, all that is described in the following discussion remains valid for proprietary data.

3. Protein Isoform Identification

The total number of genes in mammalian genomes is controversial. Leaning toward a low figure would go along with a high plasticity of macromolecules. Consequently, the challenge of detecting isoforms and variants of given proteins in the course of studying a particular cellular process in a particular tissue awaits researchers involved in understanding genome data.

To appreciate the extent of protein plasticity it is necessary to itemize possible *protein forms*—as exhaustively as possible—and relate them to each other. The relationship between a given form, for instance a splice variant, and the conditions of its production can be valuable in rationalizing the presence or the absence of this specific gene product. This information is still unevenly documented in the literature.

4. Tracking the Origin of Isoform

A variety of explanations can justify the presence of a given form of a protein. The following summarizes the events potentially generating diversity:

1. Pretranscriptional rationalization of protein forms.
   a. Duplicated genes.
   b. Polymorphism and mutations.
   c. Pseudo-genes.
2. Post-transcriptional generation of protein forms.
   a. Alternative splicing.
   b. RNA editing.
3. Post-translational generation of protein forms.
   a. Protein splicing.
   b. PTMs of amino acids.

The rationalization of one or the other event in relation to a protein identified with mass spectrometry (MS) data can only be found in the literature. However, exploring sequence data and itemizing all existing gene products is a first step toward elucidating the question. Indeed, the availability of multiple sources of data (chromosomal DNA, cDNAs, ESTs) allows such an investigation. Then, the conditions of production of data can be examined.

5. Relying on EST Data

The widespread production of ESTs, as can be seen from the millions of sequences stored in the dbEST database (www.ncbi.nlm.nih.gov/dbEST), provides a resourceful means of keeping track of variations in RNA transcripts found in various types of tissues in a given organism. Although this method may yield a number of sequence errors, it is interesting to note the existence of trends in nucleotide variations.

DbEST is a very redundant database. UniGene is a possible on-line resource in which ESTs are clustered by sequence similarity (www.ncbi.nlm.nih.gov/Unigene).

EST sequences are usually automatically clustered on the basis of nucleotide similarity. However, a careful inspection of translated sequences shows that variability at the amino acid level is also informative.

In fact, EST data are useful in proteomics in two different ways. On the one hand, translated EST sequences can either constitute or complement a protein reference database for PMF and contribute to protein *identification*. In relation to **Fig. 1**, translated ESTs represent in this case the partial or full content of the first cylinder and are thus used early on in the process. On the other hand, translated ESTs can represent the content of the second cylinder, as shown in **Fig. 2**, and thereby represent a resource for further investigation of the identified proteins, that is, *characterization*. The purpose of this chapter is mainly to illustrate the latter case and survey the former.

5.1. Translated EST for Identification

This approach is particularly helpful in the case of plant proteomics because DbEST contains a wealth of plant ESTs and so far only the *Arabidopsis thaliana* genome is available *(5,6)*. The efficiency of identification can be significantly enhanced since some translated ESTs are found neither in general-purpose protein databases nor in a dedicated complete proteome database. An increase of up to 50% in the identification possibilities is noted in published and unpublished data *(5,6)*.

Other applications were published such as in yeast *(7)*. Examples of the use of ESTs for tandem mass spectrometry (MS/MS) identification can also be found *(8,9)*.

5.2. Translated EST for Isoform Detection

When cross-checked with genomic or proteomic data, sequence variation can be associated with tissue specificity, developmental stage, or any other type of information related to conditions of expression. In the present case, the link depends *only* on sequence data. No known criteria allow the expression of a relationship between a gene and protein levels of expression.

Various situations are interpreted in **ref. 5** to complement the protein identification step.

5.3. Application

The refinement of annotation via the use of EST data is illustrated with the example of a human protein, as identified from mass data from a human sample. Using various

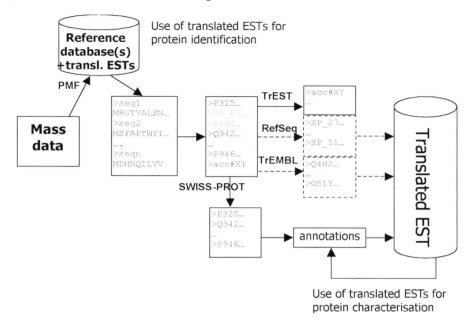

Fig. 2. The scheme described in **Fig. 1** is specified for the use of expressed sequence tag (EST) data. EST sequences can either be translated into six frames and can supplement the protein reference database for protein identification purposes, or can be considered as a cross-reference, given an identified protein to refine the protein specific annotations. PMF, peptide mass fingerprinting.

PMF engines (see options in Chap. 34) led to best match peptides to a mouse protein known as pim-3 (SWISS-PROT accession number: P58750, proto-oncogene, serine/threonine kinase). However, some matches did span a human protein known as pim-1 (P11309, another proto-oncogene, serine/threonine kinase). To appreciate the situation, the various forms of the *pim* proteins were searched in all reference databases. Three very close sequences of a serine/threonine kinase are found in human and mouse.

Here is the list of eucaryotic protein sequences used as references:

5.3.1. Pim-1

> P11309[1] = NM_002648[2] (M54915) pim-1 protein, 313 amino acids *[Homo sapiens]*.

> P06803 proto-oncogene serine/threonine-protein kinase pim-1, 313 amino acids *[Mus musculus]*.

5.3.2. Pim-2

> XP_010208 pim-2 oncogene, 334 amino acids *[Homo sapiens]*.
No mouse sequence available.

[1]SWISS-PROT entry.
[2]RefSeq entry.

5.3.3. Pim-3

> P58750 = BC017621 serine threonine kinase pim3, 326 amino acids [*Mus musculus*].

No human sequence available.

The alignment shown in **Fig. 3** outlines the very similar regions in the three gene products, with identical amino acids highlighted in light gray. It appears that the mouse pim-3 and human pim-1 are closer to each other than human pim-1 and pim-2. No publication has confirmed the existence of three genes for each organism. However, there are three major EST clusters associated with three distinct loci on the human genome:

Hs.81170 (LocusLink: 5292)
Hs.80205 (LocusLink: 11040)
Hs.5326 (LocusLink: 64840)

The first locus corresponds to pim-1, the second to pim-2, and the third to a so-called porcupine gene. The following demonstrates that Hs.5326 is wrongly annotated. This inconsistency of the EST cluster annotation was verified as follows.

Data are retrieved from the NCBI ProtEST page found in UniGene (*see* http://www.ncbi.nlm.nih.gov/UniGene/protest/) aiming at selecting two types of information from this html file:

1. The characteristics of the expressed sequence matching subject protein (EST/cDNA, Organism, Organ, tissue type, ...).
2. The alignments processed by BLASTX (BLASTX output, UniGene cluster identifier, and alignment range on subject protein).

Figure 4 shows an example of the BLASTX output where the Query is the translated EST alternatively in frame +1 and +3 and Subject is Pim-1 protein sequence. It appears that the poor quality of the data yields such complicated outputs, and the obvious question is: when is it right to change frame to keep on matching the Query and the Subject? We have developed a method to assemble the EST translations optimally and tackle the boundary problem of the sequence matches. It relies on processing the middle line of the alignment, i.e., the consensus, and exploiting the length of the consensus match as well as the amino acid composition quality in such a consensus (unpublished information). Each pairwise alignment is processed in such a way as to produce finally a global alignment of all available translated EST sequences with a given protein.

The method was used to extract ESTs of the three clusters mentioned earlier (Hs.81170, Hs.80205, and Hs.5326). The first two clearly corresponded to pim-1 and pim-2 proteins, respectively (data not shown). Although Hs.5326 is annotated as part of the porcupine gene data, all ESTs of this cluster are found in the BLASTX outputs of ProtEST, which emphasizes the inconsistency. It was, therefore, easy to retrieve sequences and to reach the conclusion of a closer relationship to the mouse pim-3. An example of information on specific amino acids brought by the alignment of translated ESTs is illustrated in **Fig. 5A**.

As a result, ESTs found in Hs.5326 could be assembled to yield the sequence of human *pim* annotated as pim-2 in SWISS-PROT (Q9P1W9, updated in June 2002). Whether there are two or three genes, including that for mouse, is still an open question. Further comments are found in **ref. *10***. **Figure 5B** provides yet another

```
Pim1_Hs (1-88)      MLLSKINSLAHLRAAPCNDLHATK-LAPG-KEKEPLESQYQVGPLLGSGGFGSVYG-IRVSDNLPVAIKHVEKDRISDWGELPNGTRVPM
Pim1_Mm (1-88)      MLLSKINSLAHLRARPCNDLHATK-LAPG-KEKEPLESQYQVGPLLGSGGFGSVYSGIRVADNLPVAIKHVEKDRISDWGELPNGTRVPM
Pim3_Mm (1-89)      MLLSKFGSLAHLCGPGGVDHLPVKILQPAKADKESFEKVYQVGAVLGSGGFGTVYAGSRIADGLPVAVKHVKERVTEWGSL-GGVAVPL
Pim2_Hs (1-82)      MLTKPLQGPP--APPG-----TPTPPPGGKDREAFEAEYRLGPLLGKGGFGTVFAGHRLTDRLQVAIKVIPRNRVLGWSPLSDSVTCPL


Pim1_Hs (89-172)    EVVLLKKVSS--GFSGVIRLLDWFERPDSFVLILERPEVQDLFDFITERGALQEELARSFFWQVLEAVRHCHNCGVLHRDIKDEN
Pim1_Mm (89-172)    EVVLLKKVSS--DFSGVIRLLDWFERPDSFVLILERPEVQDLFDFITERGALQEDLARGFFWQVLEAVRHCHNCGVLHRDIKDEN
Pim3_Mm (90-175)    EVVLLRKVGAAGGARGVIRLLDWFERPDGFLLVLERPAQDLFDFITERGALDEPLARRFFAQVLAAVRHCHNCGVVHRDIKDEN
Pim2_Hs (83-168)    EVALLWKVGAGGGHPGVIRLLDWFETQEGFMLVLERPLPAQDLFDYITEKGPLGEGPSRCFFGQVVAAIQHCHSRGVVHRDIKDEN


Pim1_Hs (173-258)   ILIDLNRGELKLIDFGSGALLKDTVYTDFDGTRVYSPPEWIRYHRYHGRSAAVWSLGILLYDMVCGDIPFEHDEIIRGQVFFRQR
Pim1_Mm (173-258)   ILIDLSRGEIKLIDFGSGALLKDTVYTDFDGTRVYSPPEWIRYHRYHGRSAAVWSLGILLYDMVCGDIPFEHDEIIKGQVFFRQT
Pim3_Mm (176-261)   LLVDLRSGELKLIDFGSGAVLKDTVYTDFDGTRVYSPPEWIRYHRYHGRSATVWSLGVLLYDMVCGDIPFEQDEILRGRLFFRRR
Pim2_Hs (169-254)   ILIDLRRGCAKLIDFGSGALLHDEPYTDFDGTRVYSPPEWISRHQYHALPATVWSLGILLYDMVCGDIPFERDQEILEAELHFPAH


Pim1_Hs (259-313)   VSSECQHLIRWCLALRPSDRPTFEEIQNHPWM-DVLLPQETAEIHLHSLSPGPSK-------------
Pim1_Mn (259-313)   VSSECQHLIKWCLSLRPSDRPSFEEIRNHPWMQGDLLPQAASEIHLHSLSPGSSK-------------
Pim3_Mn (262-326)   VSPECQQLIEWCLSLRPSERPSLDQIAAHPWM--LGTEGSVPENCDLRLCALDTDDGASTTSSS-ESL
Pim2_Hs (255-334)   VSPDCCALIRRCLAPKPSSRPSLEEILLDPWM-----QTPAEDVTPQPLQRRPCPFGLVLATLSLAWPGLAPNGQKSHPMAMSQG
```

Fig. 3. The multiple alignment of known human (Hs) and mouse (Mm) *pim* proteins. Numbers in parentheses correspond to amino acid positions. Identical residues are highlighted in light gray. Residues highlighted in darker gray emphasize the closer relationship between Mm pim-3 and Hs pim-1, as opposed to Hs pim-1 and Hs pim-2.

```
Sequence 1 BG471586 in Hs.5326 Length 972 (1-972)
Sequence 2 protein kinase (EC 2.7.1.37) pim-1 - human Length 313 (1-313)

Score = 233 bits (594), Expect = 5e-60
Identities = 118/169 (69%), Positives = 132/169 (77%)
Frame = +3

Query: 126 EPAQDLFDFITERGALDEPLARRFFAQVLAAVRHCHSCGVVHRDIKDENLLVDLRSGELK 305
           EP QDLFDFITERGAL E LAR FF QVL AVRHCH+CGV+HRDIKDEN+L+DL  GELK
Sbjct: 124 EPVQDLFDFITERGALQEELARSFFWQVLEAVRHCHNCGVLHRDIKDENILIDLNRGELK 183

Query: 306 LIDFGSGALLKDTVYTDFDGTRVYSPPEWIRYHR*HGRSSTVWSLGRASLRYGVWGHPLR 485
           LIDFGSGALLKDTVYTDFDGTRVYSPPEWIRYHR HGRS+ VWSLG    L   V G
Sbjct: 184 LIDFGSGALLKDTVYTDFDGTRVYSPPEWIRYHRYHGRSAAVWSLG-ILLYDMVCGDIPF 242

Query: 486 AGRGGSSEGRLLFRRRVSPECQQLIRWCLSLRPSERPSPGSRLRANPWM 632
               G++ FR+RVS ECQ LIRWCL+LRPS+RP+    ++ +PWM
Sbjct: 243 EHDEEIIRGQVFFRQRVSSECQHLIRWCLALRPSDRPT-FEEIQNHPWM 290

Score = 64.7 bits (156), Expect = 3e-09
Identities = 62/179 (34%), Positives = 75/179 (41%), Gaps = 15/179 (8%)
Frame = +1

Query: 7   VPLEVVLLRKVGAAGGARGVIRLLDWFERPDGFLLVLERP----------SRRRTSSTL 153
           VP+EVVLL+KV +   G  GVIRLLDWFERPD F+L+LERP          R     L
Sbjct: 86  VPMEVVLLKKVSS--GFSGVIRLLDWFERPDSFVLILERPEPVQDLFDFITERGALQEEL 143

Query: 154 SRSAAPWTSRWRAASSRRCWPPCATATAAGSCTATL----RTKICLWTCAPESSSSSTSV 321
           +RS      W+  + R   C              L   R ++ L    +    T V
Sbjct: 144 ARSFF-----WQVLEAVRHCHNCGVLHRDIKDENILIDLNRGELKLIDFGSGALLKDT-V 197

Query: 322 RVRCSRTRSTPTSTAPECTAPRSGSATTVNTGARAPCGRWAVLLYDMVCGDIPFEQDEE 498
               TR       +P         G  A       +LLYDMVCGDIPFE DEE
Sbjct: 198 YTDFDGTR--------VYSPPEWIRYHRYHGRSAAVWSLGILLYDMVCGDIPFEHDEE 247
```

Fig. 4. The classical output of a BLASTX process. Similar regions between two sequences are locally aligned. The Query sequence is a translated EST (BG471586). The Subject sequence is human pim-1. The EST is translated in six frames. In this particular case, which is quite common, a frameshift is apparent. As a result, the EST sequence matches the protein in frame +1 from matching position 86 to position 128 (in the corresponding protein) and in frame +2 from 124 position to 230. Such a common situation makes it difficult to automate the extraction of the right frame for translating an EST.

confirmation of an original sequence hidden in the EST cluster through the alignment of the C-terminal regions. The match for this region with the mouse pim-3 sequence is shown with amino acids in bold; and the regularity of the EST data tend to demonstrate the existence of a new or corrected human *pim* sequence. Noticeably, no mass data could match a peptide covering the C-terminal region of the protein.

6. Conclusions

The example detailed in this chapter emphasizes the need for careful consideration of sequence data and consistency checks of the multiple sources of data to ensure the robustness of proteome annotation. Needless to say, the correction of this erroneous and incomplete information on human *Hs.5326* sequences will be forwarded to the NCBI, and the example in this chapter will become obsolete like much of the information found

A

```
Pim1_Hs(89-172)   EVVLLKKVSS--GFSGVIRLLDWFERPDSFVLILERPEPVQDLFDFITERGALQEELARSFFWQVLEAVRHCHNCGVLHRDIKDEN
Pim1_Mm(89-172)   EVVLLKKVSS--DFSGVIRLLDWFERPDSFVLILERPEPVQDLFDFITERGALQEDLARGFFWQVLEAVRHCHNCGVLHRDIKDEN
Pim3_Mm(90-175)   EVVLRKVGAAGGARGVIRLLDWFERPDGFLLVLERPEPAQDLFDFITERGALDEPLARRFFAQVLAAVRHCHNCGVVHRDIKDEN
Pim2_Hs(83-168)   EVALLWKVGAGGGHPGVIRLLDWFETQEGFMLVLERPLPAQDLFDYITEKGPLGEGPSRCFFGQVVAAIQHCHSRGVVHRDIKDEN

AL537546_Hs5326   EVVLLRKVGTAGGARGVIRLLDWFERPDGFLLVLERPEPAQDLFDFITERGALDEPLARRFFAQVLAAVRHCHSCGVVHRDIKDEN  (brain)
BI561145_Hs5326   EVVLLRKVGTAGGARGVIRLLDWFERPDGFLLVLERPEPAQDLFDFITERGALDEPLARRFFAQVLAAVRHCHSCGVVHRDIKDEN  (testis)
```

B

```
Pim1_Hs(259-313)  VSSECQHLIRWCLALRPSDRPTFEEIQNHPWM-DVLLPQETAEIHLHSLSPGPSK-------------
Pim1_Mn(259-313)  VSSECQHLIKWCLSLRPSDRPSFEEIRNHPWMQGDLLPQAASEIHLHSLSPGSSK-------------
Pim3_Mn(262-326)  VSPECQQLIEWCLSLRPSERPSLDQIAAHPWM--LGTEGSVPENCDLRLCALDTDDGASTTSSS-ESL
Pim2_Hs(255-334)  VSPDCCALIRRCLAPKPSSRPSLEEILLDPWM----QTPAEDVTPQPLQRRPCPFGLVLATLSLAWPGLAPNGQKSHPMAMSQG

BM016405_Hs5326   VSPECQQLIRWCLSLRPSERPSLDQIAAHPWM--LGADGGAPESCDLRLCTLDP  (breast, mammary adenocarcinoma)
BI826905_Hs5326   VSPECQQLIRWCLSLRPSERPSLDQIAAHPWM--LGADGGAPESCDLRLCTLDP  (brain, medulla)
BI761657_Hs5326   VSPECQQLIRWCLSLRPSERPSLDQIAAHPWM--LGADGGAPESCDLRLCTLDP  (pooled colon, kidney, stomach)
BG681342_Hs5326   VSPECQQLIEWCLSLRPSERPSLDQIAAHPWM--LGADGGAPESCDLRLCTLEP  (skin, squamous cell carcinoma)
BG682847_Hs5326   VSPECQQLIRWCLSLRPSERPSLDQIAAHPWM--LGADGGAPESCDLRLCTLDP  (brain, neuroblastoma cell line)
BG473322_Hs5326   VSPECQQLIRWCLSLRPSERPSLDQIAAHPWM--LGADGGVPESCDLRLCTLDP  (eye, retinoblastoma)
AL543684_Hs5326   VSPECQQLIRWCLSLRPSERPSLDQIAAHPWM--LGADGGAPESCDLRLCTLDP  (placenta)
AL523928_Hs5326   VSPECQQXIRWCLXLRPSERPXLDQIAAHPWM--LGADGGAPESCDLRLCTLDP  (brain, neuroblastoma cell line)
BG739957_Hs5326   VSPECQQLIRWCLSLRPSERPSLDQIAAHPWM--LGADGGAPESCDLRLCTLDP  (skin)
BF972487_Hs5326   VSPECQQLIRWCLSLRPSERPSLDQIAAHPWM--LGADGGVPESCDLRLCTLDP  (uterus, leiomyosarcoma cell line)
```

Fig. 5. The multiple alignment of known human (Hs) and mouse (Mm) *pim* proteins along with selected translated ESTs from the Hs.5326 UniGene cluster. (A) A change of amino acid arising from examining EST sequences is boxed. (B) The C-terminal region, which regularly occurs in translated EST sequences, is aligned with the C-terminal regions of the *pim* proteins. Amino acids in bold are those matching between translated ESTs and pim-3.

in web resources. Nevertheless, the strategy will remain and hopefully may still turn out to be useful in other circumstances.

References

1. Wilkins, M. R, Williams, K. L., Appel, R. D., and Hochstrasser, D. F. (eds.) (1997) *Proteome Research: New Frontiers in Functional Genomics.* Springer, New York.
2. Pennington, S. R. and Dunn, M. J. (eds.) (2001) *Proteomics, from Protein Sequence to Function.* Springer, New York.
3. O'Donovan C., Martin, M. J., Gattiker, A., Gaseiger, E., Bairoch, A., and Apweiler, R. (2002) High-quality protein knowledge resource: SWISS-PROT and TrEMBL. *Brief. Bioinform.* **3,** 275–284.
4. Pruitt, K. D. and Maglott, D. R. (2001) RefSeq and LocusLink: NCBI gene-centered resources. *Nucleic Acids Res.* **29,** 137–140.
5. Lisacek, F. C., Traini, M. D., Sexton, D., Harry, J. L., and Wilkins, M. R. (2001) Strategy for protein isoform identification from expressed sequence tags and its application to peptide mass fingerprinting. *Proteomics* **1,** 186–193.
6. Mathesius, U., Keijzers, G., Natera, S. H., Weinman, J. J., Djordjevic, M. A., and Rolfe, B. G. (2001) Establishment of a root proteome reference map for the model legume *Medicago truncatula* using the expressed sequence tag database for peptide mass fingerprinting. *Proteomics* **1,** 1424–1440.
7. Neubauer, G., King, A., Rappsilber, J., et al. (1998) Mass spectrometry and EST-database searching allows characterization of the multi-protein spliceosome complex. *Nat. Genet.* **20,** 46–50.
8. Choudhary, J. S., Blackstock, W. P., Creasy, D. M., and Cottrell, J. S. (2001) Matching peptide mass spectra to EST and genomic databases. *Trends Biotechnol.* **9,** S17–S22.
9. Field, H. I., Fenyo, D., and Beavis, R. C. (2002) RADARS, a bioinformatics solution that automates proteome mass spectral analysis, optimises protein identification, and archives data in a relational database. *Proteomics* **2,** 36–47.
10. Chichester, C., Nikitin, F., Ravarini, J-C., Lisacek, F. (2003) Consistency checks for characterizing protein forms. *Computational Biology and Chemistry* **27,** 29–35.

18

Protein Variant Separations Using Cation Exchange Chromatography on Grafted, Polymeric Stationary Phases

Michael Weitzhandler, Dell Farnan, Nebojsa Avdalovic, and Chris Pohl

1. Introduction

Liquid chromatography is widely used in the quality control analysis of purified, formulated protein pharmaceuticals. High-resolution separations of protein variant forms provide information pertaining to product macro- and microheterogeneity that is used in assessing process consistency and product stability. High-resolution ion exchange chromatography can be used to separate and identify protein variants that may include glycosylated *(1)*, phosphorylated *(2)*, deamidated *(3)*, truncated *(4)*, and oxidized *(5)* forms. All these modifications may affect the activity or stability of the final product.

Physical-chemical characteristics associated with proteins that make the chromatographic separations of closely related variants challenging include large molecular weights and a propensity for hydrophobic interaction. The high molecular weights and large Stokes radii associated with proteins imply that they will exhibit very poor mass transfer properties even in free solution *(6)*. In conjunction with the anfractious nature of the pore structure found in conventional porous particles, these low free-solution diffusivities can lead to excessive band spreading under chromatographic conditions that would be suitable for small eluates. In recent years, an array of supports has been introduced with reported solutions to these problems *(7)*. The supports fall into three categories: macroporous, gigaporous, and gel-filled gigaporous *(8)*.

An alternative solution to the intraparticle mass transfer problem is simply to utilize polymeric particles with fluid-impervious cores in which there would not be any intraparticle mass transfer *(9,10)*. Such particle types should always result in higher efficiencies across the practical flow velocity range *(11)*. With grafted polymeric supports, the contact between the protein analyte and the base matrix support surface is markedly reduced, which reduces undesirable hydrophobic interactions between the analyte and the base matrix surface. Interesting properties observed for grafted polymeric ion exchangers are (1) reduced mass transfer resistance owing to the pellicular nature of the support, (2) marked changes in selectivity owing to the flexibility of the charge arrangement between charged sites on the protein and the functional groups located on the grafted ligands, and (3) improved mass recoveries owing to diminished nonspecific binding *(12)*.

From: *Handbook of Proteomic Methods*
Edited by: P. Michael Conn © Humana Press Inc., Totowa, NJ

In this study, we evaluated the performance of a set of grafted polymeric cation exchangers for the separation of proteins. In the described grafted supports, a rigid and impervious polymer particle has been completely coated with a base stable hydrophilic polymer. Weak (WCX) or strong (SCX) cation exchange functionalities were then grafted to the hydrophilic particle. This chemistry offers an alternative selectivity to tentacle supports attached to silica particles where silanol groups inadvertently contribute to undesirable interactions with the protein analyte. The separation of several different types of protein variants on the grafted supports was examined.

2. Materials

2.1. Apparatus

1. A DX-500 liquid chromatograph (Dionex, Sunnyvale, CA) consisting of a model GP40 gradient pump, an AD20 variable wavelength absorbance detector, and an AS3500 autosampler (Thermoseparations, San Jose, CA) equipped with a 50-μL sample loop was used throughout this work.
2. The chromatograph was controlled and data were collected using Dionex PeakNet version 5.01 software running on a Dell™ Pentium-based computer.

2.2. Materials

1. High-quality deionized water was prepared using a Millipore Milli-Q system (Bedford, MA).
2. Analytical reagent grade crystalline mono- and di-basic sodium phosphate, glacial acetic acid, and sodium acetate (J. T. Baker, Phillipsburg, NJ).
3. Analytical reagent grade sodium chloride (Fluka, Ronkonkoma, NY).
4. Samples of a humanized monoclonal, IgG1, antibody (a generous gift from a biotechnology company).
5. Carboxypeptidase B (Boehringer Mannheim, Indianapolis, IN).
6. The other proteins used in this work were from Sigma (St. Louis, MO).

3. Methods

Two types of buffer systems were used during the chromatography, sodium phosphate and sodium acetate. Two solutions were prepared for each system, one with and one without sodium chloride. Sodium phosphate buffer systems were prepared by dissolving appropriate quantities of mono- and di-basic sodium phosphate, to attain a chosen pH, and adding sodium chloride to the appropriate concentration in water. Sodium acetate solutions were prepared by titrating the solutions to the appropriate pH with glacial acetic acid. Protein samples used in this study were prepared by dissolving approx 0.5 mg of each protein per mL of the mobile phase that was used at the start of the separation.

The separations were made using the columns in the DX-500 chromatograph and operated at a flow rate of 1 mL/min. Separations were performed using methods that proportioned various linear gradients of sodium chloride in the presence of indicated sodium phosphate or sodium acetate concentrations at chosen pHs. After the required gradient for the separation was completed, the column was flushed using 750 mM NaCl for 10 column volumes, then returned to the starting conditions, and re-equilibrated for 10 column volumes before starting another analysis.

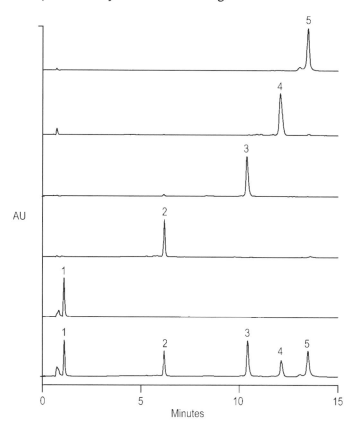

Fig. 1. Separation of five proteins on a prototype ProPac WCX column. Column: 250×4 mm. Sample: 10 μL of a solution containing approximately 0.5 mg/mL of lysozyme, ribonuclease A, cytochrome C, α-chymotrypsinogen, and myoglobin. Eluents: (A) 10 mM sodium phosphate to pH 6.0; (B) 1 M sodium chloride plus 10 mM sodium phosphate to pH 6.0. Gradient: 0–70% B for 30 min at 1 mL/min^{-1}.

The column effluent was monitored by an AD20 variable wavelength absorbance detector at a wavelength of either 254 or 280 nm. Quantification of protein peaks was accomplished using the Optimize module of the PeakNet software.

4. Results and Discussion

4.1. Mixtures of Proteins with Different pIs

Mixtures of commercially available proteins of reasonably different pIs were initially used to assess the chromatographic performance of the stationary phases. **Figure 1** is an example separation for a mixture of myoglobin, α-chymotrypsinogen, cytochrome C, ribonuclease A, and lysozyme on the WCX column using a linear sodium chloride gradient in the presence of 10 mM sodium phosphate buffer, pH 6. **Figure 2** shows that selectivity changes can be demonstrated by changing either the stationary phase, salt gradient, or eluent pH.

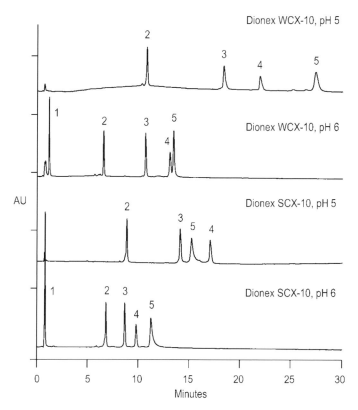

Fig. 2. Illustration of the effect of pH, salt gradient, and column type on the separation selectivity. Separation of five proteins on ProPac WCX and SCX columns. Column: 250×4 mm. Sample: 10 µL of a solution containing approximately 0.5 mg/mL of lysozyme, ribonuclease A, cytochrome C, α-chymotrypsinogen, and myoglobin. Eluents: (A) 20 mM sodium phosphate to pH 6.0; (B) 1 M sodium chloride plus 20 mM sodium phosphate to pH 6.0; (C) 20 mM sodium acetate to pH 5.0; (D) 1 M sodium chloride plus 20 mM sodium acetate to pH 5.0. (pH 6.0 gradient: 4–70% B for 30 min at 1 mL/min^{-1}; pH 5.0 gradient: 4–60% D for 30 min at 1 mL/min^{-1}).

4.2. Cytochrome C Species Variants

Baseline separations of cytochrome C variants (bovine, horse, rabbit) that differ in three amino acid residues between bovine and horse, and in four amino acid residues between bovine and rabbit, were achieved on the strong and weak cation exchangers. Chromatograms obtained on the WCX and SCX columns are shown in **Fig. 3**. Differences in selectivity between the strong and weak cation exchangers were apparent from the different retention times of the horse variant. On the WCX column, the horse cytochrome C species elutes later and much closer to the rabbit variant than on the SCX column.

4.3. Monoclonal Antibody Variants

Processing of C-terminal lysine and arginine residues of proteins isolated from mammalian cell culture has been described *(4)*. As a result of processing, C-terminal Lys or Arg residues, whose presence could be expected based on gene sequence information,

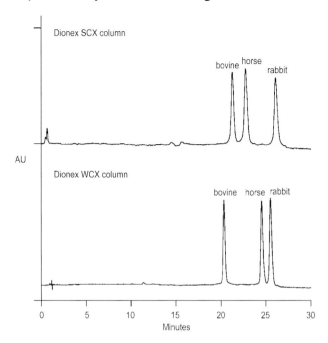

Fig. 3. Separation of three different cytochrome C species using either the WCX or SCX prototype ProPac columns. Eluents: (A) 20 m*M* sodium phosphate to pH 7.0; (B) 1 *M* sodium chloride plus 20 m*M* sodium phosphate to pH 7.0. Gradient: 40–150% B for 30 min at 1 mL/min.

are often absent in proteins isolated from mammalian cell culture. This discrepancy may result from the activity of one or more basic carboxypeptidases. Charge heterogeneity can result if the processing is incomplete. The resultant charge heterogeneity of the variant forms can be identified by cation exchange chromatography. C-terminal processing of lysine residues from heavy chains of monoclonal antibodies from a variety of sources has been reported *(13–17)*.

We assessed the performance of the grafted cation exchange columns for their ability to separate variants of a humanized IgG that was suspected of having variants that differed in the presence of lysine residues at the C-terminal of the heavy chains. As shown in **Fig. 4**, by using a shallow NaCl gradient (40–150 mM NaCl for 30 min) at neutral pH, it was possible to resolve three variant forms that differed in the occupancy of lysine at the C-terminal of the heavy chains (0, 1, or 2 lysine residues). To verify that the reason for the different retention times of the three peaks was the different content of heavy chain C-terminal lysine, the IgG preparation was treated with carboxypeptidase B, an exopeptidase that specifically cleaves C-terminal lysine residues. This treatment of the IgG preparation resulted in the quantitative disappearance of peaks 2 and 3 (containing 1 and 2 terminal lysine residues, respectively, on their heavy chains,). The decreased peak areas in peaks 1 and 2 were accompanied by a corresponding quantitative increase in peak area 1 (variant with no terminal lysine, **Fig. 4B**). This result confirmed that peaks 2 and 3 differed from peak 1 in that they contained IgG with 1 and 2 terminal heavy chain lysine residues, respectively. Further, immobilized papain-digested and Protein A-purified monoclonal antibody (MAb) Fab and Fc fragments were also

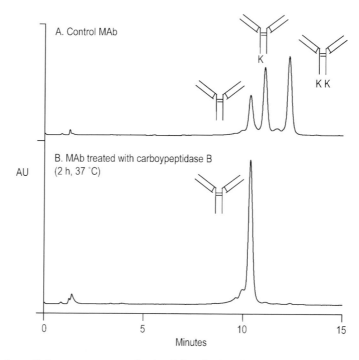

Fig. 4. Overlay of chromatograms obtained for the MAb before (**A**) and after (**B**) treatment with carboxypeptidase B for 2 h at 37°C. Column: 250 × 4 mm WC-10 cation exchanger. Sample: 10 µL of 0.5 mg of IgG per mL of eluent A. Eluents: (A) 10 m*M* sodium phosphate to pH 7.0; (B) 1 *M* sodium chloride plus 10 m*M* sodium phosphate to pH 7.0. Gradient: 4–15% B for 30 min at 1 mL/min.

separated from the intact IgG with high efficiency using a volatile buffer system on the ProPac WCX column (*18*).

4.4. Ribonuclease A Deamidation Variants

Deamidation of Asn residues or the isomerization of Asp residues occurs in a variety of protein-based pharmaceuticals including human growth hormone (*19*), tissue plasminogen activator (*20*), hirudin (*21*), MAbs (*22*), acidic fibroblast growth factor (*23*), and interleukin 1 (*24*), with varying effects on the activity or stability of the therapeutic protein. Hence, monitoring the deamidation of Asn residues in proteins is of interest to analytical and protein chemists in quality control and process departments at biotechnology and pharmaceutical companies (*25*). As described by Di Donato et al. (*26*), the selective deamidation of Asn[67] in native ribonuclease A under mild conditions states that separation of the deamidation products required cation exchange on Mono S followed by hydrophobic interaction chromatography to resolve the deamidation variants. In contrast, using only a tentacle-type weak cation exchanger, deamidation variant forms having Asp or isoAsp at Asn[67] were baseline-separated from each other and from native ribonuclease A in a single chromatographic analysis (**Fig. 5**). The baseline separation made it possible to quantify the change in amounts of each form within the mixture as a function of time. From the increase in the amount of Asn[67] deami-

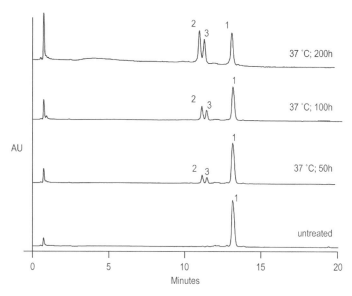

Fig. 5. Chromatograms showing the separation of ribonuclease A and its two deamidation products at several time points during the course of the forced deamidation. Column: 250 × 4 mm ID. ProPac WC-10 cation exchanger. Eluents: (A) 10 m*M* sodium phosphate to pH 6.0; (B) 1 *M* sodium chloride plus 10 m*M* sodium phosphate to pH 6.0. Gradient: 4–70% B for 30 min at 1 mL/min.

dated forms of ribonuclease A as a function of time, it was observed that the kinetics of deamidation appear to be first order, with a $t_{1/2}$ of 159 h (**Fig. 6**).

4.5. Hemoglobin Variants

Clinical laboratories frequently separate and quantify the levels of different hemoglobin variants. For the physician, the determination of glycated hemoglobin levels in the blood of a diabetic serves as an excellent indication of the average glucose level in the patient's blood during the preceding 1–2 mo *(27,28)*. Also, separating and identifying hemoglobins associated with serious hemopathies, including sickle cell, hemoglobin C, and Bart's disease, is also extremely important in the diagnosis, treatment, and counseling of afflicted children. In both cases, the availability of a new, high-resolution, cation exchange chromatographic column can make routine analyses easier and can facilitate the identification of new hemoglobin variants that may be resolved with the enhanced column performance.

In **Fig. 7**, we have demonstrated the separation of hemoglobin variants found in a sample known to contain elevated levels of glycosylated hemoglobin. The peaks are labeled in accordance with the conventions specified previously *(29)*. This sample (Sigma) had glycosylated hemoglobin at about 10% of the total hemoglobin.

The chromatogram reveals the presence of numerous glycosylated forms of hemoglobin, which is expected because such a reaction occurs nonenzymatically between the hemoglobin and sugars in the blood. In principle, nonenzymatic glycation can occur with any NH2 group in the hemoglobin protein (e.g., at the N-terminus of the protein chains or on the side chains of lysine residues) *(30)*. The major glycated component,

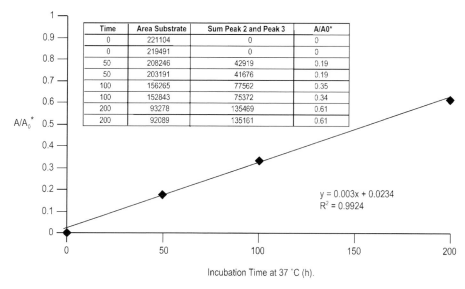

Time	Area Substrate	Sum Peak 2 and Peak 3	A/A0*
0	221104	0	0
0	219491	0	0
50	208246	42919	0.19
50	203191	41676	0.19
100	156265	77562	0.35
100	152843	75372	0.34
200	93278	135469	0.61
200	92089	135161	0.61

$y = 0.003x + 0.0234$
$R^2 = 0.9924$

Incubation Time at 37 °C (h).

Fig. 6. Fractional amount of deamidation products formed as a function of time when ribonuclease A (3 mg/mL) was incubated in 1% ammonium carbonate buffer, pH 8.2, at 37°C.

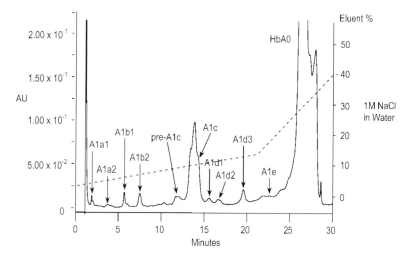

Fig. 7. Separation of hemoglobin variants using a sample known to contain elevated levels of glycated hemoglobins. Column: 250 × 4 mm ProPac SC-10 cation exchanger. Eluents: (A) 50 mM sodium phosphate and 2 mM potassium cyanide to pH 6.0; (B) 0.5 M NaCl, 50 mM sodium phosphate and 2 mM potassium cyanide to pH 6.0. Gradient: at 1 mL/min: 0 min, 3% B; 20 min, 12% B; 30 min, 40% B.

HbAc1, is formed when the N-terminals of the protein chains react with glucose, although other forms have been identified and described elsewhere (*see* e.g., **refs. 31** and **32**).

A separation of several hemoglobin sequence variants, including sickle cell hemoglobin, fetal hemoglobin, and hemoglobin C, is shown in **Fig. 8**, in which the low mass

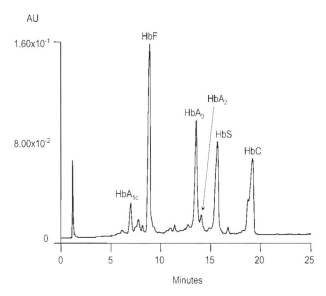

Fig. 8. Cation exchange chromatographic separation of four hemoglobin variants including fetal, sickle cell, normal, and C hemoglobins. Column: 250 × 4 mm ProPac SC-10 cation exchanger. Eluents: (A) 50 mM sodium phosphate and 2 mM potassium cyanide to pH 6.0; (B) 0.5 M NaCl, 50 mM sodium phosphate, and 2 mM potassium cyanide to pH 6.0. Gradient: at 1 mL/min: 0–50% B for 30 min.

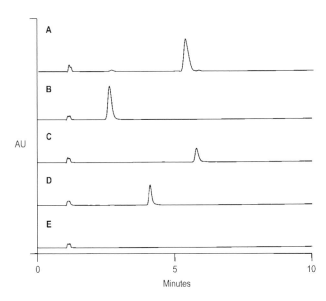

Fig. 9. N-terminal pyroglutamate removal assessed by cation exchange. (**A**) pGlu-Ser-Leu-Arg-Trp-amide peptide + pyroglutamate aminopeptidase. (**B**) pGlu-Ser-Leu-Arg-Trp-amide peptide. (**C**) Neurotensin + pyroglutamate aminopeptidase. (**D**) Neurotensin peptide (pGlu-Leu-Tyr-Gln-Asn-Lys-Pro-Arg-Arg-Pro-Tyr-Ile-Leu). (**E**) Buffer + enzyme control (0.1 mM sodium phosphate, 10 mM sodium EDTA, 5 mM DTT, 5% v/v glycerol pH 8.0.

transfer resistance associated with pellicular particles has resulted in narrow peaks and, consequently, a very high level of resolution. Also, putatively identified on the chromatogram are HbA_2 and $HbA1_c$, which were also apparently present in the sample mixture.

It is apparent from the chromatogram shown in **Figs. 7** and **8** that the SCX column demonstrates significant selectivity with regard to the hemoglobin variants and could prove to be a powerful tool in the routine clinical analysis associated with such conditions as well as affording a platform to separate hemoglobin variants from newly discovered hemopathies.

4.6 Charge Variants Arising from the Cyclization of N-terminal Glutamine

Charge variants can arise from the cyclization of N-terminal glutamine on proteins to produce pyroglutamate, another deamidation reaction. Pyroglutamate aminopeptidase is an enzyme that removes N-terminal pyroglutamate and can be used to assess the presence of N-terminal pyroglutamate in N-blocked peptides. In conjunction with the use of this enzyme, cation-exchange chromatography on the ProPac WCX-10 column can be used to assess the presence of cyclization of N-terminal glutamine **(Fig. 9)**.

5. Conclusions

The described grafted cation exchangers are expected to be most useful for high-resolution separation of protein variants usually encountered in assessments of protein macro and micro-heterogeneity. Determinations of protein micro-heterogeneity are used to assess process consistency and product stability in the production of protein therapeutics. The high-resolution separations of charge variants afforded by the cation exchange columns, as well as the opportunity of using the cation exchange columns to isolate fractions for further analysis, offer a convenient and practical alternative to isoelectric focusing gels, a manual technique often described as being tedious and of low preparative value.

Acknowledgments

The authors would like to thank Robin Higashi for her editorial assistance in the preparation of this chapter.

References

1. Hotchkiss, A., Refino, C. J., Leonard, C. K., et al. (1988) The influence of carbohydrate structure on the clearance of recombinant tissue-type plasminogen activator. *Thromb. Haemost.* **60,** 255–261.
2. Krebs, E. G. (1993) Protein-phosphorliertung und zell regulation I. *Angew. Chem.* **32** 1122–1129.
3. Caccia, J., Keck, R., Presta, L. G., and Frenz, J. (1996) Isomerization of an aspartic acid residue in the complementarity-determining regions of a recombinant antibody to human IgE: identification and effect on binding affinity. *Biochemistry* **35,** 1897–1903.
4. Harris, R. J. (1995) Processing of C-terminal lysine and arginine residues of proteins isolated from mammalian cell culture. *J. Chromatogr.* **705,** 129–134.
5. Keck, R. G. (1996) The use of t-butyl hydroperoxide as a probe for methionine oxidation in proteins. *Anal. Biochem.* **236,** 56–62.

6. Tyn, M. T. and Gusek, T. W. (1990) Prediction of diffusion coefficients of proteins. *Biotechnol. Bioeng.* **35,** 327–338.

7. Leonard, M. (1997) New packing materials for protein chromatography. *J. Chromatogr. B. Biomed. Sci. Appl.* **699,** 3–27.

8. Horvath, J., Boschetti, E., Guerrier, L., and Cooke, N. (1994) High-performance protein separations with novel strong ion exchangers. *J. Chromatogr.* **679,** 11–22.

9. Horvath, C. and Lipsky, S. (1969) Column design in high pressure liquid chromatography. *J. Chrom. Sci.* **7,** 109–116.

10. Lee, W.-C. (1997) Protein separation using non-porous sorbents. *J. Chromatogr. B. Biomed. Sci. Appl.* **699,** 29–45.

11. Frey, D. D., Schweinheim, E., and Horvath, C. (1993) Effect of intraparticle convection on the chromatography of biomacromolecules. *Biotech. Prog.* **9,** 273–284.

12. Muller, W. (1990) New ion exchangers for the chromatography of biopolymers. *J. Chromatogr.* **510,** 133–140.

13. Harris, R. J., Murnane, A. A., Utter, S. L., et al. (1993) Assessing genetic heterogeneity in production cell lines: detection by peptide mapping of a low level Tyr to Gln sequence variant in a recombinant antibody. *Biotechnology* **11,** 1293–1297.

14. Harris, R. J., Wagner, K. L., and Spellman, M. W. (1990) Structural characterization of a recombinant CD4-IgG hybrid molecule. *Eur. J. Biochem* **194,** 611–620.

15. Rao, P., Williams, A., Baldwin-Ferro, A., et al. (1991) C-terminal modification occurs in tissue culture produced OKT3. *Biopharmacology* **4,** 38–43.

16. McDonough, J. P., Furman, T. C., Bartholomew, R. M., and Jue, R. A. (1992) Method for the reduction of heterogeneity of monoclonal antibodies. U.S. Patent, 5 126 250.

17. Lewis, D. A., Guzetta, A. W., Hancock, W. S., and Costello, M. (1994) Characterization of humanized anti-TAC, an antibody directed against the interleukin 2 receptor, using electrospray ionization mass spectrometry by direct infusion, LC/MS, and MS/MS. *Anal. Chem.* **66,** 585–595.

18. Weitzhandler, M., Farnan, D., and Avdalovic, N. (2000) Monoclonal antibody variant separations using cation-exchange chromatography on grafted, polymeric stationary phases. *Am. Biotechnol. Lab.* **September,** 36–38.

19. Johnson, A. B., Shirokawa, J. M., Hancock, W. S., Spellman, M. W., Basa, L. J., and Aswad, D. W. (1989) Formation of isoaspartate at two distinct sites during in vitro aging of human growth hormone. *J. Biol. Chem.* **264,** 14,262–14,271.

20. Paranandi, M. V., Guzetta, A. W., Hancock, W. S., and Aswad, D. W. (1994) Deamidation and isoaspartate formation during in vitro aging of recombinant tissue plasminogen activator. *J. Biol. Chem.* **269,** 243–253.

21. Tuong, A., Maftouh, M., Ponthus, C., Whitechurch, O., Roitsch, C., and Picard, C. (1992) Characterization of the deamidated forms of recombinant hirudin. *Biochemistry* **31,** 8291–8299.

22. Caccia, J., Quan, C. P., Vasser, M., Sliwkowski, M. B., and Frenz, J. (1993) Protein sorting by high-performance liquid chromatography. I. Biomimetic interaction chromatography of recombinant human deoxyribonuclease I on polyionic stationary phases. *J. Chromatogr.* **634,** 229–239.

23. Tsai, P. K., Bruner, M., Irwin, J. I., et al. (1993) Origin of the isoelectric heterogeneity of monoclonal immunoglobulin h1B4. *Pharm. Res.* **10,** 1580–1586.

24. Wingfield, P. T., Mattaliano, R. J., MacDonald, H. R., et al. (1987) Recombinant-derived interleukin-1 alpha stabilized against specific deamidation. *Protein Eng.* **1,** 413–417.

25. Aswad, D. W. (1995) *Deamidation and Isoaspartate Formation in Peptides and Proteins.* CRC Series in Analytical Biotechnology. CRC Press, Boca Raton, FL.

26. Di Donato, A., Ciardiello, M. A., Nigris, M. D., Piccoli, R., Mazzarella, L., and D'Alessio, G. (1993) Selective deamidation of ribonuclease A. Isolation and characterization of the resulting isoaspartyl and aspartyl derivatives. *J. Biol. Chem.* **268,** 4745–4751.

27. The Diabetes Control and Complications Group. The effect of intensive treatment of diabetes on the development and progression of long-term complications in insulin-dependent diabetes mellitus. *N. Engl. J. Med.* **329,** 977–986.

28. Nathan, D. M., Singer, D. E., Hurxthal, K., and Goodson, J. D. (1984) The clinical information value of the glycosylated hemoglobin assay. *N. Engl. J. Med.* **310,** 341–346.

29. Frantzen, F. (1997) Chromatographic and electrophoretic methods for modified hemoglobins. *J. Chromatogr. B* **699,** 269–286.

30. Shapiro, R., McManus, M. J., Zalut, C., and Bunn, H. F. (1980) Sites of nonenzymatic glycosylation of human hemoglobin A. *J. Biol. Chem.* **350,** 3120–3127.

31. Flockiger, R. and Mortensen, H. B. (1988) Glycated haemoglobins. *J. Chromatogr. B* **429,** 279–292.

32. Biss, E., Huaman-Guillen, P., Hurth, P., et al. (1996) Heterogeneity of hemoglobin A1d: assessment and partial characterization of two new minor hemoglobins, A1d3a and A1d3b, increased in uremic and diabetic patients, respectively. *J. Chromatogr. B* **687,** 349–356.

III

SPECIFIC SYSTEMS

19

Noninvasive Imaging of Protein–Protein Interactions in Living Animals

Gary D. Luker, Vijay Sharma, and David Piwnica-Worms

1. Introduction

Protein–protein interactions are fundamental to living systems. Dynamic association and dissociation of protein complexes control most cellular functions, including cell cycle progression, signal transduction, and metabolic pathways. Although complexes of proteins form within the context of a cell, protein–protein interactions regulate signals that affect overall patterns of development, normal physiology, and pathophysiology in living animals. For example, homo- and heterodimers of different homeobox proteins control limb bud and craniofacial development *(1)*, and interactions among various cell surface receptors, scaffold proteins, and transcription factors regulate activation and trafficking of immune cells *(2)*. Complexes of transcription factors, co-repressors, and chromatin-binding proteins maintain normal cells in a quiescent state, and disruption of these protein interactions may be significant in permitting unregulated growth of cancer cells *(3)*. Additionally, protein interactions in signaling pathways are emerging as important therapeutic targets for cancer and other human diseases (reviewed in **ref. 4**). These and other similar pathways of protein interactions in specific tissues produce regional effects that cannot be investigated fully with in vitro systems.

The detection of physical interaction between two or more proteins can be facilitated if association leads to a readily observed biological or physical readout *(5)*. One such approach to produce a detectable signal involves fusion of interacting molecules to defined protein elements with specific properties or induced activities. Readout strategies might include activation of transcription, repression of transcription, activation of signal transduction pathways, or reconstitution of a disrupted enzymatic activity *(5)*. A variety of these techniques have been developed to investigate protein–protein interactions in cultured cells, and methods such as the two-hybrid system *(6,7)*, split-ubiquitin *(8,9)*, and protein fragment complementation based on β-galactosidase, dihydrofolate reductase (DHFR), or β-lactamase *(10–14)* have been used to monitor protein-protein interactions in mammalian cells. Several variations of recruitment systems have also been developed, including the Ras-recruitment system *(15,16)* and interaction traps *(17)*. Of these biotechnologies, the two-hybrid system has been used most extensively for studies of protein–protein interactions.

From: *Handbook of Proteomic Methods*
Edited by: P. Michael Conn © Humana Press Inc., Totowa, NJ

Two-hybrid systems exploit the modular nature of transcription factors, many of which can be separated into discrete DNA-binding and activation domains *(6)*. Proteins of interest are expressed as fusions with either a DNA-binding domain (BD) or activation domain (AD), creating hybrid proteins. If the hybrid proteins bind to each other as a result of interaction between the proteins of interest, then the separate BD and AD of the transcription factor are brought together within the cell nucleus to drive expression of a reporter gene. In the absence of specific interaction between the hybrid proteins, the reporter gene is not expressed because the BD and AD do not associate independently. Two-hybrid assays can detect transient and/or unstable interactions between proteins, and the technique is reported to be independent of expression of endogenous proteins *(18)*. Although the two-hybrid assay originally was developed in yeast, the system has recently been adapted for studies in bacteria and cultured mammalian cells.

Recently, we and other investigators have shown that two-hybrid systems can be used to image protein interactions in living mice with positron emission tomography (PET) *(19)* or bioluminescence imaging *(20)*. The main paradigm shift underpinning advances in molecular imaging of protein–protein interactions in vivo is the use of reporter genes that are compatible with the external imaging devices already used to image whole animals (or humans) noninvasively and repetitively. Thus, interfacing tools of molecular cell biology with the technologies of whole body imaging are at the center of research strategies to image protein–protein interactions in whole living organisms. In the following sections, we describe the multidisciplinary techniques that we use to translate the two-hybrid system from cell culture models to in vivo whole body imaging with a focus on microPET methodologies.

2. Methods

A variety of methods exist for introducing heterologous genes into living mice, including tumor xenografts with transgenes, viral or nonviral DNA vectors, and genetically engineered mice. To introduce the imaging two-hybrid system into living mice, we have used xenografts of human tumor cells that are stably transfected with the hybrid proteins and reporter gene for molecular imaging. Described below are the methods used to develop the system in cultured cells and apply the two-hybrid system to microPET imaging, a PET detection system specifically designed for high-resolution scanning of living mice *(21)*.

2.1. Reporter Gene for In Vivo Imaging

We have used thymidine kinase from herpes simplex virus 1 (HSV1-TK) as the reporter enzyme for imaging protein–protein interactions with microPET (**Fig. 1**). HSV1-TK phosphorylates nucleoside analogs, such as acyclovir, penciclovir (PCV), and 9-(4-fluoro-3-hydroxymethyl-butyl)guanine (FHBG), which are retained within cells that express the viral enzyme. Using limited mutagenesis of HSV-TK, Black et al. *(22)* derived SR39-TK, a mutant enzyme with enhanced selectivity and phosphorylation of nucleoside analogs, thereby enabling greater sensitivity for microPET imaging compared with wild-type HSV-1 TK *(23)*. By mutating a nuclear localization sequence near the N-terminus *(24,25)* in SR39-TK, we derived an enzyme (mNLS-SR39-TK) that further enhanced accumulation of PCV. We reasoned that the mutant enzyme should further improve the limits of detection for the reporter, as maximal activity was thought to be needed to image protein–protein interactions in living mice. Indeed, when analyzed by ^{3}H-PCV uptake assays, the content of radiotracer in HeLa cells expressing mNLS-SR39TK was more than 2-fold greater than HSV1-SR39TK and 12-fold greater than wild-type HSV1-TK *(19)*.

The reporter for the two-hybrid system is constructed by placing mNLS-SR39-TK under control of a Gal4 promoter. The Gal4 promoter is comprised of five copies of the consensus DNA binding sequence for the *Saccharomyces cerevisiae* Gal4 DNA-binding domain linked to a minimal TATA box promoter (BD Biosciences) **(Fig. 1)**. Use of a heterologous promoter reduces background expression of the reporter gene in mammalian cells. The Gal4 promoter can be removed from the commercially available reporter plasmid by restriction digest, or the sequence can be amplified by polymerase chain reaction (PCR) for ligation to mNLS-SR39-TK. To allow selection of cells stably transfected with the reporter plasmid, we have replaced the CMV promoter in pCDNA6-V5/HisA (Invitrogen) with the Gal4-mNLS-SR39-TK cassette. The resultant plasmid carries a blasticidin resistance marker for selecting cells stably transfected with the reporter gene. Alternatively, the Gal4-mNLS-SR39-TK reporter may be constructed in a plasmid without a selection marker, and the reporter can be cotransfected with a second plasmid that expresses a selection marker.

2.2. Construction and Validation of Hybrid Proteins

Hybrid proteins for the mammalian two-hybrid system are constructed by fusing proteins of interest to the C-termini of the Gal4 DNA BD or the VP16 transactivation domain, respectively, in plasmids pM or pVP16 (BD Biosciences). In addition to the interacting pair of proteins, a pair of mutated or closely related proteins with decreased or absent interactions are constructed as negative control hybrid proteins. For most proteins, fusions can be constructed with either Gal4 BD or VP16 **(Fig. 1)**. However, proteins with known transactivation domains should not be fused to Gal4 BD, and VP16 should not used with proteins that contain DNA binding domains. In either of these circumstances, the resultant fusion protein will reconstitute a complete transcription factor and activate the reporter for the two-hybrid system in the absence of protein–protein interactions. For expression of hybrid proteins, the interacting target proteins are cloned in the same reading frame as Gal4 BD or VP16, which is usually accomplished by PCR amplification of the target proteins. PCR products should be sequenced to verify the accuracy of amplified sequences.

After constructing pM and pVP16 plasmids containing hybrid proteins, transient transfection assays are used to test protein expression and interaction. Assays should be performed in the cell type that is to be used for tumor xenografts, thereby eliminating the potential for cell-type-specific effects on the presence or relative magnitude of interaction. Although any reporter gene regulated by a Gal4 promoter may be utilized, we recommend using the Gal4-mNLS-SR39-TK reporter because it will provide data about the signal likely to be produced by interacting hybrid proteins in vivo.

Hybrid proteins are transfected at a molar ratio of 5 : 5 : 1 for each hybrid protein and the Gal4 reporter, respectively (BD Biosciences). Separate sets of cells should be transfected with one hybrid protein and the complementary empty vector plasmid (e.g., pM-hybrid protein and pVP16 empty vector). These transfections control for hybrid proteins that activate the reporter gene in the absence of protein–protein interactions. To deter-

**1) Reverse Tetracycline
"Tet-On" Transactivator**

**2) Inducible Bidirectional
Hybrid Protein
Expression Vector**

3) Reporter Construct

4) Imaging

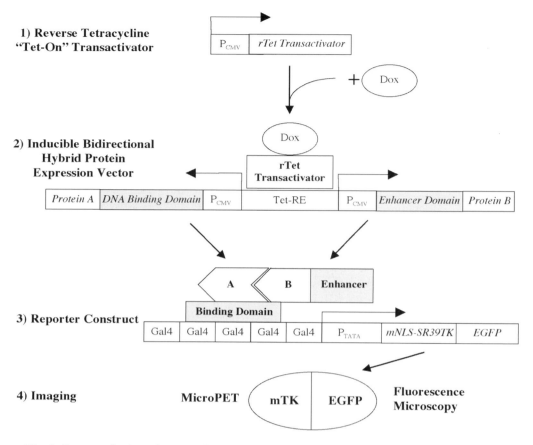

Fig. 1. Strategy for imaging protein–protein interactions in living animals in vivo. Treatment with doxycycline activates a constitutively expressed reverse tetracycline-responsive transactivator to induce bidirectional transcription of DNA-binding domain-protein A and enhancer domain-protein B hybrid proteins containing the binding partners of interest (proteins A and B). Interaction of protein A and protein B assembles the hybrid transcriptional activator on the target promoter (Gal4), inducing transcription of the mNLS-SR39-TK imaging reporter. The reporter protein can be detected by microPET imaging of living mice using TK-mediated phosphorylation and intracellular trapping of a positron-emitting nucleoside analog or, if the reporter is also fused to EGFP, by fluorescence imaging of EGFP. (Reproduced with permission from **ref. 45**).

mine background expression of the reporter gene, another set of cells is transfected with the reporter gene and empty vectors for pM and pVP16. For the Gal4-mNLS-SR39-TK reporter, assays may be performed in 24-well plates, using the number and confluency of cells appropriate for transfection with various methods. For example, we plate 35,000 HeLa cells per well and transfect DNA (150 ng each of pM and pVP16 hybrid proteins with 30 ng of reporter plasmid) at the time the cells are aliquoted into each well, following the protocol described for Fugene 6 (Roche).

1. Forty-eight hours after transfection, the function of mNLS-SR39-TK is determined by quantifying the net accumulation of ^{3}H-PCV (Moravek Biochemicals), a nucleoside analog that is phosphorylated by mNLS-SR39-TK (*19,26*). For these validation studies, ^{3}H-PCV sim-

ulates the imagable substrate wherein phosphorylation of ^{3}H-PCV by mNLS-SR39-TK traps the radiotracer within cells, whereas unphosphorylated nucleoside analog is not retained.

2. Cells are washed once with Dulbecco's modified Eagle's medium (DMEM) and then incubated with 0.2 μCi/mL ^{3}H-PCV in normal growth medium for 1 h at 37°C.
3. Medium with radiotracer is removed, and cells are washed once with DMEM.
4. Cells then are incubated in radiotracer-free growth medium for an additional hour.
5. To terminate the assay, the cells are washed twice with ice-cold phosphate-buffered saline (PBS) and then lysed with 1 mL lysis buffer [1% sodium dodecyl sulfate (SDS), 10 mM sodium borate] for at least 30 min at room temperature.
6. An aliquot (600 μL) of cell lysate is used to quantify cell content of 3H-PCV by scintillation counting, while the remainder is used to measure the concentration of protein in each well by a BCA assay (Pierce).
7. Data for cellular accumulation of ^{3}H-PCV are normalized to mg protein in each sample.

As a general guideline in evaluating the imaging constructs, positive interactions between hybrid proteins should produce at least 10-fold greater signal than non-interacting proteins or other controls in transient transfection assays. Smaller signals are unlikely to be detected in vivo with microPET imaging. Because we are using this assay solely to screen for protein interactions, we do not routinely include a reporter of transfection efficiency, such as luciferase or β-galactosidase. Also, we recommend the use of an uptake and washout protocol during this screening assay, thereby simulating the in vivo pharmacokinetics of nucleoside analogs for imaging. This arises because reporter cells in vivo are exposed to a bolus of radiolabeled compound after intravenous injection of the compound, followed by a decrease in extracellular concentration of nucleoside analog as the compound is cleared from blood. Thus, validating interacting proteins by use of a simple uptake assay in vitro may be misleading because uptake assays alone do not always distinguish phosphorylated from nonphosphorylated pools of nucleoside analog within the cells.

1. To test directly for expression of hybrid proteins by Western blotting, cells are plated in 60-mm dishes and transfected with appropriate amounts of plasmid DNA for hybrid proteins.
2. Forty-eight hours after transfection, total cell lysates are prepared by incubating each dish with 0.4 mL of RIPA buffer (1% Nonidet P-40, 0.5% sodium deoxycholate, 0.1% SDS in PBS) for 15 min at 4°C.
3. Total cell lysate (100–150 μg) is separated by SDS-polyacrylamide gel electrophoresis (PAGE) and analyzed for expression of hybrid proteins by Western blotting with antibodies specific to each hybrid protein or to the Gal4-BD or VP16 epitopes (BD Biosciences), respectively.

2.3. Stably Transfected Gal4-mNLS-SR39TK Reporter Cell Line

For use as tumor xenografts, cells transfected stably with all components of the imaging two-hybrid system must be established. We have used sequential selection of cells transfected with plasmids for the reporter gene and expression and regulation of hybrid proteins. Although sequential selection of cells requires more time than co-transfection of all plasmids, the stepwise approach optimizes each component of the system in intermediate cell lines and generates cells that can be used later in other investigations.

A number of factors should be considered when selecting a cell type to develop as the reporter cell line. The cell line should be able to grow as single colonies in tissue culture, thereby allowing isolation of clonal cell lines. Reporter cells must be able grow as tumor xenografts in immunosuppressed (e.g., *nu/nu*) mice. Importantly, transient transfection assays should demonstrate robust interactions between the hybrid proteins of interest in the selected cell line. The stably transfected reporter cell line should have the following characteristics: (1) no or minimal activation of the TK reporter in the absence of interacting proteins; and (2) high-level induction of TK activity in the presence of interacting hybrid proteins. The combination of both factors will reduce background signal and maximize the signal-to-noise ratio for in vivo detection of interacting proteins. Specifically, our initial work successfully produced expression of fusion proteins in HeLa cells, but several other cell lines, such as H1299 and MCF-7 cells, might be considered good candidates, too.

1. To establish a reporter cell line, cells are transfected with the Gal4-mNLS-SR39-TK reporter plasmid, and individual clones of stable transfectants are isolated.
2. Clones are screened for basal and induced function of the reporter protein, using accumulation of the radiolabeled nucleoside analog ^{3}H-PCV as described in **Subheading 2.2.**
3. Cells in 24-well plates are transfected with a 1:1 molar ratio of DNA for interacting or negative control hybrid proteins. Because we want to quantify potentially small differences reliably among the various clones, we account for differences in transfection efficiency by cotransfecting a constitutively expressed reporter gene, such as β-galactosidase. Parental cells without the Gal4-mNLS-SR39-TK reporter are included in the assay to account for background accumulation of ^{3}H-PCV in the absence of the reporter gene.
4. Data for the cell content of ^{3}H-PCV are normalized to relative amounts of β-galactosidase.
5. Data for accumulation of ^{3}H-PCV are normalized to β-galactosidase, and the content of radiotracer in parental cells without the reporter gene is subtracted to account for nonspecific background activity. Typically, background accumulation of radiolabeled nucleoside analog is very low, making the assay highly sensitive for quantifying the function of mNLS-SR39TK.
6. The difference between the cell content of ^{3}H-PCV in the presence or absence of positive interacting proteins is used to quantify transactivation of the reporter gene.
7. Five to 10 clones with low basal expression and high induction of mNLS-SR39TK should be maintained in culture for approximately 2 wk before repeating the ^{3}H-PCV assay.
8. Data from the two assays are compared to select a clonal line with stable induction of mNLS-SR39-TK. This cell line will be used for subsequent stable transfections with hybrid proteins.

2.4. Stable Expression of Hybrid Proteins in Reporter Cells

When selecting the method for stable expression of hybrid proteins, the effects of each protein on cellular physiology must be considered. Many proteins can be overexpressed constitutively in cells without adversely affecting cell viability. For these proteins, hybrid proteins can be stably transfected into reporter cells, using a strong, constitutive promoter such as cytomegalovirus (CMV) or elongation factor-1α (EF-1α). However, if the proteins of interest impair cell viability and/or proliferation, then an inducible system may be necessary for expression of hybrid proteins in cell culture and mice. Although more time-consuming to establish, an inducible system also allows in vivo imaging of reporter gene activity over time or in response to changes in relative amounts of hybrid proteins. Because we have used an inducible system for imaging protein–protein interactions, the following methods pertain primarily to establishing cells with regulated expression of hybrid proteins.

A variety of inducible systems have been described to regulate expression of transgenes *(27)*. Tetracycline-responsive systems are the most commonly used methods for inducible gene expression, and these systems have been well characterized in mice *(28)*. We have used the reverse tetracycline transactivator (rtTA) system, in which transcription of tetracycline-regulated genes is induced in the presence of doxycycline (tet-on). In comparison with the original tTA system (tet-off), rtTA provides faster kinetics of induction in response to doxycycline, but rtTA systems also typically have a slightly higher basal expression of regulated transgenes *(27)*.

To establish a tetracycline-inducible (tet-on) reporter cell line (**Fig. 1**), we used rtTAS-M2, a variant of rtTA that was selected for reduced background expression and greater sensitivity to doxycycline *(29)*. In some cell lines, stable expression of rtTA is enhanced by using the mammalian EF-1α promoter *(30)* or by coexpressing rtTA and a drug-resistance gene with an internal ribosomal entry site (IRES) *(31)*.

1. Based on these data, we cloned rtTA^S-M2 into pIRESneo (BD Biosciences), which uses a CMV promoter and an IRES to coexpress the transactivator and a neomycin resistant gene.
2. After the cloning of rtTA into an appropriate mammalian expression vector, Gal4-mNLS-SR39TK reporter cells are transfected with rtTA, and clones of stable transfectants are isolated. Clones that express rtTA can be identified by transient transfection of a plasmid with a tetracycline-responsive promoter driving expression of a reporter gene, such as luciferase or β-galactosidase.
3. As described for the Gal4-mNLS-TK reporter cells, clones that initially show tetracycline-inducible gene expression should be tested again after approx 2 wk to confirm stable expression of rtTA. The resultant reporter cell line, Gal4-mNLS-SR39TK-rtTA^S-M2, is then used for subsequent transfection with tetracycline-inducible hybrid proteins.
4. Pairs of hybrid proteins can be expressed from a single plasmid that contains a bidirectional, tetracycline-responsive promoter *(32)*, such as pBI (BD Biosciences).
5. DNA for each hybrid is cloned into one of the two multiple cloning sites in this vector. cDNAs should be oriented with the 5′ end at the promoter and the 3′ end at the polyadenylation site in the vector.
6. Before selecting stable transfectants, interaction of hybrid proteins in the reporter cell line is verified by a functional assay for induction of mNLS-SR39-TK.
7. Gal4-mNLS-SR39TK-rtTA^S-M2 reporter cells are transfected transiently with positive interacting or negative control pairs of hybrid proteins.
8. For tetracycline-inducible genes, parallel sets of cells are treated without or with 1 µg/mL doxycycline.
9. Forty-eight hours after transfection, accumulation of ^{3}H-PCV is quantified as described in **Subheading 2.3.** Constitutively expressed positive interacting proteins should enhance accumulation of nucleoside analog in comparison with negative control hybrid proteins or reporter cells alone. For tetracycline-regulated genes, induction of positive interacting pairs with doxycycline is anticipated to increase the cell content of ^{3}H-PCV above the other tested conditions. Accumulation of radiotracer in the absence of doxycycline treatment quantifies the basal expression ("leakiness") of the system.
10. Expression of hybrid proteins is also confirmed by Western blotting, thereby allowing comparisons of mNLS-SR39-TK activity and amounts of positive or negative control interacting proteins.
11. After verifying expression and interaction of proteins expressed from the bidirectional, tetracycline-regulated promoter, Gal4-mNLS-SR39TK-rtTA^S-M2 cells are transfected with pBI constructs for either the positive or negative interacting pairs of hybrid proteins.

12. Because pBI does not have a selection marker for mammalian cells, cells are cotransfected with a plasmid that expresses a resistance gene, such as pTK-Hyg for hygromycin (BD Biosciences). A molar ratio of 10–20:1 for pBI to pTK-Hyg is used for this transfection.
13. Clones of stably transfected cells are selected and screened for induction of hybrid proteins and activation of the mNLS-SR39-TK reporter gene following treatment with doxycycline as described above.
14. Clonal cells with absent or low basal expression and high induction of hybrid proteins in response to doxycycline are selected for in vivo use. For clones that express positive interacting proteins, the ideal cell line for xenografts will have a large (10-fold or greater) enhancement in activity of mNLS-SR39-TK in response to doxycycline, as measured by cell content of ^{3}H-PCV/mg cell protein.
15. As described for all steps in the selection, the functional assay with ^{3}H-PCV should be repeated to confirm that induction of hybrid proteins is maintained over time.

2.5. Tumor Xenografts and In Vivo Induction of Hybrid Proteins

1. To prepare tumor xenografts, stably transfected cells are incubated without selection drugs for 24 h and then washed with PBS to avoid potential toxic effects of drugs on tissues of recipient mice.
2. Cells are scraped from tissue culture flasks into DMEM, concentrated by centrifugation, resuspended at a density of 1–1.5×10^8 cells per mL DMEM, and maintained at 37°C.
3. Then 1–1.5×10^7 cells in 100 µL are injected subcutaneously into axilla of adult *nu/nu* mice (Taconic Farms). Axillary sites are selected for xenografts because [^{18}F]FHBG, the radiolabeled nucleoside analog used for microPET imaging of mNLS-SR39-TK, is excreted from the liver into the intestine. Therefore, axillary tumors will be clearly separated from background radioactivity in the abdomen.
4. With the use of stably transfected HeLa cells, 3–4-mm tumors develop by approximately 4 d after injection, and protein–protein interactions in these tumors can be detected and quantified readily by microPET imaging.

If a tet-on system is used to control hybrid proteins, doxycycline is administered to mice to induce protein expression and activate the mNLS-SR39-TK reporter. When selecting the route for delivery of doxycycline to mice, a variety of factors must be considered. Expression of hybrid interacting proteins and subsequent transcription of the reporter gene are proportional to amounts of intracellular doxycycline and time that the rtTA^S-M2 transactivator is exposed to drug. Therefore, we have found that dosing regimens producing higher concentrations of doxycycline for extended periods result in enhanced expression of hybrid proteins. Compared with oral administration of drug in drinking water, our experience has shown empirically that intraperitoneal injection of doxycycline induces higher activity of mNLS-SR39-TK in the in vivo two-hybrid system. Increased expression of hybrid proteins and reporter activity with intraperitoneal injection are probably caused by higher and/or more sustained systemic levels of doxycycline in mice. Therefore, we regulate expression of hybrid proteins for imaging by intraperitoneal injection of doxycycline (10–60 µg drug/g mouse) freshly prepared in PBS. Doses are administered every 6–12 h for 12–48 h before microPET imaging. Expression of the mNLS-SR39TK reporter increases in a time- and dose-dependent manner.

2.6. Synthesis of FHBG

Various radiolabeled nucleosides have been used as reporter probes for imaging expression of the HSV1-TK reporter gene *(33)*. Among the leading PET-based radio-

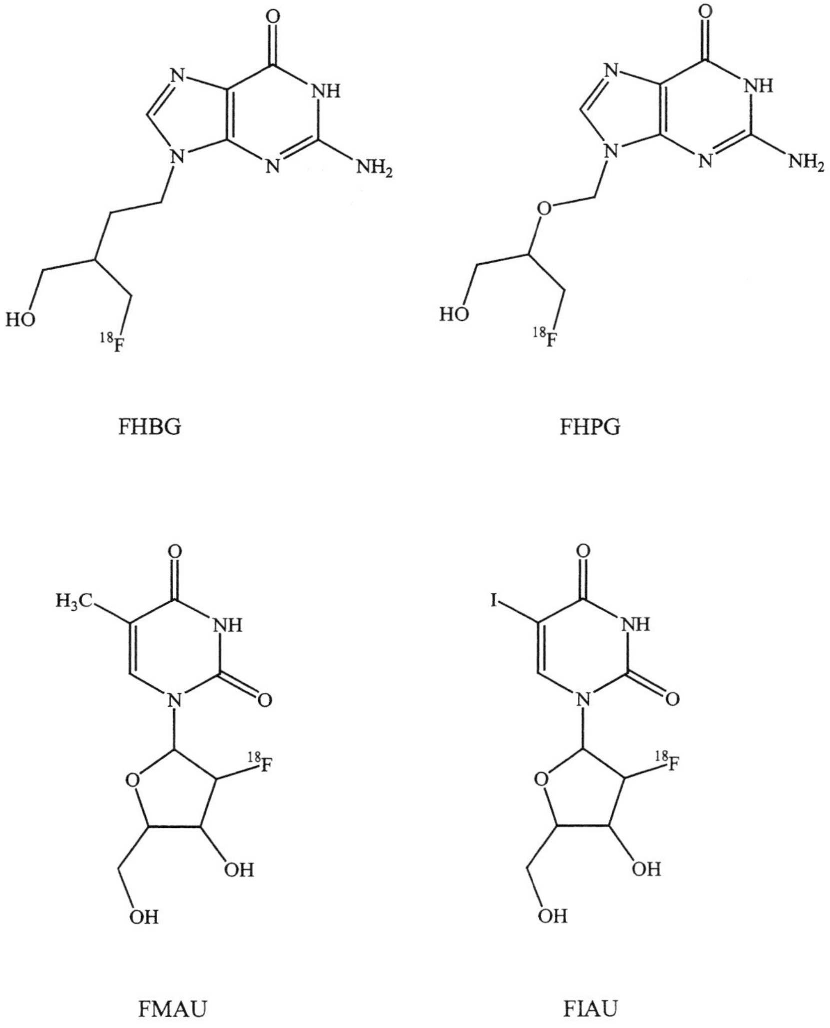

Fig. 2. Structures of several nucleoside analogs.

pharmaceuticals are the pyrimidine-based analogs, FIAU [(2'fluoro-2'-deoxy-5-iodo-1-β-D-arabinofuranosyl)uracil] *(34)* and FMAU [1-(2'-fluoro-5-methyl-1-β-D-arabinofuranosyl)uracil] *(35)* and the purine-based synthetic analogs, FGCV [8-fluorogangciclovir] *(36)*, FHBG [9-(4-fluoro-3-hydroxymethylbutyl)guanine] *(26,37,38)*, and FHPG [(9-(3-fluoro-1-hydroxy-2-propoxy)methyl)guanine] *(39)* **(Fig. 2)**. For all these PET tracers, in contrast to host thymidine kinases, HSV1-TK preferentially phosphorylates these nucleosides, thereby trapping the compounds within intracellular compartments. Therefore, to various degrees, all these radiopharmaceuticals allow selective monitoring of HSV1-TK activity in target tissues.

However, various isoforms of HSV1-TK show different enzyme kinetics between the various analogs, which translates to different levels of cellular trapping of radiolabeled compounds. Although the issue is still under active discussion within the research community, cells expressing wild-type HSV1-TK appear to show higher accumulation of pyrimidines such as FIAU *(40)*, whereas the SR39 mutant and related derivatives show enhanced accumulation selectively for purine analogs such as FHBG *(19,23)*. This would follow from the constraints of the original random sequence mutagenesis screen that produced SR39 *(22)*. Mutant HSV1-TK transformants were scored by their selective incorporation of gangciclovir and acyclovir, purine prodrugs, over thymidine, a pyrimidine. Thus, because of the ready availability of ^{18}F and because of the enhanced attributes of our SR39-based HSV1-TK mutant, we have chosen [^{18}F]FHBG for in vivo monitoring of HSV1-TK enzyme activity.

For the synthesis of [^{18}F]FHBG, the needed precursor, 9-[4-hydroxy-3-(hydroxy-methyl)but-1-yl]guanine (penciclovir) is synthesized through modification of the various steps described earlier *(41)*.

2.6.1. 2-Hydroxymethyl-Butane-1,4-Diol *(1)*

1. Triethyl-1,1,2-ethanetricarboxylate (4.6 mL, 20.0 mmol) is dissolved in t-butanol (40 mL) and treated with pellets of sodium borohydride (2 g).
2. Contents are refluxed and treated with dropwise addition of methanol (2.5 mL).
3. The solution is heated for 20 min, cooled, neutralized with HCl, filtered, and evaporated.
4. The residue is extracted with absolute ethanol (200 proof), filtered, and evaporated to yield a colorless oil *(1)* that is characterized spectroscopically and confirmed to match with literature values.

2.6.2. 5-(2-Hydroxyethyl)-2,2-Dimethyl-1,3-Dioxane *(2)*

1. **1** (1.20 g; 10 mmol) and 2,2-dimethoxypropane (1.35 g; 11 mmol) are dissolved in dry tetrahydrofuran (THF) (5.0 mL) and treated with addition of a catalytic amount of *p*-toluenesulfonic acid monohydrate (3%).
2. Contents are stirred for 30 min at room temperature and neutralized with triethylamine.
3. Volatiles are removed, the residue is purified by column chromatography, and the crude product is purified on Merck silica gel using an eluent mixture of chloroform : methanol (9 : 1).
4. Fractions containing product (R_f = 0.65) are collected and evaporated to yield purified **2**.

2.6.3. 5-(2-Bromoethyl)-2,2-Dimethyl-1,3-Dioxane *(3)*

1. **2** (600 mg; 3.8 mmol) and carbon tetrabromide (1.89 g, 5.7 mmol) are dissolved in 10 mL of dry N,N-dimethyl-formamide (DMF) (distilled over CaH$_2$).
2. The contents are stirred at 0°C for 15 min and triphenylphosphine (1.49 g, 5.7 mmol) is added to this solution.
3. The reaction mixture is stirred for 20 min, treated with sodium bicarbonate solution (10%; 5 mL), diluted with water (45 mL), and extracted with hexanes (3 × 50 mL).
4. The organic layer is separated and evaporated to yield a colorless liquid.
5. Here, we have found that analysis of spectral data often reveals significant amounts of impurities; therefore the liquid is purified on Merck silica gel using diethyl ether as an eluent, but the fractions are immediately stored in ice at 4°C.
6. The fractions are collected, combined, and evaporated to yield colorless oil **3**. Note that decomposition of **3** produces a yellowish brown oil.

2.6.4. 2-Amino-6-Chloro-9-[2-(2,2-Dimethyl-1,3-Dioxan-5-yl)-Ethyl] Purine *(4)*

1. **3** (150 mg, 0.74 mmol) is dissolved in dry DMF (10 mL) and stirred at 4°C.
2. To this solution, 2-amino-6-chloro-purine (130 mg, 0.8 mmol) and potassium carbonate (166 mg, 1.2 mmol) are added.

3. The contents are stirred overnight at 4°C, filtered, and evaporated, and the residue is purified using column chromatography.
4. A solvent mixture containing methanol:chloroform (9:1) is used to obtain purified compound **4**.

2.6.5. 9-[4-Hydroxy-3-(Hydroxymethyl)But-1-yl]Guanine **(5)**

1. **4** (0.150 g) is dissolved in HCl (2.0 *M*, 2 mL) and heated to reflux for 1 h.
2. The contents are cooled and neutralized, and the resultant precipitates are washed with water.
3. The solid **(5)** is dried through azeotropic distillation of dry acetonitrile.

2.6.6. N2-(p-Anisyldiphenylmethyl)-9-[(4-hydroxy)-3-p-Anisyldiphenyl-Methoxymethylbutyl]Guanine **(6)**

1. **5** (200 mg, 0.79 mmol), *p*-anisylchlorodiphenylmethane (600 mg, 1.94 mmol), 4-(dimethylamino) pyridine (DMAP) (10 mg), and dry triethylamine distilled over sodium (0.7 mL) are dissolved in dry DMF and heated at 50°C for 2 h.
2. The reaction mixture is cooled to room temperature and diluted with water (50 mL).
3. The aqueous solution is extracted with ethyl acetate (3 × 50 mL), and the organic layers are combined, dried with sodium sulfate, and evaporated.
4. The crude mixture containing mono-, di-, and trisubstituted trityl groups is separated through column chromatography using an eluent mixture of MeOH:CH_2Cl_2 (5:95) to obtain the purified and spectroscopically characterized product **6**.

2.6.7. N^2-(p-Anisyldiphenylmethyl)-9-[(4-tosyl)-3-p-Anisyldiphenylmethoxy-Methylbutyl]Guanine **(7)**

1. **6** (0.25 g; 0.32 mmol) is dissolved in dry pyridine (10 mL) and treated with dropwise addition of *p*-toluene sulfonyl chloride (0.7 g; 3.6 mmol) dissolved in dry pyridine (3 mL) through a cannula under argon.
2. The contents are stirred for 3 h and monitored through chromatography for disappearance of the starting precursor.
3. The reaction mixture is diluted with water (37 mL) and the aqueous layer is extracted with ethyl acetate (2 × 50 mL).
4. Combined organic extracts are washed with water (2 × 50 mL), dried over sodium sulfate, filtered, evaporated, and purified on Merck silica gel using an eluent mixture of MeOH:CH_2Cl_2 (3:97) to obtain purified **7**.

2.6.8. 9-(4-[^{18}F]-Fluoro-3-Hydroxymethylbutyl)Guanine ([^{18}F]FHBG) **(8)**

Note: This step involves the use of large quantities of radioactive isotopes; thus proper shielding, training, and radiation safety procedures are required.

1. Radiolabeled [^{18}F]FHBG is synthesized through modification of a published method *(37)*. Briefly, Kryptofix 2.2.2 (0.01 g) is added to an aliquot of resin-treated K[^{18}F] (50–100 mCi) obtained from a cyclotron and then dried by azeotropic distillation with dry acetonitrile (3 × 0.5 mL) at 115°C under a continuous stream of argon.
2. To the dried residue is added N^2-(*p*-anisyldiphenylmethyl)-9-[(4-tosyl)-3-*p*-anisyldiphenylmethoxy-methylbutyl]guanine **(7)** (300 µg) dissolved in dry acetonitrile under argon.
3. The reaction vial is sealed and heated at 135°C for 30 min.
4. Contents are then cooled to room temperature and diluted with acetonitrile (1 mL), and the mixture is passed over a C-18 Sep-pack previously activated with water (5 mL) and acetonitrile (5 mL).

5. The C-18 cartridge is then eluted with water (2 × 2 mL), and the protected ^{18}F-labeled precursor is eluted with methanol (2 mL).
6. The eluent is mixed with 1 *N* HCl (200 μL), heated at 85°C for 5 min, and neutralized with 1 *N* NaOH (200 μL), and the product is purified by high-performance liquid chromatography (HPLC) [C-18 column; isocratic solvent system: NH$_4$OAC (50 m*M*):ethanol, 93:7; flow rate, 2.25 mL/min; 8 mCi of crude product yields 450 μCi of [18F]FHBG].
7. The fraction at R_t approx 6.8 min is collected, concentrated, and used for biochemical assays and microPET imaging.

2.7. MicroPET Imaging of Protein Interactions

1. Because of the short half-life of [18]F (108 min), microPET imaging is performed as soon as possible after synthesis of [18F]FHBG in order to maximize the amount of radiolabeled compound available for injection into each animal.
2. Mice are anesthetized with isoflurane immediately before tail vein injection of 100–200 μCi of ^{18}F-FHBG.
3. Animals are placed behind lead shielding and allowed to recover for 1 h before imaging.
4. During the 1-h delay time, [18F]FHBG is cleared from blood and most tissues, allowing detection of phosphorylated, retained radiotracer in tumors that express the reporter mNLS-SR39-TK **(Fig. 3)**.
5. For microPET imaging, mice are again anesthetized with isoflurane and placed supine on the microPET scanner (Concorde Microsystems).
6. Images are acquired for 10–20 min, depending on the amount of injected radioactivity, using 1 bed position and filtered back-projection reconstruction. Isotropic image resolution is 1.8 mm.
7. To quantify accumulation of [18F]FHBG, regions of interest (ROIs) are manually drawn around each tumor image on a workstation.
8. Images are corrected for decay of ^{18}F, but attenuation correction is not performed routinely. Given the small size of mice and the superficial location of tumor xenografts, attenuation of positrons in the animal is minimal.
9. AnalyzePC 3.0 software is used to convert ROI counts on images to radioactive counts per g of tissue (nCi/g), assuming a tissue density of 1 g/mL.
10. Standard uptake values (SUVs) for each tumor are calculated by dividing counts per g of tissue by injected dose of radioactivity per g of animal weight *(42)*. SUV data from ROI analyses have been shown to correlate highly with amounts of radioactivity in extracted tissues.
11. If no additional imaging studies are planned, tumors may be excised and assayed directly for radioactivity in a gamma counter for further correlation with images (Cobra II, Beckman). Radiotracer content is expressed as percent injected dose per g of tissue (%ID/g) *(43)*.
12. The tumor content of [18F]FHBG is normalized to accumulation of radiotracer in heart, a representative background organ that does not significantly accumulate [18F]FHBG *(38)*. For both SUV measurements and assays of excised tumors, the tumor content of [18F]FHBG has been shown to be proportional to the activity of HSV1-TK *(33,44)*.

3. Concluding Remarks

Recent advances in imaging technology have allowed in vivo studies of cellular and molecular signaling processes. Although in vitro assays can provide important data about interacting proteins, these studies do not reproduce the complex regulatory path-

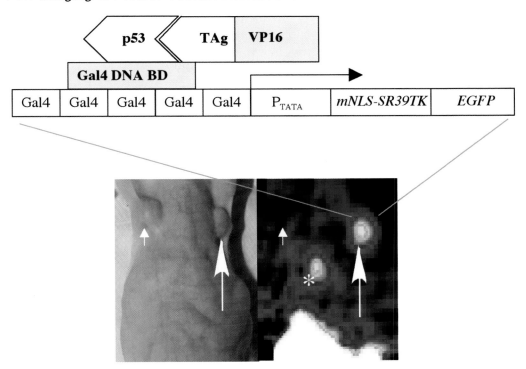

Fig. 3. Molecular imaging of protein-protein interactions in vivo using a PET reporter. In this example, tumor suppressor p53 and transforming factor SV40 large T-antigen (TAg), two interacting proteins known to be involved in transforming cells, are fused to the Gal4 DNA binding domain and HSV-1-VP16 enhancer domain, respectively **(upper panel)**. A similar fusion protein was also engineered comprising polyoma virus coat protein (CP), a protein known to lack interactions with p53, fused to HSV-1-VP16. The activated reporter can be detected by PET imaging of living mice using the positron-emitting radiopharmaceutical [^{18}F]FHBG. Photograph of the anterior thorax of a mouse with axillary xenograft tumors of reporter HeLa cells stably transfected with the noninteracting p53-CP control proteins (small arrow) or interacting p53-TAg proteins (large arrow) **(bottom, left panel)**. Coronal microPET image of the same mouse showing accumulation of [^{18}F]FHBG only in the tumor expressing the interacting p53-TAg proteins (large arrow) **(bottom, right panel)**. Gallbladder (asterisk) and intestinal activity from normal hepatobiliary clearance of the radiotracer is observed in the lower portion of the image. (Reproduced with permission from **ref. 45**.)

ways that exist in living animals. As demonstrated by the imaging two-hybrid system, investigations of protein–protein interactions in living animals require researchers to integrate knowledge about the biological hypothesis with molecular imaging reporters and pharmacokinetics of contrast agents or tracers. Further research is necessary to increase the sensitivity of this imaging two-hybrid system and others for monitoring protein interactions in real time and thus expanding the biological questions that can be studied in vivo. In addition, translating other techniques from cell culture to living animals will provide complementary methods for imaging protein interactions that are not compatible with two-hybrid systems, such as complexes of membrane proteins or transcription factors. The development and application of technologies for molecular imaging of protein–protein interactions in vivo will allow researchers to determine how intrinsic binding specificities of proteins are regulated during normal development and disease progression. In addition, evaluation of the pharmacokinetics and targeting of drugs that promote or disrupt these interactions will allow novel aspects of drug action to be studied directly within the living organism.

Acknowledgments

We thank colleagues of the Molecular Imaging Center for insightful discussions and excellent technical assistance. This work was supported by grants from the National Institutes of Health (P50 CA94056) and Department of Energy (DE FG02 94ER61885).

References

1. Zhang, H., Hu, G., Wang, H., et al. (1997) Heterodimerization of Msx and Dlx homeoproteins results in functional antagonism. *Mol. Cell. Biol.* **17,** 2920–2932.
2. Stark, G., Kerr, I., Williams, B., Silverman, R., and Schreiber, R. (1998) How cells respond to interferons. *Annu. Rev. Biochem.* **67,** 227–264.
3. Ogawa, H., Ishiguro, S., Gaubatz, S., Livingston, D., and Nakatani, Y. (2002) A complex with chromatin modifiers that occupies E2F- and Myc-responsive genes in G0 cells. *Science* **296,** 1132–1136.
4. Heldin, C. (2001) Signal transduction: multiple pathways, multiple options for therapy. *Stem Cells* **19,** 295–303.
5. Toby, G. and Golemis, E. (2001) Using the yeast interactioin trap and other two-hybrid-based approaches to study protein-protein interactions. *Methods* **24,** 201–217.
6. Fields, S. and Song, O. (1989) A novel genetic system to detect protein-protein interaction. *Nature* **340,** 245–246.
7. Serebriiskii, I., Khazak, V., and Golemis, E. (2001) Redefinition of the yeast two-hybrid system in dialogue with changing priorities in biological research. *Biotechniques* **30,** 634–636.
8. Johnsson, N. and Varshavsky, A. (1994) Split ubiquitin as a sensor of protein interactions in vivo. *Proc. Natl. Acad. Sci. USA* **91,** 10,340–10,344.
9. Stagljar, I., Korostensky, C., Johnsson, N., and te Heesen, S. (1998) A genetic system based on split-ubiquitin for the analysis of interactions between membrane proteins *in vivo. Proc. Natl. Acad. Sci. USA* **95,** 5187–5192.
10. Rossi, F., Blakely, B., and Blau, H. (2000) Interaction blues: protein interactions monitored in live mammalian cells by beta-galactosidase complementation. *Trends Cell. Biol.* **10,** 119–122.
11. Rossi, F., Charlton, C., and Blau, H. (1997) Monitoring protein-protein interactions in intact eukaryotic cells by beta-galactosidase complementation. *Proc. Natl. Acad. Sci. USA* **94,** 8405–8410.
12. Remy, I. and Michnick, S. (1999) Clonal selection and *in vivo* quantitation of protein interactions with protein-fragment complementation assays. *Proc. Natl. Acad. Sci. USA* **96,** 5394–5399.
13. Remy, I., Wilson, I., and Michnick, S. (1999) Erythropoietin receptor activation by a ligand-induced conformation change. *Science* **283,** 990–993.
14. Galarneau, A., Primeau, M., Trudeau, L.-E., and Michnick, S. (2002) β-Lactamase protein fragment complementation assays as *in vivo* and *in vitro* sensors of protein-protein interactions. *Nat. Biotechnol.* **20,** 619–622.

15. Aronheim, A., Zandi, E., Hennemann, H., Elledge, S., and Karin, M. (1997) Isolation of an AP-1 repressor by a novel method for detecting protein-protein interactions. *Mol. Cell. Biol.* **17,** 3094–3102.
16. Broder, Y., Katz, S., and Aronheim, A. (1998) The Ras recruitment system, a novel approach to the study of protein-protein interactions. *Curr. Biol.* **8,** 1121–1124.
17. Eyckerman, S., Verhee, A., Van der Heyden, J., et al. (2001) Design and application of a cytokine-receptor-based interaction trap. *Nat. Cell Biol.* **3,** 1114–1119.
18. von Mering, C., Krause, R., Snel, B., et al. (2002) Comparative assessment of large-scale sets of protein-protein interactions. *Nature* **471,** 399–403.
19. Luker, G., Sharma, V., Pica, C., et al. (2002) Noninvasive imaging of protein-protein interactions in living animals. *Proc. Natl. Acad. Sci. USA* **99,** 6961–6966.
20. Ray, P., Pimenta, H., Paulmurugan, R., et al. (2002) Noninvasive quantitative imaging of protein-protein interactions in living subjects. *Proc. Natl. Acad. Sci. USA* **99,** 2105–3110.
21. Cherry, S., Shao, Y., Silverman, R., et al. (1997) MicroPET: a high resolution PET scanner for imaging small animals. *IEEE Trans. Nucl. Sci.* **44,** 1161–1166.
22. Black, M., Newcomb, T., Wilson, H., and Loeb, M. (1996) Creation of drug-specific herpes simplex virus type 1 thymidine kinase mutants for gene therapy. *Proc. Natl. Acad. Sci. USA* **93,** 3525–3529.
23. Gambhir, S., Bauer, E., Black, M., et al. (2000) A mutant herpes simplex virus type 1 thymidine kinase reporter gene shows improved sensitivity for imaging reporter gene expression with positron emission tomography. *Proc. Natl. Acad. Sci. USA* **97,** 2785–2790.
24. Degreve, B., Esnouf, R., De Clerq, E., and Balzarini, J. (1999) Characterization of multiple nuclear localization signals in herpes simplex virus type 1 thymidine kinase. *Biochem. Biophys. Res. Commun.* **264,** 338–342.
25. Ponomarev, V., Dubrovin, M., Beresten, T., Balatoni, J., Blasberg, R., and Tjuavev, J. (2000) Optimization of a reporter gene for dual-modality imaging of transgene expression. *J. Nucl. Med.* **41,** 81P.
26. Luker, G., Luker, K., Sharma, V., et al. (2002) *In vitro* and *in vivo* characterization of a dual-function green fluorescent protein-HSV1-thymidine kinase reporter gene driven by the human elongation factor 1α promoter. *Mol. Imaging* **1,** 65–73.
27. Ryding, A., Sharp, M., and Mullins, J. (2001) Conditional transgenic technologies. *J. Endocrinol.* **171,** 1–14.
28. Lewandoski, M. (2001) Conditional control of gene expression in the mouse. *Nat. Rev. Genet.* **2,** 743–755.
29. Urlinger, S., Baron, U., Thellmann, M., Hasan, M., Bujard, H., and Hillen, W. (2000) Exploring the sequence space for tetracycline-dependent transcriptional co-activators: novel mutations yield expanded range and sensitivity. *Proc. Natl. Acad. Sci. USA* **97,** 7963–7968.
30. Gopalkrishnan, R., Christiansen, K., Goldstein, N., DePinho, R., and Fisher, P. (1999) Use of the human EF-1alpha promoter for expression can significantly increase success in establishing stable cell lines with consistent expression: a study using the tetracycline-inducible system in human cancer cells. *Nucleic Acids Res.* **27,** 4775–4782.
31. Yu, J., Zhang, L., Hwang, P., Rago, C., Kinzler, K., and Vogelstein, B. (1999) Identification and classification of p53-regulated genes. *Proc. Natl. Acad. Sci. USA* **96,** 14,517–14,522.
32. Baron, U., Freundlieb, S., Gossen, M., and Bujard, H. (1995) Co-regulation of two gene activities by tetracycline via a bidirectional promoter. *Nucleic Acids Res.* **23,** 3605–3606.
33. Tjuavev, J. G., Stockhammer, G., Desai, R., et al. (1995) Imaging the expression of transfected genes in vivo. *Cancer Res.* **55,** 6126–6132.
34. Tjuavev, J., Finn, R., Watanabe, K., et al. (1996) Noninvasive imaging of herpes virus thymidine kinase gene transfer and expression: a potential method for monitoring clinical gene therapy. *Cancer Res.* **56,** 4087–4095.

35. Conti, P., Allauddin, M., Fissekis, J., and Wanatabe, K. (1999) Synthesis of [F-18]2-fluoro-5-methyl-1-beta-D-arabinofuranosyluricil ([F-18]FMAU). *J. Nucl. Med.* **40,** 83P.

36. Gambhir, S., Barrio, J., Phelps, M., et al. (1999) Imaging adenoviral-directed reporter gene expression in living animals with positron emission tomography. *Proc. Natl. Acad. Sci. USA* **96,** 2333–2338.

37. Alauddin, M. and Conti, P. (1998) Synthesis and preliminary evaluation of 9-(4-[^{18}F]-fluoro-3-hydroxymethylbutyl)guanine ([^{18}F]FHBG): a new potential imaging agent for viral infection and gene therapy using PET. *Nucl. Med. Biol.* **25,** 175–180.

38. Alauddin, M., Shahinian, A., Gordon, E., Bading, J., and Conti, P. (2001) Preclinical evaluation of the penciclovir analog 9-(4-[^{18}F]fluoro-3-hydroxymethylbutyl)guanine for in vivo measurement of suicide gene expression with PET. *J. Nucl. Med.* **42,** 1682–1690.

39. Alauddin, M., Conti, P., Mazza, S., Hamzeh, F., and Lever, J. (1996) 9-[(3-[18F]-fluoro-1-hydroxy-2-propoxy)methyl]guanine ([^{18}F-FHPG): a potential imaging agent of viral infection and gene therapy using PET. *Nucl. Med. Biol.* **23,** 787.

40. Tjuvajev, J., Doubrovin, M., Akhurst, T., et al. (2002) Comparison of radiolabeled nucleoside probes (FIAU, FHBG, and FHPG) for PET imaging of HSV1-*tk* gene expression. *J. Nucl. Med.* **43,** 1072–1083.

41. Harnden, M., Jarvest, R., Bacon, T., and Boyd, M. (1987) Synthesis and antiviral activity of 9-[4-hydroxy-3-(hydroxymethyl)but-1-yl]purines. *J. Med. Chem.* **30,** 1636–1642.

42. Yeung, H., Sanches, A., Squire, O., Macapinlac, H., Larson, S., and Erdi, Y. (2002) Standardized uptake value in pediatric patients: an investigation to determine the optimum measurement parameter. *Eur. J. Nucl. Med. Mol. Imaging* **29,** 61–66.

43. Sharma, V., Beatty, A., Wey, S.-P., et al. (2000) Novel gallium(III) complexes transported by *MDR1* P-glycoprotein: potential PET imaging agents for probing P-glycoprotein-mediated transport activity in vivo. *Chem. Biol.* **7,** 335–343.

44. Gambhir, S., Barrio, J., Wu, L., et al. (1998) Imaging of adenoviral-directed herpes simplex virus type 1 thymidine kinase reporter gene expression in mice with radiolabeled ganciclovir. *J. Nucl. Med.* **39,** 2003–2011.

45. Luker, G., Sharma, V., and Piwnica-Worms, D. (2003) Visualizing protein-protein interactions in living animals. *Methods* **29,** 110–122.

20

Strategies in Clinical Proteomics

Eric T. Fung

1. Introduction

Clinical proteomics attempts to address clinical questions using a proteomics approach, with the hypothesis that proteins can provide either diagnostic or therapeutic solutions. Although there are numerous proteomics techniques from which to choose, the overriding determinant of the success of a proteomics program is the choice of clinical question followed by careful study design and implementation. The underlying clinical question will drive the decision of which proteomics technique to use, the success criteria, how many samples to examine, how to analyze the data, and, ultimately, whether the clinical proteomics program is a success. Although this chapter focuses on how ProteinChip® technology can be applied to clinical proteomics, many of the principles should be used in conjunction with any technology. This chapter will guide the reader through the steps of defining the clinical question, sample collection, sample analysis, and data analysis. It is not intended to describe the theory behind ProteinChip technology, nor is it intended to review the various applications other than protein expression profiling that can be performed using the technology; for these, the reader is referred to several reviews listed in the reference section (1–6).

2. Defining the Clinical Question

Any clinical proteomics program begins with the clinical question. The more specific the clinical question, the more readily one can develop inclusion and exclusion criteria for the clinical study. For example, although it is tempting to ask whether one can develop a screening test for breast cancer, a closer look reveals that the breadth of the question requires a large sample population to study. This is because the outcome of the study must be applicable to the population at large—the goal is to be able to distinguish the women with breast cancer from the rest of the tested population, which includes healthy women, women with other neoplasms, women with non-neoplastic diseases, and so on. The group of women with breast cancer is itself a heterogeneous population, since some women have a family history of breast cancer, some possess BRCA1 or BRCA2 genotype, and so on. Although no one doubts the value of creating a screening test for breast cancer, to develop one would require a large population-based study and therefore would be an enormous undertaking. A more defined question might be "among BRCA1-positive individuals, can I predict who will get breast cancer?" This

From: *Handbook of Proteomic Methods*
Edited by: P. Michael Conn © Humana Press Inc., Totowa, NJ

question can be further refined to ask questions about the relationship of BRCA1 to age, family history, age at first pregnancy, and so on. These refinements more precisely define the requirements for patient enrollment in the study and therefore make it more likely that a statistically meaningful difference can be ascertained because the population noise will decrease with each additional criterion for patient inclusion or exclusion. Obviously, the more limited the population on which an outcome is based, the more limited the population the outcome can be applied to, unless additional work is performed to demonstrate the validity of the biomarkers on a more general population. Therefore, the initial task for the researcher is to balance the advantages of simplifying the clinical study with the practical utility of the outcome.

Another parameter that should be considered in defining the clinical question is the criterion for success. Typically, the clinical proteomics problem is one of classification: can one predict whether a woman will get breast cancer (predictive medicine), can one classify a woman as having breast cancer (diagnostics), can one predict whether a woman with breast cancer will have a favorable outcome (prognostics), or can one predict if a woman will respond to tamoxifen (theranostics). There are two types of incorrect prediction: predicting that an individual has the disease (or a favorable outcome, or will respond to tamoxifen) when the individual does not, or predicting that an individual does not have the disease when she in fact does. These errors are encapsulated in specificity (the former error decreases specificity) and sensitivity (the latter error decreases sensitivity). Another commonly used statistical measure is the positive predictive value: how likely it is that a positive result on a test represents a true positive. These parameters, sensitivity, specificity, and positive predictive value, are commonly used by clinicians to determine the utility of a classification test (diagnostic test). Therefore, success criteria for a clinical proteomics question are dependent on how the available gold standard for clinical decision making performs by these metrics and, in the instances in which no gold standard exists, the statistical requirement for clinicians to adopt a diagnostic test. Establishing the success criteria at the beginning of the study helps to determine the feasibility of answering the clinical question with utility.

Once the clinical question has been asked and specific inclusion and exclusion criteria are established, the next step in developing the clinical proteomics project is to define the number of samples to be studied. Generally, 30 is the lower limit for the number of samples in each classification group (e.g., disease and healthy or treated and untreated). This number of samples is in theory enough to give greater than 90% statistical confidence in a single marker with a p value greater than 0.01. (Note that data analysis should not be limited to univariate or single marker analysis; see below for a discussion on multivariate analysis.) Because this sample set size is relatively small, inherent biological variability can confound the ability to conclude that differences seen are consequences specific to the disturbance being studied. Therefore, it is imperative that the study include well-chosen samples (e.g., patients of the same age group or all of a single sex) and, equally important, appropriately chosen controls. Confounding variables should be considered and their importance determined in the context of the disease process and study goals. Is age important? Smoking history? Menopause? Number of children? For some diseases these parameters should be controlled, and for others, they are irrelevant. Finally, it should be remembered that the number of samples required becomes larger as the inclusion and exclusion criteria become less defined.

3. Study Implementation

Now that the number and types of samples have been determined, the next consideration is sample acquisition. Many researchers have been collecting serum, tissue, or other biological samples for many years and storing them for proteomics analysis. Before thawing these samples, the researcher should go through the patient records and ensure that the samples already available indeed satisfy the specific inclusion and exclusion criteria set forth. It is better to take the time to collect new samples than to rush in and do a "freezer analysis," which often leads to inconclusive results. Beyond the patient profiles, other characteristics of the samples should be considered. They should have been handled identically, and care should have been taken to minimize the number of freeze–thaw cycles. Generally, storage should be at –80°C in a frost-free environment with temperature protection backup.

1. If starting with fresh serum, the most frequently profiled specimen, the serum should be prepared from whole blood and then aliquotted into four tubes, each containing 25 µL, and, if material is available, four additional tubes, each containing 100 µL. The 25-µL tubes are used for profiling, whereas the 100-µL tubes can be used for purification and identification of protein biomarkers.

2. If starting with frozen serum, samples should be thawed on ice at the time of profiling, and aliquots should be made of unused material.

3. Urine should be centrifuged and the supernatant aliquotted into 0.5-mL aliquots and frozen.

4. For any body fluid, the color should be noted. If hemolysis has occurred, the fluid will be pink or red, and these samples are best avoided. The hemoglobin in these samples will be readily detected and can inhibit the visualization of other proteins.

5. Studies that involve tissue require close communication between the surgeon (or other physician acquiring the sample) and the pathologist who processes the sample. A specific protocol should be drawn up so that samples intended for a study do not sit in the gross room all day waiting to be accessioned and processed.

6. Biopsy specimens that will be used for protein profiling should not be placed in formalin, since the formalin crosslinks proteins and therefore makes protein analysis difficult. Instead, tissue should be rinsed with saline to remove blood on the surface, minced with a clean razor blade (a new blade for each sample) into pieces no longer than 0.5 cm in greatest dimension, immediately frozen in liquid nitrogen, and then stored.

7. It is recommended that, if possible, 50 mg of tissue be made available.

8. Once the samples are ready to be profiled, they can be processed either directly on the arrays with minimal preparation or prefractionated using small spin columns. The decision should be based on the amount of sample available and the complexity (number of proteins) of the sample. Limited sample availability obviously limits the amount of prefractionation one can perform, and simpler proteomes (e.g., urine) often do not need much prefractionation.

9. When preprocessing is not necessary or not possible, we recommend that the samples be denatured (we typically use 9 *M* urea/2% CHAPS in a suitable buffer) in order to disrupt noncovalent protein-protein interactions, which increases the reproducibility of the assay.

10. The denatured sample should then be diluted into the appropriate chip binding buffer and a suitable amount (e.g., 10 µg) of protein incubated with the chip.

11. When there is adequate material (for serum, 20 µL of volume), we typically perform some form of fractionation prior to any ProteinChip Array binding procedure. From our experience, this fractionation step significantly increases the number of peaks visualized and therefore increases the likelihood that biomarkers will be found.

12. For serum, we use anion exchange fractionation, whereas for cells or tissues we often perform subcellular fractionation followed by size exclusion or anion exchange fractionation.

13. Specific applications might make other types of fractionation worthwhile. For example, if one is specifically interested in assaying calcium-binding proteins, immobilized metal affinity chromatography fractionation should be performed.

14. Following fractionation, each fraction is profiled under a series of ProteinChip Array assay conditions, which can include different permutations of array surface chemistries, choice of energy-absorbing molecules, and laser energies. Consequently, each sample in the study generates multiple spectra and therefore data analysis and reduction must be manageable while taking advantage of the plethora of information made available.

15. In addition to the samples being tested, it is always a good idea to run standard samples. Typically, to monitor the experiments, we use a standard sample that closely mimics the test sample (e.g., commercially available pooled human serum), and we process these samples side by side with the test samples.

16. The most rigorous use of the standard sample is to have one spot on each chip contain the standard; this spot should vary from chip to chip. In fact, all the samples (standards, control, and disease) should be processed so that they are distributed uniformly throughout the chips. One should never put all cancer samples on one set of chips and all control samples on another or put all cancer samples on spots A–D and all control samples on spots E–H.

17. During the discovery phase it is prudent to test as many of the available arrays as possible under low-stringency conditions to obtain the largest number of peaks.

18. Once the samples are bound to the chips and the chips washed, matrix is applied. The most commonly used matrices are cyano-4-hydroxycinnaminic acid (CHCA) and sinapinic acid (SPA), although specialized applications may require the use of others.

4. Data Acquisition and Analysis

1. Once the arrays have been processed, they are read in the ProteinChip reader. The performance of the ProteinChip reader should always meet a baseline specification using a reproducible standard. For example, a chip containing immunoglobulin can be used to ensure that instrument performance for sensitivity is adequate.

2. Once instrument performance has been checked, the next task is to identify a combination of laser energy and detector sensitivity that produces good protein profiles. This is accomplished by performing a prequalification run, which consists of spotting a standard sample (generally the same one used for monitoring of the project) onto a series of chips and reading these chips at a range of laser energies and detector sensitivities.

3. From the spectra, two laser energies to read the arrays are used. A low laser energy allows peaks in the low mass range (2–20 kDa) to be well visualized, whereas a high laser energy allows peaks in the high mass range (>20 kDa) to be well visualized.

4. The intensity and shape of the peaks should be noted. Peaks with flat tops or with un-normalized, baseline-subtracted laser intensities greater than 60 generally are unreliable, since individual laser shots were probably off-scale.

5. The array reading spot protocol generally consists of taking 15 shots at approx 5-unit intervals along the spot surface; this results in close to 200 shots, which are then averaged to generate a final spectrum for that spot.

4.1. Data Analysis: Preprocessing

Data analysis consists of two phases: preprocessing and postprocessing. Preprocessing ensures that all the data can be compared together and so includes baseline subtraction, mass calibration, and total ion current normalization.

1. We typically use a baseline subtraction setting of eight times the fitting width.
2. For mass calibration, the All-in-One protein mixture works well, although for best accuracy, separate calibration equations should be calculated for different mass ranges.
3. Data for calibration should be collected at the various time lag focusing settings that match the actual time lag focusing settings used to read the chips containing the samples.
4. The total ion current (TIC) normalization employed in ProteinChip software is analogous to the normalization employed in gene array analysis. An average ion current (the ion current can be thought of as the combined signal and noise) is calculated for each spectrum, and the overall average of the average ion current is calculated across all the spectra to be normalized together; this generates the normalization factor, which is used to calculate a normalization coefficient for each spectrum. Each spectrum's intensity is then adjusted by its normalization coefficient.
5. TIC normalization parameters should be set to exclude the matrix signal, since the matrix region is not quantitative.
6. Once these preprocessing steps have been performed, the true data analysis begins—finding peaks and determining their value in classifying samples.
7. The Biomarker Wizard is an automated method of peak detection. This algorithm begins by establishing a threshold for determining a signal to be a peak; by convention, a peak is defined as a signal having three times the intensity of the noise (a signal-to-noise ratio of 3).
8. In some instances, one may set the threshold lower and allow quality control procedures to remove errantly selected peaks. Because the baseline tends to be noisy, automated peak detection algorithms remain imperfect, and so setting a signal-to-noise ratio of 3 as a threshold often misses obvious peaks.
9. One of the mechanisms by which signals errantly labeled as peaks are removed is to establish the threshold number of spectra in which a called peak must appear to be considered a true peak. If a called peak appears in less than this threshold number of spectra, it is removed from consideration. Errantly called peaks often fall into this category. The percentage can be adjusted at the user's discretion. For example, if one hypothesized that all the disease samples should contain a peak and none of the controls should, then the percentage should be set as the percentage of disease samples in the study. This of course is highly unlikely, and so it is usually best to use a lower setting.
10. Once a peak has been defined based on the signal-to-noise ratio and percentage of spectra thresholds, the intensity of that peak is recorded for every sample. If a sample does not contain that peak, a value corresponding to the local noise is still recorded.

4.2. Data Analysis: Postprocessing

Once an intensity value has been obtained for each peak for each sample, statistics can be applied to determine the peaks that best answer the clinical question. Analyzing data from ProteinChip Arrays is analogous to analyzing data from gene arrays, and therefore some of the same techniques can be applied. Broadly speaking, there are two types of data analysis: univariate and multivariate. Univariate analysis consists of tests of means and medians. Parametric tests such as *t*-tests or analysis of variance (ANOVA) assume gaussian distribution of the data, whereas nonparametric tests such as Mann-Whitney and Kruskal-Wallis do not. For most of the sample set sizes analyzed in these small clinical proteomics studies, the nonparametric tests are preferred. These tests are easily performed and provide a *p* value for each peak.

1. A threshold for a useful *p* value should be set, and the peaks that qualify should be examined more closely.

2. The spectra should be re-examined to ensure that the peaks satisfy the user's definition of a peak.
3. The distribution of the data should be re-examined to look for outliers and qualities that might indicate subgroups of patients.
4. Unfortunately, p values alone are not always indicative of the usefulness of a peak for predictive purposes. A highly predictive diagnostic test relies on minimizing the overlap between patients and controls in distribution in values for a parameter. The overlap in distribution leads to the inverse correlation between sensitivity and specificity, which can be graphed in a plot called the receiver operator characteristic (ROC) curve. Since a p value does not take this overlap into account, a peak can have an outstanding p value and still have an overlap that would obviate its value as a diagnostic. It is therefore advisable to examine the ROC curves for peaks that have a good p value and determine whether they satisfy the success criteria set forth at the initiation of the study.

Because univariate analysis often leads to peaks with favorable p values that do not provide adequate diagnostic accuracy, multivariate analysis is often employed. Multivariate analysis is an intuitive solution, since most diagnoses are made based on multiple variables (i.e., patient history, signs and symptoms, laboratory tests). For ProteinChip Array data (and many other types of data) analysis, multivariate analysis falls into two classes: unsupervised learning and supervised learning. The former are often called clustering techniques, whereas the latter are often called classification techniques.

Clustering techniques do not require prior knowledge of the class (e.g., healthy or disease) of the samples, which is why they are described as unsupervised learning techniques. Commonly used clustering techniques include principal component analysis, hierarchal clustering, self-organized maps, and k-means clustering. All these techniques generally attempt to use the data to distribute the samples as far from each other as possible in n-dimensional space. The distances between individual samples are calculated using any of a variety of distance metrics, which leads to the clustering of similar samples close to each other and of dissimilar samples far from each other. These clusters are then compared with the predefined classes to determine how well the data describe the predefined classes. Ideally, healthy samples will cluster together and disease samples will cluster together. In some situations, subgroups can be identified, and this can lead to the discovery of a heretofore-unrecognized phenotype, which is one of the great strengths of unsupervised learning. How these techniques might actually be used from the diagnostic standpoint is still uncertain. Unsupervised learning does not by itself provide "rules" to apply to a sample of unknown class; such rules would have to be derived from the clusters that arise during the unsupervised learning process.

Classification techniques do require prior knowledge of the class of the samples, which is why they are described as supervised learning techniques. The modeling of data consists of two steps: training and testing. During the training phase, the data along with the sample identity are provided to the classification algorithm to generate models and determine the quality of the models. During the testing phase, data that the algorithm has never seen are applied to the best model generated during the training phase. Supervised learning techniques include decision trees, neural networks, and support vector machines. All supervised learning techniques have the ability to overfit the data, in which a model is built that can classify the samples in the training set extremely well but that classifies test samples poorly. Almost all models generated by any technique will perform better on the training set than on the test set; the difference in performance on the two datasets describes the level of overfitness of the model. More complex models are more likely to be overfit, because they tend to look at an increasing number of variables simultaneously.

The implementation of any of these algorithms, although trivial at the operational level, requires thought and care. Some datasets are inappropriate for specific types of analyses. For example, while small sample set sizes are often analyzed using unsupervised learning techniques, the same set of data would be prone to being overfit when being analyzed using supervised learning techniques. Although decision tree analysis is a good approach toward solving two class problems (healthy vs disease), it is not ideal for solving certain types of multiclass problems such as timecourse changes. Another consideration in the use of these tools is the number of peaks vs the number of samples. If one looks at the data in spreadsheet format with each column representing a peak and each row representing a sample, then each cell in that spreadsheet corresponds to the peak intensity value for a given peak for a given sample. If there are more samples than numbers of peaks, the spreadsheet is longer than it is wide, which is a manageable task for multivariate analysis techniques. If there are more peaks than number of samples, the spreadsheet is wider than it is long, which is a less desirable situation. If this discrepancy is very large, then it is quite possible that the algorithm will find a solution through pure chance. This solution will of course not be applicable to new data and is a mechanism by which an irrelevant model might be generated. Feature selection tools should therefore be applied to the data prior to use of the supervised learning algorithm.

5. Biomarker Validation

Once the data have been analyzed, there is usually enough information to determine whether the success criteria have been fulfilled. If not, the entire process should be re-examined. Was the clinical question approachable? Were the samples the right ones to address the clinical questions? Were the success criteria set too high? Were enough experimental conditions assayed? Probing of these questions will often reveal approaches with which one can salvage the project. In the more pleasant scenario that the initial discovery project can be called a success, the next step is to expand the study in several ways to validate the initial markers. This would include studying new samples from the same clinic, studying samples from additional clinics, and studying samples that address additional clinical subgroups to widen the applicability of the findings. In general, for initial studies, samples are procured from a single clinical center. Procuring samples from multiple centers is required to ensure that candidate biomarkers are not particular to one site, and so single-center studies should always be followed by multicenter studies (at least in terms of sample acquisition, even if all the samples are processed in the same laboratory). In the multicenter study, it is important that control and disease samples be equally distributed between the centers. There have been situations in which all the controls have been collected at one site and the patients at another, and the "markers" that were revealed could not rigorously be attributed to disease state, since slight variations in sample handling at each site can lead to different proteomic profiles.

The start of the validation phase is an ideal time to develop a high-throughput type of assay that may either be purely chromatographically based, antibody based, or some combination of the two. Recall that in the discovery phase, many experimental conditions were assayed: serum was fractionated, and each fraction was applied to multiple array chemistries, multiple matrix types were used, and various laser intensity settings were used. This level of detail is not feasible or necessary when trying to validate the findings on many samples. A purely chromatographic assay might therefore rely on using only one array type that binds the markers that give high enough diagnostic accuracy. In contrast, an antibody assay would consist of a multiplexed antibody capture assay performed on a ProteinChip Array. A natural consideration might be to transfer the antibody assays to a traditional immunoassay [e.g., enzyme-linked immunosorbent assay (ELISA) or radioimmunoassay (RIA)]. This may be useful for a single analyte assay, but for complex diagnostic patterns this is not practical. Moreover, detection of antigens using a ProteinChip Array has the advantage of revealing the relative amounts of protein variants. For example, the amyloid-β ProteinChip Array antibody assay reveals the relative quantities of the various secretase products, which is important since certain fragments are more amyloidogenic than others. An ELISA would not be able to provide such information.

The decision to pursue the chromatographic versus antibody assay depends mostly on practical issues. A pattern of peaks that provides high diagnostic accuracy may itself be used as an assay. Consequently, no time needs to be invested in purifying and identifying the components of the pattern and then developing and characterizing antibodies. However, if one wants to understand the mechanism of disease, or wants to characterize the biomarker as a candidate drug target, then identification of the relevant biomarkers is necessary.

6. Protein Purification and Identification

Protein purification has traditionally been the tedious and arduous task of taking a large amount of starting material, running column A, analyzing the fractions on a gel, choosing the appropriate fractions and running column B, analyzing the fractions on a gel, and performing this iterative process until the protein was purified to homogeneity. Although in practice it is not too different from the traditional methods, the combined use of spin columns and ProteinChip Arrays can accelerate this process substantially.

1. As an example, to purify a marker from serum, we usually start with approx 500 µL of serum and perform our first fractionation on an anion exchange spin column (or filtration plate in conjunction with an automated liquid handler) using a pH gradient.
2. The fractions are assayed on ProteinChip Arrays, and the appropriate fraction is then buffered to pH 4 and fractionated on a cation exchange spin column using a salt gradient.
3. The fractions are again assayed on ProteinChip Arrays, and, if necessary, a reversed-phase column is employed.
4. Variations of this technique may include the use of size-exclusion chromatography, various types of affinity chromatography, and lectin chromatography.
5. Once a protein has been purified to a satisfactory level of homogeneity, it is digested with a site-specific protease such as trypsin, and the products of the digest are analyzed in the ProteinChip reader with internal mass calibration.
6. The masses of the digest fragments are used to search a database of theoretical digests products for all cDNAs, and a match is determined based on the number of peptides in the actual digest that match peptides that would appear in a theoretical digest of the candidate protein.

7. A caveat for these types of searches is that post-translational modifications, if they do exist, will alter the mass of the peptide in the actual digest but will not be predicted in the database. Therefore, in situations in which a high probability match is not obtained using the ProteinChip reader, one can analyze the digest products using the ProteinChip interface with a tandem mass spectrometer to obtain the actual protein sequence.

7. Concluding Remarks

The successful clinical proteomics endeavor begins with the clinical question. The clinical question should be asked with the endpoint in mind—how would the answer change medical practice? What are the performance specifications that the biomarkers need to satisfy? This ensures proper study design and execution. Data analysis requires the application of the appropriate univariate and multivariate statistical algorithms. A well-executed clinical proteomics program will therefore integrate the knowledge of clinicians, biochemists, and statisticians for an optimal chance of success and the discovery of biomarkers with diagnostic and therapeutic applications.

References

1. Petricoin, E. F., Zoon, K. C., Kohn, E. C., Barrett, J. C., and Liotta, L. (2002) Clinical proteomics: translating benchside promise into bedside reality. *Nat. Rev. Drug Discovery* **1,** 683–695.
2. Santambien, P., Brenac, V., Schwartz, W., Boschetti, E., and Spencer, J. (2002) Bioprocess tutorial: rapid "on-chip" protein analysis and purification. *Genet. Eng.* **22.**
3. Srinivas, P. R., Verma, M., Zhao, Y., and Srivastava, S. (2002) Proteomics for cancer biomarker discovery. *Clin. Chem.* **48,** 1160–1169.
4. Fung, E. T. and Enderwick, C. (2002) ProteinChip® clinical proteomics: computational challenges and solutions. *Biotechn. Comput. Proteomics Suppl.* **32,** S34–S41.
5. Leung, H., Leung, S., and Karavanov, A. (2001) Solid-phase profiling of proteins, in *Current Protocols in Protein Science* (Coligan, J. E., Dunn, B. M., Ploegh, H. L., Speicher, D. W., and Wingfield, P. T., eds.), Wiley, Hoboken, NJ.
6. Weinberger, S. R., Dalmasso, E. A., and Fung, E. T. (2001) Current achievements using ProteinChip® Array technology. *Curr. Opin. Chem. Biol.* **6,** 86–91.

21

Proteomic Profiling of the Cancer Microenvironment

Vladimir Knezevic and Michael R. Emmert-Buck

1. Introduction

Cancer is the result of a disturbed biological equilibrium that results in uncontrolled proliferation of malignant cells accompanied by invasion and destruction of surrounding normal tissue. Tumorigenesis is initiated by genetic susceptibility (DNA mutation inherited from parents), epigenetic influences (environmental stimuli to which the organism is exposed), or a combination of both. As the tumor develops over time, additional mutations are acquired in the genome that lead to increasingly aggressive behavior. However, it is important to note that, in addition to the molecular alterations that occur in the malignant cells, the process of tumor progression is also significantly influenced by tumor–host cell interactions. There is a body of literature that supports the critical importance of stromal and other host cells in the development of cancer *(1–6)*. Investigators who are interested in determining the mechanisms that underlie tumor progression in vivo must be aware of and account for the role that each individual cell type plays in the process. Therefore, depending on the goals of the study, it may be important to select methodologies that permit selective analysis of both normal and neoplastic cells.

Extracting proteins from a whole tissue sample is a rapid and simple method to collect enough material for analysis in the laboratory. However, in addition to neoplastic cells, a bulk tissue sample contains multiple cellular components (blood, neurological, connective, fat) Together, these normal host cells may represent 50–80% of the cellular content of the tissue specimen. After sample homogenization, proteins from all the cell types are admixed together, making it difficult to detect and measure changes accurately that are occurring in any individual cell population.

There are two approaches to analysis of proteins in specific compartments of tissue: *in situ* and microdissection (**Table 1**). With the *in situ* approach, proteins are detected directly inside the tissue section. Technologies available for such studies include immunohystochemistry and layered expression scanning (LES) *(7)*. In the microdissection-based approach, cells of interest are removed from the tissue and subsequently analyzed *(8–10)*. Microdissection can be performed manually or via commercially available laser-based technologies (*see* **Table 1** for details). There are several molecular techniques that can be utilized to analyze protein samples collected by microdissection this way. They include traditional Western blotting *(11,12)*, enzyme-linked

From: *Handbook of Proteomic Methods*
Edited by: P. Michael Conn © Humana Press Inc., Totowa, NJ

Table 1
Approaches to Proteomic Profiling of Tissue Samples

	Type of sample collection	Type of analysis	Advantages	Disadvantages
In situ approach	None	Imunohistochemistry	Preserves tissue morphology	Hard to quantitate Low throughput No single cell resolution
		Layered expression scanning (LES) (*7*)	Preserves tissue morphology High throughput	
Microdissection	Manual	Western blotting (*11,12*)	Quantitative	Requires large amount of material Low throughput Does not retain tissue morphology
	Laser capture microdissection (http://www.arctur.com)	ELISA (*13*)	Quantitative	Does not retain tissue morphology
	Laser pressure catapulting (http://www.palm-mikrolaser.com)	Protein arrays (*14,15*)	High throughput	Does not retain tissue morphology
	LMD laser microdissection System (www.leica microsystems.com)	SELDI (*16*)	Good tool for profiling	Does not retain tissue morphology Expensive
	Laser-based cell isolation (http://www.micros copy.bio-rad.com)			

ELISA, enzyme-linked immunosorbent assay; SELDI, surface-enhanced laser desorption/ionization.

immunosorbent assay (ELISA) *(13)*, protein arrays *(14,15)*, and surface-enhanced laser desorption/ionization (SELDI) *(16)*. It is the purpose of this chapter to provide detailed information on the techniques that are useful for measuring protein levels in tissue specimens, including methods not covered in the literature. For the interested reader, we also provide references for standard techniques.

2. *In Situ* Protein Detection
2.1. Immunohistochemistry

For almost five decades, immunohistochemistry (IHC) has been the primary method for visualizing the distribution of proteins directly in tissue samples. The technique is performed by mounting a thin tissue section on a glass surface (in other words, preparing a standard histology slide) and incubating it with an antibody directed against the protein of interest. The antibody-protein complex is typically visualized by applying a second antibody bound to a reporter enzyme, for example, an enzyme capable of catalyzing a chromogenic substrate. Direct detection of proteins in tissue (using labeled primary antibodies) has also been employed. A number of commercially available automated systems can be used to perform this analysis. (Additional information can be found at http://www.ventanamed.com, http://ca.us.dakocytomation.com, http://www.labvision.com, and http://www.meyerinst.com.)

2.2. Layered Expression Scanning

LES[1] is a new approach that allows high-throughput proteomics profiling on tissue samples while preserving histology. This technology is being codeveloped by 20/20 Gene Systems, Inc. and scientists at the National Cancer Institute. Layered membranes are the basis of this novel approach to molecular diagnosis of tissues and/or cell samples that allows for multiplex examination of biomolecules in a specimen. The membranes are 10-μm thick and specially constructed to achieve unique protein binding characteristics that allow them to be stacked such that proteins can be transferred simultaneously bound to multiple layers (**Fig. 1**). Following transfer of the sample, individual proteins of interest can be detected and visualized on each membrane. Importantly, the ability to measure multiple analytes in the same sample conserves material and facilitates the analysis of whole biochemical pathways as opposed to single individual proteins. Relative to standard immunohistochemical techniques, this represents a significant step forward in the age of high-throughput molecular profiling. (Refer to **Table 1** for a summary of the advantages and disadvantages of each approach.)

2.3. Protocol: LES of Soluble Cytoplasmic Proteins in Frozen Tissue Sections

2.3.1. Materials

1. Tissue LES pack (20/20 Gene Systems, Inc.).
2. 70% ethanol.

---

[1]This method and certain materials described herein are covered by one or more patent applications owned by or licensed to 20/20 Gene Systems, Inc. (Rockville, MD). No license to this intellectual property may be implied by this publication. For license information please contact 20/20 Gene Systems, Inc., 9700 Great Seneca Highway, Rockville, MD 20852, Attention: Director of Business Development.

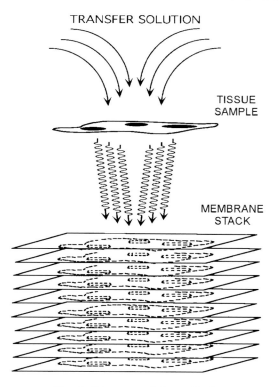

Fig. 1. Diagram of tissue LES.

3. Tris-buffered solution (TBS; 50 mM Tris, pH 8.0, 150 mM sodium chloride).
4. Tris-buffered solution with Tween-20 (TBST; 50 mM Tris-HCl, pH 8.0, 150 mM sodium chloride, 0.05% Tween-20).
5. Primary antibody.
6. Secondary antibody conjugated to horseradish peroxidase (HRP).
7. ECL detection reagent.
8. X-ray film.

2.3.2. Methods

1. Cut 10-μm frozen section and place it on the tissue carrier membrane from the Tissue LES pack.
2. Fix in 70% ethanol for 10 min.
3. Set up transfer assembly by putting the sample stand into the buffer tray and filling the buffer tray with TBS **(Fig. 2)**.
4. Lay filter paper wicks onto the sample stand. Make sure that the lower portion of the wick is immersed into the buffer.
5. Place tissue carrier membrane with the tissue section onto the sample stand.
6. Cover with one filter paper prewetted in buffer.
7. Place a stack of dry filter papers onto the section. Apply light weight.

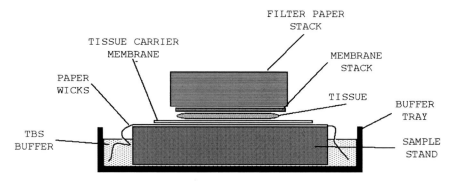

Fig. 2. Diagram of tissue LES transfer assembly.

8. Transfer until all filter papers are wet.
9. Remove the membrane stack and wash it in TBST buffer for 5 min at room temperature.
10. At this point membranes can be allowed to air dry and can be stored at 4°C or put into primary antibody solution diluted in TBST with 0.5% bovine serum albumin (BSA) for 8–12 h at 4°C. *Note:* Dried membranes should be moistened with TBST before they are put into antibody solution.
11. Wash membranes with TBST buffer 3 × 5 min.
12. Incubate in secondary antibody conjugated to the reporter enzyme diluted in TBST with 0.5% BSA for 30 min at room temperature.
13. Wash membranes with TBST buffer 3 × 5 min.
14. Visualize signal by enhanced chemiluminescence reaction and exposure to X-ray film.

3. Microdissection

Several techniques are available to dissect selected cells of interest from tissue **(Table 1)**. Samples collected in this manner contain a relatively pure population of cells obtained directly from the tissue of interest, without an intermediate step of culturing the cells in vitro. Thus, investigators can be more confident that the expression measurements of each tissue compartment are closely reflective of the native molecular state of the cells in vivo.

3.1. Laser Capture Microdissection

One of the newer methods for tissue microdissection is laser capture microdissection (LCM). The following lists basic methods for LCM, including protocols for slide preparation, procurement of cells, and labeling of proteins for subsequent array analysis. Additional information on technical improvements and protocol modifications related to tissue microdissection can be found at:

http://cgap-mf.nih.gov
http://www.arctur.com/
http://dir.nichd.nih.gov/lcm/lcm.htm
http://dir.niehs.nih.gov/dirlep/lcm.html
http://cancer.ucsd.edu/molpath.

3.1.1. Protocol: Preparation of Frozen Tissue Section for LCM

3.1.1.1. MATERIALS

1. Regular, clean glass slides.
2. 100% ethanol.
3. Mayer's hematoxylin.
4. Eosin Y.
5. Xylenes.

3.1.1.2. METHODS

1. Cut 5-µm frozen sections and place them on clean glass slides. At this point, sections can be stored at –20°C (although a lower temperature is preferred).
2. Melt the section gently (e.g., on the back of the hand) for approx 30 s after removal from the freezer. This will create a "rougher" tissue surface and allow for better adhesion to the LCM cap.
3. Place the sections in the following solutions:
   a. 95% ethanol, 15 s
   b. 70% ethanol, 15 s
   c. Deionized water, 15 s
   d. Mayer's hematoxylin, 30 s. Use the minimal amount of staining to visualize the tissue for microdissection. This will significantly improve protein recovery. For example, hematoxylin can be used at 10% of the standard concentration.
   e. Deionized water. Rinse (×2), 15 s.
   f. 70% ethanol, 15 s
   g. Eosin Y, 5 s. Use the minimal amount of staining to visualize the tissue for microdissection. This will significantly improve protein recovery. For example, eosin can be used at 10% of its standard concentration.
   h. 95% ethanol, 15 s.
   i. 95% ethanol, 15 s.
   j. 100% ethanol, 15 s.
   k. 100% ethanol, 15 s.
   l. Xylenes (to ensure dehydration of the section), 60 s.
4. Air-dry for approx 2 min or gently use an air gun to remove xylenes completely. The tissue is now ready for LCM.
5. Once the tissue has been properly processed, sectioned, and stained, it is important to microdissect samples within 20–30 min. This will maximize the amount and the quality of the protein collected.
6. The tissue is first visualized under the microscope, and an initial road map image as well as predissection, postdissection, and cap images are taken to document the histology, the steps of microdissection, and the microdissected cells, respectively. See the website and protocols available on the Arcturus Engineering website for more details on this aspect of the procedure (http://www.arctur.com/).
7. The laser beam size may be adjusted to allow microdissection of groups of cells, or, alternatively, single cells depending on the needs of the investigator.
8. It is essential that there are no irregularities in the tissue surface in or near the area to be microdissected. Wrinkles in the tissue will elevate the LCM cap away from the surface and decrease the membrane contact during laser activation. Occasionally, there are subtle irregularities of the tissue surface (under the LCM cap) that cannot be visually appreciated; however, these can be noted by a decrease in the laser activation spot size. This can be partially

Table 2
Commercial Suppliers of Protein Arrays

Supplier	Array type	Web address
BioCat	Antibody array	http://www.biocat.de
BioChain	Total protein array	http://www.biochain.com
BioNova Cientifica	Antibody array	http://www.bionova.es
Bio-Rad	Total protein array	http://www .bio-rad.com
Cashmere Biotech	Antibody array	http://www.cashmere.com
Clontech	Antibody array	http://www.clontech.com
Perkin Elmer	Total protein array	http://www.perkinelmer.com
Pierce	Antibody array	http://www.piercenet.com
Zyomyx	Antibody array	http://www.zyomyx.com
	Ligand array	

or completely alleviated by adding an extra weight to the cap support arm, or temporarily increasing the laser power.

9. Use an adhesive pad after microdissection to remove cells that may have attached nonspecifically to the LCM cap. Place the cap on the adhesive pad three separate times and then view it microscopically to ensure that all the nonspecific material has been removed.

10. A cap-alone control is recommended for each experiment to ensure that nonspecific transfer is not occurring during microdissection. This is best performed by placing an LCM cap on the tissue section being dissected and aiming and firing the laser at regions where there are no cells or structures present, e.g., lumens of large vessels, cystic structures, and so on. (Alternatively, one can place a portion of the LCM cap "off" the tissue and target this region.)

11. The control cap should be similarly processed through the buffer and analysis methodology utilized in the study.

3.2. Preparation of Proteins from LCM Sample for Protein Array Analysis

Protein arrays are a new and exciting methodology for proteomic studies. The basic principle of this technology is similar to that of cDNA microarrays. Briefly, the technology is based on anchoring target molecules (typically antibodies) with known identities on solid surfaces in an array format. Each target thus has a unique "address" based on its X-Y location in the array. A complex protein sample is then ubiquitously labeled and applied to the array. Molecules that do not bind are washed off, and those that do bind are visualized and measured. The identity of the bound proteins is determined based on their location in the two-dimensional grid.

Over the last few years, protein arrays have become commercially available, minimizing the need for researchers to make them "in-house." **Table 2** provides information on suppliers of arrays. Although each supplier provides specific information useful for their particular array, most of the protocols do not address the issue of using small amounts of protein sample, such as that recovered by LCM. The following section provides guidelines for labeling of proteins collected by LCM for subsequent analysis on protein arrays. This protocol is based on labeling proteins by tagging them with biotin molecules.

3.2.1. Protocol: Labeling of Proteins for Protein Array Studies

3.2.1.1. MATERIALS

1. CENTRISEPT spin columns (Princeton Separations).
2. TBS (50 m*M* Tris, pH 8.0, 150 m*M* sodium chloride).
3. TBST (50 m*M* Tris, pH 8.0, 150 m*M* sodium chloride, 0.05% Tween-20).
4. Phosphate-buffered solution (100 m*M* sodium phosphate, 150 m*M* sodium chloride).
5. EZ-Link Sulfo-NHS-Biotin (Pierce).

3.2.1.2. METHODS

1. Hydrate the CENTRISEPT spin columns with TBS for at least 1 h at room temperature. The number of spin columns should equal the number of samples to be analyzed.
2. Place 55.0 µL of PBS with 1% sodium dodecyl sulfate (SDS) in a 0.5-mL tube. The number of tubes prepared should equal the number of samples to be analyzed.
3. Shave off polymer-containing samples from the bottom of the LCM cap with a sharp blade. Place immediately inside the tube and spin down. It is important to ensure that the polymer is immersed completely in the buffer.
4. Vortex the tube for 2 min and place in a 50°C heating block for 10 min. For best results use a shaking heating block.
5. Remove the tube and spin down at 14,000*g* to collect liquid at the bottom of the tube. Transfer the liquid portion of the sample into a clean tube.
6. Use 1.0–3.0 µL of the sample to measure the concentration of the protein. Since the protein yield is usually relatively low, we recommend using CBQCA Protein Quantitation Kit (Molecular Probes). This method is particularly good if the total expected yield of protein is less then 1.0 µg.
7. Heat up to 49.0 µL of protein sample (containing up to 1.0 µg of protein) at 95°C for 5 min. If the volume of the sample is less then 49.0 µL, add PBS with 1% SDS to adjust to the volume. Spin down at 14,000*g* to collect liquid at the bottom of the tube.
8. Dissolve EZ-Link Sulfo-NHS-Biotin in ddH$_2$O for the final concentration of 10 mg/mL and immediately add 1.0 µL of this solution to 49.0 µL of the sample. Mix well by vortexing!
9. Incubate for 30 min at room temperature.
10. Five minutes before the end of the biotinylation reaction, start preparing spin columns to receive the samples. It is important to use the columns immediately after they are prepared.
11. Carefully apply the sample in the middle of the column and spin down at 750*g* for 2 min. Recovered volume should be 40.0–45.0 µL.
12. If necessary, adjust the volume with PBS containing 1% SDS to 45.0 µL and mix well by vortexing.
13. Take 5.0 µL of each sample and separate by use of a denaturing 10% Tris-HCl polyacrylamide gel electrophoresis (PAGE) gel. Include a marker lane for later determination of protein size.
14. Electroblot gel onto a polyvinylidene fluoride (PVDF) membrane.
15. Block the membrane for 30 min in TBST with 5% BSA at room temperature with rocking.
16. Incubate membrane in streptavidin conjugated to a reporter enzyme diluted in TBST with 0.5% BSA. We use Vectastain ABC-AmP Kit (Vector Laboratories).
17. Wash membrane 3 × 5 min in TBST.
18. Incubate the membrane in developing reagent. We recommend use of the ECL detection system followed by exposure of the membrane to X-ray film. However, using alkaline phosphatase as a reporter and developing with BCIP/NBT substrate is also an option.

The biotinylated sample should appear as a smear ranging from 200 to 30 kDa in size. If only small molecular weight proteins are found, one has to consider that the sample may have been degraded. When multiple labeled samples are compared with each other, their appearance on the SDS-PAGE gel should be similar. Look for any major differences in sample amount. Protein samples are now ready to be applied to the protein array, followed by colorimetric, ECL, or fluorescent detection.

References

1. Tuxhorn, J. A., Ayala, G. E., and Rowley, D. R. (2001) Reactive stroma in prostate cancer progression. *J. Urol.* **166,** 2472–2483.
2. Sausville, E. A. (2001) The challenge of pathway and environment-mediated drug resistance. *Cancer Metastasis Rev.* **20,** 117–122.
3. Ruiter, D. J., van Krieken, J. H., van Muijen, G. N., and de Waal, R. M. (2001) Tumour metastasis: is tissue an issue? *Lancet Oncol.* **2,** 109–112.
4. Ruiter, D., Bogenrieder, T., Elder, D., and Herlyn, M. (2002) Melanoma-stroma interactions: structural and functional aspects. *Lancet Oncol.* **3,** 35–43.
5. Pupa, S. M., Menard, S., Forti, S., and Tagliabue, E. (2002) New insights into the role of extracellular matrix during tumor onset and progression. *J. Cell. Physiol.* **192,** 259–267.
6. Cooper, C. R., Chay, C. H., and Pienta, K. J. (2002) The role of alpha(v)beta(3) in prostate cancer progression. *Neoplasia* **4,** 191–194.
7. Englert, C. R., Baibakov, G. V., and Emmert-Buck, M. R. (2000) Layered expression scanning: rapid molecular profiling of tumor samples. *Cancer Res.* **60,** 1526–1530.
8. Cor, A., Vogt, N., and Malfoy, B. (2002) Microdissection techniques for cancer analysis. *Folia Biol. (Praha)* **48,** 3–8.
9. Best, C. J. and Emmert-Buck, M. R. (2001) Molecular profiling of tissue samples using laser capture microdissection. *Expert Rev. Mol. Diagn.* **1,** 53–60.
10. Eltoum, I. A., Siegal, G. P., and Frost, A. R. (2002) Microdissection of histologic sections: past, present, and future. *Adv. Anat. Pathol.* **9,** 316–322.
11. Jones, M. B., Krutzsch, H., Shu, H., et al. (2002) Proteomic analysis and identification of new biomarkers and therapeutic targets for invasive ovarian cancer. *Proteomics* **2,** 76–84.
12. Ornstein, D. K., Gillespie, J. W., Paweletz, C. P., et al. (2000) Proteomic analysis of laser capture microdissected human prostate cancer and in vitro prostate cell lines. *Electrophoresis* **21,** 2235–2242.
13. Simone, N. L., Remaley, A. T., Charboneau, L., et al. (2000) Sensitive immunoassay of tissue cell proteins procured by laser capture microdissection. *Am. J. Pathol.* **156,** 445–452.
14. Knezevic, V., Leethanakul, C., Bichsel, V. E., et al. (2001) Proteomic profiling of the cancer microenvironment by antibody arrays. *Proteomics* **1,** 1271–1278.
15. Paweletz, C. P., Charboneau, L., Bichsel, V. E., et al. (2001) Reverse phase protein microarrays which capture disease progression show activation of pro-survival pathways at the cancer invasion front. *Oncogene* **20,** 1981–1989.
16. Wulfkuhle, J. D., McLean, K. C., Paweletz, C. P., et al. (2001) New approaches to proteomic analysis of breast cancer. *Proteomics* **1,** 1205–1215.

22

Identification of Determinants of Sensitivity to Antitumor Drugs

Paola Perego, Giovanni Luca Beretta, and Laura Gatti

1. Tumor Cell Lines as Model Systems

In the postgenomic era, new technologies have been designed to define the molecular context that confers sensitivity to antitumor drugs or to identify the appropriate target for pharmacological intervention (*1*). In this context, the significance of cellular studies of transcriptional responses/protein expression modulation by microarray/proteomic technologies often depends on the choice of the appropriate concentration of the cytotoxic insults to which tumor cell lines are exposed. Thus, although easy to perform, chemosensitivity assays are not trivial, and they represent the starting point of more complex experiments in which a reasonable drug concentration has to be used. For example, a concentration for defining the profile of protein expression modulation should not be too low or too high. In the former case, lack of biological effect is the most likely result, whereas in the latter it is possible that the observed effects are more linked to "nonspecific" phenomena (i.e., cell lysis by osmotic shock) rather than to a drug-related, possibly genetically driven, mechanism (e.g., cell death by apoptosis). Besides, exposure to high drug concentrations could result in shutdown of transcription and as a consequence of protein expression. Moreover, exposure to high drug concentrations has a questionable pharmacological meaning because it is unlikely to be reached in vivo. Chemosensitivity of tumor cells can be analyzed by using a variety of techniques, which provide distinct information. Thus, a precise understanding of all chemosensitivity assays is required before making a choice.

1.1. Growing Tumor Cell Lines for Chemosensitivity Studies

Tumor cell lines grow as monolayers or in liquid depending on the specific tumor type. Standard growth conditions are in incubators at 37°C with a 5% CO_2 level. Several media are available for the growth, the most commonly used being RPMI-1640. Medium is usually supplemented with 10% fetal calf serum, although some cell lines might require 15% for optimal growth. The doubling time of established cell lines varies among different cell lines and is a critical parameter when running chemosensitivity studies as, after drug treatment, the cells must be kept in drug-free medium to provide time for the damage to be processed. This time should be two or more doublings.

From: *Handbook of Proteomic Methods*
Edited by: P. Michael Conn © Humana Press Inc., Totowa, NJ

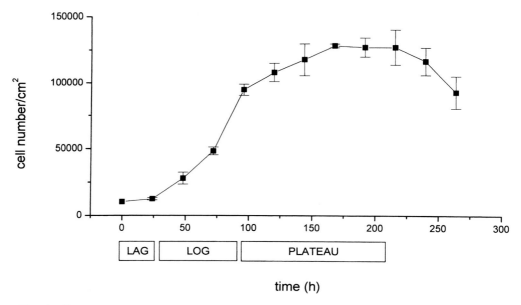

Fig. 1. Growth curve of a human tumor cell line. The growth of the human osteosarcoma U2-OS cell line is shown. The different growth phases are indicated below the *x* axis.

The growth curve can be divided into three phases (**Fig. 1**):

1. The lag phase. After subculture, cells can display a period of adaptation to the fresh medium. Adherent cells during this time re-establish the connections lost after trypsin-EDTA treatment for detachment from substrate.
2. The log phase. This is the time of exponential growth. The length depends on the seeding density. The doubling time has to be calculated in this phase.
3. The plateau phase. This is also called the stationary phase. The cell fraction still growing is below 10%.

When used for chemosensitivity assays, cells have to be in the log phase. Seeding number has to be defined in such a way that cells remain in the log phase during the entire experiment time. Cells should not be exposed to drug in the lag or plateau phases unless specifically required by the experimental design.

1.2. Testing Cellular Sensitivity of Tumor Cell Lines

A variety of assays are available for testing cellular sensitivity to drugs. The best measure of chemosensitivity is provided by colony-forming assays in which cell survival/proliferative capacity after drug exposure is determined *(2–4)*. A good approximation of cell sensitivity is obtained with growth-inhibition assays in which cell number provides an idea of cell growth/proliferative capacity after exposure to the cytotoxic agent. Microtiter assays can be employed to measure inhibition of cell growth. In such cases, cell growth can be estimated as a function of the amount of cell protein content or as a function of viable cells metabolizing a defined substrate (*see* **Subheading 1.2.2.**) *(5–7)*.

1.2.1. Inhibition of Colony Formation

1. Grow up cells in a T25 flask and harvest them during the log phase of growth. For adherent cells, wash the monolayer with saline, add 0.5 mL of prewarmed trypsin-EDTA (37°C) to cover the layer, and remove the excess. Incubate the cells at 37°C and check for detachment within minutes. Resuspend in 5 mL of medium and count using the emocytometer or the cell counter. Prepare a 1 : 100 diluted suspension and further dilute this for reaching the desired cell number.

   Work in triplicates and seed 3 control dishes and 15 dishes for exposing to five drug concentrations. To have an idea of the concentrations, use 1 : 10 dilutions of the tested drug and define the appropriate range. For example, if testing cisplatin for 1 h start with 100 μM and go down. In subsequent experiments the number of seeded cells/dishes exposed to high drug concentrations has to be increased (e.g., 3–10 times more cells than control) to allow reasonable counting at the end of the experiment.

2. Seed cells in 5-cm diameter dishes in 5 mL of medium. The number of cells to seed depends on the capability of the cell line to form colonies. Preliminary experiments have to be performed to define the seeding number. For several cell lines, around 300 cells/dish is appropriate. In this case cell concentration will be around 15 cells/cm^2.

3. Forty-eight hours after seeding, expose cells to the drug for 1 h or longer times if required.

4. Remove the drug-containing medium and wash the monolayer twice with saline.

5. Add 5 mL of fresh medium and place the dishes in the incubator, usually for 10–14 d.

6. When single colonies (consisting of at least 30 cells) can be observed under the microscope in control dishes, remove medium and proceed with fixation and staining.

7. Add 4 mL methanol/dish and keep at room temperature for 1 h.

8. Remove methanol and stain the colonies by adding 4 mL of 2% crystal violet in 70% methanol for 1 h.

9. Wash with water.

10. Count colonies (consisting of at least 30 cells) using an inverted microscope possibly equipped with an imaging system that allows automatic counting.

11. Average values from triplicates and calculate the surviving fraction (colony number in drug-treated dishes/colony number in controls). Plot the obtained values as a function of concentration and calculate the median inhibitory concentration (IC$_{50}$) by interpolation of values providing a linear relationship. A linear scale for the y axis and a log scale for the x axis usually gives a nice curve for all survival levels, although the reverse is more indicated for studying high levels of inhibition. The IC$_{50}$ is defined as the drug concentration that inhibits the cell survival by 50%.

The protocol cannot be used for cells growing in suspension, for which colony formation has to be measured in soft agar (*2*).

1.2.2. Growth-Inhibition Assays

Inhibition of cell growth can be measured by cell counting or by employing microtiter colorimetric assays **(Fig. 2)**.

1.2.2.1. CELL COUNTING

1. Grow up cells in a T75 flask and harvest them during the log phase of growth. For adherent cells, wash the monolayer with saline, add 1.0 mL of prewarmed (37°C) trypsin-EDTA to cover the layer and remove the excess. Incubate the cells at 37°C and check for detach-

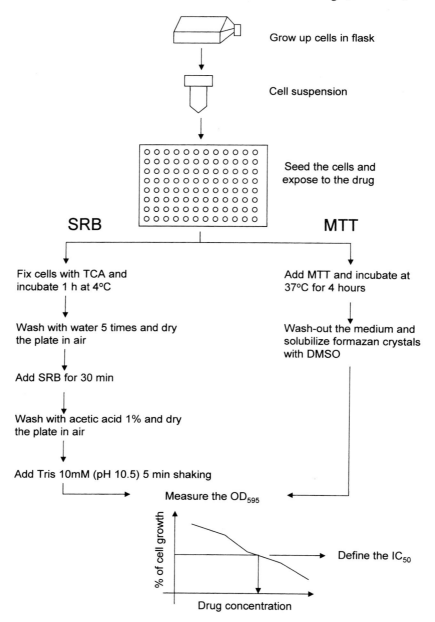

Fig. 2. Flow chart of two microtiter colorimetric assays commonly used in antitumor pharmacology studies. MTT, 3-(4,5-dimethylthiazol-2-yl)-2,5-diphenyltetrazolium bromide; SRB, sulforhodamine B; TCA, trichloroacetic acid; DMSO, dimethylsulfoxide; IC_{50}, median inhibitory concentration.

ment within minutes. Resuspend in 10 mL of medium and count using the emocytometer or the cell counter.

2. Seed cells in 6-well plates (9.6 cm² diameter) in triplicates in 2 mL/well. For adherent cells, cell number usually ranges between 1500 and 17,000 cells/cm² depending on the cell line.

3. Twenty-four to 48 h after seeding, add the drug to the medium. If cells are already attached, drug treatment can be done 24 h after seeding; if not, wait until 48 h.

4. Expose cells to the drug for 1 h (in case of short-term exposure) or longer (24 or 72 h) in case of long-term exposure.

5. Remove the drug-containing medium and wash the monolayer twice with saline.

6. Add 2 mL of fresh medium and incubate for 72 h in drug-free medium in case of 1- or 24-h drug exposure. Proceed directly with **step 7** for the 72-h exposure.

7. Remove medium, wash with saline, and add 300 µL of trypsin-EDTA/well.

8. Harvest the detached cells in 10 mL of counting fluid and count, using a cell counter.

9. Average values from triplicates and calculate the percent of cell growth (cell number of drug-treated cells/cell number in controls). Plot the obtained values as a function of concentration and calculate the IC_{50} by interpolation of values providing a linear relationship. A linear scale for the y axis and a log scale for the x axis usually gives a nice curve for all growth levels, although the reverse is more appropriate for studying high levels of inhibition. The IC_{50} is defined as the drug concentration that inhibits the cell growth by 50%.

A similar protocol can be used for cells growing in suspension for which the following modifications are performed: (1) harvesting is performed by centrifugation; and (2) short-term treatment is in 12-mL tubes (5 mL volume) and is followed by seeding in 6-well plates. For long-term treatment (72 h), the experiment is in 6-well plates.

1.2.2.2. THE MTT ASSAY

This is a colorimetric assay based on the fact that 3-(4,5-dimethylthiazol-2-yl)-2,5-diphenyltetrazolium bromide (MTT) is converted to a colored product only in the mitochondria of viable cells that are capable of reducing it. Therefore, the number of viable cells/well is directly proportional to the production of formazan, which after solubilization can be measured spectrophotometrically. The original MTT assay, described by Mosmann (*6*), has been modified to improve the level of formazan generation, solubility, and stability of MTT formazan (*7*). Although both short- and long-term drug exposures can be employed in principle with this assay, it is recommended for long-term exposure, as the former schedule, which requires washing of small cell numbers after drug treatment (with concomitant serendipitous removal of cells by aspiration), can result in high variability. The protocol below refers to adherent cells.

1. Grow up cells in a T25 flask and harvest them during the log phase of growth as described for inhibition of colony formation.

2. Seed the cells in 96-well plates in 200 µL medium using a repeating pipet. Do not fill the whole plate as peripheral wells (usually eight) have to be filled with medium to provide blanks. The number of seeded cells is dependent on the cell line (range to test: 500–10,000 cells/well). Preliminary experiments have to be carried out to define optimal seeding number.

3. After 24 h, add 10 µL of 20× drug and incubate for 72–96 h.

4. Add to each well 10 µL of MTT (5 mg/mL) dissolved in Hanks' buffer and incubate at 37°C for 4 h.

5. Remove medium by pouring the plate into a basin and wipe residual MTT-containing medium with napkins.

6. Add dimethylsulfoxide (100 µL/well) to solubilize the reduced tetrazolium salts.

7. Measure OD_{550nm} in a microtiter plate reader equipped with the proper filter.

8. IC_{50} is defined as the drug concentration causing a 50% decrease of absorbance over that of control cells.

For cells growing in suspension, the protocol is similar **(steps 2–4)** except that at the end of long-term incubation with the drug, the plates have to be centrifuged for around 20 min to spin down the cells using an appropriate rotor. After centrifugation, remove the medium by gentle aspiration using a needle that has been connected to a pump, and then proceed with **step 6**.

1.2.2.3. THE SRB ASSAY

Sulforhodamine B (SRB) is a bright pink aminoxanthene dye with two sulfonic groups, which under mild acidic conditions binds to protein basic amino acid residues in trichloroacetic acid (TCA)-fixed cells *(3)*. Therefore, SRB staining provides a sensitive index of cellular protein content. The dye is then extracted from the cells and solubilized to measure optical density through mild bases. The SRB assay provides a sensitive index of cellular protein content that is linear over a cell density range of at least 2 orders of magnitude.

1. Seed the cells in 96-well microtiter plates (flat bottomed) in 200 µL volume. The number of cells can vary depending on cell lines and experimental setting (10–200 thousands cells/well). If working with adherent cells, preincubate them for 12–24 h before drug treatment to allow attachment to substrate. Cells growing in suspension can be treated just after seeding. Do not seed cells in the first column of wells (8 wells), in which only medium (200 µL) has to be added to allow subtraction of background staining at the end of the experiment during analysis of results.
2. Expose the cells to drugs for 48 h.
3. Fix cells with TCA by adding 50 µL cold 50% TCA so that final TCA concentration will be 10% and incubating the plates for 1 h at 4°C.
4. Wash each plate five times with water to remove TCA, growth medium, low molecular weight metabolites, and serum proteins.
5. Let the plates dry before proceeding with the SRB staining. At this step, plates can be stored at room temperature for weeks.
6. Add 100 µL/well SRB 0.4% (w/v) in acetic acid 1% for 30 min.
7. Remove SRB and quickly wash the plates (four times) with 1% acetic acid to remove the unbound dye. The best way is to pour the acetic acid directly into the plates so that washings can be fast without causing loss of protein-bound dye.
8. Let the plates dry in the air.
9. Solubilize the bound SRB using 100 µL/well Tris-HCl 10 m*M* (not buffered, pH 10.5); put the plates for 5 min on an orbital shaker before proceeding with absorbance reading.
10. Measure OD_{550} using a microtiter plate reader equipped with the proper filter. Microplate readers are usually connected to a computer endowed with a specific software providing an output that can be edited in programs commonly used for calculations.
11. Proceed with data analysis by making the average for each data point and then subtracting the background staining. Calculate the cell growth percent by dividing the absorbance values corresponding to drug-treated cells by those of control cells. Plot cell growth percent as a function of drug concentrations.
12. A linear scale for the *x* axis and log scale for the *x* axis is suggested when testing a wide range of drug concentrations (half a log dilutions). For studying cell survival at high concentrations, the reverse is more appropriate. IC_{50} is defined as the drug concentration causing a 50% decrease of absorbance over that of control cells.

2. Yeast as a Model Organism

The fission yeast *Schizosaccharomyces pombe* and the budding yeast *Saccharomyces cerevisiae* have become valuable tools for the study of basic cellular functions of eukaryotic cells *(8)*. The utility of the yeast in elucidating complex eukaryotic processes such as DNA replication, cell proliferation, and gene expression is well established. This is in part owing to the well-defined genetics of this eukaryotic microrganism, including the availability of DNA vectors that can be efficiently introduced and stably maintained. These tools facilitate the expression of heterologous proteins and the study of mutant protein function. The relatively high frequency of homologous recombination in yeast also allows integration of DNA sequences into specific chromosomal loci. It is therefore possible to construct isogenic yeast strains that differ only at one genetic locus and to assess the phenotypic consequences of specific genetic alterations. The yeast life cycle allows the propagation and maintenance of stable haploids or diploids. Yeast strains of different mating types can be mated and induced to undergo meiosis, and the individual products of a single meiosis can be analyzed. Thus, the effect of specific mutations in a given gene sequence can be easily assessed in any genetic background. The availability of the entire yeast (both fission and budding yeasts) genome and polymerase chain reaction (PCR)-based technologies simplify the genetic analysis of complementing DNA sequences derived from any genetic screen.

Since the major signaling pathways and cellular processes involved in cellular response to cytotoxic agents are conserved between yeast and mammalian cells, these simple eukaryotic systems are excellent models for identification of molecular/cellular mechanisms of sensitivity to antitumor drugs. Yeast can be used as a model for identification of factors conferring sensitivity or resistance as well as for establishing the mechanism of action of protein-targeted antitumor drugs *(9)*. In the latter case, the organism has proved useful in: (1) finding the cellular target of a given compound; and (2) defining the relevance of specific gene mutations on drug sensitivity. For example, when sensitivity to a drug requires the expression of a specific target gene (e.g., DNA topoisomerase I), deletion of this gene (if compatible with cell viability) renders the cells drug-resistant. Conversely, when the gene is reintroduced into these cells on a plasmid, drug sensitivity is restored. In this view it is possible to study the effects of mutant target proteins to identify the protein domains and the specific amino acid residues that are required for the productive interaction with the drug. Information on the distribution of residues essential for drug action is necessary for understanding the possible mechanisms of drug resistance and how analogs may be designed to overcome drug resistance. In addition, altered expression or mutations in genes other than gene-target (genes involved in checkpoint control, nucleotide excision repair, recombinational repair, membrane transport) that render the cells drug-resistant or -sensitive can be investigated.

Here we mainly focus on techniques for drug screening and only in part deal with other methods, for which we refer the reader to more detailed books.

2.1. Strain Preservation

Fission and budding yeast strains can be stored for short periods at 4°C on appropriate medium in Petri dishes or in vials. Yeast strains can be stored indefinitely in medium supplemented with 15% (v/v) glycerol at –80°C. The strains can be grown on plates, and the colonies can be picked up with sterile applicator sticks and suspended in the glycerol solution. Alternatively, yeast can be grown in liquid and 300 μL of 80% glycerol can be added to 700 μL of culture/cryovial. The yeast can be revived by transferring a small part of the frozen sample to a medium plate or in liquid culture.

2.2. Growth of Yeast

Fission and budding yeasts are usually grown at 30°C on complete or minimal medium. For experimental purposes, it is important that cultures be maintained in the exponential phase of growth. The optical density of a culture can be used to measure the concentration of the cells. It is important to establish a relationship between optical density and cell concentration. This relationship will vary between strains, from one spectrophotometer to another, and among growth conditions. For example: for the *S. pombe* 972 *h⁻* strain, an OD_{595} of 0.1 is equivalent to 2×10^6 cells/mL; for *S. cerevisiae* strains, an OD_{595} of 0.1 is usually equivalent to 7×10^6 cells/mL.

To generate cultures in midexponential growth, use a fresh patch of a strain of checked phenotype to inoculate 3 mL of medium in a 50-mL sterile tube and incubate for 1–2 d at the appropriate temperature until cells are in early stationary phase. At this point the preculture can be used to inoculate a large culture. It takes one generation for cells to recover from the stationary phase and re-enter exponential growth. Gentle shaking is required to maintain a uniform growth condition.

2.3. Yeast Plasmids

Budding and fission yeast plasmids consist of a bacterial origin of replication and selectable marker, a yeast-selectable marker, and an equivalent to an autonomous replication sequence (ARS), which is responsible for a high frequency of transformation *(10)*. Only two general types of vectors are used for transformation of yeast, designated YEp (yeast episomal plasmid) and YCp (yeast centromeric plasmid). The yeast-selectable marker is routinely a cloned yeast gene for which strains with nonrevertible null alleles are available. The most common *S. cerevisiae* markers are LEU2, HIS3, URA3, and TRP1 *(11)*. Budding yeast markers also used in *S. pombe* are the LEU2 and URA3 genes. Plasmids containing these markers complement the *S. pombe* mutations *leu1⁻* and *ura4⁻*. The URA3 gene is expressed very poorly in *S. pombe* and does not rescue the *ura4⁻* mutation when it is present as a single copy. Fission yeast markers commonly used are the *ura4⁺* and *sup3–5 (12)*. The ARS elements have been shown to be yeast origins of DNA replication, and the presence of these elements on a vector promotes high-frequency transformation of yeast, generally at a level of several thousands transformants per microgram of vector DNA *(13)*. Plasmids derived from the ones described above have been used to increase the expression of certain gene products. Genes linked to the SV40 early promoter are expressed at moderate levels, whereas more powerful promoters are those of GAL1, GAL10, and ADH1 genes of *S. cerevisiae* as well as the *S. pombe adh1⁺* promoter. In addition to vectors for constitutive gene expression, plasmids containing inducible/repressible promoters have been also developed.

2.4. Transformation of Yeast

A prerequisite for molecular biology manipulation of any organism is a reliable and efficient means for introducing exogenous DNA into the cell. Various techniques are available for transforming budding and fission yeasts including lithium acetate and electroporation transformation *(14–17)*. Although fast and simple, lithium acetate transformation provides only a low efficiency of DNA transfer, whereas the electroporation technique is extremely efficient. The latter comprises the three steps described below.

1. Preparation of electrocompetent cells. An overnight culture is grown with vigorous shaking at 30°C to an OD_{595} of 1.3–1.5, diluted the next morning, and grown (around 40 mL of culture/sample) until the yeast enters the exponential growth phase. Cells are then spun down, washed in sterile water to concentrate the cells (1000-fold) and reduce the conductivity of the culture, and resuspended in ice-cold 1 *M* sorbitol.
2. Electroporation. An aliquot (40 μL) of yeast electrocompetent cells is transferred to a sterile tube. The DNA (≤100 ng) is added to the cell suspension, mixed, and incubated on ice for about 5 min. The DNA should be dissolved in a small volume (e.g., 5 μL) of a low ionic strength buffer (10 m*M* Tris-HCl, 1 m*M* EDTA, pH 8.0) or water. Transfer to a 0.2-cm sterile electroporation cuvette. Pulse at 1.5 kV, 25 μF, 200 Ω. The apparatus and the cuvettes are commercially available. Immediately add 1 mL cold 1 *M* sorbitol to the cuvette, and mix by gentle pipeting. Transfer to a culture tube.
3. Plating. Aliquots of the transformed cells are plated on selective medium. No incubation is required after resuspension and transfer of the reaction from the cuvette; plate as soon as possible.

The transformation efficiency can be determined by calculating the number of transformants in 1 mL of resuspended cells per 1 μg plasmid per 10^8 cells. For example, if the transformation of 1.0×10^8 cells with 100 ng plasmid results in 500 colonies on a plate spread with 1 μL of suspension, then:

$$\text{Transformation efficiency} =$$
$$500 \times 1000 \text{ (plating factor)} \times 10 \text{ (plasmid factor)} \times 1 \text{ (cell/transformation} \times 10^8)$$

$$\text{Transformation efficiency} = 5 \times 10^6 \text{ transformants/1.0 μg plasmid/}10^8 \text{ cells.}$$

Do not use too much DNA, as transformation efficiency declines as plasmid concentration is increased *(15,18)*.

The protocol just described refers to *S. cerevisiae*, but a similar method can be employed for *S. pombe (14)*.

2.5. Sensitivity of Yeast Strains to Drugs

The assay described below is the corresponding of the measurement of inhibition of colony formation in tumor cells. The protocol **(Fig. 3)** refers to *S. cerevisiae*, but it can be employed for other yeasts as well.

1. Thaw yeast strains in appropriate liquid medium at 30°C. Alternatively, yeasts are first grown on the surface of solid medium plate at 30°C, and then a single colony is picked up with sterile applicator sticks, suspended in 5 mL of liquid medium, and shaken for 1–2 d at the appropriate temperature until cells are in early stationary phase.
2. Dilute the culture to have the appropriate ratio between OD_{595} and cell concentration. One generation is necessary for cells to recover from stationary phase and re-enter exponential growth.
3. Three milliliters of exponentially growing cells are treated with the drug at different concentrations for a fixed time (concentration-dependent cytotoxicity) or for varying times with a fixed concentration (time-dependent cytotoxicity).
4. Plate control and treated cells on solid medium (in triplicates, 10-cm diameter dishes) taking care to dilute the culture to have about 100 cells/plate (for example, dilute the sample at

10–100 cells/μL in sterile water and plate the necessary volume to have 100, 500, 1000, or 5000 cells/plate depending on drug dose used). Dishes are maintained at the appropriate temperature until colonies are formed. The appropriate time depends on the yeast strain. Longer time is required in minimal medium. Usually 2–3 d are enough in rich medium.

5. Count the grown colonies. The number of colonies/dish is used to calculate the effect of the drug that is expressed as the fraction of colonies grown in treated sample compared with those grown in control samples (surviving fraction). The IC_{50} is defined as the drug concentration that inhibits the cell survival by 50%.

2.6. Microtiter Assays

Panels of yeast strains carrying specific mutations can be a valuable tool for screening drugs in the attempt to find an agent more cytotoxic for the mutant than for the wildtype yeast as well as for defining the biological/molecular background in which a certain drug exerts optimal effects, in clarifying the contribution of a specific gene in regulating drug sensitivity/resistance and providing novel approaches for identification of new drug targets *(19–21)*.

An antiproliferative assay performed in microtiter plates can be used to evaluate the cytotoxic effect of antitumor drugs **(Fig. 3)**. It is important to perform preliminary experiments to verify the linearity of the relation between cell number and absorbance at 595 nm and to make sure that the drug tested does not interfere with the employed wavelength.

1. Yeast cell cultures (3 mL in 50-mL tubes) are grown overnight in appropriate liquid medium until midlog phase.
2. Twelve thousand cells (in 100 μL 2× medium) are then seeded in 96-well microtiter plates, and 100 μL of 2× drug is added (each sample is treated in triplicate wells). These conditions might need to be adjusted depending on the drug solubility/strain sensitivity.
3. Plates are then incubated at 30°C for the established time (usually 48 h), and OD_{595} is measured. The IC_{50} is defined as the drug concentration causing a 50% decrease of absorbance over that of control cells.

2.7. New Emerging Techniques for Drug Screening

In addition to the classic techniques of drug screening, the recently developed synthetic lethal screening and genetic selection of peptide inhibitors are methodologies that can be employed for drug screening. Synthetic lethal screening is used in yeast to identify mutations that are not lethal *per se* but are lethal in combination with another mutation. This technology has been proposed as a tool for new drug targets *(20)*. Thus, a specific inhibitor of a cellular pathway that contributes to cell survival could be used in a cell deficient in another pathway concomitantly participating in maintaining cell survival. The procedure could be useful in killing cells with a specific defect, because it provides other drug targets in addition to a specific mutation whose inactivation may produce an advantage in killing the cell. Genetic selection of peptide inhibitors has been performed in *S. cerevisiae* to identify cellular pathways that are potential targets for drug discovery *(21)*. This strategy consists of (1) selection of peptides whose binding to unknown targets produces a phenotype, like mutations produce phenotypes by inactivating genes; and (2) identification of putative targets for the inhibitors by a combination of a two-hybrid system and genetic dissection of the target pathways. The technique allows screening of different peptamers that are presented inside cells on the surface of

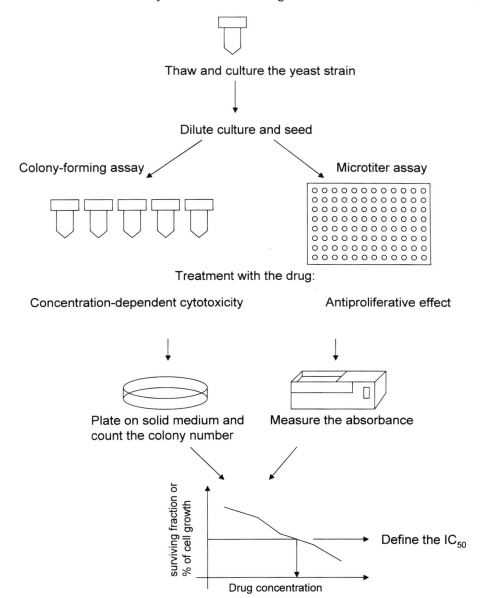

Fig. 3. Schematic representation of methods used to assess the drug sensitivity of yeast strains. IC_{50}, median inhibitory concentration.

an inert carrier protein. Target proteins inhibited by peptamers could be similarly inhibited with small organic molecules including drugs. The identification of peptamers with different potency could provide useful information about the correlation between structure and activity of a drug.

References

1. Lockhart, D. J. and Winzeler, E. A. (2000) Genomics, gene expression and DNA arrays. *Nature* **405,** 827–836.

2. Hamburger, A. W. and Salmon, S. E. (1977) Primary bioassay of human tumor stem cells. *Science* **197,** 461–463.
3. Sikic, B. I. and Taber, R. L. (1981) Human tumor clonogenic assays. An overview. *Cancer Chemother. Pharmacol.* **6,** 201–203.
4. Weisenthal, L. M. and Lippman, M. E. (1985) Clonogenic and nonclonogenic in vitro chemosensitivity assays. *Cancer Treat. Rep.* **69,** 615–632.
5. Skehan, P., Storeng, R., Scudiero, D., et al. (1990) New colorimetric cytotoxicity assay for anticancer-drug screening. *J. Natl. Cancer Inst.* **82,** 1107–1112.
6. Mosmann, T. (1983) Rapid colorimetric assay for cellular growth and survival: application to proliferation and cytotoxicity assays. *J. Immunol. Methods* **65,** 55–63.
7. Alley, M. C., Scudiero, D. A., Monks, A., et al. (1988) Feasibility of drug screening with panels of human tumor cell lines using a microculture tetrazolium assay. *Cancer Res.* **48,** 589–601.
8. Perego, P., Jimenez, G. S., Gatti, L., Howell, S. B., and Zunino, F. (2000) Yeast mutants as a model system for identification of determinants of chemosensitivity. *Pharmacol. Rev.* **52,** 477–491.
9. Beretta, G. L., Binaschi, M., Zagni, E., Capuani, L., and Capranico, G. (1999) Tethering a type IB topoisomerase to a DNA site by enzyme fusion to a heterologous site-selective DNA-binding protein domain. *Cancer Res.* **59,** 3689–3697.
10. Moreno, S., Klar, A., and Nurse, P. (1991) Molecular genetic analysis of fission yeast *Schizosaccharomyces pombe. Methods Enzymol.* **194,** 795–823.
11. Rose, M. D. and Broach, J. R. (1991) Cloning genes by complementation in yeast. *Methods Enzymol.* **94,** 195–230.
12. Hottinger, H., Pearson, D., Yamao, F., et al. (1982) Nonsense suppression in *Schizosaccharomyces pombe:* the *S. pombe* Sup3-e tRNASerUGA gene is active in *S. cerevisiae. Mol. Gen. Genet.* **188,** 219–224.
13. Brewer, B. J. and Fangman, L. (1987) The localization of replication origins on ARS plasmids in *S. cerevisiae. Cell* **51,** 463–471.
14. Ito, H., Fukada, Y., Murata, K., and Kimura, A. (1983) Transformation of intact yeast cells treated with alkali cations. *J. Bacteriol.* **153,** 163–168.
15. Gietz, R. D. and Woods, R. A. (2002) Transformation of yeast by lithium acetate/single-stranded carrier DNA/polyethylene glycol method. *Methods Enzymol.* **350,** 87–93.
16. Uno, I., Fukami, K., Kato, H., Takenawa, T., and Ishikawa, T. (1988) Essential role for phosphatidylinositol 4,5-bisphosphate in yeast cell proliferation. *Nature* **333,** 188–190.
17. Prentice, H. L. (1991) High efficiency transformation of *Schizosaccharomyces pombe* by electroporation. *Nucleic Acids Res.* **20,** 621.
18. Becker, D. M. and Guarente, L. (1991) High-efficiency transformation of yeast by electroporation. *Methods Enzymol.* **194,** 182–187.
19. Perego, P., Zunino, F., Carenini, N., Giuliani, F., Spinelli, S., and Howell, S. B. (1998) Sensitivity to cisplatin and platinum-containing compounds of *Schizosaccharomyces pombe* rad mutants. *Mol. Pharmacol.* **54,** 213–219.
20. Hartwell, L. H., Szankasi, P., Roberts, C. J., Murray, A. W., and Friend, S. H. (1997) Integrating genetic approaches into the discovery of anticancer drugs. *Science* **278,** 1064–1068.
21. Norman, T. C., Smith, D. L., Sorger, P. K., et al. (1999) Genetic selection of peptide inhibitors of biological pathways. *Science* **285,** 591–595.

23

Application of Proteomics to the Discovery of Serological Tumor Markers

Terence C. W. Poon and Philip J. Johnson

1. Introduction

As life expectancy rises throughout the world, so cancer, predominantly a disease of aging populations, becomes an ever greater public health problem. Currently, between one-fourth and one-third of Western populations will die of cancer. Most cancers have a prolonged subclinical phase, and it probably takes many years for them to become symptomatic or visible by imaging techniques. Indeed, for the great majority of a tumor's lifetime, it will be microscopic. To date, surgical resection is the most widely used approach to effect a cure. Unfortunately, most of the cancers are diagnosed at an advanced stage when they are no longer suitable for surgical removal either because of their physical size or, more often, because the tumor has already undergone metastatic spread. This metastatic spread is the major problem in cancer therapeutics. Once metastatic spread has occurred, local therapies, such as surgery or radiotherapy, are no longer appropriate, and the administration of systemic chemotherapeutic drugs is usually the only option. Such drugs are relatively nonspecific in their effects on cancer cells, and hence they are often toxic to normal, noncancerous tissues; despite some notable exceptions, only a small percentage of patients is cured by this approach. Until more effective treatment for metastatic disease becomes available, prevention or early detection of cancer remains the most effective ways to reduce mortality.

Effective serological tumor markers may be one way of achieving early diagnosis of the disease in a noninvasive manner. In the last decade many potential serological tumor markers have been discovered and investigated experimentally, but few have entered into routine clinical practice, and there is little consensus on their use. More sensitive and specific markers are urgently needed. The development of high-throughput profiling of proteome contents may provide a useful approach for the systematic discovery of new and more effective serological tumor markers.

2. General Aspects of Tumor Markers

Tumor markers are broadly defined as biomolecules that can be measured quantitatively in body tissues or fluids; they are clinically useful in patients with cancer. In addition to early diagnosis, clinical usage includes screening, staging, prognosis, and monitoring treatment response and detection of tumor recurrence *(1)*.

From: *Handbook of Proteomic Methods*
Edited by: P. Michael Conn © Humana Press Inc., Totowa, NJ

Hundreds of potential tumor markers have been investigated over the last 50 years but few of these have entered routine oncology practice. Nonetheless, the successes should be acknowledged and recognized as "proof of principle" that serological tumor markers can have a real clinical impact. Thus, monoclonal immunoglobulins in myeloma, estimation of human chorionic gonadotrophin (β-HCG) in patients with gestational trophoblastic disease, and α-fetoprotein (AFP) in patients with nonseminomatous germ cell tumor and endodermal sinus tumors all permit early diagnosis of disease and disease relapse, and have sufficient specificity to warrant initiation of treatment. Although such markers approach the ideal, they are applicable to only a minority of tumors.

Tumor markers for more common cancers have proved less successful either because of low specificity or low sensitivity. Plasma prostate-specific antigen (PSA) estimation is widely used in screening for prostate cancer, but there is a high frequency of false positive results. Others that have failed in the setting of screening the normal population for specific cancers, such as carcinoembryonic antigen (CEA; in colorectal cancer), are more widely used to detect recurrence after curative surgery, but again there is little consensus about the efficacy of this approach. Although AFP is widely used for the diagnosis of hepatocellular carcinoma, its level frequently rises in patients with testicular or ovarian cancers or patients with benign liver disease. Serum CEA levels may increase in patients other than those with colorectal cancer, such as lung and breast cancers. Obviously, these markers are far from the ideal that should be highly specific for a particular cancer type and not associated with nonmalignant diseases.

The development of "conventional" tumor markers, CEA and AFP, for example, was driven largely by the introduction of new methods [immunodiffusion techniques and subsequently radioimmunoassays and enzyme-linked immunosorbent assays (ELISAs)] for quantifying small amounts of circulating proteins that were relatively specific for certain types of cancer. In an attempt to increase the pool of possible serological molecular markers, several new approaches, including proteomics, are now being applied to constituents of plasma and serum *(2)*.

From the practical point of view, a serological tumor marker is more versatile than a marker that can only be detected in tumor tissues. The former can be readily measured by simple chemistry, such as immunoassay, in a noninvasive manner. In contrast, tumor biopsies are not always available for histological analysis and certainly cannot be undertaken serially. Furthermore, serological markers are particularly important for early detection and screening. In these two situations, a tumor may be too small to be visualized by imaging methods or for biopsy. An ideal tumor marker for diagnosis or screening should be detectable in biological fluids before the tumor has developed the capacity to metastasize. We also need tumor markers that can accurately reflect the tumor load. Indeed, there is a most striking imbalance between resources that are invested in developing new treatment and those invested in the development of methods for determining the efficacy of these new treatments. Unfortunately, very few of the existing tumor markers can fulfil the above requirements.

There is, of course, no guarantee that an ideal serological tumor marker exists for each cancer type. The transformation of a normal cell into a malignant one is a slow, complex process involving multiple steps. Both exogenous and endogenous factors contribute simultaneously or sequentially. Conventionally, cancer is diagnosed and classified according to morphological changes. However, somatic mutation of different protooncogenes or tumor-suppressor genes may result in the same morphological change. This may explain why a newly discovered marker is usually later proved to be applicable only to a limited subset of patients with a particular cancer type. If ideal tumor markers do not exist for the majority of cancers, will it be possible to improve the diagnosis or increase the predictive power by combined use of multiple serological markers? How many markers are required in order to achieve 100% accuracy in the diagnosis or outcome prediction? What are these markers? How can we effectively identify these markers? Proteomics may be able to help us find the answers to some or all of these questions.

3. Proteomic Technologies and Identification of Serological Tumor Markers

Proteomics is the study of protein properties (expression level, post-translational modification, interactions, and so on) on a large scale that results in a global, integrated view of disease processes, cellular programs, and networks at the protein level *(3)*. Proteomic studies can be classified into two major types—protein structure analysis and quantitative regulation measurement. The former aims to identify the important structural features of proteins, especially their functional domains, and to identify proteins that interact and form complexes. This is important in identifying the protein–protein interaction pathways. Quantitative regulation studies aim to measure the expression of the proteins within a cell or tissue and to examine the quantitative changes under different conditions. An example of the latter might be the study of the differences between healthy and diseased states. It is the proteomic technologies developed for the quantitative approaches that are particularly useful in the discovery of tumor markers because both are related to protein expression levels *(4)*.

3.1. Two-Dimensional Polyacrylamide Gel Electrophoresis

Proteomics first became feasible with the combined use of two-dimensional polyacrylamide gel electrophoresis (2-D-PAGE) and mass spectrometry (MS) for resolving the proteome content and protein identification, respectively. Using 2-D-PAGE, proteins can be separated according to their isoelectric points (pI values) and apparent molecular weights, and they appear as discrete spots. The protein identity can then be obtained by tryptic peptide fingerprinting. An excised protein spot is first digested with trypsin. The masses of the resultant set of peptides are then measured by matrix-assisted laser desorption/ionization time-of-flight (MALDI-TOF) MS and compared with the theoretical mass spectrum of different proteins in publicly available databases, such as the EMBL and SwissProt *(5,6)*. Postsource decay analysis can then be performed to obtain the amino acid sequence information of one or more tryptic peptides to confirm the protein identity. Quadrupole-TOF (Q-TOF) MS *(7)* and TOF-TOF MS *(8)* are contemporary MS technologies that make amino acid sequencing much easier by collision-induced dissociation analysis. Using the molecular mass and the partial amino acid sequence information of a peptide, one can search public databases that contain incomplete gene information, e.g., the expressed sequence tags (ESTs) database *(9)* to obtain the protein identity. If the peptide sequence is novel, the whole protein may be sequenced.

The 2-D-PAGE approach has been successfully applied to the analysis of the proteomes of various cancers including lung cancer *(10)*, breast cancer *(11,12)*, prostate cancer *(13,14)*, esophageal cancer *(15)*, hepatocellular carcinoma (HCC) *(16)*, bladder cancer *(17)*, renal cell cancer (RCC) *(18)*, and ovarian cancer *(19)*. By comparing the proteomes of cancer and normal tissues, cancer-associated proteins were identified in some of these studies.

For example, the comparison between primary lung adenocarcinomas and normal tissues showed that a protein of unknown identity (35 kDa, pI 5.5) was present only in the cancer tissue *(10)*. Compared with normal breast epithelial cells, 2-D-PAGE analysis revealed that 14-3-3σ was strongly downregulated in the human breast cancer cell lines MCF-7 and MDA-MB-231 and in primary breast carcinomas *(11)*. In the case of prostate cancer, NEDD8, calponin, and folistatin-related protein were found only in normal prostate tissue and not in malignant tissue *(13)*. Another 2-D-PAGE study revealed six proteins that were only seen in microdissected malignant prostate cells and two proteins that were only seen in microdissected benign epithelium *(14)*. 2-D-PAGE images of the microdissected normal squamous esophageal epithelium and corresponding tumor cells were almost (98%) identical, but 17 proteins showed tumor-specific alterations, including 10 that were uniquely present in the tumors and 7 that were observed only in the normal epithelium *(15)*. Two of the altered proteins were identified as cytokeratin 1 and annexin I. Comparison of normal and tumor kidney tissues revealed that ubiquinol cytochrome c reductase and mitochondrial NADH-ubiquinone oxido-reductase complex I were present only in normal tissues and not in RCCs *(18)*. The downregulation of these proteins may suggest that mitchondrial dysfunction plays an important role in the carcinogenesis of RCC.

Such studies, aimed primarily at the identification of the cancer-associated proteins with a view to understanding the carcinogenesis process, clearly demonstrate that proteins, relatively specific for individual cancers, are present in cell/tissue extracts and can be detected by proteomic techniques. However, before such proteins can be used as tumor markers, further studies will be required to determine whether they can also be detected in biological fluids, such as plasma or serum.

3.2. Detection of Cancer-Related Proteins in Serum

Autoantibodies against tumor antigens, such as p53 *(20)*, have been shown to be present in the sera of cancer patients. On the basis of this phenomenon, an approach has been developed to identify proteins that commonly induce an autoantibody response in patients with lung cancer *(21,22)*, breast cancer *(23)*, and RCC *(24)*. Serum samples were collected from a cancer patient group and a noncancer control group. The proteome of a cancer cell line or cancer tissue was first separated by 2-D-PAGE, followed by Western blot analysis in which individual sera were tested for primary antibodies. The patterns of the stained spots between the two study groups were compared to identify the autoantibody response specifically associated with the cancer. The corresponding antigens in the 2-D-PAGE were then identified by tryptic peptide fingerprinting. Using this approach, autoantibodies against the protein gene product 9.5 (PGP 9.5) *(21)* and annexins I and II *(22)* were found in the lung cancer patients. PGP 9.5 was found in the sera of the cancer patients, but not in sera of the noncancer control subjects. By screening for IgG antibodies to proteins from the breast cancer cell line SUM-44, antibodies against a novel oncogenic protein RS/DJ-1 that regulates RNA–protein interaction were identified in the sera of the breast cancer patients *(23)*. RS/DJ-1 was later shown to be present in sera from 37% of newly diagnosed patients

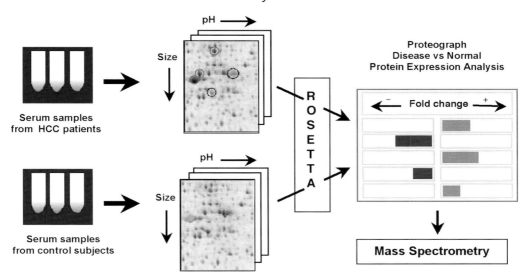

Fig. 1. An outline of the strategy for the identification of serological tumor marker proteins using hepatocellular carcinoma (HCC) as a model tumor. 2-D-PAGE images from HCC samples were placed in the foreground and those from cirrhosis and normal samples in the background. Proteomic features common to (occurrence ≥50%) individual images in the foreground and those common to (occurrence (50%) individual images in the background were identified. These two sets of proteomic features were compared with respect to identity (MCI number) and intensity. Features differing between the two sets by a relative quantitative difference of equal to or greater than two times were used to generate a Proteograph. For each MCI, the p value between HCC and liver cirrhosis cases was calculated. Those giving a p value <0.05 upon Student's t-test were considered significant *(23)*.

with breast cancer. Using a similar approach, antibodies against smooth muscle protein 22-α and carbonic anhydrase I (CA I) were identified in RCC patients, but not in healthy subjects *(24)*. All these findings suggest that this autoantibody-proteomic approach may be useful in the identification of serum antigens and/or antibodies for cancer diagnosis. However, this approach cannot detect circulating tumor antigens that do not trigger an autoimmune response or tumor markers that are downregulated.

Another approach is to subject serum samples directly to 2-D-PAGE analyses. By comparing the 2-D-PAGE images of serum proteomes of the cancer and control groups, proteins specifically released by tumors may be identified. Because this approach is noninvasive, it is easy to collect large amounts of patient and control samples for investigation. We have attempted to use this direct approach to identify novel serological tumor markers for the diagnosis of HCC **(Fig. 1)**. The proteins in individual sera from HCC patients, cirrhotic patients, and healthy normal subjects were first separated by 2-D-PAGE and then stained with fluorescent dye OGT MP 17 to allow quantification of each protein spot from the picogram to milligram level. The gel images were processed with a custom version of MELANIE II (Bio-Rad). Individual resolved protein features were enumerated and quantified on the basis of fluorescence signal intensity. To normalize the experimental variations in gel loading and staining, the quantity of individual spots

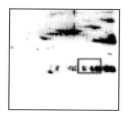

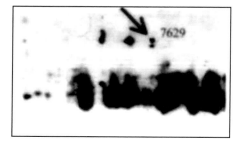
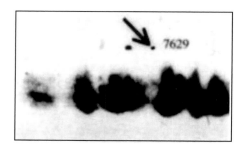

Hepatocellular   Liver cirrhosis
carcinoma

Fig. 2. A typical example of a gel image region that contains proteome features upregulated in HCC *(23)*.

("proteomic features") was expressed as a fractional fluorescence intensity of that feature, relative to the sum of all detected features in that sample. Each protein spot present in a master gel was given a numerical label [molecular cluster index (MCI)]. In other words, each MCI corresponded to a particular protein spot in the master gel with a specific pI value and relative molecular weight.

Rosetta (Oxford GlycoScience), a gel image analysis software, was used to match the protein spots in other gels to the protein spots in the master gel, and assigned MCIs. Protein spots occurring in more than 50% of the HCC samples and those occurring in more than 50% of cirrhosis samples were identified and compared with respect to the MCI number and intensity. Spots differing between the two sets by a relative quantitative difference of ≥ two times were used to generate a Proteograph. Those giving a *p*-value <0.05 upon Student's *t*-test were considered significant. The identified spots were subjected to protein identification by tryptic peptide fingerprinting with MALDI-TOF MS, amino acid sequencing with Q-TOF MS, and searching of the SwissProt database. Using this direct approach, more than 20 proteomic features were found to be either upregulated or downregulated in HCC (**Fig. 2**). Complement factor 4 and ceruloplasmin are two examples of the upregulated features, which have previously been reported in HCC patients *(25)*. This approach can also be used to verify newly discovered markers by comparing the proteomes of pre- and postsurgical resection sera from HCC patients. For example, in the case of the HCC-specific AFP isoform, its corresponding MCI candidate on the 2-D-PAGE image disappeared after the resection of the tumor mass (**Fig. 3**).

Despite the successful application of 2-D-SDS-PAGE technology in cancer research, this method of protein or peptide separation is not without its problems, some of which

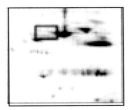

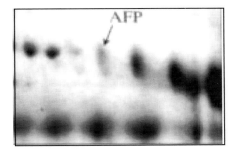 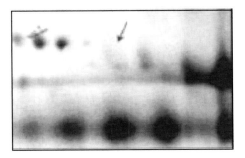

Before resection   After

Fig. 3. The identification of α-fetoprotein (AFP) in the proteome of sera collected from the same HCC patient before tumor resection and after resection. After tumor resection, the quantity of a proteome feature corresponding to an AFP isoform was greatly reduced in the serum *(23)*.

impact directly on protein tumor marker discovery. Some are technical, such as maintaining reproducibility (so that valid comparisons can be made between tumor and normal proteomes). More seriously, the "dynamic range" of 2-D-SDS-PAGE is usually limited, spot densities reflecting a range of concentration of only 10^2–10^3, whereas we know that the true range in serum is more than 10^6. We also know that only a fraction of proteins that are expressed can be detected because of low abundance. Proteins present in very low concentrations may still be important tumor markers.

3.3. Surface-Enhanced Laser Desorption/Ionization Time-of-Flight Mass Spectrometry

Surface-enhanced laser desorption/ionization (SELDI)-TOF MS is a non-electrophoresis-based proteomic technology, introduced by Hutchens and Yip *(26)*. In this technology, proteins are bound on a solid-phase chromatographic surface (i.e., the ProteinChip Array) and subsequently ionized and detected by TOF MS. There are several types of ProteinChip Arrays with different chromatographic properties, such as reverse phase, anion exchange, cation exchange, and metal affinity to allow the retention of different subsets of proteins in a complex protein mixture. When serum proteomes are analyzed, neat or diluted serum samples can be directly applied to the chromatographic surface. After washing and addition of an energy-absorbing molecule (EAM), usually α-cyano-4-hydroxycinnamic acid (CHCA) or sinapinic acid, the mass/charge *(m/z)* profile of the retained proteins can be obtained with the ProteinChip Reader. Because most of the proteins are single-charged during the EAM-assisted MS analysis, each peak in mass spectrum usually corresponds to a single peptide/protein with a molecular weight equivalent to the *m/z* value. The peak intensity is directly proportional to the serum level. This SELDI-TOF MS technology has been applied to identify potential serological diagnostic markers for various cancers including ovarian cancer *(27)*, prostate cancer *(28,29)*, breast cancer *(30)*, and HCC *(31)*.

The experimental procedure of this technology is much simpler than that of 2-D-PAGE. In the study by Petricoin et al. *(27)*, C16 reverse-phase ProteinChip arrays were first equilibrated with acetonitrile, followed by the addition of 1 µL of serum. After air-drying, washing with deionized water, and addition of 0.5 µL of EAM CHCA, the chips were analyzed with the ProteinChip Reader to obtain the mass spectra of *m/z* range 0–20,000 to profile the low molecular weight proteins/peptides (<20 kDa). Using a genetic algorithm and self-organizing cluster analysis, the comparison of mass spectra between 50 unaffected women and 50 patients with ovarian cancer identified five masses (534, 989, 2111, 2251, and 2465 Daltons), which formed a unique cluster pattern for discriminating cancer from noncancer subjects. The discriminatory pattern identified unseen ovarian cancer cases with a sensitivity of 100% and specificity of 95% *(27)*.

When prostate cancer is studied (using sera from patients with prostate cancer, benign prostate hyperplasia and healthy men), immobilized metal affinity capture ProteinChip IMAC3 that was preloaded with Cu^{2+}, was found to give the best results. Serum samples were first diluted with urea solution containing 3-[(3-cholamidopropyl)dimethylammonio]-1-propane-sulfonic acid, then applied to the IMAC3-Cu ProteinChip. After washing with deionised water and adding EAM sinapinic acid, the chips were analysed with the ProteinChip Reader to obtain the mass spectra of *m/z* range 2000–40,000. A decision tree classification algorithm was used to identify the polypeptide peaks that could classify the cases. Nine masses (4475, 5074, 5382, 7024, 7820, 8141, 9149, 9507, and 9656 Daltons) were identified to form a "serum protein fingerprinting" for the detection of prostate cancer with a sensitivity of 83% and specificity of 97% *(28)*. Besides Cu^{2+}, IMAC3 ProteinChip can be preloaded with Ni^{2+}. In the study of Li et al. *(30)*, IMAC3-Ni ProteinChip was used to analyze the sera from patients with breast cancer, patients with benign breast diseases, and healthy women. Three peaks (4.3, 8.1, and 8.9 kDa) were identified as the potential diagnostic markers for early detection of breast cancer at a sensitivity of 93% and specificity of 91%.

Besides focusing on peptides and low molecular weight proteins, it is also possible to apply SELDI-TOF MS technology to obtain proteomic profiles from 0.5 to 200 kDa. We have recently used the IMAC3-Cu ProteinChip and weak cation exchange ProteinChip WCX2, in a combination of anion-exchange fractionation to analyze and compare the serum proteomic profiles of patients with HCC and those of noncancer control subjects with chronic liver disease (CLD) *(31,32)*. Because the chemistries of the binding surface of the IMAC3-Cu ProteinChip Array and that of the WCX2 are different, we can identify more potential serum tumor markers within the same mass range **(Fig. 4)**. In this study, 2384 common peaks of mass from 0.5 to 200 kDa were observed in the SELDI-TOF mass spectra. Among them, 250 peaks were identified as a panel of potential markers forming a proteomic signature for the detection of HCC. These 250 potential markers were used to generate artificial neural networks (ANNs) for calculation of a diagnostic score for the detection of HCC. The cross-validation analysis showed that the ANN diagnostic score was useful in the differentiation between HCC and CLD cases at a sensitivity of 92% and specificity of 90%. The ANN diag-

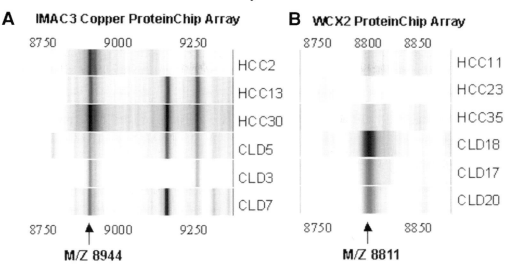

Fig. 4. Gel views of the mass spectra of the serum sample illustrating the higher intensities of M/Z 8944 polypeptide and lower intensities of M/Z 8811 polypeptide in the serum proteomes of the HCC cases, compared to CLD cases. The M/Z 8944 polypeptide was identified with the IMAC3 copper ProteinChip Array (**A**), whereas the M/Z 8811 polypeptide was identified with the WCX2 ProteinChip Array (**B**) (from **ref. *32*.**)

nostic score was useful in the identification of HCC cases with nondiagnostic serum AFP levels. Differentiation of HCC with nondiagnostic AFP levels from CLD is particularly important in Southeast Asian countries including Hong Kong and China, as more than 80% of HCC cases arise in patients with pre-existing liver cirrhosis.

All these studies indicate that the SELDI-TOF technology is a useful high-throughput tool for the identification of a panel of potential diagnostic markers by analyzing and comparing the serum proteomes of the cancer and noncancer patients. When each of these potential markers was used alone to set the diagnostic criterion, the diagnostic accuracy was far from satisfactory. However, combined use of multiple markers as a diagnostic signature or fingerprint could achieve a diagnostic accuracy of more than 90%. These studies strongly suggest that a single, ideal, serological tumor marker does not exist for most cancers; combined use of multiple tumor markers is likely to be the best way forward.

It is noteworthy that the protein identities of all these potential markers are unknown. However, even without knowing their identities, they can be unambiguously detected and quantified in patient sera by the SELDI-TOF ProteinChip system because each protein has a unique *m/z* value and biochemical properties. However, the fact that the identities of the tumor markers remain unknown means that the possibility of measuring them, using other technology platforms, such as microarray-based multiplex ELISA, is limited *(33)*.

3.4. Isotope-Coded Affinity Tags

Isotope-coded affinity tags (ICATs) are reagents developed to allow simultaneous quantification and identification of protein components in a complex protein mixture by MS/MS *(34,35)*. This technology is based on comparison of the MS signal intensities of hallmark peptides from individual proteins between the test sample and the reference sample. The proteins under investigation and the reference samples are labeled at the cysteine residues with different biotin affinity tags comprised of heavy (d8-ICAT) and light (d0-ICAT) isotopes, respectively, and the two samples are mixed before further analysis. After trypsinization, multidimensional high-performance liquid chromatography (HPLC) separation (strong cation exchange chromatography, biotin affinity chromatography, and reverse-phase liquid chromatography) is performed to purify and separate the ICAT-labeled hallmark peptides into fractions to reduce the complexity of the hallmark peptides for subsequent MS analyses *(35)*. In the deisotoped mass spectrum of each fraction, hallmark peptides of the same sequence belonging to the test and reference samples will appear as a pair of peaks with a mass difference of 8 Daltons (for peptides containing one cysteine residue) or 16 Daltons (for peptides containing two cysteine residues). Fragmentation analysis such as postsource decay or collision-induced dissociation can be used to identify the protein to which the hallmark peptide belongs. The ratio of the peptide peaks is equivalent to the concentration of this particular protein in the test sample relative to the reference sample. This technology has been successfully used to obtain comprehensive proteomic profiles of the membrane proteins of naïve and in vitro differentiated myeloid leukemia (HL-60) cells *(34)*. It is reasonable to expect that this ICAT technology will be readily applicable to serum proteomic profiling.

Compared with the SELDI-TOF technology, it is obvious that this ICAT technology can provide the protein identity for each MS peak, but the SELDI-TOF technology cannot. Knowing the protein identity is particularly important when we are investigating biological questions. In the case of serum marker discovery, knowing the protein identities of the potential tumor markers can allow us to measure these markers using other technology platforms, hence increasing their practical value. However, advantages of the SELDI-TOF technology are that the experimental procedures are simpler and it is applicable to the profiling of small peptides, whereas the ICAT technology is not, because small peptides usually do not contain cysteine residues.

3.5. Tagless Extraction-Retentate Chromatography

Tagless extraction-retentate chromatography is a recently developed technology that facilitates quantitative proteomic profiling by SELDI-TOF technology *(36)*. A protein mixture is first digested with a protease (e.g., trypsin) and alkylated. The peptides containing methionine residues are extracted by covalent attachment of the methionine residues to bromoacetyl reactive groups tethered to the surface of glass beads packed in small reaction vessels. After washing away nonspecifically bound peptides, β-mercaptoethanol is used to release the captured methionine-containing peptides in their nascent state. The recovered methionine-containing peptides are then profiled by SELDI-TOF technology. MS/MS can be used later to obtain the protein identity of the methionine-containing peptides.

An analogy could be made between DNA microarray experiments and quantitative proteomic profiling experiments. The ICAT approach is analogous to the cDNA microarray approach, whereas the tagless extraction-retentate chromatography approach is analogous to the Affymetrix oligonucleotide microarray approach. In both the ICAT and cDNA microarray approaches, one needs a pair of test and reference samples that are differentially labeled. The concentration of an analyte will be obtained as a ratio relative to the concentration of the same analyte in the reference samples. In the case of the tagless extraction-retentate chromatography and Affymetrix oligonucleotide microarray approaches, only the test sample is needed. The concentration of an analyte will be obtained as an absolute signal intensity, which is later normalized with the internal calibration standard(s) or the total signal intensity. Although this tagless technology appears to be an attractive alternative to the ICAT technology, its effectiveness and limitations await further evaluation.

4. Bioinformatic Issues

Bioinformatic tools are needed to assist in the identification of potential serological tumor markers from 2-D-PAGE images or proteomic profiles and to integrate the clinical parameters and multiple markers for outcome prediction. Because proteomic profiling methods are high-throughput technologies, values of large numbers of variables (i.e., concentrations of various proteins) will be generated simultaneously. These data are of the high-dimensional type. The number of observations on the data variables depends on the number of cases in the study groups. In practice, the number of variables is usually much larger than the number of observations, and it is difficult to extract meaningful patterns from a small number of observations on a large number of variables.

One of the simplest experimental aims is to identify proteins the levels of which are significantly different between two conditions. In a tumor marker discovery experiment, one may have sets of 2-D-PAGE images or proteomic profiles for the cancer patient and noncancer subject groups. For example, we may have serum samples collected from 50 cases for each group and there may be 1000 common proteomic features observed for each serum sample. In clinical chemistry studies, Student's *t*-test or the Mann-Whitney U test are commonly used to compare the serum levels of a marker between two study groups. Suppose we use the Mann-Whitney U test to compare the levels of each proteomic feature between the two groups to identify the potential tumor markers. The commonly accepted probability level of 0.05 is used to reject the null hypothesis. Following this test rule, we will have 1 chance out of 20 of wrongly rejecting the null hypothesis. After performing the Mann-Whitney U test 1000 times to compare the levels of each of the proteomic features between the two study groups, we will incorrectly reject the null hypothesis 50 times. In other words, we will identify 50 differential proteomic features that are "statistically significant," even though none of the differential proteomic features are genuinely different between the two groups. This is known as the problem of multiple comparisons. Furthermore, statistic tests such as Student's *t*-test and the Mann-Whitney U test do not consider the relationships among the proteomic data. Analysis of the proteomic data thus needs sophisticated statistical interpretation before we can identify valid tumor markers or tumor-specific proteomic signatures.

The multidimensional nature of the data and the problem of multiple comparisons have been recognized and addressed in the analysis of gene expression data generated from microarray experiments. For example, significant analysis of microarray (SAM) is a statistical technique designed to find significant genes in a set of microarray experiments *(37)*. The algorithm uses repeated permutations of the data to determine whether the expression of any gene/protein is significantly related to the response. The cutoff for significance is determined by the user, based on the false discovery rate (FDR). The FDR is the number of falsely called genes/proteins divided by number of differential genes/proteins in original data. In other words, FDR provides information on the percentage of nonsignificant genes one can expect to find in the resulting list of differentially expressed genes. SAM has been successfully used to identify differentially expressed genes *(37)* or to filter out the probable irrelevant gene expression data *(38)* and thereby facilitate further analysis.

We have found that SAM is also applicable to serum proteomic data *(32)*. After filtering the insignificant proteomic data using SAM, two-way hierarchical clustering analysis can then be performed to classify the cases according to the significant proteomic data. Other more complex approaches, including genetic algorithms *(27)*, self-organizing cluster analysis *(27)*, discrete wavelet transform analysis *(39)*, ANNs *(40)*, and unified maximum separation analysis *(30)*, have been used to compare the SELDI-TOF MS data between cancer and noncancer groups in order to identify tumor-specific proteomic features.

Once a panel of serological tumor markers has been identified, the next issue is how to integrate the clinical values of these data to achieve high sensitivity and specific diagnosis. Boosted decision tree analysis *(29)* and calculation of Euclidean distance vector *(27)* have been used to interpret the tumor-specific proteomic patterns in cancer detection. Recently, we have shown that ANNs are useful in the interpretation of a panel of serological liver markers to improve the detection of HCC *(32,41)*. They may be also applicable to the interpretation of tumor-specific proteomic patterns.

5. Future Prospects

Advances in the field of MS such as the development of TOF-TOF MS are dramatically improving our ability to resolve highly complex proteomic profiles and are making protein identification much quicker and easier. Although recent technologies such as SELDI-TOF MS and ICAT may appear more powerful than the 2-D-PAGE technology, the 2-D-PAGE technology still provides some advantages over the current gel-free MS technologies. Thus many serum proteins carry post-translational modification such as glycosylation or phosphorylation, and, by focusing on a specific glycoform, the clinical value of a tumor marker can be enhanced, as in the case of AFP glycoforms *(42)*. 2-D-PAGE can readily resolve proteins differing in post-translation modifications, but SELDI-TOF MS and ICAT approaches cannot. Furthermore, 2-D-PAGE is a relatively cheap technology affordable by most laboratories. As proteomic technologies advance rapidly, it seems likely that the availability of good bioinformatic tools may become the major bottleneck. Compared with gene expression microarray data, the computational analysis tools for proteomic data are still underdeveloped.

To date, proteomic technologies have only been applied to the identification of tumor markers for diagnosis. In the future the identification of serological proteomic fingerprint/signatures may help us to address more difficult clinical problems, such as the prediction of a patient's treatment response or adverse effects to chemotherapy.

Finally, it is noteworthy that circulating nucleic acids form a new category of tumor markers, which is expanding rapidly and which may provide specific and sensitive markers for many tumors *(43)*. Ultimately, combined use of multiple circulating nucleic acid markers and protein/glycoprotein markers may allow outcome prediction at an accuracy close to 100%.

References

1. Diamandis, E. P., Fritsche, H. A., Lilja, H., Chan, D. W., and Schwartz, M. W. (eds.) (2002) *Tumor Markers—Physiology, Pathobiology, Technology, and Clinical Applications.* AACC Press, Washington, DC.
2. Johnson, P. J. (2001) A framework for the molecular classification of circulating tumor markers. *Ann. NY Acad. Sci.* **945,** 8–21.
3. Blackstock, W. P. and Weir, M. P. (1999) Proteomics: quantitative and physical mapping of cellular proteins. *Trends Biotechnol.* **17,** 121–127.
4. Alaiya, A. A., Franzen, B., Auer, G., and Linder, S. (2000) Cancer proteomics. *Electrophoresis* **21,** 1210–1217.
5. Fenyo, D. (2000) Identifying the proteome: software tools. *Curr. Opin. Biotechnol.* **11,** 391–395.
6. Shevchenko, A., Jensen, O. N., Podtelejnikov, A. V., et al. (1996) Linking genome and proteome by mass spectrometry: large-scale identification of yeast proteins from two dimensional gels. *Proc. Natl. Acad. Sci. USA* **93,** 14440–14445.
7. Andersen, J. S. and Mann, M. (2000) Functional genomics by mass spectrometry. *FEBS Lett.* **480,** 25–31.
8. Medzihradsky, K. F., Campbell, J. M., Baldwin, M. A., et al. (2000) The characteristics of peptide collision-induced dissociation using a high performance MALDI-TOF/TOF tandem mass spectrometer. *Anal. Chem.* **72,** 552–558.
9. Fenyo, D. (2000) Identifying the proteome: software tools. *Curr. Opin, Biotechnol.* **11,** 391–395.
10. Okuzawa, K., Franzen, B., Lindholm, J., et al. (1994) Characterization of gene expression in clinical lung cancer materials by two-dimensional polyacrylamide gel electrophoresis. *Electrophoresis* **15,** 382–390.
11. Vercoutter-Edouart, A. S., Lemoine, J., Le Bourhis, X., et al. (2001) Proteomic analysis reveals that 14-3-3 sigma is down-regulated in human breast cancer cells. *Cancer Res.* **61,** 76–80.
12. Page, M. J., Amess, B., Townsend, R. R., et al. (1999) Proteomic definition of normal human luminal and myoepithelial breast cells purified from reduction mammoplasties. *Proc. Natl. Acad. Sci. USA* **96,** 12589–12594.
13. Meehan, K. L., Holland, J. W., and Dawkins, H. J. (2002) Proteomic analysis of normal and malignant prostate tissue to identify novel proteins lost in prostate cancer. *Prostate* **50,** 54–63.
14. Ornstein, D. K., Gillespie, J. W., Paweletz, C. P., et al. (2000) Proteomic analysis of laser capture microdissected human prostate cancer and in vitro prostate cell lines. *Electrophoresis* **21,** 2235–2242.
15. Emmert-Buck, M. R., Gillespie, J. W., Paweletz, C. P., et al. (2000) An approach to proteomic analysis of human tumors. *Mol. Carcinog.* **27,** 158–165.
16. Seow, T. K., Liang, R. C. M. T., Leow, C. K., and Chung, M. C. M. (2001) Hepatocellular carcinoma: from bedside to proteomics. *Proteomics* **1,** 1249–1263.
17. Celis, J. E., Wolf, H., and Ostergaard, M. (2000) Bladder squamous cell carcinoma biomarkers derived from proteomics. *Electrophoresis* **21,** 2115–2121.
18. Sarto, C., Marocchi, A., Sanchez, J. C., et al. (1997) Renal cell carcinoma and normal kidney protein expression. *Electrophoresis* **18,** 599–604.
19. Jones, M. B., Krutzsch, H., Shu, H., et al. (2002) Proteomic analysis and identification of new biomarkers and therapeutic targets for invasive ovarian cancer. *Proteomics* **2,** 76–84.

20. Marx, D., Frey, M., Zentgraf, H., et al. (2001) Detection of serum autoantibodies to tumor suppressor gene p53 with a new enzyme-linked immunosorbent assay in patients with ovarian cancer. *Cancer Detect. Prev.* **25,** 117–122.

21. Brichory, F., Beer, D., Le Naour, F., Giordano, T., and Hanash, S. (2001) Proteomics-based identification of protein gene product 9.5 as a tumor antigen that induces a humoral immune response in lung cancer. *Cancer Res.* **61,** 7908–7912.

22. Brichory, F. M., Misek, D. E., Yim, A. M., et al. (2001) An immune response manifested by the common occurrence of annexins I and II autoantibodies and high circulating levels of IL-6 in lung cancer. *Proc. Natl. Acad. Sci. USA* **98,** 9824–9829.

23. Le Naour, F., Misek, D. E., Krause, M. C., et al. (2001) Proteomics-based identification of RS/DJ-1 as a novel circulating tumor antigen in breast cancer. *Clin. Cancer Res.* **7,** 3328–3335.

24. Klade, C. S., Voss, T., Krystek, E., et al. (2001) Identification of tumor antigens in renal cell carcinoma by serological proteome analysis. *Proteomics* **1,** 890–899.

25. Poon, T. C. and Johnson, P. J. (2001) Proteome analysis and its impact on the discovery of serological tumor markers. *Clin. Chim. Acta* **313,** 231–239.

26. Hutchins, T. W. and Yip, T. T. (1993) New desorption strategies for the mass spectrometric analysis of macromolecules. *Rapid Commun. Mass Spectrom.* **7,** 576–580.

27. Petricoin, E. F. III, Ardekani, A. M., Hitt, A. B., et al. (2002) Use of proteomic patterns in serum to identify ovarian cancer. *Lancet* **359,** 572–577.

28. Adam, B. L., Qu, Y., Davis, J. W., et al. (2002) Serum protein fingerprinting coupled with a pattern-matching algorithm distinguishes prostate cancer from benign prostate hyperplasia and healthy men. *Cancer Res.* **62,** 3609–3614.

29. Qu, Y., Adam, B. L., Yasui, Y., et al. Boosted decision tree analysis of surface-enhanced laser desorption/ionization mass spectral serum profiles discriminates prostate cancer from noncancer patients. *Clin. Chem.* **48,** 1835–1843.

30. Li, J., Zhang, Z., Rosenzweig, J., Wang, Y. Y., and Chan, D. W. (2002) Proteomics and bioinformatics approaches for identification of serum biomarkers to detect breast cancer. *Clin. Chem.* **48,** 1296–1304.

31. Poon, T. C. W., Yip, T.-T., Yip, C., et al. (2001) Application of the surface-enhanced laser desorption ionisation (SELDI) ProteinChip® biomarker system to the discovery of serological biomarkers associated with hepatocellular carcinoma. *Proc. AACR* **12,** 1976.

32. Poon, T. C. W., Yip, T. T., Chan, A. T. C., et al. (2003) Comprehensive proteomic profiling identifies serum proteomic signatures for detection of hepatocellular carcinoma and its subtypes. *Clin. Chem.* **49(5),** in press.

33. Wiese, R., Belosludtsev, Y., Powdrill, T., Thompson, P., and Hogan, M. (2001) Simultaneous multianalyte ELISA performed on a microarray platform. *Clin. Chem.* **47,** 1451–1457.

34. Han, D. K., Eng, J., Zhou, H., and Aebersold, R. (2001) Quantitative profiling of differentiation-induced microsomal proteins using isotope-coded affinity tags and mass spectrometry. *Nat. Biotech.* **19,** 946–951.

35. Gygi, S. P., Rist, B., Griffin, T. J., Eng, J., and Aebersold, R. (2002) Proteome analysis of low-abundance proteins using multidimensional chromatography and isotope-coded affinity tags. *J. Proteome Res.* **1,** 47–54.

36. Weinberger, S. R., Viner, R. I., and Ho, P. (2002) Tagless extraction-retentate chromatography: a new global protein digestion strategy for monitoring differential protein expression. *Electrophoresis* **23,** 3182–3192.

37. Tusher, V. G., Tibshirani, R., and Chu, G. (2001) Significance analysis of microarrays applied to the ionizing radiation response. *Proc. Natl. Acad. Sci. USA* **98,** 5116–5121.

38. Xu, Y., Selaru, F. M., Yin, J., et al. (2002) Artificial neural networks and gene filtering distinguish between global gene expression profiles of Barrett's esophagus and esophageal cancer. *Cancer Res.* **62,** 3493–3497.

39. Srinivas, P. R., Srivastava, S., Hanash, S., and Wright, G. L. Jr. (2001) Proteomics in early detection of cancer. *Clin. Chem.* **47,** 1901–1911.
40. Ball, G., Mian, S., Holding, F., et al. (2002) An integrated approach utilizing artificial neural networks and SELDI mass spectrometry for the classification of human tumours and rapid identification of potential biomarkers. *Bioinformatics* **18,** 395–404.
41. Poon, T. C. W., Chan, A. T. C., Zee, B., et al. (2001) Application of classification tree and neural network algorithm to the identification of serological liver marker profiles in the diagnosis of hepatocellular carcinoma. *Oncology* **61,** 275–283.
42. Poon, T. C., Mok, T. S., Chan, A. T., et al. (2002) Quantification and utility of monosialylated alpha-fetoprotein in the diagnosis of hepatocellular carcinoma with nondiagnostic serum total alpha-fetoprotein. *Clin. Chem.* **48,** 1021–1027.
43. Johnson, P. J. and Lo, Y. M. (2002) Plasma nucleic acids in the diagnosis and management of malignant disease. *Clin. Chem.* **48,** 1186–1193.

24

Infectomic Analysis of Microbial Infections Using Proteomics

Sheng-He Huang, Ambrose Jong, and James T. Summersgill

1. Introduction

Infectious diseases caused by various microbial pathogens have held considerable importance for medicine, being a major cause of deaths and disabilities throughout the world despite the availability of effective antimicrobial agents and vaccines over the last 50 years *(1,2)*. The continual emergence of previously undescribed new infectious diseases and the re-emergence of old pathogens will certainly heighten the global impact of infectious diseases in the 21st century. Another significant problem in medicine is the development of microbial resistance to antimicrobial drugs, owing to the widespread and often inappropriate use of these antimicrobials. As clinical practice exhibits a trend toward greater use of invasive interventions and with patients living longer, there is a continually growing proportion of older and immunocompromised patients. These groups are predisposed to opportunistic infections caused by nonpathogenic microbes such as yeast. In addition, how to deal with and prevent bioterrorism is becoming a very serious issue in the 21st century. The development of new anti-infective agents against various microbial pathogens and in favor of the host defense has emerged as an urgent issue in modern medicine. The combination of new (e.g., genomics, proteomics, glycomics, and bioinformatics) and traditional approaches [e.g., cloning, polymerase chain reaction (PCR), gene knockout and knockin, antisense] will lead to a solution for the challenges we are facing today *(3)*.

1.1. Overview of Infectomics

The term "infectomics" was recently coined and was defined as an integrative omic approach to study microbial infections globally *(3)*. The most important omic approaches include genomics and proteomics *(3)*. Recently, glycomics has become a valuable addition to the general omic approaches *(4,5)*. Each omics embraces two essential aspects: structural and functional studies. The fundamental issue of infectious diseases is how to dissect the interactions between microbial pathogens and their hosts holistically and integratively by using infectomics.

1.1.1. Infectomes of Microbial Pathogens

Microbial pathogens are exposed to multifactoral and dynamic environmental conditions during an infection cycle and have to regulate their gene expression globally at

From: *Handbook of Proteomic Methods*
Edited by: P. Michael Conn © Humana Press Inc., Totowa, NJ

both RNA and protein levels (infectomes) accordingly. Presumably, different microbial pathogens have their distinct infectomes that are induced in their hosts during microbial infections. The same pathogen may have distinct infectomes induced in different tissues of the same host. These characteristic features of infectomes can be used for globally dissecting microbial infection. To date, most of the advances in our understanding of human infectious diseases have come from analysis of a single, or at most, a small number of genes. Recently, DNA microarray and proteomic analyses have been used to monitor microbial gene expression profiles globally under certain environmental conditions or in their hosts.

1.1.2. Infectomes of the Host

The infectomic changes, including mRNA and protein expression profiles, in the host infected by pathogens are believed to be patterned and stereotyped. These infectomes can be used to distinguish among different infectious agents or different pathogenic mechanisms. It has been proposed that the host gene expression signatures can be used as potential diagnostic markers for infectious diseases (3). DNA and protein microarray techniques permit us to obtain global information on gene expression profiles inside cells. Human cDNA microarrays have been used to monitor globally the host response to various microbial pathogens including viruses, bacteria, fungi, and parasites (3). It is anticipated that the global monitoring of host infectomes induced by pathogens will be much more specific than the detecting of the traditional markers of inflammation, such as cytokines.

1.1.3. Applications of Infectomics in Infectious Diseases

1.1.3.1. BASIC RESEARCH

Basic infectomics is essential for globally dissecting the pathogenesis of microbial infections. The key fundamental issue of infectious diseases is how to understand globally and integratively the interactions between microbial pathogens and their hosts contributing to infection by using infectomics. On the one hand, we need to identify and characterize holistically the virulence factors, targets for vaccines and antimicrobials, and in vivo survival mechanisms of the invading micro-organism; on the other hand, we have to dissect globally the components of the host response that will lead to the elimination of the invading pathogen and resolution of disease.

1.1.3.2. DIAGNOSTIC INFECTOMICS

Infectomes of both microbial pathogens and the hosts can be used for diagnosis. Almost all currently available diagnostic tests for identification of microbial pathogens depend on the techniques in which the microorganisms can be obtained, manipulated, and analyzed in the laboratory (6). However, a number of recent studies suggest that there are certain inherent limitations in current diagnostic methods (7). These include the lack of a genome-wide survey of infectious agents and host responses, as well as ignorance of the host environmental conditions that are required by certain microbial pathogens and cannot be duplicated in the laboratory. The availability of complete human and numerous microbial genome sequences has profoundly revolutionized the ways of prediction, prognosis, and diagnosis of diseases. We assume that genotypic and phenotypic infectomes contributing to microbial infections are encoded by the genomes of microbial pathogens and their hosts and can be used as diagnostic indications to identify specific microbial infections. Genome-wide approaches to genotyping and phenotyping facilitate a rapid translation of molecular discoveries to diagnostics.

1.1.3.3. PREVENTIVE AND THERAPEUTIC INFECTOMICS

The development of microbial infections is determined by the nature of host–microbe relationships. These include host–pathogen, host–commensal, and pathogen–commensal interactions. A holistic balance of these relationships is essential to our health. However, this balance remains poorly defined, and little attention has been paid to commensal microbes that may be beneficial to the host defense systems. The availability of genomic and proteomic approaches allows for global study of preventive and therapeutic infectomics that will lead to holistic solutions to infectious diseases. For example, recent advances in genomics and proteomics have provided a tremendous opportunity in illuminating the present crisis of antibiotic resistance and expanding the range of potential antimicrobial targets. Global approaches based on microbial genomes have also facilitated a fundamental shift from the conventional methods of vaccine development and direct antimicrobial screening programs toward rational and genome-wide target-based strategies *(3)*. A combination of the strengths and advantages of both the genome-wide and the conventional screening strategies will greatly revolutionize vaccine developments and drug discoveries.

1.2. Mathematical Studies of Infectious Diseases

The application of mathematical approaches to infectious diseases dates back at least as far as Daniel Bernoulli's mathematical theory on smallpox control in 1760 *(8)*. By far the most important result to come out of mathematical studies of infectious diseases is the mathematical epidemic theory *(8)*. Mathematical models have been essential in analyzing the transmission dynamics of infectious diseases and informing decision making *(8,9)*. However, as in most of the biomedical sciences, mathematical and computational methods have not heretofore played a significant role in infectious diseases. The melding of the biomedical sciences (including infectious diseases) with mathematical and computational approaches is desperately needed in the postgenomic era. It is impossible to monitor and dissect infectomes holistically without the aid of mathematical and computational methods.

The central issues in infectious diseases involve the interplay between host resistance and microbial infectivity. The most promising hope for the future for dissecting both of these issues is network information embedded in microbial and host's genomes. Recently, genomic and proteomic analyses have generated mountains of data. Information for thousands of gene products can be produced in a single experiment. Therefore, mathematical or computational analysis of biological data (bioinformatics) is becoming an essential part of biomedical sciences. Bioinformatics is an integral part of proteomics *(10)*. Currently, the most common statistical approach for gene and protein expression data analyses is to cluster genes and proteins that are coregulated. There are two types of clustering algorithms, supervised versus unsupervised. In unsupervised clustering, no predefined set of classes is required. If a coregulated class of genes or proteins is known, supervised clustering algorithms are used to assign uncharacterized

Table 1
Selected Web Sources for Proteomics and Infectious Diseases

Host and model organism genomes
  Human Genome Project: http://www.ncbi.nlm.nih.gov/genome/seq/
  Mouse Genome Project: http://www.nih.gov/science/models/mouse
Microbial pathogen genomes
  Microbial genomes (general): http://www-fp.mcs.anl.gov/~gaasterland/genomes.html
  Pathogen genomics (NIAID): http://www.niaid.nih.gov/dmid/genomes/
  Parasite genomes: http://www.ebi.ac.uk/parasites/parasite-genome.html
  WHO-TDR Parasite Genome Committee: http:www.who.ch/tdr/workplan/genome.htm
  HIV database: http.//hiv-web.lanl.gov/
  Virus database: http://life.anu.edu.au
Protein microarrays and proteomics
  A Human Pathogen Microarray: http://www.noabdiagnostics.com/ANMicroarrays.htm
  Nagayama Protein Array: http:// www.jst.go.jp/erato/project/nts_P/nts_P.html
  Biochain's Protein Array: http:// www.biochain.com/proteinarray.htm
  Protein Microarray Instruments: http://www.biocompare.com/spotlight.asp
General 2-DGE Databases
  2D-gel databases: http://proteomewww.glycob.ox.ac.uk/OGPlnk2D.html
  Meta-database 2DWG: http://www.lecb.ncifcrf.gov/ips-databases.html
  WORLD-2DPAGE: http://us.expasy.org/ch2d/2d-index.html
  SIENA-2DPAGE: http://www.bio-mol.unisi.it/2d/2d.html
Cell/organism-specific 2-DGE databases
  Bacterial pathogens: http://www.mpiib-berlin.mpg.de/2D-PAGE/
  Human and mouse cells: http://proteome.tmig.or.jp/2D/
  Meningitic bacteria: http://www.abdn.ac.uk/~mmb023/2dhome.htm
  Mycobacterium tuberculosis: http://www.ssi.dk/en/forskning/tbimmun/tbhjemme.htm
  Parasite host cell: http://www.gram.au.dk/2d/2d.html
  Toxoplasma gondii-2D map: http://www-public.rz.uni-duesseldorf.de/~hfischer/
Organ/tissue-specific 2-DGE databases
  Human cancer 2-D-PAGE: http://proteomics.cancer.dk/
  Human placenta: http://www.ludwig.edu.au/jpsl/Databases.asp

molecules to that class. So far, most proteomics study has focused on two-dimensional gel electrophoresis (2-DGE)-based technologies. One of the first analysis steps in proteomics begins with identification of protein spots on 2-DGE gels. Subsequently, these identified spots are linked to the images of the gels. Proteins can be further identified by matrix-assisted laser desorption/ionization (MALDI) and mass spectrometry (MS). A number of 2-DGE image databases have been developed *(10)*. There are two kinds of 2-DGE databases, general versus cell/organism- and organ/tissue-specific **(Table 1)**.

1.3. Proteomic Approaches to Infectious Diseases

The ultimate goal of genomics is the global elucidation of the functional partners of genes and genomes (proteins and proteome) *(3)*. The proteome represents the functional status of a cell in response to environmental stimuli and thus provides more direct information on functional changes. Therefore, as an alternative and complementary approach to genomic-based technologies, proteomics is essential for the identification and validation of proteins and for the global monitoring of infectomic changes in protein expression and modification during infections. The major approach to proteomics to date is based on 2-DGE with a combination of MS and computer technologies. This makes it possible to separate and detect several thousands of protein spots in a good gel. Commercial robots are now available for staining gels, spot excision, and subsequent proteolysis before MS. MALDI, MS, and special softwares such as Melanie 3 (http://www.expasy.ch/melanie/) can be used to characterize proteins of interest further.

Proteomics has been successfully used for comparative analysis of protein expression profiles of pathogens and the infected host cells. For example, the protein expression patterns of virulent *Mycobacterium tuberculosis* strains and attenuated vaccine strains were compared using a combination of 2-DGE and MS *(3)*. Among the 1800 protein spots isolated, six new gene products, not previously predicted by the genomic study of *M. tuberculosis*, were identified with proteomics. A combination of proteomics and genomics has been used to identify unknown regulons and proteins in *Bacillus subtilis* and Gram-positive pathogens *(3)*. Recently, proteomics has been a powerful tool for infectomic studies of infectious diseases *(11–19)*. For example, infectomic analyses of the *Chlamydia pneumonia* elementary body (EB) in HEp-2 cells and *Chlamydia trachomatis* reticulate body (RB) in HeLa 229 cells have been performed by using MALDI-time-of-flight (TOF) MS *(20–22)*. Several novel features of *C. pneumonia*, such as energy-producing enzymes and type III secretion proteins, have been implicated in its proteomics maps *(20)*. A novel 7-kDa RB protein was identified in *C. trachomatis* but was not found in *C. pneumonia (22)*. However, there are a number of limitations in using the 2-DGE MS approach *(23)*. These include labor-intensive and time-consuming procedures and the inability to detect low-abundance proteins (*see* **Notes 1–3**). To overcome these limitations, a number of approaches have been developed to bypass gel electrophoresis. Recently, isotope-coded affinity tag (ICAT) has emerged as one of the most interesting alternative techniques to 2-DGE *(23)* (*see* **Note 2**). In the following sections of this chapter, the focus is placed on how to detect infectomes of *C. pneumonia* in infected HEp-2 cells by using the major proteomic approach, 2-DGE-MS *(21)* (**Fig. 1**).

2. Materials

1. Microbial pathogens. *C. pneumoniae* A-03 was previously isolated from an atheroma of a patient with coronary artery disease *(24)*.
2. Host cells and medium. HEp-2 cells (ATCC CCL-23) were cultured in RPMI-1640 medium (Cellgro, Herndon, VA) as described previously *(25)*. HEp-2 medium for propagation of *C. pneumoniae* A-03: RPMI-1640 medium with 10% fetal bovine serum, 2 m*M* L-glutamine, 1% (vol/vol) nonessential amino acids, 10 m*M* HEPES, 10 μg gentamicin per mL, and 25 μg vancomycin per mL.
3. Radioisotopes. [^{35}S] methonine/cysteine (ICN Biochemicals, Irvine, CA) was used for labeling proteins.
4. Buffers for protein extractions:
   a. PE buffer 1: 50 m*M* Tris-HCl (pH 7.0) containing 1% sodium dodecyl sulfate (SDS).
   b. PE buffer 2: 7 *M* urea, 2 *M* thiourea, 4% CHAPS [3-[(3-cholamidopropyl)-dimethylammonio]-1-propanesulfonate], 40 m*M* Tris base, 65 m*M* dithiothreitol (DTT), and 2% Pharmalyte 3-10 (Amersham Pharmacia, Piscataway, NJ).
   c. PE mixture: 50 m*M* MgCl$_2$, 476 m*M* Tris-HCl, 24 m*M* Tris base, 1 mg DNase I per mL, and 0.250 mg RNase A per mL (pH 8.0).

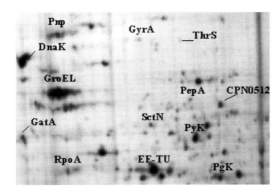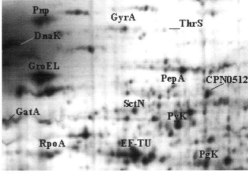

0 U/ml INF-γ   **50 U/ml INF-γ**

Fig. 1. Protein infectomes of *C. pneumoniae* in infected HEp-2 cells in the the absence or presence of interferon-γ (IFN-γ). HEp-2 cells were infected with *C. pneumoniae* A-03 and incubated for 24 h without (**left**) or with (**right**) IFN-γ (50 U/mL). Chlamydial proteins were labeled for 2 h with [^{35}S]methionine/cysteine, and 2-D electrophoresis was performed as described in the text. (Reprinted with permission from **ref. *21*.**)

5. Supplies and Buffers for 2-DGE:
   a. Investigator 2-D Electrophoresis System (Genomic Solutions, Ann Arbor, MI).
   b. A SenSys camera system (Photometrics, Tucson, AZ) was used to scan silver-stained gels and X-ray films.
   c. Immobilized pH gradient (IPG) rehydration buffer: 8 *M* urea, 4% CHAPS, 0.04 *M* Tris base, 0.065 *M* DTT, 0.01% bromophenol blue.
   d. 2-D buffer: 6 *M* urea, 2% DTT, 30% glycerol, and 1× Tris-acetate.
   e. Cathode buffer: 200 m*M* Tricine.
   f. Anode buffer: 0.4% SDS plus 625 m*M* Tris-acetate (pH 8.3).
   g. Software for image analysis. Bioimage 2D Analyzer software (Genomic Solutions) was used to estimate M_r and pI coordinates of proteins and to match spots of proteins localized in the EB with those of proteins expressed intracellularly.
   h. MALDI-TOF MS (Micromass TofSpec 2E). This machine is a time-of-flight mass spectrometer fitted with a MALDI source. The instrument can be operated in the linear or reflectron modes and possesses delayed ion extraction for increased resolution.

3. Methods

3.1. Infection

1. HEp-2 cells were plated in 6-well plates at 10^6 cells per well and allowed to adhere overnight.
2. Cells were subsequently infected with 20 inclusion-forming units (IFU) per cell of *C. pneumoniae* A03, centrifuged at 800*g* for 1 h at 4°C, and incubated at 37°C in 5% CO_2 for 30 min. Infected cells were cultivated in RPMI medium for 24 h (*26*).

3.2. Pulse-Labeling

1. To trace chlamydial protein synthesis, pulse labeling was performed in infected HEp-2 cells for 2 h in methionine/cysteine-free RPMI medium containing 100 µCi of [^{35}S]methionine/cysteine per mL (ICN Biomedicals) and 3 mg of cycloheximide per mL. Uninfected

HEp-2 cells were labeled as controls to determine potential incorporation of [^{35}S]methionine/cysteine into host cell proteins. At the end of the 2-h labeling period, the cells were washed in cold phosphate-buffered saline (PBS), scraped with a rubber policeman, and pelleted by centrifugation at 16,000g. Proteins were then extracted as described in **Subheading 3.4.** for EB preparations.

3.3. Bacterial Isolation

C. pneumoniae is an obligate intracellular pathogen. The infectious extracellular form of the bacterium is known as EB.

1. HEp-2 cells were cultured in HEp-2 medium and used for propagation of *C. pneumoniae* A-03 as described previously *(25)*.
2. EBs were isolated and purified by disruption of HEp-2 cell monolayers with sterile glass beads, sonication, and centrifugation over a Renografin density gradient as described previously *(27)*.
3. EB suspensions were stored in sucrose-phosphate-glutamic acid buffer at –70°C at concentrations of 400 µg of total protein per mL, as determined by a Bio-Rad (Hercules, CA).

3.4. Protein Extraction

1. Bacterial proteins were extracted as described previously *(21)*.
2. Briefly, the pelleted EB preparations were resuspended in 30 µL of PE buffer 1, sonicated for 15 s, and boiled for 5 min. Samples were then treated with 2 µL of PE mixture on ice for 10 min. Finally, samples were suspended in 90 µL of PE buffer 2 and frozen at –70°C.

3.5. 2-DGE

1. 2-DGE was performed with the Investigator 2-D Electrophoresis System *(21)*.
2. For isoelectric focusing, protein extracts were mixed in IPG rehydration buffer in a final volume of 400 µL.
3. One hundred micrograms of protein extracts was used for generating an electrophoretic map of purified *C. pneumoniae* EBs.
4. To detect intracellular *C. pneumoniae* protein expression, 300,000 cpm of ^{35}S-labeled chlamydial proteins was mixed with the IPG buffer.
5. The prepared samples were loaded onto IPG strips and allowed to swell overnight.
6. The rehydrated strips were run with the following focusing parameters: 5000 V of maximum voltage, 124 V of holding voltage, 80 µA of maximum current per gel, 100,000 V•h, and a duration of 24 h.
7. After the first dimension was completed, IPG strips were equilibrated in 2-D buffer.
8. 2-DGE was carried out with SDS-polyacrylamide gel electrophoresis (PAGE) (10% polyacrylamide) at 4°C for 4.5 h under a 500-V maximum voltage and 20,000 mW per gel.
9. After the 2-DGE, gels were fixed in 45% methanol/7.5% acetic acid for 1 h.
10. Gels containing purified EB proteins were stained with silver and used for MS, whereas gels containing radiolabeled chlamydial proteins were treated for 30 min with Amplify fluorographic reagent (Amersham Pharmacia, Buckinghamshire, UK), vacuum dried, and exposed to Kodak Biomax MR films.

3.6. 2-D Image Analysis

1. Both silver-stained gels and X-ray films were scanned under a SenSys camera system.
2. Using Bioimage 2D Analyzer software, M_r and pI coordinates of proteins were determined, and spots of proteins localized in the EB were matched with those of proteins

expressed intracellularly *(21)*. This software was also used to estimate the volumes of radiolabeled protein spots.

3. Statistical analysis of *C. pneumoniae* protein expression profiling under control or persistent infection conditions was carried out by using data from a minimum of five experiments and performing a Mann-Whitney test with GraphPad software (http://graphpad.com/instatman/instat3.htm).

3.7. MALDI-TOF MS Analysis and Database Fitting

1. Protein spots from *C. pneumoniae* EBs were excised from silver-stained gels and incubated with 20 µg/mL of trypsin in 50 m*M* ammonium bicarbonate at 37°C overnight *(21)*.

2. Two microliters of supernatant were mixed in an equal volume of saturated-cyano-4-hydroxycinnamic acid, 50% (vol/vol) acetonitrile, and 0.1% trifluoroacetic acid (TFA), and 0.8 µL of the resulting solution was applied to the MALDI MS template. MALDI MS measurements were performed on a Micromass TOF-Spec 2E mass spectrometer. Patterns of measured masses were matched against theoretical masses of proteins found in the annotated databases SWISS-PROT and TREMBL, which are accessible in the ExPASy Molecular Biology server (http://expasy.cbr.nrc.ca/). Database fitting was performed with the MS-FIT server (http://prospector.ucsf.edu/) with restrictions to proteins from 1 to 100 kDa and mass tolerances of 150 ppm.

4. Notes

1. The resolution of the 2-DGE MS approach is limited when a total cell lysate is analyzed *(23)*. In the crude extract, it is difficult to detect the low-copy proteins, as the most abundant proteins may dominate the gel. It was demonstrated in yeast crude extracts that no low-copy proteins were detected by 2-DGE *(23)*. To overcome this limitation, it is better to use 2-DGE for dissecting subproteomes in partially purified protein complexes instead of whole proteomes in a total cell lysate. For example, the analysis of subproteomes for pathogens (such as stimulons, regulons, and genetic or pathogenicity islands) and hosts (e.g., caveolae, endosomes, phagosomes, and mitochondria) not only makes the work more manageable but frequently provides more detailed information that is lost if a total cell lysate is used.

2. Another limitation of 2-DGE is that the procedure for protein analysis is labor-intensive and time-consuming. A number of approaches have been developed to bypass gel electrophoresis. Protein and carbohydrate microarrays and ICAT have emerged as the most interesting alternative techniques *(4,5,23)*. The microarray technologies are undergoing rapid development for various applications *(4,28)*. Proteins or carbohydrates are immobilized by being covalently attached to glass microscope slides, and the microarrays are shown to be able to interact with their partner molecules *(4,28)*.

3. Protein phosphorylation is an important regulatory mechanism contributing to microbial pathogenesis. MS is one of the most powerful tools for detecting phosphoproteomes *(23)*. The use of in vivo labeling of proteins with inorganic ^{32}P is a common approach to studying protein phosphorylation. The infectomes of cells that differ in protein phosphorylation can be analyzed by culturing normal and infected cells in medium containing inorganic ^{32}P and preparing cell lysates. Alterations in the phosphorylation state of proteins can then be detected by 2-DGE and autoradiography. Proteins of interest are isolated from the gel and then microsequenced by MS *(23)*.

References

1. Fauci, A. S. (2001) Infectious diseases: considerations for the 21st century. *Clin. Infect. Dis.* **32,** 675–685.

2. Huang, S. H. and Jong, A. (2001) Cellular mechanisms of microbial proteins contributing to invasion of the blood-brain barrier. *Cell. Microbiol.* **3**, 277–287.
3. Huang, S. H., Triche, T., and Jong, A. Y. (2002) Infectomics: genomics and proteomics of microbial infections. *Funct. Integr. Genomics* **1**, 331–344.
4. Gronow, S. and Brade, H. (2001) Lipopolysaccharide biosynthesis: which steps do bacteria need to survive? *J. Endotoxin Res.* **7**, 3–23.
5. Wang, D., Liu, S., Trummer, B. J., Deng, C., and Wang, A. (2002) Carbohydrate microarrays for the recognition of cross-reactive molecular markers of microbes and host cells. *Nat. Biotechnol.* **20**, 275–281.
6. Casadevall, A. and Pirofski, L. A. (2000) Host-pathogen interactions: basic concepts of microbial commensalism, colonization, infection, and disease. *Infect. Immun.* **68**, 6511–6518.
7. Relman, D. A. (1999) The search for unrecognized pathogens. *Science* **292**, 1308–1310.
8. Levin, S. A. (2002) New directions in the mathematics of infectious disease, in *Mathematical Approaches for Emerging and Reemerging Infectious Diseases* (Castillo-Chavez, C., et al., eds.), Springer, New York, pp. 1–5.
9. Levin, S. A., Grenfell, B., Hastings, A., and Perelson, A. S. (1997) Mathematical and computational challenges in population biology and ecosystems science. *Science* **275**, 334–343.
10. Vihinen, M. (2001) Bioinformatics in proteomics. *Biomol. Eng.* **18**, 241–248.
11. Jungblut, P. R. (2001) Proteome analysis of bacterial pathogens. *Microbes Infect.* **3**, 831–840.
12. Shah, H. N., Keys, C. J., Schmid, O., and Gharbia, S. E. (2002) Matrix-assisted laser desorption/ionization time-of-flight mass spectrometry and proteomics: a new era in anaerobic microbiology. *Clin. Infect. Dis.* **35**, S58–S64.
13. Klade, C. S. (2002) Proteomics approaches towards antigen discovery and vaccine development. *Curr. Opin. Mol. Ther.* **4**, 216–223.
14. Nilsson, C. L. (2002) Bacterial proteomics and vaccine development. *Am. J. Pharmacogenomics* **2**, 59–65.
15. Smith, R. D., Anderson, G. A., Lipton, M. S., et al. (2002) The use of accurate mass tags for high-throughput microbial proteomics. *OMICS* **6**, 61–90.
16. Cash, P. (2000) Proteomics in medical microbiology. *Electrophoresis* **21**, 1187–1201.
17. VanBogelen, R. A., Schiller, E. E., Thomas, J. D., and Neidhardt, F. C. (1999) Diagnosis of cellular states of microbial organisms using proteomics. *Electrophoresis* **20**, 2149–2159.
18. Washburn, M. P. and Yates, J. R. 3rd (2000) Analysis of the microbial proteome. *Curr. Opin. Microbiol.* **3**, 292–297.
19. O'Connor, C. D., Adams, P., Alefounder, P., et al. (2000) The analysis of microbial proteomes: strategies and data exploitation. *Electrophoresis* **21**, 1178–1186.
20. Vandahl, B. B., Birkelund, S., Demol, H., et al. (2001) Proteome analysis of the *Chlamydia pneumoniae* elementary body. *Electrophoresis* **22**, 1204–1223.
21. Molestina, R. E., Klein, J. B., Miller, R. D., Pierce, W. H., Ramirez, J. A., and Summersgill, J. T. (2002) Proteomic analysis of differentially expressed *Chlamydia pneumoniae* genes during persistent infection of HEp-2 cells. *Infect. Immun.* **70**, 2976–2981.
22. Shaw, A. C., Larsen, M. R., Roepstorff, P., Christiansen, G., and Birkelund, S. (2002) Identification and characterization of a novel *Chlamydia trachomatis* reticulate body protein. *FEMS Microbiol. Lett.* **212**, 193–202.
23. Graves, P. R. and Haystead, T. A. (2002) Molecular biologist's guide to proteomics. *Microbiol. Mol. Biol. Rev.* **66**, 39–63.
24. Ramirez, J. A., Ahkee, S., Summersgill, J. T., et al. (1996) Isolation of *Chlamydia pneumoniae* from the coronary artery of a patient with coronary atherosclerosis. *Ann. Intern. Med.* **125**, 979–982.

25. Molestina, R. E., Dean, D., Miller, R. D., Ramirez, J. A., and Summersgill, J. T. (1998) Characterization of a strain of *Chlamydia pneumoniae* isolated from a coronary atheroma by analysis of the *omp1* gene and biological activity in human endothelial cells. *Infect. Immun.* **66,** 1370–1376.

26. Pantoja, L. G., Miller, R. D., Ramirez, J. A., Molestina, R. E., and Summersgill, J. T. (2001) Characterization of *Chlamydia pneumoniae* persistence in HEp-2 cells treated with interferon-gamma. *Infect. Immun.* **12,** 7927–7932.

27. Caldwell, H. D., Kromhout, J., and Schachter, J. (1981) Purification and partial characterization of the major outer membrane protein of *Chlamydia trachomatis*. *Infect. Immun.* **31,** 1161–1176.

28. Walter, G., Bussow, K., Lueking, A., and Glokler, J. (2002) High-throughput protein arrays: prospects for molecular diagnostics. *Trends Mol. Med.* **8,** 250–253.

25

Toward a Complete Proteome of *Bacillus subtilis*

Cytosolic, Cell Wall-Associated, and Extracellular Proteins

Haike Antelmann, Jan Maarten van Dijl, and Michael Hecker

1. Introduction

Bacillus subtilis is widely regarded as a model organism for the functional genome analysis of Gram-positive bacteria. This is based on two factors: first, the genome sequence that predicts about 4100 open reading frames was completed in 1997 *(1)* and second, *B. subtilis* strain 168 is highly amenable to genetic manipulation. Thus, systematic programs have been initiated to elucidate the functions of all the genes with previously unknown functions. These projects are based on the systematic construction of a mutant library that is used for the genome-wide functional analysis to expand our knowledge of *B. subtilis* physiology and regulatory mechanisms. Data resulting from these programs are compiled in specific databases (e.g., SubtiList, Micado, Sub2D, SubScript, and SPID) including databases for the genome sequence, functional analysis, proteomics, transcriptomics, and two-hybrid protein interactions *(2)*.

Currently, the transcriptome analysis using cDNA arrays has proved to be a powerful tool and the method of choice to uncover unknown regulons and stimulons in *B. subtilis (3–11)*. However, there are several examples in which this technique alone is not sufficient for the complete characterization of biological systems. For instance, mRNA abundance often does not correlate with protein abundance *(12–16)*. Thus, transcriptome analyses should be accompanied by proteomic approaches to verify mRNA expression profiles and to detect post-translational modifications, protein stability, and degradation.

Another important area of proteomics is the verification of predictions for protein localization by the definition of subproteomes. This is much more complicated for multicellular eukaryote systems than for prokaryotes. Thus, microorganisms may serve as good model systems to study protein localization. Compared with eukaryotes, *B. subtilis* consists of a limited number of subcellular compartments: the cytoplasm is surrounded by a single cytoplasmic membrane, which in turn is separated from the extracellular space by a thick Gram-positive cell wall consisting of peptidoglycan and covalently attached anionic polymers such as teichoic or teichuronic acids *(17)*. In contrast to Gram-negative bacteria, *B. subtilis* is lacking an outer membrane. Thus, the thick cell wall is thought to perform some roles of the periplasm of Gram-negative bacteria

From: *Handbook of Proteomic Methods*
Edited by: P. Michael Conn © Humana Press Inc., Totowa, NJ

(18). Wall-binding proteins are retained in the cell wall owing to the presence of specific wall-binding domains (CWB) in addition to the N-terminal signal peptides *(19,20)*. In contrast to these noncovalently linked cell wall proteins, many surface proteins of pathogen Gram-positive bacteria (e.g., protein A or fibronectin binding protein of *Staphylococcus aureus*) are anchored to the cell wall by a mechanism that requires a C-terminal sorting signal with a conserved L-P-X-T-G motif that is cleaved and anchored to the peptidoglycan-chains by the sortase *(21)*. Because two sortases are present on the *B. subtilis* genome encoded by the *yhcS* and *ywpE* genes, some as yet unknown surface proteins could be covalently linked to the *B. subtilis* cell wall *(19,20)*.

The lack of an outer membrane simplifies the secretion pathway: *B. subtilis* is able to secrete large amounts of proteins directly into the surrounding growth medium. Thus, *B. subtilis* is also used as an industrial host for the large-scale production of enzymes like proteases, amylases, or lipases. In contrast to most of the pathogenic bacteria, the mechanisms that allow secretion of proteins are well understood in *B. subtilis* *(19,20)*. Based on predictions for signal peptides, the composition of the *B. subtilis* secretome was defined including both the secreted proteins and the protein secretion machineries *(19,20)*. In summary, 300 exported proteins were predicted for *B. subtilis*, and most of them should follow the major "Sec" pathway for protein secretion. In contrast, the alternative twin-arginine translocation "Tat" pathway *(22,23)*, a pseudopilin export pathway for competence development, and certain ATP-binding cassette (ABC) transporters contribute only selectively to protein secretion in *B. subtilis* *(19,20)*.

During or shortly after translocation of the (pre-) proteins across the membrane, the N-terminal signal peptide is cleaved by the signal peptidases (SPases). Two major classes of signal peptides can be distinguished based on the SPase recognition sequence *(19,20)*. The first class consists of 180 Sec-type signal peptides that are cleaved by one of the five type I SPases (SipS-W) of *B. subtilis* *(24,25)*. The second class of signal peptides is present in 114 (putative) pre-lipoproteins, which are cleaved by the lipoprotein-specific type II SPase (LspA) of *B. subtilis* *(26)*. Signal peptides of pre-lipoproteins are characterized by the "lipobox" containing an invariable cysteine residue that is lipid modified by the diacylglyceryl transferase (Lgt), prior to precursor cleavage by LspA. The exported modified lipoproteins remain attached to the cytoplasmic membrane by their N-terminal lipid anchors.

In regard to verification of predicted subcellular localizations this review focuses on the characterization of several subproteomic fractions of *B. subtilis*. The cytoplasmic proteome that includes the majority of proteins with housekeeping functions reflects the main and specific metabolic pathways that can be used to complement transcriptome analysis. The extracellular proteome analysis has shown that around 50% of the extracellular proteins are unpredicted because these lack signal peptides (phage-related, flagella-related, and cytoplasmic proteins) or possess retention signals in addition to the signal peptide (membrane proteins, cell wall proteins, and lipoproteins). The cell wall proteome based on the LiCl extraction procedure includes mainly CWB proteins resulting from the WapA and WprA precursor processing. In addition, the changes in the extracellular and cell wall proteome in response to mutations affecting regulation, export pathways, protein modification, and stability are described in this review.

2. Definition of Subproteomes for *B. subtilis*

2.1. The Cytoplasmic Proteome of B. subtilis in Complete Medium

1. Cytoplasmic protein extracts from *B. subtilis* were prepared by breaking the cells with a French press and removing the cell waste by repeated centrifugation.
2. The resulting crude cytoplasmic protein extract was dissolved in the reswelling solution containing 2 *M* thiourea, 8 *M* urea, 1% Nonidet P-40, 20 m*M* dithiothreitol (DTT), and 0.5% Pharmalyte (pH 3–10). This sample containing reswelling solution was used for in-gel rehydration of the immobilized pH gradient (IPG) strips.
3. A silver-stained cytoplasmic proteome of *B. subtilis* grown in minimal medium has been published previously *(27)*, which reflects the standard pI range of 4–7 and a narrow pI range of 4.5–5.5. Since two-dimensional (2-D) gels stained with silver nitrate are not quantitative, we now stain the 2-D gels with Sypro Ruby, which is a highly sensitive quantitative dye with a high dynamic range.
4. A Sypro Ruby-stained cytoplasmic proteome in the standard pI range 4–7 that has been defined in complete LB medium is shown in **Fig. 1**. This cytoplasmic proteome was defined from samples harvested during the stationary phase because many proteins are induced on entry into the stationary phase.
5. Of the visible protein spots, 300 spots could be identified, most of which perform important housekeeping functions and are involved in the metabolism of carbohydrates, amino acids, and nucleotides as well as in the translation process *(27)*. However, the cytoplasmic proteome in complete medium depicts only that subset of proteins that is currently expressed at the defined condition. To visualize the complete proteome for cytoplasmic proteins, several different physiological conditions must be choosen that occur in the natural environment of *B. subtilis*. For example, cytoplasmic proteome maps have been defined in response to heat, cold, salt, oxidative stress, or starvation for glucose, phosphate, amino acids, oxygen, and nitrogen *(4,28–32)*.

2.2. The Extracellular Proteome of B. subtilis in Complete Medium

Because of the lack of an outer membrane, *B. subtilis* is able to secrete large amounts of extracellular proteins directly into the growth medium. The highest level of protein secretion is observed when cells of *B. subtilis* are grown in rich medium, in particular during the stationary phase.

1. Thus, the extracellular proteome was defined in L-broth medium during exponential growth as well as 1 h after entry into the stationary phase (**Fig. 2A** and **B**).
2. For preparation of extracellular protein extracts, cells were separated from the growth medium by centrifugation, and the extracellular proteins secreted into the medium were precipitated with 10% trichloroacetic acid (TCA) overnight.
3. After centrifugation, the extracellular proteins were scraped from the wall of the centrifuge tube and washed several times with 96% ethanol *(33)*.
4. The dried pellet was resolved in a solution containing 2 *M* thiourea and 8 *M* urea, and the insoluble material was removed by centrifugation.
5. The extracellular protein sample of 200 µg was adjusted to 360 µL with the urea/thiourea solution and filled up with a 10-fold concentrated reswelling solution.
6. This sample containing reswelling solution was used for in-gel rehydration of IPG strips in the pH range of 3–10. This pH range was found to separate optimal extracellular proteins from *B. subtilis* that are located in the acidic as well as highly alkaline range.

Fig. 1. The cytoplasmic proteome of *B. subtilis* 168. Cells of *B. subtilis* 168 were grown in L-broth and harvested 1 h after entry into the stationary phase. Cytoplasmic proteins were separated by 2-D-PAGE and stained with Sypro Ruby as described in **Subheading 2.1.**

The extracellular proteome of *B. subtilis* consists of 82 identified proteins including 75 proteins identified in complete medium and 7 additional proteins that are secreted under the conditions of phosphate starvation *(29)*. Most of these exported proteins are mainly involved in the degradation of carbohydrates, proteins, nucleotides, lipids, and phosphate as well as in cell wall metabolism. Other extracellular proteins are lipopro-

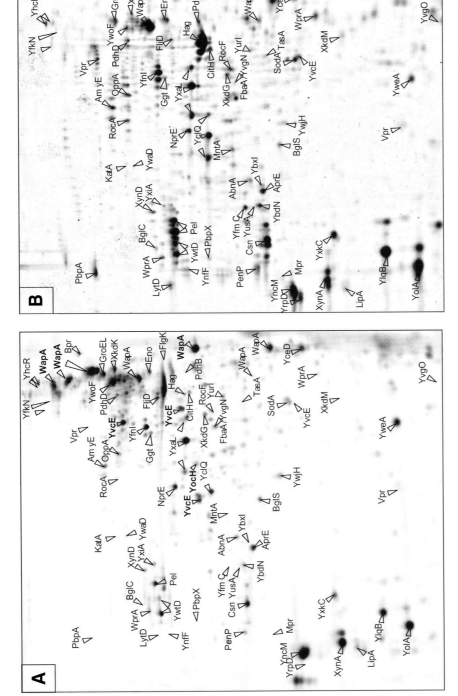

Fig. 2. The extracellular proteome of *B. subtilis* 168. Cells of *B. subtilis* 168 were grown in L-broth, and proteins in the growth medium were harvested during exponential growth (OD₅₄₀ = 0.8) (**A**) and 1 h after entry into the stationary phase (**B**). After precipitation with TCA, the extracellular proteins were separated by 2-D-PAGE and stained with Sypro Ruby as described in **Subheading 2.2.** The large WapA processing products, the unprocessed YvcE and YocH, are in bold in (**A**) because these disappeared during the stationary phase.

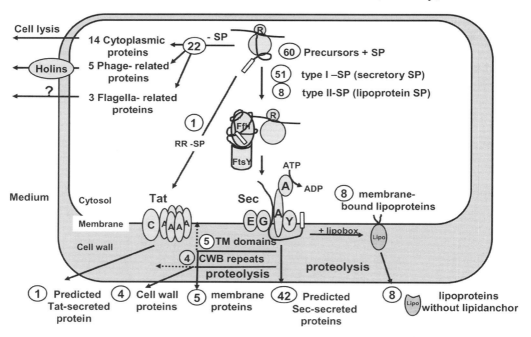

Fig. 3. Verification of predictions for the identified extracellular proteins of *B. subtilis* 168. A total number of 82 extracellular proteins was identified on the extracellular proteome in complete and phosphate starvation medium. Extracellular proteins were classified as "predicted" to be secreted or "unpredicted" based on the presence of signal peptides according to Tjalsma et al. *(19)*. Consequently, 42 extracellular proteins were predicted to be targeted to the Sec translocase in the membrane via the SRP-like Ffh/FtsY complex and secreted into the medium because these possess N-terminal type I signal peptides (SP) and lack retention signals *(19,33)*. Only the PhoD protein is transported via the alternative Tat pathway *(22)*. The remaining 39 extracellular proteins are unpredicted because these either lack SP (14 cytoplasmic proteins, 5 phage-related proteins, and 3 flagella-related proteins) or have retention signals in addition to the SP (8 lipoproteins, 5 transmembrane proteins, and 4 cell wall proteins) *(33)*.

teins that are substrate-binding components of various transport systems, proteins involved in detoxification, flagella-related functions, and phage-related functions *(33)*. A total of 43 identified extracellular proteins was predicted to be secreted because of the presence of the amino-terminal type I SPase cleavage site and no retention signals *(19)*. The remaining extracellular proteins are unpredicted to be secreted because of the absence of signal peptides (14 cytoplasmic proteins, 5 phage-related proteins, and 3 flagella-related proteins) or the presence of specific retention signals in addition to a signal peptide (8 lipoproteins, 4 CWB proteins, and 5 membrane proteins) **(Fig. 3)** *(33)*.

As shown previously, most of these extracellular proteins are secreted at higher amounts during the stationary phase (**Fig. 2A** and **B**) *(33,34)*. However, there are some CWB proteins that are exclusively present on the extracellular proteome during the exponential growth phase and that disappeared on entry into the stationary phase. These include the large WapA processing products, the D,-L-endopeptidase homologue YvcE, and the cell wall protein YocH. Interestingly, the murein hydrolase YvcE was found during exponential growth at its estimated position at 50 kDa on the 2-D gel as well as at two further positions at 30 kDa. Upon entry into the stationary phase, a smaller degradation product of YvcE accumulated besides the SodA spot. This suggests that WapA as well as YvcE might be subjected to proteolytic degradation because extracellular proteases are induced during the stationary phase. In addition, the strong secretion of the large WapA and YvcE processing products during the exponential growth might be caused by the high cell wall turnover during growth that is reduced during the stationary phase *(34)*.

2.3. The Cell Wall Proteome of B. subtilis in Complete Medium

1. Non-covalently linked CWB proteins were extracted from whole cells according to the previously described LiCl extraction procedure *(35)*.
2. For this purpose *B. subtilis* cells were grown in L-broth, and samples of 250 mL were harvested by centrifugation during the transition into the stationary phase (t_0), or 1 h after the transition into the stationary phase (t_1).
3. Cells were washed twice in 10 mL 10 mM Tris-HCl, pH 8.0.
4. The cell pellet was resuspended in 10 mL 1.5 M LiCl, 25 mM Tris-HCl, pH 8.0, and after incubation for 10 min on ice, the cell suspension was centrifuged again (20,000g, 10 min, 4°C).
5. The supernatant containing the CWB proteins was collected, and then the wall proteins were precipitated with ice-cold 10% (w/v) TCA. Because strong cell lysis occurred if we tried to extract CWB proteins from exponentially growing cells, this method is rather useful for the extraction of the cell wall fraction from stationary phase cultures. The cell wall proteome extracted from stationary phase cells of the wild-type strain *B. subtilis* 168 is shown in **Fig. 4** *(34)*.
6. Using matrix-assisted laser desorption/ionization time-of-flight (MALDI-TOF) mass spectrometry 12 of these cell wall proteins were identified. These include most abundantly CWBP105 and CWBP62, resulting from the WapA processing, and CWBP52 and CWBP23 as the processing products of the major cell wall-associated protease WprA. Interestingly, CWBP52 appeared at two different positions in the 2-D gel that showed the same N-terminal sequence starting with amino acid 414 (ANDIQY). Additionally, the major amidase LytC as well as its modifier protein LytB could be extracted from the cell wall. Besides these known CWB proteins, we also identified flagellin (Hag), YwsB, and YqgA as putative CWB proteins.

3. Influence of Mutations on Expression, Secretion Pathways, Modification, and Stability of Extracellular and Cell Wall-Associated Proteins

3.1. Regulation of Extracellular Proteins: Hyperproduction of Degradative Extracellular Enzymes in the degU32(hy) Mutant

B. subtilis is used as an industrial host for the large-scale production of enzymes such as proteases, amylases, or lipases. Such production strains often harbor pleiotropic mutations in central regulatory genes like the *degU32(hy)* mutation, resulting in hyperproduction of degradative enzymes (hy-phenotype) as well as in repression of motility and competence *(36–38)*.

The comparison of the extracellular proteome of the wild type with that of the *degU32(hy)* strain revealed an upregulation of 13 degradative enzymes in the *degU32(hy)* strain **(Fig. 5)** *(32,39)*. Consistently, earlier studies showed that the genes encoding the proteases AprE and NprE are positively regulated by DegU-P *(38)*.

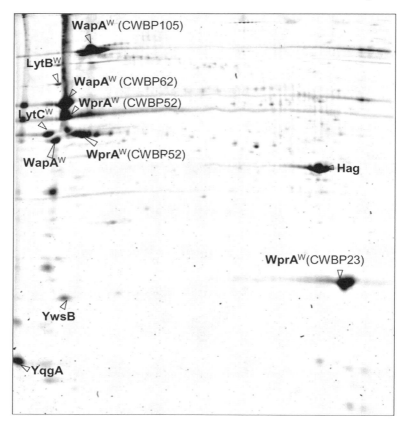

Fig. 4. The cell wall proteome of the *B. subtilis* wild-type 168. Cells of *B. subtilis* 168 grown in L-broth were harvested 1 h after entry into the stationary phase, and the cell wall proteins were extracted using an LiCl extraction procedure, precipitated with TCA, separated by 2-D-PAGE, and stained with Sypro Ruby. Cell wall proteins with wall binding repeats are indicated with a superscript W. Processing products of WapA and WprA are also indicated as a CWBP (cell wall binding protein), according to their molecular masses.

Different ratios could be calculated for the extracellular proteins. Whereas the relative amounts of AmyE, Ggt, Vpr, and YnfF were only two to three times higher in the mutant, ratios between 10 and 33 were obtained for AprE, BglC, BglS, Bpr, Mpr, NprE, PelB, YfkN, YwaD, and YurI *(39)*. In addition, the amounts of eight extracellular proteins involved in motility and chemotaxis were found to be strongly reduced in the *degU32(hy)* mutant **(Fig. 5)**. Of the downregulated proteins, the autolysins LytD and YwtD, the flagellin Hag, the flagellar hook-associated proteins FlgK and FliD, and the cell wall-associated protein WapA are encoded by SigmaD-dependent genes.

3.2. Protein Export Pathways: Determination of Sec-Dependent Secretion of Extracellular Proteins

Of the identified extracellular proteins, 60 are synthesized with N-terminal signal peptides, most of which should be translocated via the general secretion (Sec) pathway in an unfolded conformation. Cytoplasmic chaperones and targeting factors (like the Ffh

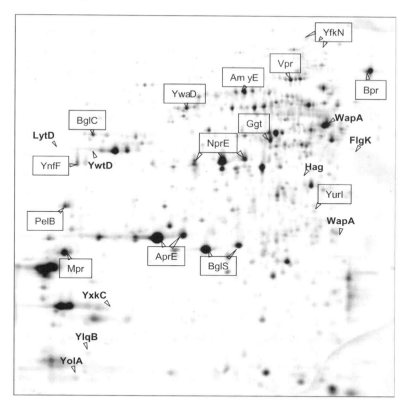

Fig. 5. The extracellular proteome of *B. subtilis degU32(hy)* mutant. Extracellular proteins were separated by 2-D-PAGE and stained with Sypro Ruby. Extracellular proteins secreted at higher amounts in the *degU32(hy)* mutant are boxed, and those that are repressed in the mutant are in bold.

protein homologous to the 54-kDa subunit of the mammalian signal recognition particle (SRP) and the FtsY protein homologous to the mammalian SRP receptor α-subunit) facilitate targeting of the preproteins to the translocase in the membrane *(40,41)*.

The composition of the extracellular proteome has also been determined by Hirose and co-workers *(42)*. In this study, 23 extracellular proteins were identified, most of which were verified by our extracellular proteome analysis. A temperature-sensitive *secA* mutant and a conditional *ffh* mutant were used to determine the number of proteins that are translocated via the SRP-Sec pathway *(42)*. Hirose and co-workers found that all identified extracellular proteins that are synthesized with secretory signal peptides are absent from the extracellular proteome in the temperature-sensitive *secA* strain after the upshift to the higher temperature as well as after depletion for Ffh. Because *secA*, *ffh*, and *ftsY* are essential for growth of *B. subtilis*, we used a different approach, based on the inhibition of SecA activity with sodium azide to visualize the number of extracellular proteins that are translocated via the Sec pathway.

1. After cells reached the stationary phase, they were washed twice with fresh medium and then resuspended in L-broth with or without 15 m*M* sodium azide addition (**Fig. 6A** and **B**).
2. The proteins secreted during the next 10–20 min were subjected to extracellular proteome analysis. Because only *de novo* secretion can be visualized after washing of the cells, only

those proteins could be analyzed that are secreted at high levels and normally accumulate during the stationary phase.

3. Of the 26 identified *de novo* synthesized proteins that are secreted during 20 min without SecA inhibition by sodium azide, a subset was secreted at strongly reduced levels in the presence of azide. These azide-sensitive proteins include the most strongly secreted proteins Csn, LipA, WapA, XynA, YolA, YvcE, YweA, and YxaL. While the secretion of LipA, WapA, YolA, YvcE, YweA and YxaL was completely blocked by azide, the secretion of Csn and XynA was only weakly reduced (**Fig. 6A** and **B**).

Interestingly, three of these nine Sec-dependent proteins are synthesized with predicted RR/KR signal peptides that could be secreted via the alternative Tat pathway *(23)*. However, extracellular proteome comparison between the wild type and a *total-tat* mutant strain revealed that only PhoD was secreted in a strictly Tat-dependent manner *(22,23)*. Thus, the Tat pathway makes a highly selective contribution to the extracellular proteome of *B. subtilis (23)*.

3.3. Lipid Modification: Shedding of Lipoproteins and Autolysins by a Diacyl-Glyceryl Transferase Mutant

One hundred fourteen extracellular proteins of *B. subtilis* are predicted lipoproteins that are retained in the cytoplasmic membrane because of the N-terminal lipid anchor *(26)*. The diacylglyceryl transferase Lgt of *B. subtilis* is responsible for the diacylglyceryl modification of the N-terminal cysteine residue of pre-lipoproteins *(43)*. As shown in **Fig. 2B**, the lipoproteins MntA, OppA, YclQ, and YfmC were shown to be present on the extracellular proteome in complete medium in the wild type *(33)*. To explore further the factors required for lipoprotein processing and retention in the cell, the composition of the extracellular proteome of an *lgt* mutant was analyzed, which is defective in the lipid modification of lipoproteins *(43)*. As shown in **Fig. 7**, the amounts of 13 lipoproteins were significantly increased on the extracellular proteome of the *lgt* mutant in complete medium including 9 lipoproteins that were not present on the extracellular proteome of the wild-type strain. In addition, the extracellular levels of the autolytic enzymes YvcE, LytD, XepA, and XlyA were significantly increased by the *lgt* mutation. These observations show that cells lacking the diacyl-glyceryl transferase shed lipoproteins and autolysins into their growth medium. Because the released lipoproteins have lost their N-terminal cysteine residue, it might be possible that pre-lipoproteins are released into the medium after an alternative proteolytic shaving by LspA or other proteases *(33)*. The fact that transcription of the *mntA* or *oppA* genes is not increased in the absence of Lgt further confirmed that the increased levels of lipoproteins in the medium are caused by the redistribution of lipoproteins from the membrane to the extracellular space *(33)*.

3.4. Protein Stability: Stabilization of Secreted and Wall-Associated WapA Processing Products in the Absence of Extracellular Proteases as well as in an sigD Mutant

As shown in **Fig. 2**, extracellular proteases are secreted at higher amounts during the stationary phase. To verify whether extracellular proteases are involved in the processing of those cell wall proteins that are degraded upon entry into the stationary phase, the extracellular proteome of the wild-type was compared with that of a multiple exoprotease-deficient strain lacking seven extracellular proteases (AprE, Bpr, Epr, Mpr,

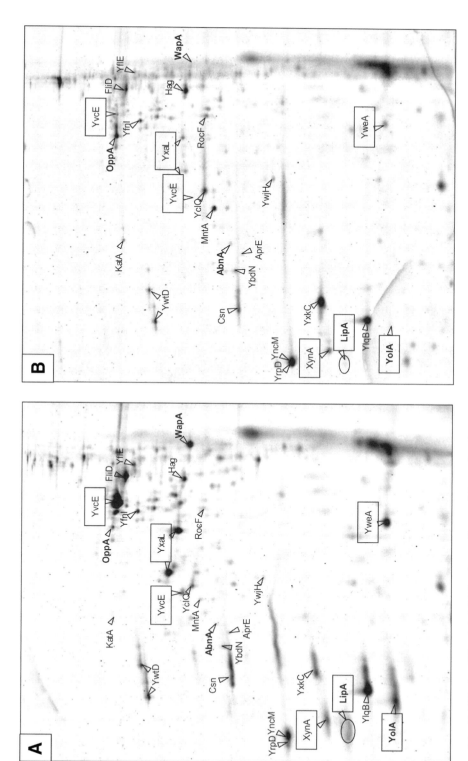

Fig. 6. Effects of SecA inhibition by sodium azide on the *de novo* secretion of extracellular proteins of *B. subtilis* 168. Cells of *B. subtilis* grown into the stationary phase were washed twice in prewarmed medium and then resuspended in L-broth supplemented without (**A**) or with 15 m*M* sodium azide (**B**). After 20 min of growth, the extracellular proteins were separated by 2-D-PAGE and stained with Sypro Ruby. Proteins with predicted RR/KR signal petides are in bold, and proteins present at decreased amounts after sodium azide addition are boxed.

Fig. 7. The extracellular proteome of diacylglyceryl transferase mutant cells of *B. subtilis*. Cells of *B. subtilis* Δ*lgt*, which lack the diacylglyceryl transferase, were grown in L-broth, and proteins in the growth medium were harvested 1 h after entry into the stationary phase. After precipitation with TCA, the extracellular proteins were separated by 2-D-PAGE and stained with Sypro Ruby. The proteins identified by mass spectrometry are indicated. Non-lipoproteins that are present at elevated levels in the medium of the Δ*lgt* strain are in bold; lipoproteins that are released into the medium of the Δlgt strain are printed in bold and marked with a superscript lipo.

NprB, NprE, and Vpr) *(34,44)*. As shown in **Fig. 8A**, the large WapA processing products and the 50-kDa YvcE spot were observed in the extracellular proteome of the protease-depleted strain during both exponential growth and the stationary phase. Thus, the WapA processing products as well as the 50-kDa YvcE protein seem to be stabilized in the protease-deficient strain during the stationary phase. In addition, the level of the WapA processing product CWBP105 was significantly increased and the amount of the WapA processing product CWBP62 was decreased in the cell wall proteome of the protease-deficient strain **(Fig. 9)**. A detailed extracellular and cell wall proteome analysis of an *epr* mutant showed that Epr might be the candidate protease responsible

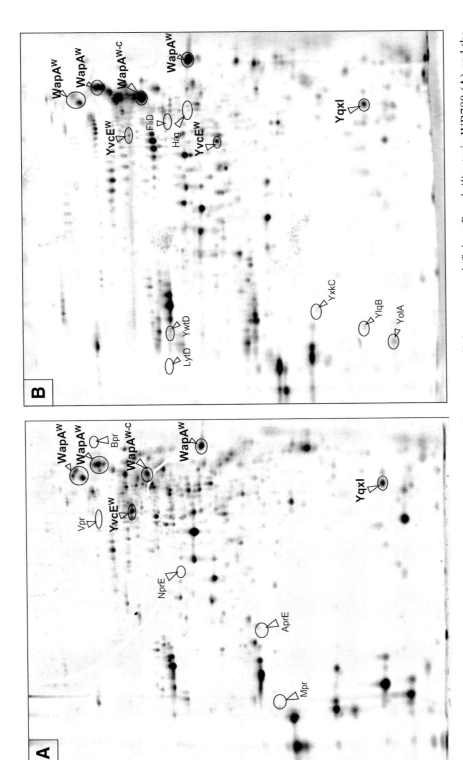

Fig. 8. The stationary phase extracellular proteome of the multiple extracellular protease-deficient *B. subtilis* strain WB700 (**A**) and the *sigD* mutant (**B**). Extracellular proteins were separated by 2-D-PAGE and stained with Sypro Ruby. Secreted cell wall proteins that are stabilized in the protease mutant as well as in the *sigD* mutant during the stationary phase mutant are in bold.

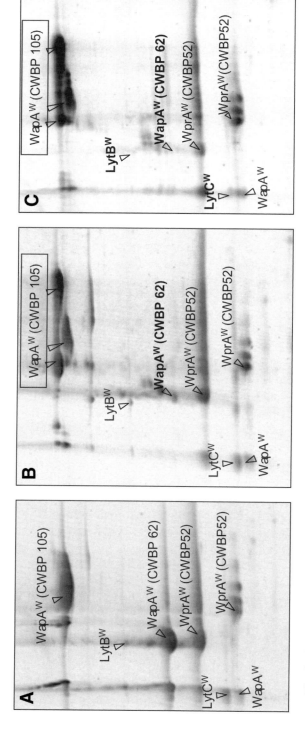

Fig. 9. Comparison of the cell wall proteome between the wild-type (**A**), the multiple extracellular protease-deficient *B. subtilis* strain WB700 (**B**), and the *sigD* mutant (**C**). Cell wall-associated proteins were separated by 2-D-PAGE and stained with Sypro Ruby; the alkaline high molecular weight region is shown. The WapA processing product CWBP105 is increased and the WapA processing product CWBP62 is decreased in the protease mutant as well as in the *sigD* mutant.

for the processing of cell wall proteins during the stationary phase. Thus, the extracellular serine protease Epr might be involved in the degradation of secreted CWB\ proteins and in the processing of the cell wall-associated product CWBP105 during the stationary phase.

Previous studies have shown that an artificial CWB lipase was stabilized in a *sigD* mutant lacking the alternative sigma factor SigmaD, which is involved in motility and chemotaxis as well as in cell wall turnover *(45)*. To test the involvement of the alternative sigma factor SigmaD in the stability of cell wall proteins, a *sigD* mutant was subjected to extracellular and cell wall proteome analysis. Consistent with the protease-deficient strain, increased amounts of the secreted WapA processing products on the extracellular proteome and CWBP105 on the cell wall proteome were detected in *sigD* mutant cells during the stationary phase (**Figs. 8B** and **9**). These results confirm the previous finding that a *sigD* mutation increased the stability of cell wall proteins. The increased stability of cell wall proteins in the absence of *sigD* might be the result of an impaired cell wall turnover owing to the loss of the SigmaD-regulated autolysins.

4. Concluding Remarks

Prior to the proteomic approach, subcellular fractionation provided an attractive addition to the protein separation techniques commonly used in proteome analysis. The advantage lies in the enrichment of compartment-specific proteins dealing with specific subcellular functions. Most importantly, proteomics is the only technique to verify predicted sucellular localizations. Thus, we have defined three subproteomes for the Gram-positive bacterium *B. subtilis* (the cytoplasmic, extracellular, and cell wall proteome) and have shown that a specific set of proteins is expressed in the different compartments. Moreover, not all of these proteins were expected to be localized in these specific cellular fractions. These subproteomes can be used to monitor changes in response to mutations that affect expression, targeting, modification, or stability of extracellular or cell wall proteins that have escaped the global transcriptome analysis. Future proteome analyses of *B. subtilis* will focus on the definition of the membrane proteome and the detection of post-translational modifications, for example, protein phosphorylation or glycosylation.

Acknowledgments

We thank Karin Binder, Sebastian Grund, and Doreen Kliewe for expert technical assistance and the members of the European *Bacillus* Secretion Groups (*see* http://www.ncl.ac.uk/ebsg) for stimulating discussions. This work was supported in part by Quality of Life and Management of Living resources grants QLK3-CT-1999-00413 and QLK3-CT-1999-00917 from the European Union (to H.A., J.M.v.D., and M.H.), grants from the Deutsche Forschungsgemeinschaft (DFG), the Bundesministerium für Bildung, Wissenschaft, Forschung, und Technologie (BMFT), the Fonds der Chemischen Industrie (to M.H.), and Genencor International (Palo Alto, CA).

References

1. Kunst, F., Ogasawara, N., Moszer, I., et al. (1997) The complete genome sequence of the Gram-positive bacterium *Bacillus subtilis. Nature* **390,** 249–256.
2. Harwood, C. R. and Moszer, I. (2002) From gene regulation to gene function: regulatory networks in *Bacillus subtilis. Comp. Funct. Genom.* **3,** 37–41.

3. Caldwell, R., Sapolsky, R., Weyler, W., Maile, R. R., Causey, S. C., and Ferrari, E. (2001) Correlation between *Bacillus subtilis* scoC phenotype and gene expression determined using microarrays for transcriptome analysis. *J. Bacteriol.* **183,** 7329–7340.

4. Eymann, C., Homuth, G., Scharf, C., and Hecker, M. (2002) *Bacillus subtilis* functional genomics: global characterization of the stringent response by proteome and transcriptome analysis. *J. Bacteriol.* **184,** 2500–2520.

5. Fawcett, P., Eichenberger, P., Losick, R., and Youngman, P. (2000) The transcriptional profile of early to middle sporulation in *Bacillus subtilis*. *Proc. Natl. Acad. Sci. USA* **97,** 8063–8068.

6. Kobayashi, K., Ogura, M., Yamaguchi, H., et al. (2001) Comprehensive DNA microarray analysis of *Bacillus subtilis* two-component regulatory systems. *J. Bacteriol.* **183,** 7365–7370.

7. Moreno, M. S, Schneider, B. L., Maile, R. R., Weyler, W., and Saier, M. H. Jr. (2001) Catabolite repression mediated by the CcpA protein in *Bacillus subtilis*: novel modes of regulation revealed by whole-genome analyses. *Mol. Microbiol.* **39,** 1366–1381.

8. Petersohn, A., Brigulla, M., Haas, S., Hoheisel, J. D., Völker, U., and Hecker, M. (2001) Global analysis of the general stress response of *Bacillus subtilis*. *J. Bacteriol.* **183,** 5617–5631.

9. Price, C. W, Fawcett, P., Ceremonie, H., et al. (2001) Genome-wide analysis of the general stress response in *Bacillus subtilis*. *Mol. Microbiol.* **41,** 757–774.

10. Ye, R. W., Tao, W., Bedzyk, L., Young, T., Chen, M., and Li, L. (2000) Global gene expression profiles of *Bacillus subtilis* grown under anaerobic conditions. *J. Bacteriol.* **182,** 4458–4465.

11. Yoshida, K., Kobayashi, K., Miwa, Y., et al. (2001) Combined transcriptome and proteome analysis as a powerful approach to study genes under glucose repression in *Bacillus subtilis*. *Nucleic Acids Res.* **29,** 683–692.

12. Gygi, S. P., Rochon, Y, Franza, B. R., and Aebersold, R. (1999) Correlation between protein and mRNA abundance in yeast. *Mol. Cell. Biol.* **19,** 1720–1730.

13. Anderson, L. and Seilhammer, J. (1997) A comparison of selected mRNA and protein abundances in human liver. *Electrophoresis* **18,** 533–537.

14. Futcher, B., Latter, G. I., Monardo, P., McLaughlin, C. S., and Garrels, J. I. (1999) A sampling of the yeast proteome. *Mol. Cell. Biol.* **19,** 7357–7368.

15. Griffin, T. J., Gygi, S. P., Ideker, T., et al. (2002) Complementary profiling of gene expression at the transcriptome and proteome levels in *Saccharomyces cerevisiae*. *Mol. Cell. Proteomics* **1,** 323–333.

16. Bernhardt. J., Weibezahn, J., Scharf, C., and Hecker (2003) *Bacillus subtilis* during feast and famine: Visualization of the overall regulation of protein synthesis during glucose starvation by proteome analysis. *Genom. Res.* **13,** 224–237.

17. Archibald, A. R., Hancock, I. C., and Harwood. C. R. (1993) Cell wall structure, synthesis and turnover, in *Bacillus subtilis and other Gram-Positive Bacteria* (Sonenshein, A. L., Hoch, J. A., and Losick, R., eds.), American Society for Microbiology, Washington, DC, pp. 381–410.

18. Pooley, H. M., Merchante, R., and Karamata, D. (1996) Overall protein content and induced enzyme components of the periplasm of *Bacillus subtilis*. *Microb. Drug. Resist.* **2,** 9–15.

19. Tjalsma, H., Bolhuis, A., Jongbloed, J. D. H., Bron, S., and van Dijl, J. M. (2000) Signal peptide-dependent protein transport in *Bacillus subtilis*: a genome-based survey of the secretome. *Microbiol. Mol. Biol. Rev.* **64,** 515–547.

20. van Dijl., J. M., Bolhuis, A, Tjalsma, H., Jongbloed, J. D. H., de Jong, A., and Bron, S. (2001) Protein transport pathways in *Bacillus subtilis*: a genome-based road map, in *Bacillus subtilis and Its Closest Relatives* (Sonenshein, A. L., Hoch, J. A., and Losick, R., eds.), ASM Press, Washington, DC, pp. 337–355.

21. Navarre, W. W. and Schneewind, O. (1999) Surface proteins of Gram-positive bacteria and mechanisms of their targeting to the cell wall envelope. *Microbiol. Mol. Biol. Rev.* **63,** 174–229.
22. Jongbloed, J. D. H., Martin, U., Antelmann, H., et al. (2000) TatC is a specificity determinant for protein secretion via the twin-arginine translocation pathway. *J. Biol. Chem.* **275,** 41,350–41,357.
23. Jongbloed, J. D. H., Antelmann, H., Hecker, M., et al. (2002) Selective contribution of the twin arginine translocation pathway to protein secretion in *Bacillus subtilis. J. Biol. Chem.* **277,** 44,068–44,078.
24. Tjalsma, H., Noback, M. A., Bron, S., Venema, G., Yamane, K., and van Dijl, J. M. (1997) *Bacillus subtilis* contains four closely related type I signal peptidases with overlapping substrate specificities. Constitutive and temporally controlled expression of different sip genes. *J. Biol. Chem.* **272,** 25,983–25,992.
25. Tjalsma, H., Bolhuis, A., van Roosmalen, M. L., et al. (1998) Functional analysis of the secretory precursor processing machinery of *Bacillus subtilis*: identification of a eubacterial homolog of archaeal and eukaryotic signal peptidases. *Genes Dev.* **12,** 2318–2331.
26. Tjalsma, H., Kontinen, V. P., Pragai, Z., et al. (1999) The role of lipoprotein processing by signal peptidase II in the Gram-positive eubacterium *Bacillus subtilis*. Signal peptidase II is required for the efficient secretion of alpha-amylase, a non-lipoprotein. *J. Biol. Chem.* **274,** 1698–1707.
27. Büttner, K., Bernhardt, J., Scharf, C., et al. (2001) A comprehensive two-dimensional map of cytosolic proteins of *Bacillus subtilis*. *Electrophoresis* **22,** 2908–2935.
28. Antelmann, H., Bernhardt, J., Schmid, R., Mach, H., Völker, U., and Hecker, M. (1997) First steps from a two-dimensional protein index towards a response-regulation map for *Bacillus subtilis*. *Electrophoresis* **18,** 1451–1463.
29. Antelmann, H., Scharf, C., and Hecker, M. (2000) Phosphate starvation-inducible proteins of *Bacillus subtilis*: proteomics and transcriptional analysis. *J. Bacteriol.* **182,** 4478–4490.
30. Bernhardt, J., Büttner, K., Scharf, C., and Hecker, M. (1999) Dual channel imaging of two-dimensional electropherograms in *Bacillus subtilis*. *Electrophoresis* **20,** 2225–2240.
31. Mäder, U., Homuth, G., Scharf, C., Büttner, K., Bode, R., and Hecker, M. (2002) Transcriptome and proteome analysis of *Bacillus subtilis* gene expression modulated by amino acid availability. *J. Bacteriol.* **184,** 4288–4295.
32. Graumann, P., Schröder, K., Schmid, R., and Marahiel, M. A. (1996) Cold shock stress-induced proteins in *Bacillus subtilis*. *J. Bacteriol.* **178,** 4611–4619.
33. Antelmann, H., Tjalsma, H., Voigt, B., Ohlmeier, S., Bron, S., van Dijl, J. M., and Hecker, M. (2001) A proteomic view on genome-based signal peptide predictions. *Genome Res.* **11,** 1484–1502.
34. Antelmann, H., Yamamoto, H., Sekiguchi, J., and Hecker, M. (2002) Stabilization of cell wall proteins in *Bacillus subtilis*: a proteomic approach. *Proteomics* **5,** 591–602.
35. Rashid, M. H., Sato, N., and Sekiguchi, J. (1995) *FEMS Microbiol. Lett.* **132,** 131–137.
36. Henner, D. J., Yang, M., and Ferrari, E. (1988) Localization of *Bacillus subtilis sacU(Hy)* mutations to two linked genes with similarities to the conserved procaryotic family of two-component signalling systems. *J. Bacteriol.* **170,** 5102–5109.
37. Kunst, F., Debarbouille, M., Msadek, T., et al. (1988) Deduced polypeptides encoded by the *Bacillus subtilis sacU* locus share homology with two-component sensor-regulator systems. *J. Bacteriol.* **170,** 5093–5101.
38. Msadek, T., Kunst, F., and Rapoport, G. (1995) A signal transduction network in *Bacillus subtilis* includes the DegS/DegU and ComP/ComA two-component systems, in *Two-Component Signal Transduction* (Hoch, J. A. and Silhavy, T. J., eds.), American Society for Microbiology, Washington, DC, pp. 447–471.

39. Mäder, U., Antelmann, H., Buder, T. Dahl, M. K., Hecker, M., and Homuth, G. (2002) *Bacillus subtilis* functional genomics: genome wide analysis of the DegS-DegU regulon by transcriptomics and proteomics. *Mol. Genet. Genom.* **268,** 455–467.

40. Honda, K., Nakamura, K., Nishiguchi, M., and Yamane, K. (1993) Cloning and characterization of a *Bacillus subtilis* gene encoding a homolog of the 54-kilodalton subunit of mammalian signal recognition particle and *Escherichia coli* Ffh. *J. Bacteriol.* **175,** 4885–4894.

41. Ogura, A., Kakeshita, H., Honda, K., Takamatsu, H., Nakamura, K., and Yamane, K. (1995) *srb*: a *Bacillus subtilis* gene encoding a homologue of the alpha-subunit of the mammalian signal recognition particle receptor. *DNA Res.* **2,** 95–100.

42. Hirose, I., Sano, K., Shioda, I., Kumano, M., Nakamura, K., and Yamane, K. (2000) Proteome analysis of *Bacillus subtilis* extracellular proteins: a two-dimensional protein electrophoretic study. *Microbiology* **146,** 65–75.

43. Leskelä, S., Wahlström, E., Kontinen V. P., and Sarvas, M. (1999) Lipid modification of prelipoproteins is dispensable for growth but essential for efficient protein secretion in *Bacillus subtilis*: characterization of the *lgt* gene. *Mol. Microbiol.* **31,** 1075–1085.

44. Ye, R., Yang, L. P., and Wong, S.-L. (1996) Construction of protease deficient *Bacillus subtilis* strains for expression studies: inactivation of seven extracellular proteases and the intracellular LonA protease, in *Proceedings of the International Symposium on Recent Advances in Bioindustry*, pp. 160–169.

45. Kobayashi, G., Toida, J., Akamatsu, T., Yamamoto, H., Shida, T., and Sekiguchi, J. (2000) Accumulation of an artificial cell wall-binding lipase by *Bacillus subtilis* wprA and/or sigD mutants. *FEMS Microbiol. Lett.* **188,** 165–169.

26

Renal and Urinary Proteomics

Visith Thongboonkerd, Elias Klein, and Jon B. Klein

1. Introduction

Genetic information can be successfully used to explain some diseases at the gene level. This is also true for the case of genetic disorders primarily involving the kidneys. However, genomic analysis may not be able to explain mechanisms in normal physiology and diseases that are directly represented by proteins, not genes. Additionally, genomic analysis is limited by dynamic processes, which occur during or after translation, especially post-translational modifications (PTMs). Previously, studies of proteins were mainly performed using immunoblotting techniques that are, however, limited because specific antibodies must be available, and only a few proteins can be tested in a single experiment. Moreover, Western blot analysis requires a prior assumption from previous studies about the protein to be identified.

Recent advances in proteomic technologies allow the simultaneous analysis of a large number of proteins. Proteomic analysis is thus a promising tool for studying normal renal physiology and renal pathophysiology and for discovering biomarkers for diseases. Proteomics in nephrology can be classified as *renal proteomics* and *urinary proteomics* by differences in protein extraction and/or isolation methods and data interpretation. Extraction techniques used for renal parenchymal proteins are similar to methods used in various tissues and cells, with some special considerations. Protein isolation from urine is completely different because the proteins have already been solubilized. Because of low urinary protein concentration compared with contaminating salts and urea, the proteins need to be isolated. Strategies used in protein isolation from urine are similar to and can be applied to protein isolation from other body fluids. The source of urinary proteins should be considered because the proteins can be from renal tubules, urinary passages, and prostatic secretion. Additionally, interpretation of alterations in overall urinary protein excretion is a crucial issue. Urinary protein excretion is influenced by changes in their production or by changes in their excretion.

The earliest reference to the application of *renal proteomics* may have been in 1995, when Witzmann and colleagues *(1)* used two-dimensional polyacrylamide gel electrophoresis (2-D-PAGE) to identify renal stress proteins as biomarkers for chemical toxicity. *Urinary proteomics* probably began in 1996, when Marshall and Williams *(2)* identified urinary proteins on a 2-D gel using a dye precipitation method. During the late 1990s, only a few proteomic studies in nephrology were conducted because of the limited ability to identify proteins. More rapid progress began with the new era in mass

From: *Handbook of Proteomic Methods*
Edited by: P. Michael Conn © Humana Press Inc., Totowa, NJ

spectrometric (MS) analyses, when Patterson et al. *(3)* identified urinary proteomes using liquid chromatography tandem MS (LC MS/MS) *(4)*, and Knepper et al. *(5,6)* applied immunologically based methods for targeted proteomic profiling of renal Na⁺ transporters. More recently, the present authors have established novel protein isolation/extraction techniques for urinary proteins and for integral membrane proteins *(7,8)*. Additionally, we have reported several renal and urinary proteomic studies *(7,9–12)*. Our objectives were to determine differential protein expression in renal cortex vs medulla; in the urine after acute sodium loading; in the kidney during hypoxia-induced hypertension; and in the kidney of mice with diabetic nephropathy. A longer term objective is to create complete proteome maps for rat, mouse, and human urine and kidneys.

Recent applications of proteomics in nephrology can be divided into three categories: *normal physiology*, *pathophysiology* of diseases, and *biomarker* discovery. A major limitation in these studies had been the difficulty of identifying proteins by methods available at that time. Another problem was the availability of sample preparation protocols to isolate a comprehensive profile of protein components. Most of the studies were expression proteomic analyses that determined alterations in protein expression without functional study. However, the data from these studies indicated the usefulness and trends of proteomics in renal medicine. Improvements in protein isolation, separation, and identification will provide a more complete image of kidney and urinary proteins in normal and disease states. When the experimental protocols are optimized and highly sensitive identification methods are applied, proteomics applied to nephrology will be more powerful and fruitful.

2. Proteomics and Renal Physiology

To understand renal physiology better, a global approach toward renal protein analyses is necessary. Although Sarto and colleagues *(13)* identified only 47 proteins in the kidney proteome map, they demonstrated the magnitude of protein present in the whole kidney because more than 2000 spots were visualized on 2-D gels. More recently, Arthur and Klein et al. *(9)* determined differential protein expression between renal cortex and medulla. Renal proteins precipitated from DNA/RNA free whole cell homogenates (*see protocol I*, **Subheading 5.1.**) were separated by 2-D-PAGE and identified by matrix-assisted laser desorption/ionization time-of-flight (MALDI-TOF) MS followed by peptide mass fingerprinting (PMF). A total of 1095 protein spots were visualized in renal cortex, and 885 protein spots were visualized in renal medulla. Of these visualized spots, 72 proteins were identified in the proteome map. Ten identified proteins were expressed to a greater extent in the cortex, whereas six identified proteins were expressed to a greater extent in the medulla (**Table 1**).

Differential expression of kidney cortex versus medulla was also studied previously by Witzmann et al. *(14)* but in cytosolic, not whole cell homogenates. They resolved 727 protein spots in the cortex and 716 in the medulla on the 2-D gel. A total of 127 spots were found to differ in abundance in the two locales. Twenty of these spots representing 14 different proteins and variants were identified. Although the data from cytosolic and whole cell homogenates are not the same, they are complementary to each other. Approximately 100 of the differentially expressed proteins between cortex and medulla from these two studies remained unidentified. Eventual identification of these

Table 1
Differential Expression of Proteins in Renal Cortex and Medulla

Cortex-predominant proteins
　3-Mercaptopyruvate sulfurtransferase
　α-2u globulin
　α-enolase
　Aldehyde dehydrogenase
　Contraception-associated protein 1
　Heat shock protein 60 kDa (HSP60)
　Isocitrate dehydrogenase
　NADH-ubiquinone oxidoreductase
　Ornithine aminotransferase
　Retinol binding protein
Medulla-predominant proteins
　Aflatoxin b1 aldehyde reductase
　Albumin
　α-B crystallin
　BH3 interacting domain death agonist
　Fatty acid binding protein
　Glucose-regulated protein precursor 78

differentially expressed proteins may be able to explain various responses to physiological stimuli in the cortex and the medulla.

Urinary proteomics is becoming an important field by virtue of its noninvasiveness, for studies involving human subjects. Urinary proteins generally reflect mostly normal tubular physiology, but various interventions can be applied to determine tubular or glomerular response to physiological stimuli. Although Marshall and Williams *(2)* created a 2-D image of normal human urinary proteins using dye precipitation to isolate the proteins, no protein structure was identified in their study. Heine et al. *(15)* performed mapping of peptides and protein fragments in fractionated human urine using high-performance liquid chromatography electrospray ionization (HPLC ESI) MS. They identified 34 peptides and protein fragments from the urine. These identified components, however, were limited to only major abundant proteins. Targeted proteomics was applied by Knepper et al. *(16)* to identify type 3 Na-H exchanger (NHE3), bumetanide-sensitive Na-K-2Cl cotransporter (NKCC2), and thiazide-sensitive Na-Cl cotransporter (NCC) in immunoblotting experiments. Although this technique is more sensitive than visualization of the spots on the 2-D gel, it is limited by availability of the specific antibody.

Spahr et al. *(4)* applied LC MS/MS to identify protein components from unfractionated normal human urine. A total of 124 proteins or translations of expressed sequence tags (ESTs) were identified. However, HPLC MS/MS requires instrumentations, and MS analyses are necessary for all experiments. Construction of a urinary proteome map on 2-D gel may be a useful technique for most laboratories, because recent 2-D-PAGE methodology provides a consistent protein spot pattern *(17,18)*. Recently,

Thongboonkerd et al. *(7)* created a 2-D-PAGE proteome map of normal human urinary proteins isolated by acetone precipitation or ultracentrifugation (*see protocols VII and VIII*, **Subheadings 7.1.** and **7.2.**). These two isolation methods are complementary to each other for identification of urinary protein components with different physical and chemical properties. A total of 67 protein forms of 47 unique proteins were identified, including transporters, adhesion molecules, complement, chaperones, receptors, enzymes, serpins, cell signaling proteins, and matrix proteins. Acetone precipitates more acidic and hydrophilic proteins, whereas ultracentrifugation isolates those that are more basic, hydrophobic, and, perhaps, fragments of membrane. Interestingly, proteins in the normal urine are expressed as multiple series of protein spots with modest changes of pI and Mw that appears to match PTM patterns, as determined by bioinformatics-based analyses. These findings are consistent with findings in other body fluids such as plasma and cerebrospinal fluid (CSF) *(19)*, indicating an important role of various PTMs in normal physiology.

More recently, we have created a proteome map for normal rat urinary proteins isolated by ultracentrifugation and also studied alterations in excretion of urinary proteins that occur during acute sodium loading *(11,12)*. A total of 111 protein forms representing 57 unique proteins were identified and, again, several PTMs were detected using bioinformatics-based analyses. Some proteins, for example diphor-1, growth-associated protein 43, which is regulated by transforming growth factor-β (TGF-β), and l-myc, which is an oncogene product, have not previously been identified in the urine by other techniques. After acute sodium loading, several proteins including solute carrier family 3, diphor-1, meprin 1-α, neutral endopeptidase, H^+-ATPase, and ezrin were decreased in quantity, whereas kidney aminopeptidase, albumin, transferrin, and α-2u globulin were increased. The role of these proteins in renal sodium regulation has not previously been established. However, functional studies on these changes are necessary to explain the novel mechanisms. An exploration by proteomic analyses may lead to rethinking of the physiology of renal sodium handling.

3. Proteomics and Renal Pathophysiology

3.1. Hypertension and the Kidney

Proteomic analysis in hypertension studies may define novel pathways of blood pressure regulation. Thongboonkerd et al. *(10)* determined a role of the renal kallikrein pathway in hypoxia-induced hypertension using a proteomic approach. Renal expression of kallistatin, a potent vasodilator, was downregulated by both episodic hypoxia (EH) and sustained hypoxia (SH). Renal expression of α_1-antitrypsin (A1AT), an inhibitor of kallikrein activation, was upregulated in EH, in which all animals were hypertensive, but downregulated in SH, in which all animals were normotensive (**Fig. 1**).

B2-bradykinin receptor (B2R) was upregulated in normotensive animals but remained unchanged in hypertensive EH animals, indicating role of B2R in a compensatory mechanism. The proteomic data indicate a significant role of the renal kallikrein pathway in hypertension associated with EH that mimics obstructive sleep apnea syndrome (OSAS) (**Fig. 2A** and **B**). A confirmation of the proteomic data was performed by introducing the human kallikrein (hKLK1) gene to the animals. All the animals overexpressed with human kallikrein were prevented from EH-induced hypertension (**Fig. 2C** and **D**). This is an example that demonstrates the power of proteomic analysis in the

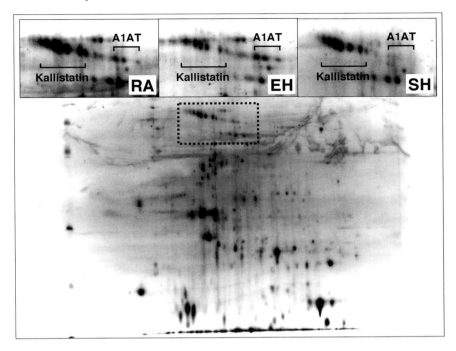

Fig. 1. Proteomic analysis of the entire kidneys of rats exposed to hypoxia. Renal proteins were extracted by using the three buffers method (*see Protocol I*). A total of 100 μg of renal proteins was separated by 2-D-PAGE and visualized by Sypro Ruby staining. Renal proteomes of the animals exposed to episodic hypoxia (EH) and sustained hypoxia (SH) were compared with the control [room air (RA)]. Expression of kallistatin was downregulated by both EH and SH. Expression of α_1-antitrypsin (A1AT) was upregulated in EH, in which all animals were hypertensive, but downregulated in SH, in which all animals were normotensive. (Adapted from **ref. *10*.**)

generation of a confirmable new hypothesis and in the exploration of previously undetermined disease pathways.

3.2. Diabetic Nephropathy

Although much progress has been made in the study of the pathophysiology of diabetic nephropathy, the cellular mechanisms leading to diabetes-induced renal damage are incompletely understood. Recently, Thongboonkerd et al. (ms. submitted), compared a proteomic analysis of renal tissue in 120-d-old OVE26 transgenic mice, which display characteristics of early-onset insulin-dependent diabetes mellitus (IDDM), with normal renal tissue. Because there was no available mouse kidney proteome map, we have created a renal proteome map for normal mouse kidney with 92 identified protein forms representing 65 unique proteins. Thirty identified proteins were regulated in diabetic kidney, including proteases, protease inhibitors, apoptotic-induced proteins, regulating proteins for oxidative tolerance, calcium-binding proteins, transport regulators, cell signaling proteins, and smooth muscle contractile elements. Nineteen of the altered proteins have previously been shown to be regulated during diabetes, whereas the roles of another 11 altered proteins have not been established, indicating the novel mechanisms of diabetic nephropathy. Most of the altered proteins play important roles in apoptosis,

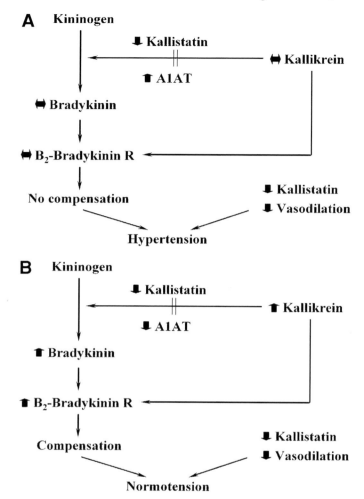

Fig. 2. Alterations in the renal kallikrein system during hypoxic-induced hypertension. The data from the proteomic analysis in **Fig. 1** were used to generate a new hypothesis in episodic hypoxia (EH)-induced hypertension and a compensatory mechanism in sustained hypoxia (SH). The hypothesis was confirmed by an experiment in TG (hKLK1) animals, which overexpressed human tissue kallikrein. Overexpression of renal tissue kallikrein prevented the animals from experiencing EH-induced hypertension. BD, blood pressure; A1AT, α_1-antitrypsin. (Adpated from **ref. 10**.)

vascular thrombosis and remodeling, glomerulopathy, and fibrogenesis and are closely related to each other in the renal protein trafficking of diabetic nephropathy **(Fig. 3)**.

We found an increase of the expression of an unknown protein. Using bioinformatics-based analyses, the sequence of the unknown protein was homologous to that of elastase inhibitor. We confirmed the structure by an immunoblotting method. We have proposed a new hypothesis, that its substrate, elastase, is downregulated and that elastin, a substrate of elastase, is upregulated as a consequence. The hypothesis has been confirmed by an immunohistochemistry method. Again, this is another example that demonstrates the usefulness of proteomic analysis to generate a confirmable hypothesis for

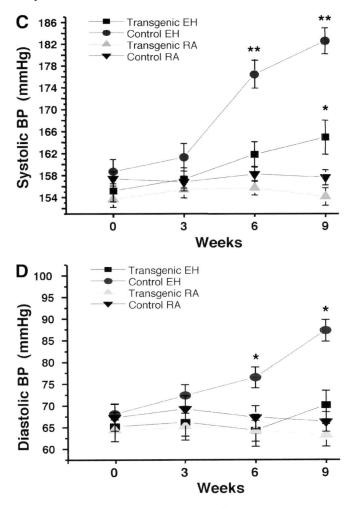

Fig. 2. *(continued)*.

the disease mechanisms. These data indicate the complexity in pathogenic mechanisms of diabetic nephropathy, leading to a coherent understanding of the pathophysiology of this renal complication. We have found some changes in type 2 diabetic nephropathy that are similar to those of type 1 and some changes that are opposite to those of the type 1 model (unpublished data). This has allowed us to plan functional studies in an attempt to understand these complex mechanisms in diabetic nephropathy better.

3.3. Nephrotoxicity

Proteomic analysis for nephrotoxicity was first introduced by Aicher et al. *(20)* to determine the association between renal protein changes and cyclosporin A (CsA) nephrotoxicity in renal transplant patients. Calbindin-D (28 kDa), an antagonist for cell apoptosis, was decreased in most of the patients with CsA toxicity. Although expression study shows this important information, functional studies are still needed. Cutler et al. *(21)* used an integrated proteomic approach to determine the renal effects of

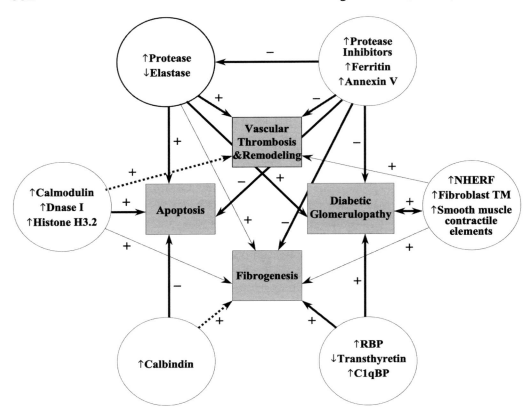

Fig. 3. The renal protein trafficking in type 1 diabetic nephropathy obtained by proteomic and bioinformatic analysis of OVE26 diabetic mice kidneys. +, activation or stimulation; −, inhibition. NHERF, NA+/+++ exchanger regulatory factor; TM, tropomyosin; RBP, retinol-binding protein.

puromycin aminonucleoside (PAN) administration, which generally induces ultra-structural glomerular changes and a nephrotic syndrome mimicking minimal-change nephropathy. At baseline, low molecular weight rat urinary proteins were present with α-2u globulin and glial fibrillary acidic proteins as the major components. Administration of PAN caused shifting of urinary protein spot patterns to the higher molecular sizes range, including albumin, transferrin, and vitamin D-binding protein. These changes disappeared after 672 h following a single PAN administration. These findings may be applied for identification of urinary biomarkers for glomerular diseases in the near future.

Lead nephrotoxicity has been evaluated with proteomic analysis by Witzmann et al. *(22)*. Expression of 76 protein spots on a 2-D gel was altered in the cortex, whereas only 13 spots were changed in the medulla. Unfortunately, of these altered proteins, only 11 proteins were identified by MALDI MS/ESI MS, immunological methods, and gel matching. Expression of aflatoxin B1 aldehyde reductase, α₂-microglobulin, aldose reductase, and glutathione S-transferase P1 (GSTP1) were altered both in the cortex and, to a lesser extent, in the medulla. Expression of transferrin, transketolase, heat shock protein 90 kDa (HSP90), calbindin, argininosuccinate synthase, calcineurin, and sorbitol dehydrogenase was changed only in the cortex. These findings could have explained the differential vulnerability between cortex and medulla to nephrotoxic substances. Witzmann et al. *(23,24)* also used proteomic analysis to determine renal effects of simulated military occupational jet fuel (JP-8) exposure. Only numerical, but not statistically significant, differences were observed in renal expression of glutathione-S-transferase and 10-formyltetrahydrofolate dehydrogenase. These data supported a possibility of the renal effect from JP-8 occupational exposure in the military. However, supporting data such as functional studies are needed.

More recently, Hampel et al. *(25)* applied surface-enhanced laser desorption/ionization (SELDI) ProteinChip array TOF MS to determine urinary protein changes after radiocontrast medium administration. In this method, ProteinChip arrays contained 2-mm-diameter adsorptive target spots with either chemical (e.g., hydrophilic or hydrophobic) or immunoadsorptive (antibody) surfaces. The adsorptive surfaces were designed to capture the widest range of complementary proteins. After removal of unbound proteins and interfering substances, molecular masses of the retained proteins were identified by TOF MS. Proteins with molecular weights of 9.75, 11.75, 23.5, and 66.4 kDa were either increased or decreased in their expression. The protein at 11.75 kDa that was upregulated was identified as β_2-microglobulin by an immunological method. These changes were observed in all patients with or without renal complications after radiocontrast administration but were more persistent in patients with renal effects. Although the SELDI ProteinChip technique is rapid, reproducible, and highly sensitive, it is limited by the difficulty in protein identification, and other confirmatory methods are needed *(26,27)*.

4. Proteomics and Biomarkers

4.1. Urological Carcinoma

Urological carcinomas, especially bladder cancer, have been studied with proteomic methods. Rasmussen et al. *(28)* identified 124 polypeptides (with several repeated variants) in the urine of patients with bladder squamous cell carcinoma (SCC). The identified protein components, however, were limited to only major abundance groups. Among the identified proteins, psoriasin was expressed only in the urine from SCC patients. The same group *(29)* used immunoblotting to confirm the results: psoriasin was present only in SCC urine, not in transitional cell carcinoma or normal urine. These data are consistent with the knowledge that psoriasin is a major abundant protein in human keratinocytes and would be expected to be present in SCC also. Psoriasin alone, or in combination with other markers, is therefore a promising biomarker to detect bladder SCC. In addition to conventional methods to determine biomarkers for renal cell carcinoma (RCC), Sarto et al. *(30)* demonstrated that renal tissue in RCC patients expressed two multimeric and five monomeric forms of Mn-superoxide dismutase (SOD), whereas normal kidney expressed only two monomeric Mn-SOD spots without a multimeric form. These findings could apparently be from a particular PTM. To date, there is no validated and unique biomarker for detecting and monitoring RCC evolution and treatment.

4.2. Glomerular Diseases

Detection of biomarkers for various glomerular diseases may become one of the most fruitful and useful applications of proteomics in nephrology. To our knowledge, there is no available reference using proteomic analysis to determine biomarkers for the glomerular diseases. Our group has performed a preliminary study to detect biomarkers from the urine of patients with various glomerular diseases. The preliminary data show the obvious differences between urinary proteome patterns in glomerular diseases and the normal and among various glomerular defects *(31)*. We are continuing the work to complete this project.

The techniques used to perform proteomic analyses for the study of renal and urinary abnormalities are essentially the same. Four main steps are involved in such studies: protein extraction (or isolation), protein separation, protein identification, and bioinformatics-based analyses. In this chapter, we focus only on the methodology used for extraction/isolation of proteins from the kidneys and the urine for 2-D-PAGE.

5. Protein Extraction Methods for the Entire Kidney

Studies of renal protein expression can be divided into two levels: *the entire kidney* and *the individual intrarenal structure*. The study of the entire kidney is simply performed by removal of the kidneys from the animals, removal of renal capsules, and flushing the kidneys with saline to wash out contaminating blood. Study of individual intrarenal structure is more complicated because each structure has to be isolated before protein extraction. At the beginning phase of renal proteomics, analysis of the entire kidney is necessary to create a global image of the renal proteome. A limitation of proteomic analysis of the entire kidney is that it cannot provide information about the source and location of the proteome identified. The locations of the identified renal proteins can be determined either by *subproteome analysis* or by other localizing techniques, especially immunohistochemistry. However, proteomic analysis of the entire kidney is still very useful to screen alterations in the renal proteomes by various physiological conditions, pathophysiological states, and diseases. Proteomic information from the entire kidney is also helpful to hint at or guide the direction and selection for the future subproteome study. In this section, we provide various protocols that we use in our laboratory to extract proteins from the entire kidney.

The principles in these protocols are to solubilize and preserve the proteins to be ready for and compatible with isoelectric focusing (IEF) separation. Detergents used for solubilizing the proteins include sodium dodecyl sulfate (SDS), urea, thiourea, Triton-X, and zwitterionic detergents such as CHAPS (3-[(3-cholamidopropyl)dimethylammonio]-1-propane-sulfonate) and Zwittergent. Protease inhibitors such as phenylmethylsulfonyl fluoride (PMSF), aprotinin, and leupeptin are used to preserve the proteins. Phosphatase inhibitors such as okadaic acid, calyculin A, and Na_3VO_4 are applied to preserve phosphoproteins, especially when phosphorylation study is needed. Excess DNA and RNA components that interfere with the IEF protein separation are removed by DNAse and RNAse or by precipitation with ampholytes. The main function of carrier ampholytes is to establish or condition a pH gradient increasing from the anode to the cathode. Excess salts and ions are removed by chelating with EDTA (ethylenediaminetetraacetic acid) and EGTA [ethylene glycol-bis(2-aminoethylether)-N,N,N′,N′-tetraacetic acid] or by precipitating proteins using trichloroacetic acid (TCA), acetone, or other organic solvents. For immobilized pH gradient (IPG) IEF separation, the protein solution from each protocol after the final step is subjected to mixing with standard rehydration buffer. Two standard rehydration buffers are used in our laboratory. One is the urea rehydration buffer containing 8 M urea, 2% CHAPS, 0.01 M dithiothreitol (DTT), 2% ampholytes, and bromophenol blue. Another is urea/thiourea rehydration buffer containing 7 M urea, 2 M thiourea, 2% CHAPS, 0.65 mM DTT, 1% Zwittergent, 0.8% ampholytes, and bromophenol blue.

5.1. Protocol I: Three Buffers Technique

See **Fig. 1**.

1. After removal of the renal capsule and flushing of the excess blood, the kidney is excised into thin slices and the slices are washed in phosphate-buffered saline (PBS). The kidney slices are then rapidly frozen in liquid nitrogen. The frozen tissue is then ground up into powder using a prechilled mortar and pestle.
2. The weight of tissue is measured. Preheated buffer I (2 mL/g of tissue) is added and the mixture is incubated at 100°C for 5 min.
3. The mixture is then placed on ice for 5 min. Buffer II (100 µL/mL of mixture) is added and further incubated on ice for 10 min.
4. The mixture is then centrifuged at 13,000g for 5 min. The supernatant is saved and 50% TCA is added to supernatant for a concentration of 10%. The sample is incubated on ice again for 10 min.
5. The sample is then centrifuged at 13,000g for 2 min. The pellet is saved on ice.
6. The pellet is then washed with 95% acetone (resuspended in 0.5 mL by using a 21-G needle). The sample is centrifuged at 13,000g for 5 min. The pellet is saved. This washing step is performed twice.
7. The pellet is left to dry at room temperature for 5 min. Then, 1 mL of buffer III is added and the pellet is resuspended by using a 21-G needle.
8. Acetone 4 mL is added to the sample, which is further incubated at room temperature for 5 min.
9. The mixture is then centrifuged at 13,000g for 5 min. The pellet is saved and allowed to air dry at room temperature for 5 min.
10. Then buffer III (200 µL) is added and the mixture is resuspended. The protein concentration is measured (*see* **Note 1**).
    Buffer I: 0.3 % SDS, 200 mM DTT, 28 mM Tris-HCl, and 22 mM Tris base in 18 MΩ water.
    Buffer II: 476 mM Tris-HCl, 24 mM Tris base, 50 mM MgCl$_2$, 1 mg/mL DNAse I, and 0.25 mg/mL RNAse A in 18 MΩ water.
    Buffer III: 7.92 M urea, 0.06% SDS, 1.76% ampholytes, 120 mM DTT, 3.2% Triton X-100, 22.4 mM Tris-HCl, and 17.6 mM Tris base.

5.2. Protocol II: Simple Lysis

1. After removal of the renal capsule and flushing of the excess blood, the kidney is excised into thin slices and the slices are washed in PBS. The kidney slices are then rapidly frozen in liquid nitrogen. The frozen tissue is then ground up into powder using a prechilled mortar and pestle.
2. The weight of the tissue is measured and lysis buffer is added (2.5 µL/mg of tissue).
3. The mixture is then incubated at 4°C on a shaker for 30–60 min and centrifuged at 13,000g for 5 min.

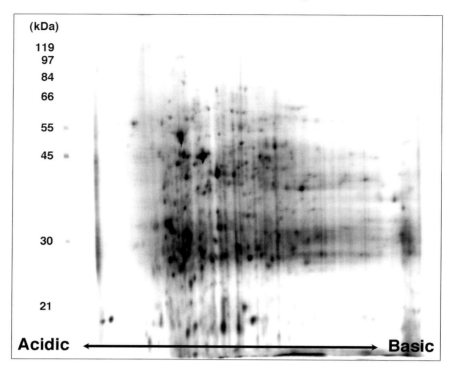

Fig. 4. Renal proteome of the entire rat kidney extracted by the urea/thiourea method (150 μg of total proteins, IPG strip, linear pH 3–10, Sypro Ruby staining).

4. The supernatant is saved and the protein concentration is measured.

   Lysis buffer: 20 m*M* Tris-HCl (pH 7.4), 150 m*M* NaCl, 1% Triton X-100, 1% NP-40, 1 m*M* EDTA, 1 m*M* EGTA, 1 m*M* Na$_3$VO$_4$, 20 m*M* NaF, 5 m*M* PMSF, 21 μg/mL aprotinin, and 5 μg/mL leupeptin.

   Note: 2-D-PAGE and MS analysis showed that the proteins identified using this technique were mostly hydrophilic and cytosolic proteins.

5.3. Protocol III: Urea/Thiourea Method

See **Fig. 4**.

1. After removal of the renal capsule and flushing of the excess blood, the kidney is excised into thin slices and the slices are washed in PBS. The kidney slices are then rapidly frozen in liquid nitrogen. The frozen tissue is then ground up into powder using a prechilled mortar and pestle.
2. The weight of tissue is measured and urea/thiourea buffer is added (2.5 μL/mg of tissue).
3. The mixture is then incubated on a shaker at 4°C for 30–60 min and centrifuged at 13,000*g* at 4°C for 5 min.
4. The supernatant is saved and the protein concentration is measured (*see* **Note 2**).

   Urea/thiourea buffer: 9 mg/mL DTT, 40 mg/mL CHAPS, 0.42 g/mL urea, 0.15 g/mL thiourea, and 4.45% (v/v) carrier ampholytes, pH 3–10 (protease inhibitors can be added up to: 10 μg/mL aprotinin, 20 μg/mL leupeptin, and 0.5 m*M* PMSF).

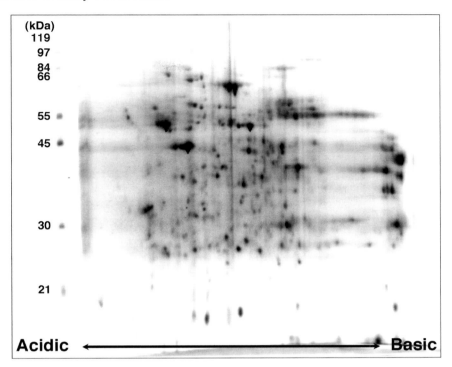

Fig. 5. Renal proteome of the entire rat kidney extracted by the SDS/Cymal method (150 µg of total proteins, IPG strip, linear pH 3–10, Sypro Ruby staining).

5.4. Protocol IV: SDS/Cymal Method

See **Fig. 5**.

One major limitation in proteomic study of proteins separated by 2-D-PAGE is the inability to isolate integral membrane proteins using available extraction protocols. Even the efficacy of recently developed extraction buffers containing thiourea is unsatisfactory when applied to hydrophobic transmembrane proteins. We have developed a new extraction technique using SDS followed by dialysis in the presence of cyclo alkyl substituted maltosides (Cymal) to remove SDS and to maintain dispersion of the proteins.

1. After removal of the renal capsule and flushing of the excess blood, the kidney is excised into thin slices and the slices are washed in PBS. The kidney slices are then rapidly frozen in liquid nitrogen. The frozen tissue is then ground up into a powder using a prechilled mortar and pestle.
2. The weight of tissue is measured and 10% SDS (in 18 MΩ water) is added [2.5 g of SDS per g of tissue dry weight (dry weight is approx 30% of the powder weight in **Step 1**)].
3. The mixture is then incubated at room temperature on the shaker for 30–60 min, heated at 100°C for 5 min, and then centrifuged at 13,000*g* for 5 min.
4. The supernatant is saved and SDS is removed by dialysis against dialysate containing 0.2% 6-cyclohexyl-1-hexyl-β-D-maltoside (Cymal-6; Anatrace, Maumee, OH) in 18 MΩ water using a 0.5-cm-diameter cellulose dialysis tube with a nominal molecular weight cutoff at 6 kDa. The dialysis is performed at room temperature for 18 h. (Note that the sample volume will be increased after completion of dialysis.)
5. Protein concentration is measured (*see* **Note 3**).

5.5. Protocol V: Sequential Extractions

Because of differences in physical and chemical properties of proteins, each extraction method is not perfect for all protein components expressed in the tissues. Techniques used for hydrophilic proteins are not suitable in the case of hydrophobic proteins. In contrast, techniques used for hydrophobic protein extraction may not work well in the case of hydrophilic proteins. Remaining unsolubilized proteins can be observed as the pellet left after completion of the extraction by all available techniques, without any exception. One strategy to solve this problem is using sequential extraction procedures to get a more complete extraction. Although this technique is laborious, it may be necessary in some cases. The exact procedure still needs to be developed.

6. Protein Extraction Methods for Isolated Intrarenal Structures and Cultured Cells

Several protocols are used for isolation of individual intrarenal structures including cortex, medulla, glomeruli, mesangial cells, podocytes, tubules, brush border membrane, vessels, and vascular endothelia. Basically, extraction of proteins from isolated intrarenal structures can be performed as for the entire kidney. Protein extraction from the cultured cells can also be performed using the same techniques as those for the entire kidney and the isolated intrarenal structures. Extraction protocols I–V can be performed after the cultured cells are harvested. The following protocol is an additional technique for the cultured cells.

6.1. Protocol VI: Extraction Protocol for Cultured Cells

See **Fig. 6**.

1. Confluent cells in T-75 flasks (75 cm^2) are obtained and the culture media is removed.
2. The flasks are washed three times with 10 mL of ice-cold PBS for 1 min. After the final wash, PBS should be completely removed to eliminate the excess salt and water problem.
3. Then 240 µL of preheated (100°C) buffer I is added to the flask. The flask is held at a 30° angle and the cells are scraped using a cell scraper. The lysate and buffer are mixed thoroughly and placed into a 1.5-mL Eppendorf tube and heated at 100°C for 5 min.
4. The sample is placed on ice for 5 min, then 24 µL of buffer II is added to the tube, and the mixture is mixed.
5. The sample is incubated on ice for additional 8 min and the contents are mixed again and split into two aliquots (132 µL each).
6. Then 80% v/v of acetone (528 µL) is added to each tube and the samples are incubated on ice for additional 20 min.
7. The samples are then centrifuged at 13,000*g* for 10 min and the supernatants are discarded.
8. The pellets are allowed to air dry for 5 min and resuspended with 200 µL of buffer III using a 21-G syringe needle.
9. Protein concentration is measured.
   Buffer I: 0.3% SDS, 200 m*M* DTT, 28 m*M* Tris-HCl, and 22 m*M* Tris base in 18 MΩ water.
   Buffer II: 476 m*M* Tris-HCl, 24 m*M* Tris base, 50 m*M* MgCl$_2$, 1 mg/mL DNAse I, and 0.25 mg/mL RNAse A in 18 MΩ water.
   Buffer III: 7.92 *M* urea, 0.06% SDS, 1.76% ampholytes, 120 m*M* DTT, 3.2% Triton X-100, 22.4 m*M* Tris-HCl, and 17.6 m*M* Tris Base.

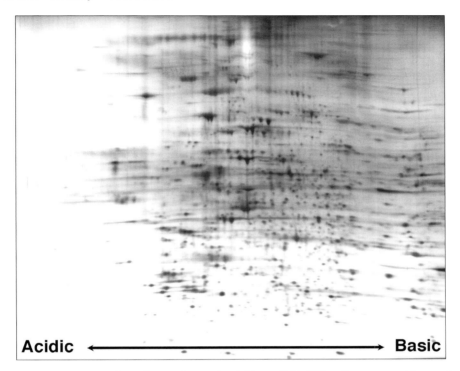

Acidic ←——————————————→ **Basic**

Fig. 6. The proteome of Madin-Darby canine kidney (MDCK) cells extracted by methods in protocol VI (80 µg of total proteins, IEF-mobilized tube gel, pH 3–10, EMBL silver staining).

7. Protein Isolation Methods for Urine

Because protein concentration in the urine is low and salts and urea contents in the urine may affect IEF separation, proteins need to be concentrated or isolated from the urine. Several techniques can be applied to isolate the proteins. Precipitation with organic solvents is a common method, whereas ultracentrifugation is a new method we use to isolate urinary proteins. Following are the successful methods we use in our laboratory to isolate urinary proteins.

7.1. Protocol VII: Acetone Precipitation

See **Fig. 7**.

1. A urine sample of 10 mL is collected in a polyethylene tube containing 1 mL of protease inhibitors (0.1 mg/mL leupeptin, 0.1 mg/mL PMSF, and 1 mM sodium azide in 1 M Tris-HCl, pH 6.8) and centrifuged at 500g for 5 min.
2. The supernatant is passed through 0.34-mm Whatman chromatography paper.
3. Acetone 10 mL is added to the samples, and the mixture is incubated for 10 min and centrifuged at 13,000g for 5 min.
4. The pellet is allowed to air dry for 5 min and then resuspended with 1 mL of buffer containing 7.92 M urea, 0.06% SDS, 1.76% ampholytes, 120 mM DTT, 3.2% Triton X-100, 22.4 mM Tris-HCl, and 17.6 mM Tris base (using a 21-G syringe needle).
5. Protein concentration is then measured (*see* **Note 4**).

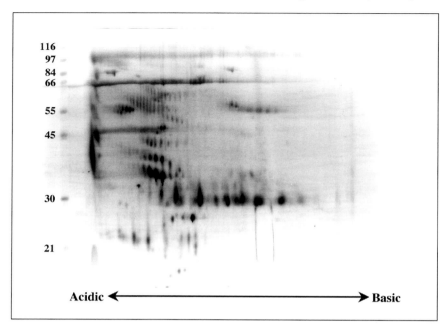

Fig. 7. Normal human urinary proteome isolated by acetone precipitation (150 µg of total proteins, IPG strip, linear pH 3–10, EMBL silver staining). (Adapted from **ref. 7**.)

7.2. Protocol VIII: Ultracentrifugation

See **Fig. 8**.

1. A urine sample of 10 mL is collected in a polyethylene tube containing 1 mL of protease inhibitors (0.1 mg/mL leupeptin, 0.1 mg/mL PMSF, and 1 mM sodium azide in 1 M Tris-HCl, pH 6.8) and centrifuged at 500g for 5 min.
2. The supernatant is passed through 0.34-mm Whatman chromatography paper.
3. The sample is centrifuged at 200,000g for 120 min at 4°C. The supernatant is then removed. (The residue remaining after this step is only a thin film of urinary proteins.)
4. The pellet (or protein film) is resuspended with 100–200 µL of buffer containing 7.92 M urea, 0.06% SDS, 1.76% ampholytes, 120 mM DTT, 3.2% Triton X-100, 22.4 mM Tris-HCl, and 17.6 mM Tris base.

7.3. Protocol IX: Centrifugal Filtration

See **Fig. 9**.

1. A urine sample of 10 mL is collected in a polyethylene tube containing 1 mL of protease inhibitors (0.1 mg/mL leupeptin, 0.1 mg/mL PMSF, and 1 mM sodium azide in 1 M Tris-HCl, pH 6.8) and centrifuged at 500g for 5 min.
2. The supernatant is passed through 0.34-mm Whatman chromatography paper.
3. The sample is filtrated through a centrifugal filter device (Centricon Plus-20, M$_w$ cutoff at 5 kDa; Millipore, Bedford, MA) at 3500g until the urine volume decreases to 0.2–1.0 mL.
4. The remaining urine is collected and the protein concentration is measured.

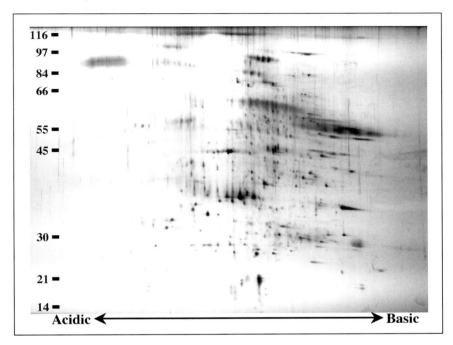

Fig. 8. Normal human urinary proteome isolated by ultracentrifugation (150 µg of total proteins, IEF mobilized tube gel pH 3–10, EMBL silver staining). (Adapted from **ref. 7**.)

7.4. Protocol X: Sequential Isolations and/or Combination

As in the case of protein extraction of the entire kidney, physical and chemical properties of urinary proteins are important to select an isolation method. There is no single perfect technique that can be used to study all protein components expressed in the urine. Sequential isolation, for example, ultracentrifugation followed by acetone precipitation, or vice versa, should be kept in mind. Additionally, a combination of the two techniques can be applied for two aliquots of the same samples.

7.5. Protocol XI: Lyophilization

1. Fresh urine is collected and debris and nuclei are removed as for the other methods by using low-speed spin and filtration.
2. The urine is frozen at –20 to –70°C and then lyophilized.
3. The lyophilized pellet is resuspended in buffer containing 7.92 M urea, 0.06% SDS, 1.76% ampholytes, 120 mM DTT, 3.2% Triton X-100, 22.4 mM Tris-HCl, and 17.6 mM Tris base.

8. Concerns

8.1. Differential Expression of Proteins Extracted from the Entire Kidney

Protein expression of the entire kidney represents overall expression of proteins from all intrarenal structures. To determine differential expression of proteins that are extracted from the entire kidneys, we are not able to specify the location of changes. On another hand, we may not obtain differential expressed proteins in case changes in

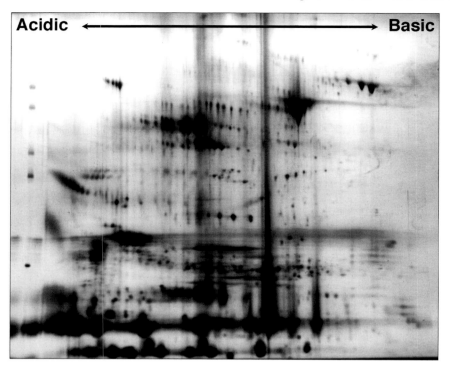

Fig. 9. Normal rat urinary proteome isolated by centrifugal filtration (50 μg of total proteins, IEF mobilized tube gel pH 3–10, EMBL silver staining).

each intrarenal structure are not in the same direction. We only examine a final product of those changes in individual structures. If expression of a protein is upregulated in one structure but is downregulated in other structures, the measuring level of protein expression in the entire kidney may be unchanged. This may not be a frequent case, but it should be kept in mind. Proteomic analysis of an isolated intrarenal structure and immunohistochemistry are the potential solutions for this concern.

8.2. Sources of the Urinary Proteins

Proteins in the urine mainly originate from the renal tubules in normal physiological status and in tubular defects. However, urinary proteins found in the normal urine and urine from the patients can also originate from other sources. Urinary proteins are obtained from the plasma in case of capillary leakage, especially in glomerular diseases and hematuria from other causes. Additionally, proteins can be secreted from the prostate in male normal urine. To interpret the data, this issue should be considered.

8.3. Alterations in Urinary Protein Excretion

Changes in excretion of proteins into the urine can be from two different mechanisms: alterations in their production in the renal tubules or alterations in their secretion into the urine. For example, a decrease in excretion of a specific protein can be from a decrease in its production in the renal tubules, or from a normal production but less secretion because the renal tubules need to keep this protein in response to a physiological stimulus to maintain the normal physiology.

8.4. Total Amount of Proteins Loaded for 2-D-PAGE Separation

We have found that optimal protein concentration for a consistent 2-D spot pattern is 2–15 µg/µL. The amount of total proteins should be 100–300 µg per gel in order to get the best spot resolution without streaking and background interference. However, overloading the samples is very useful to increase efficacy of protein identification by MS analysis.

9. Notes

1. 2-D-PAGE and MS analysis showed that the proteins identified using this technique were mostly hydrophilic and cytosolic proteins.
2. 2-D-PAGE and MS analysis showed that the proteins identified using this technique were mostly hydrophilic and cytosolic proteins. However, a few hydrophobic and integral membrane proteins could be identified.
3. Using this technique, several typical integral membrane proteins can be identified by 2-D-PAGE separation and MS analysis. Predicted or computed transmembrane regions of proteins identified by this technique were significantly greater than by other methods. However, expression of hydrophilic and cytosolic proteins on 2-D-PAGE was less than in other methods by equal total protein loading.
4. This technique precipitates mostly hydrophilic and acidic proteins. All the procedures should be performed at 4°C. Other resuspension buffers that can be used are IEF rehydration buffer and 250 m*M* sucrose with 10 m*M* triethanolamine. The urine volume in this protocol is suitable for nonproteinuric donors. For proteinuric patients, the urine volume should be 2–5 mL for 1 mL of resuspension buffer. The urine with protease inhibitors can be kept at –20 to –70°C for several months before protein isolation is performed. The processed samples can be stored at –20 to –70°C for several months without any effect on 2-D-PAGE.
5. This method isolates most basic proteins and some hydrophobic proteins. All the procedures should be performed at 4°C. Other resuspension buffers that can be used are IEF rehydration buffer and 250 m*M* sucrose with 10 m*M* triethanolamine. Volume of resuspension can be adjusted based on the pellet size or thickness of the protein film, especially in case of proteinuria. The urine with protease inhibitors can be kept at –20 to –70°C for several months before performing protein isolation. The processed samples can be stored at –20 to –70°C for several months without any effect on 2-D-PAGE.
6. Proteins can be stuck on the filter. Thus, this method may not be suitable for determination of differential protein expression by various physiological stimuli and renal diseases.

References

1. Witzmann, F., Clack, J., Fultz, C., and Jarnot, B. (1995) Two-dimensional electrophoretic mapping of hepatic and renal stress proteins. *Electrophoresis* **16**, 451–459.
2. Marshall, T. and Williams, K. (1996) Two-dimensional electrophoresis of human urinary proteins following concentration by dye precipitation. *Electrophoresis* **17**, 1265–1272.
3. Davis, M. T., Spahr, C. S., McGinley, M. D., et al. (2001) Towards defining the urinary proteome using liquid chromatography-tandem mass spectrometry. II. Limitations of complex mixture analyses. *Proteomics* **1**, 108–117.

4. Spahr, C. S., Davis, M. T., McGinley, M. D., et al. (2001) Towards defining the urinary proteome using liquid chromatography-tandem mass spectrometry. I. Profiling an unfractionated tryptic digest. *Proteomics* **1,** 93–107.

5. Brooks, H. L., Sorensen, A. M., Terris, J., et al. (2001) Profiling of renal tubule Na$^+$ transporter abundances in NHE3 and NCC null mice using targeted proteomics. *J. Physiol.* **530,** 359–366.

6. Knepper, M. A. and Masilamani, S. (2001) Targeted proteomics in the kidney using ensembles of antibodies. *Acta Physiol. Scand.* **173,** 11–21.

7. Thongboonkerd, V., McLeish, K. R., Arthur, J. M., and Klein, J. B. (2002) Proteomic analysis of normal human urinary proteins isolated by acetone precipitation or ultracentrifugation. *Kidney Int.* **62,** 1461–1469.

8. Thongboonkerd, V., Klein, J. B., and Klein, E. (2002) A new protein extraction technique that permits identification of integral membrane proteins by two-dimensional gel electrophoresis, submitted.

9. Arthur, J. M., Thongboonkerd, V., Scherzer, J. A., Cai, J., Pierce, W. M., and Klein, J. B. (2002) Differential expression of proteins in renal cortex and medulla: a proteomic approach. *Kidney Int.* **62,** 1314–1321.

10. Thongboonkerd, V., Gozal, E., Sachleben, L. R., et al. (2002) Proteomic analysis reveals alterations in the renal kallikrein pathway during hypoxia-induced hypertension. *J. Biol. Chem.* **277,** 34,708–34,716.

11. Thongboonkerd, V., Klein, J. B., and Arthur, J. M. (2001) Proteomic identification of urine proteins. *J. Am. Soc. Nephrol.* **12 (Suppl.),** 50A.

12. Thongboonkerd, V., Klein, J. B., Pierce, W. M., Jevans, A. W., and Arthur, J. M. (2003) Sodium loading changes urinary excretion: a proteomic analysis. *Am. J. Physiol. Renal Physiol.* **284,** in press.

13. Sarto, C., Marocchi, A., Sanchez, J. C., et al. (1997) Renal cell carcinoma and normal kidney protein expression. *Electrophoresis* **18,** 599–604.

14. Witzmann, F. A., Fultz, C. D., Grant, R. A., Wright, L. S., Kornguth, S. E., and Siegel, F. L. (1998) Differential expression of cytosolic proteins in the rat kidney cortex and medulla: preliminary proteomics. *Electrophoresis* **19,** 2491–2497.

15. Heine, G., Raida, M., and Forssmann, W. G. (1997) Mapping of peptides and protein fragments in human urine using liquid chromatography-mass spectrometry. *J. Chromatogr. A* **776,** 117–124.

16. McKee, J. A., Kumar, S., Ecelbarger, C. A., Fernandez-Llama, P., Terris, J., and Knepper, M. A. (2000) Detection of Na(+) transporter proteins in urine. *J. Am. Soc. Nephrol.* **11,** 2128–2132.

17. Blomberg, A., Blomberg, L., Norbeck, J., et al. (1995) Interlaboratory reproducibility of yeast protein patterns analyzed by immobilized pH gradient two-dimensional gel electrophoresis. *Electrophoresis* **16,** 1935–1945.

18. Witzmann, F. A. and Li, J. (2002) Cutting-edge technology. II. Proteomics: core technologies and applications in physiology. *Am. J. Physiol. Gastrointest. Liver Physiol.* **282,** G735–G741.

19. Wait, R., Gianazza, E., Eberini, I., et al. (2001) Proteins of rat serum, urine, and cerebrospinal fluid: VI. Further protein identifications and interstrain comparison. *Electrophoresis* **22,** 3043–3052.

20. Aicher, L., Wahl, D., Arce, A., Grenet, O., and Steiner, S. (1998) New insights into cyclosporine A nephrotoxicity by proteome analysis. *Electrophoresis* **19,** 1998–2003.

21. Cutler, P., Bell, D. J., Birrell, H. C., et al. (1999) An integrated proteomic approach to studying glomerular nephrotoxicity. *Electrophoresis* **20,** 3647–3658.

22. Witzmann, F. A., Fultz, C. D., Grant, R. A., Wright, L. S., Kornguth, S. E., and Siegel, F. L. (1999) Regional protein alterations in rat kidneys induced by lead exposure. *Electrophoresis* **20,** 943–951.

23. Witzmann, F. A., Bauer, M. D., Fieno, A. M., et al. (2000) Proteomic analysis of the renal effects of simulated occupational jet fuel exposure. *Electrophoresis* **21,** 976–984.

24. Witzmann, F. A., Carpenter, R. L., Ritchie, G. D., Wilson, C. L., Nordholm, A. F., and Rossi, J. III (2000) Toxicity of chemical mixtures: proteomic analysis of persisting liver and kidney protein alterations induced by repeated exposure of rats to JP-8 jet fuel vapor. *Electrophoresis* **21,** 2138–2147.

25. Hampel, D. J., Sansome, C., Sha, M., Brodsky, S., Lawson, W. E., and Goligorsky, M. S. (2001) Toward proteomics in uroscopy: urinary protein profiles after radiocontrast medium administration. *J. Am. Soc. Nephrol.* **12,** 1026–1035.

26. Petricoin, E. F., Ardekani, A. M., Hitt, B. A., et al. (2002) Use of proteomic patterns in serum to identify ovarian cancer. *Lancet* **359,** 572–577.

27. Verma, M., Wright, G. L. Jr., Hanash, S. M., Gopal-Srivastava, R., and Srivastava, S. (2001) Proteomic approaches within the NCI early detection research network for the discovery and identification of cancer biomarkers. *Ann. NY Acad. Sci.* **945,** 103–115.

28. Rasmussen, H. H., Orntoft, T. F., Wolf, H., and Celis, J. E. (1996) Towards a comprehensive database of proteins from the urine of patients with bladder cancer. *J. Urol.* **155,** 2113–2119.

29. Celis, J. E., Wolf, H., and Ostergaard, M. (2000) Bladder squamous cell carcinoma biomarkers derived from proteomics. *Electrophoresis* **21,** 2115–2121.

30. Sarto, C., Deon, C., Doro, G., Hochstrasser, D. F., Mocarelli, P., and Sanchez, J. C. (2001) Contribution of proteomics to the molecular analysis of renal cell carcinoma with an emphasis on manganese superoxide dismutase. *Proteomics* **1,** 1288–1294.

31. Thongboonkerd, V., Klein, J. B., and McLeish, K. R. (2002) Proteomic identification of biomarkers for glomerular diseases. *J. Am. Soc. Nephrol.* **13 (Suppl.),** 120A.

27

Proteomics in Endocrinology

Jan W. A. Smit and Johannes A. Romijn

1. Introduction

Endocrinology focuses on the understanding of (patho)physiological processes arising from communication between chemical messengers and their receptors, postreceptor signaling cascades and the translation of these signals into effector molecules, and the ultimate effects of the interaction between these factors. The techniques of endocrine research have evolved impressively, from interventions aimed at elucidating the function of endocrine organs to innovative molecular biological approaches at the subcellular level within endocrine organs and target tissues. Endocrine research has revealed numerous important discoveries, which have been adopted by other scientific fields (e.g., immunology, oncology, and infectious diseases). The introduction of high-throughput genomics and proteomics will considerably increase the understanding of molecular endocrine principles and consequently of (patho)physiology in general.

Traditionally, endocrine research was based on hypotheses derived from clinical or experimental observations of excess or deficient hormone production and sensitivity. The era of genomics has allowed a new approach in endocrinology in which the starting point of research is related to molecular changes in genes or gene function, which is referred to as reverse endocrinology. This chapter focuses on the impact of structural genomics and proteomics on endocrine research, preceded by a historical perspective of endocrine research.

2. Traditional Endocrine Research

Traditionally, clinical and experimental observations of phenotypes caused by excessive or insufficient secretion of a particular hormone allowed the establishment of fundamental endocrine principles. Exemplary for this approach were the experiments of Berthold (1) in the middle of the 18th century, who demonstrated that implantation of the testes into the peritoneal cavity of castrated roosters prevented atrophy of the cock's comb. These experiments indicated that the effects of the testis on the comb were not explained by neural mechanisms but most likely involved humoral substances. The definition of hormones was introduced by Starling in 1905 (2). In this definition, a hormone was transported by the blood from the endocrine gland to the target tissue. In the following decades, identification of endocrine organs permitted identification and purification of the hormone(s) produced by the respective organs. For instance, the dis-

From: *Handbook of Proteomic Methods*
Edited by: P. Michael Conn © Humana Press Inc., Totowa, NJ

covery of insulin in 1921 was rapidly followed by the industrial production of insulin from pancreatic tissues obtained from cows and pigs. The synthesis of the particular hormones could be accomplished by traditional biochemical and physicochemical techniques for relatively uncomplicated molecules such as oligopeptides or thyroid or steroid hormones; thyroid hormone and steroids have been synthesized artificially and applied as substitutes or drugs for many decades. However, for complex polypeptides, like insulin, purification of the hormones from the endocrine glands has been the main source until the era of recombinant DNA techniques. After the identification of endocrine glands and their products, methods to determine hormones and hormonal effects were established of which the introduction of the radioimmunoassay was a milestone, allowing the measurement of hormones in picomolar concentrations.

In addition to organ–organ interactions through hormones, the same principles of chemical interaction appeared to be present in the paracrine communication between cells within organs, e.g., within the pituitary gland *(3)* and later even within cells in the form of autocrine communication *(4)*. Although receptors for hormones were postulated early on to exist, their identification was only possible many decades after the discovery of the hormones. For instance, the insulin receptor was cloned only in 1985, more than 60 years after the discovery of insulin *(5)*. Moreover, the determination of the structure of receptors is even more complex and is unresolved for many receptors. The three-dimensional (3-D) structure of insulin and the insulin receptor was proposed only recently in 1999 *(6,7)*.

Investigation of the postreceptor effects of hormones reveals a high degree of complexity that is increasing because of an ever increasing amount of details on these cascades. Nonetheless, the study of these intracellular signal transduction chains opened new horizons in endocrine research and the understanding of hormone action.

3. Molecular Genetics
3.1. The Single-Gene Approach

After the introduction of molecular genetic techniques, new strategies became available to identify hormones; it became possible to translate phenotypes or physiological events directly to the genes involved and, once these genes were identified, to characterize the related proteins, the endocrine glands involved, and the receptors. The modern equivalents of the classical experiments involving resection and/or implantation of endocrine organs are knockout and transfection experiments *(8)*. In addition, the identification of genes encoding hormones also obviated the need for animal or human endocrine tissues as a source of hormones. The introduction of cDNA clones into biological expression systems made it possible to produce recombinant hormones in pharmacological quantities. For instance, growth hormone and insulin have been produced by recombinant genetic techniques since the 1980s *(9,10)*. Human thyrotropin, a dimeric polypeptide, is another example of a hormone with a highly complex structure that can now be produced as recombinant hormone and is used to increase radioiodine uptake in thyroid cancer *(11)*. These molecular approaches are still largely aimed at limited numbers of genes involved in well-defined physiological processes based on clinical or experimental observations.

3.2. High-Throughput Genetic Techniques

In general, the hormones have multiple effects, but each effect is controlled by multiple hormones. It appears that the effects of hormones are much more complex than previously thought. For instance, each hormone affects the expression of multiple genes. Hormones seem to act on gene expression, like a conductor directing different instruments within a large symphony orchestra. New molecular techniques have allowed us for the first time to integrate these complex effects of hormones into gene expression profiles.

Large-scale molecular analyses became possible with the introduction of high-throughput techniques, including serial analysis of gene expression (SAGE) and DNA microarrays. These techniques have identified relevant genes in physiological processes that could never have been found by the single-gene approach. Examples are the identification of proteins in endocrine glands such as the thyroid *(12)*, the adrenal gland *(13)*, adipose tissue (a virtual endocrine gland) *(14)*, the pituitary *(15)*, and the pancreas *(16)*. Other applications are the study of how hormones work: the understanding of the effects and working mechanisms of "common" hormones is limited, as for thyroid hormone *(17–19)*, growth hormone *(20)*, estradiol *(21)*, and androgens *(22)*. In addition, the role of hormones is elucidated in pathological conditions like male infertility *(23)*, the skeletal muscle and adipose tissue conditions in type 2 diabetes mellitus *(24,25)*, and pituitary adenomas *(26)*. Examples of the clinical relevance of DNA array studies are the identification of prognostic markers in breast carcinoma *(27)* and of diagnostic markers in thyroid carcinoma *(28)*. High-throughput techniques have also been applied in the development of drugs or vaccines *(29)* and the understanding of their working mechanism *(30)*. The introduction of high-throughput genetics has created new challenges. In contrast to the single-gene approach, in which the experimental capacity was the limiting factor, in high-throughput techniques, the output of a single DNA array provides thousands of data points, thus necessitating the development of new bioanalytical algorithms. Algorithms have been developed to cluster and classify genes according to their expression profiles in different tissues, organs, or organisms or in relation to their response to specific interventions *(31)*.

4. Genomics and Proteomics

4.1. Definitions

Although the aforementioned techniques are innovative, their starting point is still the elucidation of a physiological process leading to a biological endpoint. In other words, endocrine research (and that of many other sciences) has traditionally been hypothesis-based: appropriate techniques are developed to answer the questions that arise from observations in populations, organisms, organs, and cells and at the subcellular level. This approach is opposed to another scientific approach in which not one particular hypothesis is the starting point of research. Instead, a complete and systematic inventory of characteristics of a defined population is made. This inventory is used as a base to develop hypotheses, which in turn may form the base for experimental research projects. Although this scientific approach is being associated nowadays with genetic mapping projects, it has long been used in other disciplines; many epidemiological cohorts have been established without an underlying hypothesis only to provide data, from which hypotheses are derived.

The avalanche of new molecular genetic techniques and discoveries has provided numerous new approaches for endocrine research that focus on the systematic inventory of all genes, gene products, and their functions in an organism. This process is generally called genomics. However, the term genomics is better used to indicate the process of *mapping* and *sequencing* of all genes of an organism. *Structural genomics* has been used in the past as a synonym to genomics but is defined nowadays as the process in which the three-dimensional structure of all gene *products* are characterized. It would be better to use the term *structural proteomics* for this process, as structural genomics also involves the structural characterization of other, nonpeptide macromolecules, such as RNA molecules (which in turn is called ribonomics). In addition, *functional genomics* defines the process in which the function of macromolecules is characterized. A better term to characterize the function of proteins is *functional proteomics*.

4.2. Genomics

The first published whole genome was that of *Hemophilus influenzae* in 1995 *(32)*, followed in 1998 by the first complete mapping of the genome of a multicellular organism, *Caenorhabtitis elegans (32)*. Only 3 years passed between the publication of the whole genome of *C. elegans* and the completion of the first draft of the human genome *(34)*. The International Human Genome Sequencing Consortium estimates that there are approx 30,000–40,000 protein-encoding genes in humans *(35)*.

4.3. Reverse Endocrinology

In contrast to the aforementioned classical approach in endocrine research, reverse endocrinology involves the use of the genomic inventory of an organ or organism as a starting point to discover new receptors, new hormones, and, consequently, new physiological pathways. This approach is illustrated by the discovery of a new class of orphan nuclear receptors (*see* **Subheading 5.**). In addition, more than 400 G-protein-coupled receptors have been identified *(36)*, which can be visualized at http://protomap. cornell.edu/Amino/Clusters.

4.4. Proteomics

However, mapping and sequencing of genes is a beginning rather than an endpoint of reverse endocrinology. Relatively little is still known about the products of these genes. In *C. elegans*, the function of only 50% of the genes is known. In eukaryotes, including humans, this percentage is much lower. Understanding the function of the genes and gene products is limited by the fact that the approx 40,000 genes encode 370 million proteins. Several mechanisms may be involved. Intracellularly, variations in reading frames within one gene lead to different mRNAs. Posttranscriptional processing includes alternative splicing, which may lead to up to 100 different proteins from one mRNA *(37)*. The coupling of nonprotein groups (phosphorylation, glycation, prenylation) and oligomerization may lead to structurally and functionally different proteins. After secretion of the protein by the cell, proteolytic enzymes may convert the native protein into functionally different molecules. An important example is the conversion of plasminogen into plasmin and angiostatin, which have biologically different functions.

Consequently, a particular DNA sequence cannot simply be translated into a protein sequence, let alone protein structure or function. An additional problem is that there is no straight relationship between the primary amino acid sequence and protein function: proteins with a comparable primary structure may have entire different functions. On the other hand, proteins with a comparable tertiary structure, but with a different sequence, may have comparable functions, such as myoglobin and hemoglobin *(38)*. The importance of protein structures is also illustrated by the fact that knowledge of the tertiary structures of hormones and their receptors is required for the design of hormone analogs with enhanced activity or with superior pharmacokinetic profiles. Examples of the importance of the structure–function relationship are the development of short-acting designer insulin analogs, which are a major improvement in the treatment of diabetes mellitus *(39)* or the development of somatostatin analogs, which are used to treat various neuroendocrine and endocrine tumors *(40)*. Often, attempts to design superagonists result in the discovery of antagonists, as is the case in the growth hormone receptor antagonist pegvisomant, which was introduced recently as an alternative therapy for growth hormone excess *(41)*. These examples indicate that although exact knowledge of the molecular structure of hormones and their receptors has not always been necessary to develop therapies for endocrine disorders, assessment of structure–function relationships is important to improve existing therapies and to develop tailor-made molecules with enhanced activity compared with natural ligands. Moreover, analysis of structural homologies between hormones or receptors with known and unknown function is an important tool to predict the function of newly discovered molecules.

Thus, although the Human Genome Project has offered enormous amounts of data on the primary sequences of proteins, the ultimate goal of characterizing protein function can only be accomplished by additional strategies. All proteins characterized so far are registered in the Protein Data Bank (PDB). The current state of the PDB can be viewed at www.rcsb.org/pdb *(42)*. To date, approx 19,000 protein structures have been deposited. Other important proteomic databases are the one maintained by the Swiss Institute of Bioinformatics at the University of Geneva, which maintains the ExPASy (Expert Protein Analysis System) proteomics server (http://www.expasy.org/), a source connected to many other informative databases. These database interfaces have tools designed to overcome the enormous computational challenges associated with proteome analysis.

4.4.1. Structural Proteomics

The premise of structural proteomics is that for the identification of protein function it is necessary to identify the 3-D structure. As the 3-D structure cannot be derived merely from gene sequences, this requires a major effort, more complicated than the Human Genome Project. The determination of 3-D protein structures requires high-throughput X-ray and nuclear magnetic resonance (NMR) crystallography *(43,44)*, which is much more complicated than the linear sequencing of DNA. Two strategies have been initiated to achieve these goals.

The first strategy is to characterize all possible protein *structures*, as reviewed in **ref. 45**. The Structural Genomics Project aims at determination of the 3-D structure of all proteins. This can be achieved in four steps: organize known protein sequences into families, select family representatives as targets, solve the 3-D structure of targets by X-ray crystallography or NMR spectroscopy, and build models for other proteins by homology to solved 3-D structures.

Protein structures are determined by folds. Although the number of human proteins is estimated to be approx 370,000, the number of folds may be about 10,000–30,000 *(46,47)*. The hypothesis is that when all folds are known, protein structures can be predicted from this database without the need to determine the 3-D structure of each individual protein. About 800 folds are deposited in the PDB. According to the presence of specific folds, proteins can be categorized into several families. In this approach, the purpose is to identify new folds. To realize this, genes are chosen with unknown functions and without homology with known genes or gene products. cDNAs of these genes are expressed in bacteria, and the 3-D structures of the gene products are characterized in high-throughput X-ray and NMR systems. Several consortia have been formed to perform this enormous task. Their approaches and their achievements can be viewed at www.structuralgenomics.org, www.rcsb.org/pdb, and www.ncbi.nlm.nih.gov/Structure/MMDB/mmdb.shtml.

Another strategy is not to aspire to complete the structure of all human proteins, but instead to target potentially relevant structures. In contrast to the first approach, relevant genes are not genes with an unknown function or homology, but genes that may be involved in physiological processes of interest. Admittedly, this approach may be partially based on hypotheses regarding the potential involvement of genes in biological processes and as such does not entirely fulfil the definition of genomics. However, this approach has already proved successful in the identification of novel proteins as targets for drug development. One example is the development of anti-HIV drugs *(48)*. After whole genome mapping of HIV, genes potentially encoding proteases have been expressed and the 3-D structures of their products determined. Once the critical regions in these structures were identified, protease inhibitors could be designed that have proved to be an enormous breakthrough in AIDS. Major examples of the significance of reverse endocrinology are the discovery of orphan nuclear receptors and the subsequent development of drugs serving as ligands for these receptors, as is reviewed in **Subheading 5.**

4.4.2. Functional Proteomics

Once the structure of proteins is determined in structural proteomics, their functions need to be elucidated (functional proteomics). This can be achieved by the following strategies:

> *Homology searches.* When comparable folds are present in a protein with a known function, the protein of interest is probably involved in a comparable physiologic system. Alternatively, the protein can be screened systematically in ligand binding assays, in enzymatic assays, or in protein hybridization arrays.
>
> *Phylogenetic profile searches.* Proteins and protein folds that have been conserved through generations in different species may have important functional relationships. For example, the neuropeptide Y family of peptides consists of neuropeptide Y (NPY), which is expressed in the central and peripheral nervous systems, as well as peptide YY (PYY) and pancreatic polypeptide (PP), which are gut endocrine peptides. NPY and PYY are present in all vertebrates, whereas PP probably arose as a copy of PYY in an early tetrapod ancestor. NPY is one of the most conserved peptides during evolution. PYY is more

variable, whereas PP may be the most rapidly evolving neuroendocrine peptide among tetrapods *(49)*. Another example is gonadotrophin-releasing hormone (GnRH) and its receptors, of which multiple variants have been identified in vertebrates *(50)*.

Rosetta stone linkage. Proteins that are separate molecules in one species but one molecule in other species are likely to be functionally related. An example in endocrinology is adrenocorticotropic hormone (ACTH) in humans, ACTH also contains melanocyte-stimulating hormone (MSH) sequences, whereas MSH is secreted as a separate hormone in other species *(51)*.

Gene-neighbor profiles. Genes that are neighbors in different genomes may have related functions.

All these methods are used to designate protein function. If a functional relationship is suggested by more than one method, the probability is high that this relationship exists in reality. Thus, functional proteomics can best be regarded as an integrative process, combining the data from structural proteomics with data collected from statistical and experimental approaches. This integrative process may yield functional protein networks, as illustrated by Eisenberg et al. *(52)* and by GenMapp, which provides a graphic representation of the relative expression of selected groups of genes categorized by their known positions in physiological pathways (http://www.genmapp.org).

5. An Example of Reverse Endocrinology: Orphan Nuclear Receptors

Nuclear receptors have traditionally been assigned to thyroid hormone, steroids (glucocorticoids, mineralocorticoids, estradiol, progesterone, testosterone, vitamin D) and all-*trans*-retinoic acid (ATRA) *(53)*. The potential importance of nuclear receptors led to an effort in which DNA libraries were screened for homologs to DBD or LBD sequences of known nuclear receptors. In 1988, the first cDNA clones were identified, encoding proteins that were structurally related to nuclear receptors *(54)*.

To date, more than 50 nuclear receptor-like proteins have been identified in several species, as a result of the genome projects. The superfamily of nuclear receptors is given at http://protomap.cornell.edu/Amino/Clusters. One of the first results was the identification of multiple receptors for already known hormones: two thyroid hormone receptors *(55)*, two estrogen receptors, and three retinoic acid receptors (RARs) appeared to be present *(56)*. However, for multiple nuclear receptor-like proteins, the ligands and, consequently, the physiological role of most of these nuclear receptors were not known. Thus they were referred to as orphan nuclear receptors.

As the classical nuclear receptors are involved in important biological processes, the discovery of orphan nuclear receptors led to the assumption that novel physiological pathways could be identified, with subsequent opportunities to clarify the pathogenesis of so far unexplained disorders and to identify new targets for drug design.

Many research groups have followed these leads, which has revealed novel hormonal pathways and led to the development of new categories of drugs. The first example of reverse endocrinology leading to a new hormone was the identification of the retinoid x-receptors (RXRs). These receptors were found after screening libraries for homologs to DBD of RAR-α *(57)*. RXR appeared to play an important role in the signaling pathways of nuclear receptors. Most nuclear receptors can form heterodimers with RXR, which provides additional routes to activate and modulate nuclear receptor-induced gene transcription. Later on, 9-cis-retinoic acid (9-cisRA), a metabolite of ATRA, was found to be the ligand for this receptor *(58)*. Thus, 9-cisRA was the first hormone found by reverse endocrinology. This discovery may have clinical implications: synthetic RXR-specific activators have been developed that have beneficial in vitro effects in breast cancer *(59)*.

Another remarkable class of orphan nuclear receptors, leading to the discovery of a new class of drugs, is the peroxisome proliferator-associated receptors (PPAR) (reviewed in **ref. 60**). These genes have also been found by screening cDNA libraries with DBD sequences. The nomenclature is the result of the finding that a class of compounds called peroxisome proliferators could induce their activity. Four different PPARs have been identified: α, β, γ, and δ. In subsequent experiments, the distribution of the expression of the PPAR genes in different organs was established. For example, the expression of PPAR-α in the liver, heart, and brown adipose tissue suggested an involvement in fatty acid metabolism. This was proved by several experimental strategies including PPAR-α knockout mice, which develop hepatic triglyceride accumulation owing to their inability to oxidize fatty acids and (interestingly) to develop insulin resistance *(61)*. PPAR-γ is mainly distributed in white and brown adipose tissue, indicating a role in adipocyte physiology. PPAR-γ is involved in the differentiation of fibroblasts to adipocytes and probably also in the differentiation of other cell types, including macrophages *(62,63)*.

Ligands for PPAR receptors were identified in cells transfected with the chimeric PPAR-glucocorticoid receptor gene. The first ligands discovered were fibric acid derivatives, drugs used to lower cholesterol and triglycerides, a mechanism of action that was not previously understood. Fibric acid derivatives appeared to be strong activators of PPAR-α. Thereafter, several species of fatty acids and fatty acid derivatives were found to be activated by PPAR-α and -γ. This is a new concept in endocrinology, because, according to the definition of hormones, fatty acids can now be regarded as hormones! Other classes of PPAR-α and -γ activators are eicosanoids and prostaglandin analogs, which are mediators in inflammation. Probably the most important ligands for PPAR-γ are drugs with antidiabetic and hypolipidemic properties, called thiazolidinedione (TZD) derivatives *(64)*. These drugs are currently prescribed to a large number of patients with type 2 diabetes mellitus, indicated by the fact that these drugs gained a multibillion dollar market in the United States. Finally, it appeared that these PPAR nuclear receptors are involved not only in metabolism (of fatty acids) but also in many other fundamental processes like cell differentiation and inflammation. The integration of traditionally distinct physiological processes in one class of receptors offers important perspectives for concepts dealing with the pathogenesis and treatment of complex disorders associated with these systems, including diabetes mellitus, atherosclerosis, and cancer. From a historical perspective, it is interesting that the drugs acting through these orphan receptors were developed before the natural ligands were identified.

6. Conclusion

The era of genomics and proteomics has provided endocrinology with powerful tools to dissect the molecular (patho)physiology that is potentially involved in health and disease. Meanwhile, traditional endocrine research will remain strong for several reasons: traditional endocrine approaches provide the tools to evaluate the discoveries of reverse endocrinology. In addition, structural proteomics has methodological and conceptual limitations: it is still difficult to isolate membrane-bound proteins for structural analyses *(65)*. As membrane-bound receptors are important players in many endocrine pathways, this is an important problem. Another limitation is that many hormones are nonpeptide hormones. The structure and function of complex nonpeptide hormones can therefore not be clarified by genomics alone: the mirror image of a protein receptor is necessary to delineate their structure. The reiterative combination of innovative genomics and proteomics with classical endocrine approaches will enable us to maintain endocrinology as a front runner in biological research and to develop innovative therapeutic approaches in a continuing interaction between bed and bench.

References

1. Berthold, A. A. (1848) Transplantation der Hoden. *Arch. Anat. Physiol. Wiss. Med.* 42–46.
2. Starling, E. H. (1905) On the chemical correlation of the functions of the body. *Lancet* **2,** 339–341.
3. Brokken, L. J., Scheenhart, J. W., Wiersinga, W. M., and Prummel, M. F. (2001) Suppression of serum TSH by Graves' Ig: evidence for a functional pituitary TSH receptor. *J. Clin. Endocrinol. Metab.* **86,** 4814–4817.
4. Kulkarni, R. N., Bruning, J. C., Winnay, J. N., Postic, C., Magnuson, M. A., and Kahn, C. R. (1999) Tissue-specific knockout of the insulin receptor in pancreatic beta cells creates an insulin secretory defect similar to that in type 2 diabetes. *Cell* **96,** 329–339.
5. Ebina, Y., Ellis, L., Jarnagin, K., et al. (1985) The human insulin receptor cDNA: the structural basis for hormone-activated transmembrane signalling. *Cell* **40,** 747–758.
6. Ebina, Y., Edery, M., Ellis, L., et al. (1985) Expression of a functional human insulin receptor from a cloned cDNA in Chinese hamster ovary cells. *Proc. Natl. Acad. Sci. USA* **82,** 8014–8018.
7. Luo, R. Z., Beniac, D. R., Fernandes, A., Yip, C. C., and Ottensmeyer, F. P. (1999) Quaternary structure of the insulin-insulin receptor complex. *Science* **285,** 1077–1080.
8. Kulkarni, R. N., Winnay, J. N., Daniels, M., et al. (1999) Altered function of insulin receptor substrate-1-deficient mouse islets and cultured beta-cell lines. *J. Clin. Invest.* **104,** R69–R75.
9. Frasier, S. D. (1983) Human pituitary growth hormone (hGH) therapy in growth hormone deficiency. *Endocr. Rev.* **4,** 155–170.
10. Home, P. D. and Alberti, K. G. (1982) The new insulins. Their characteristics and clinical indications. *Drugs* **24,** 401–413.
11. Weintraub, B. D. and Szkudlinski, M. W. (1999) Development and in vitro characterization of human recombinant thyrotropin. *Thyroid* **9,** 447–450.
12. Moreno, J. C., Pauws, E., van Kampen, A. H., Jedlickova, M., de Vijlder, J. J., and Ris-Stalpers, C. (2001) Cloning of tissue-specific genes using serial analysis of gene expression and a novel computational substraction approach. *Genomics* **75,** 70–76.
13. Rainey, W. E., Carr, B. R., Wang, Z. N., and Parker, C. R. J. (2001) Gene profiling of human fetal and adult adrenals. *J. Endocrinol.* **171,** 209–215.
14. Halvorsen, Y. D., Wilkison, W. O., and Briggs, M. R. (2000) Human adipocyte proteomics—a complementary way of looking at fat. *Pharmacogenomics* **1,** 179–185.
15. Flores-Morales, A., Stahlberg, N., Tollet-Egnell, P., et al. (2001) Microarray analysis of the in vivo effects of hypophysectomy and growth hormone treatment on gene expression in the rat. *Endocrinology* **142,** 3163–3176.
16. Webb, G. C., Akbar, M. S., Zhao, C., and Steiner, D. F. (2001) Expression profiling of pancreatic beta-cells: glucose regulation of secretory and metabolic pathway genes. *Diabetes* **50 (Suppl. 1),** S135–S136.
17. Clement, K., Viguerie, N., Diehn, M., et al. (2002) In vivo regulation of human skeletal muscle gene expression by thyroid hormone. *Genome Res.* **12,** 281–291.

18. Feng, X., Jiang, Y., Meltzer, P., and Yen, P. M. (2000) Thyroid hormone regulation of hepatic genes in vivo detected by complementary DNA microarray. *Mol. Endocrinol.* **14,** 947–955.

19. Viguerie, N., Millet, L., Avizou, S., Vidal, H., Larrouy, D., and Langin, D. (2002) Regulation of human adipocyte gene expression by thyroid hormone. *J. Clin. Endocrinol. Metab.* **87,** 630–634.

20. Thompson, B. J., Shang, C. A., and Waters, M. J. (2000) Identification of genes induced by growth hormone in rat liver using cDNA arrays. *Endocrinology* **141,** 4321–4324.

21. Watanabe, H., Suzuki, A., Mizutani, T., et al. (2002) Genome-wide analysis of changes in early gene expression induced by oestrogen. *Genes Cells* **7,** 497–507.

22. Waghray, A., Feroze, F., Schober, M. S., et al. (2001) Identification of androgen-regulated genes in the prostate cancer cell line LNCaP by serial analysis of gene expression and proteomic analysis. *Proteomics* **1,** 1327–1338.

23. Olesen, C., Larsen, N. J., Byskov, A. G., Harboe, T. L., and Tommerup, N. (2001) Human FATE is a novel X-linked gene expressed in fetal and adult testis. *Mol. Cell Endocrinol.* **184,** 25–32.

24. Nadler, S. T., Stoehr, J. P., Schueler, K. L., Tanimoto, G., Yandell, B. S., and Attie, A. D. (2000) The expression of adipogenic genes is decreased in obesity and diabetes mellitus. *Proc. Natl. Acad. Sci. USA* **97,** 11371–11376.

25. Sreekumar, R., Halvatsiotis, P., Schimke, J. C., and Nair, K. S. (2002) Gene expression profile in skeletal muscle of type 2 diabetes and the effect of insulin treatment. *Diabetes* **51,** 1913–1920.

26. Evans, C. O., Young, A. N., Brown, M. R., et al. (2001) Novel patterns of gene expression in pituitary adenomas identified by complementary deoxyribonucleic acid microarrays and quantitative reverse transcription-polymerase chain reaction. *J. Clin. Endocrinol. Metab.* **86,** 3097–3107.

27. van't Veer, L., Dai, H., van de Vijver, M. J., et al. (2002) Gene expression profiling predicts clinical outcome of breast cancer. *Nature* **415,** 530–536.

28. Huang, Y., Prasad, M., Lemon, W. J., et al. (2001) Gene expression in papillary thyroid carcinoma reveals highly consistent profiles. *Proc. Natl. Acad. Sci. USA* **98,** 15,044–15,049.

29. Klade, C. S. (2002) Proteomics approaches towards antigen discovery and vaccine development. *Curr. Opin. Mol. Ther.* **4,** 216–223.

30. Albrektsen, T., Frederiksen, K. S., Holmes, W. E., Boel, E., Taylor, K., and Fleckner, J. (2002) Novel genes regulated by the insulin sensitizer rosiglitazone during adipocyte differentiation. *Diabetes* **51,** 1042–1051.

31. Quackenbush, J. (2001) Computational analysis of microarray data. *Nat. Rev. Genet.* **2,** 418–427.

32. Fleischmann, R. D., Adams, M. D., White, O., et al. (1995) Whole-genome random sequencing and assembly of *Haemophilus influenzae* Rd. *Science* **269,** 496–512.

33. Anonymous (1998) Genome sequence of the nematode *C. elegans*: a platform for investigating biology. The *C. elegans* Sequencing Consortium. *Science* **282,** 2012–2018.

34. Lander, E. S., Linton, L. M., Birren, B., et al. (2001) Initial sequencing and analysis of the human genome. *Nature* **409,** 860–921.

35. Baltimore, D. (2001) Our genome unveiled. *Nature* **409,** 814–816.

36. Howard, A. D., McAllister, G., Feighner, S. D., et al. (2001) Orphan G-protein-coupled receptors and natural ligand discovery. *Trends. Pharmacol. Sci.* **22,** 132–140.

37. Black, D.L. (2000) Protein diversity from alternative splicing: a challenge for bioinformatics and post-genome biology. *Cell* **103,** 367–370.

38. Eisen, M. B., Spellman, P. T., Brown, P. O., and Botstein, D. (1998) Cluster analysis and display of genome-wide expression patterns. *Proc. Natl. Acad. Sci. USA* **95,** 14863–14868.

39. Barnett, A. H. and Owens, D. R. (1997) Insulin analogues. *Lancet* **349,** 47–51.

40. Harris, A. G. (1994) Somatostatin and somatostatin analogues: pharmacokinetics and pharmacodynamic effects. *Gut* **35,** S1–S4.

41. Trainer, P. J., Drake, W. M., Katznelson, L., et al. (2000) Treatment of acromegaly with the growth hormone-receptor antagonist pegvisomant. *N. Engl. J. Med.* **342,** 1171–1177.

42. Berman, H. M., Bhat, T. N., Bourne, P. E., et al. (2000) The Protein Data Bank and the challenge of structural genomics. *Nat. Struct. Biol.* **7 (Suppl.),** 957–959.

43. Montelione, G. T., Zheng, D., Huang, Y. J., Gunsalus, K. C., and Szyperski, T. (2000) Protein NMR spectroscopy in structural genomics. *Nat. Struct. Biol.* **7 (Suppl.),** 982–985.

44. Yee, A., Chang, X., Pineda-Lucena, A., et al. (1902) An NMR approach to structural proteomics. *Proc. Natl. Acad. Sci. USA* **99,** 1825–1830.

45. Linial, M. and Yona, G. (2000) Methodologies for target selection in structural genomics. *Prog. Biophys. Mol. Biol.* **73,** 297–320.

46. Montelione, G. T. (2001) Structural genomics: an approach to the protein folding problem. *Proc. Natl. Acad. Sci. USA* **98,** 13,488–13,489.

47. Moult, J. and Melamud, E. (2000) From fold to function. *Curr. Opin. Struct. Biol.* **10,** 384–389.

48. Lin, J. H., Ostovic, D., and Vacca, J. P. (1998) The integration of medicinal chemistry, drug metabolism, and pharmaceutical research and development in drug discovery and development. The story of Crixivan, an HIV protease inhibitor. *Pharm. Biotechnol.* **11,** 233–255.

49. Larhammar, D. (1996) Evolution of neuropeptide Y, peptide YY and pancreatic polypeptide. *Regul. Pept.* **62,** 1–11.

50. Neill, J. D. (2002) Mammalian gonadotropin-releasing hormone (GnRH) receptor subtypes. *Arch. Physiol. Biochem.* **110,** 129–136.

51. Asa, S. L. and Ezzat, S. (1999) Molecular determinants of pituitary cytodifferentiation. *Pituitary* **1,** 159–168.

52. Eisenberg, D., Marcotte, E. M., Xenarios, I., and Yeates, T. O. (2000) Protein function in the post-genomic era. *Nature* **405,** 823–826.

53. Evans, R. M. (1988) The steroid and thyroid hormone receptor superfamily. *Science* **240,** 889–895.

54. Giguere, V., Yang, N., Segui, P., and Evans, R. M. (1988) Identification of a new class of steroid hormone receptors. *Nature* **331,** 91–94.

55. Thompson, C. C., Weinberger, C., Lebo, R., and Evans, R. M. (1987) Identification of a novel thyroid hormone receptor expressed in the mammalian central nervous system. *Science* **237,** 1610–1614.

56. Mosselman, S., Polman, J., and Dijkema, R. (1996) ER beta: identification and characterization of a novel human estrogen receptor. *FEBS Lett.* **392,** 49–53.

57. Mangelsdorf, D. J., Ong, E. S., Dyck, J. A., and Evans, R. M. (1990) Nuclear receptor that identifies a novel retinoic acid response pathway. *Nature* **345,** 224–229.

58. Levin, A. A., Sturzenbecker, L. J., Kazmer, S., et al. (1992) 9-Cis retinoic acid stereoisomer binds and activates the nuclear receptor RXR alpha. *Nature* **355,** 359–361.

59. Wu, K., Kim, H. T., Rodriquez, J. L., et al. (2000) 9-cis-Retinoic acid suppresses mammary tumorigenesis in C3(1)-simian virus 40 T antigen-transgenic mice. *Clin. Cancer Res.* **6,** 3696–3704.

60. Desvergne, B. and Wahli, W. (1999) Peroxisome proliferator-activated receptors: nuclear control of metabolism. *Endocr. Rev.* **20,** 649–688.

61. Kersten, S., Seydoux, J., Peters, J. M., Gonzalez, F. J., Desvergne, B., and Wahli, W. (1999) Peroxisome proliferator-activated receptor alpha mediates the adaptive response to fasting. *J. Clin. Invest.* **103,** 1489–1498.

62. Auwerx, J. (1999): PPARgamma, the ultimate thrifty gene. *Diabetologia* **42,** 1033–1049.

63. Kersten, S., Desvergne, B., and Wahli, W. (2000) Roles of PPARs in health and disease. *Nature* **405,** 421–424.
64. Olefsky, J. M. (2000) Treatment of insulin resistance with peroxisome proliferator-activated receptor gamma agonists. *J. Clin. Invest.* **106,** 467–472.
65. Quick, M. and Wright, E. M. (2002) Employing Escherichia coli to functionally express, purify, and characterize a human transporter. *Proc. Natl. Acad. Sci. USA* **99,** 8597–8601.

28

Proteomics in Plant Biology

Christina Mihr and Hans-Peter Braun

1. Introduction

As in microbiology, zoology, and medicine, proteomics in plant biology has become more and more important for investigating molecular physiology. Proteomics is based on (1) protein extraction, (2) high-resolution protein separations, (3) efficient protein identification, and (4) bioinformatics for data processing. Several prerequisites for proteome analyses were very much improved during the last years, for example, the introduction of immobilized pH gradient gels for isoelectric focusing (IEF) and protein mass spectrometry (MS). Furthermore, the recent completion of genome sequencing projects for *Arabidopsis thaliana (1)* and rice *(2,3)* is an important step in facilitating efficient protein identifications. Proteome analyses have been used successfully for several physiological and genetic studies in plants (reviewed in **refs.** *4* and *5*).

Although different procedures have been developed for protein separations (such as various types of chromatography, capillary electrophoresis, and direct MS), two-dimensional gel electrophoresis (2-DGE) using isoelectric focusing and sodium dodecyl sulfate polyacrylamide gel electrophoresis (SDS-PAGE) still form the most important separation technology for proteome analyses. However, sample preparation for 2-D resolution of plant proteins is difficult owing to special features of plant cells. In contrast to several other eukaryotic lineages, plant cells have a very robust cell wall, plastids, and large acidic vacuoles that contain secondary plant metabolites. The nature and the concentration of these metabolites depend on the genotype but also on the tissue and the age of a plant. Furthermore, pigments and phenolic compounds are characteristic metabolites in plant cells. Like nucleic acids and lipids, all these substances can interfere with IEF and lead to high background and streaking on 2-D gels, which make image analysis complicated. Nevertheless, some procedures were developed that give good results for 2-DGE of complex protein mixtures from plant tissues. On the other hand, it is very difficult to provide general guidelines for sample preparation of proteins from plant cells because of the great diversity of plant tissues. Therefore, optimization of procedures for different plant tissue is absolutely necessary for good 2-DGE results.

In this chapter, we describe our experience with 2-D separations of protein extracts from two tissues of rapeseed (5-d-old etiolated rapeseed seedlings and rapeseed flower buds) and potato (2-wk-old dark grown potato seedlings and potato leaves). We used modified versions of two formerly described extraction procedures, which show different suitability for different tissues: (1) protein precipitation with 10% trichloroacetic

From: *Handbook of Proteomic Methods*
Edited by: P. Michael Conn © Humana Press Inc., Totowa, NJ

acid (TCA) in acetone *(6)* and (2) protein precipitation with 0.1 M ammonium acetate in methanol after phenol extraction *(7)*.

2. Methods
2.1. Disruption of Plant Cells

Several different methods of cell disruption are used in biology. More rigorous methods with respect to shear forces are necessary for tissues with robust cell walls, e.g., sonication, grinding, or use of a French press or a Waring blender. We preferred disruption of plant material frozen in liquid nitrogen by hand grinding with a mortar and pestle or grinding of frozen material with a mixer mill. Tissue is ground into a fine powder using a prechilled mortar and pestle or using the nitrogen-chilled steel cylinders of a mixer mill. If small tissue amounts have to be prepared, this kind of pulverization is fast and easy to perform.

2.2. Extraction and Precipitation of Protein Fractions from Plant Cells

Protein precipitation is an optional step in sample preparation for 2-DGE. It is also possible to homogenize plant material directly in lysis buffer for IEF. 2-DGE was successfully carried out with directly lysed rice shoots and *Arabidopsis* seeds *(8,9)*. However, depending on the kind of plant tissue and the composition of the lysis buffer, there is some risk in using large amounts of nonproteinous material, which interferes with IEF. Direct lysis of *Arabidopsis* cells from suspension cell cultures gave unsatisfactory results (Mihr and Braun, unpublished data). We therefore preferred precipitation of proteins before suspension of plant proteins in lysis buffer for IEF. Two procedures gave very good results in combination with 2-DGE: TCA/acetone precipitation and ammonium acetate/methanol precipitation after phenol extraction.

2.2.1. TCA/Acetone Precipitation

The precipitation of proteins from plant extracts by TCA/acetone was previously described by Darmerval et al. *(6)* and Tsugita and Kamo *(10)*.

1. First, 500 mg of pulverized rapeseed or potato tissue is suspended in 5 mL TCA precipitation solution [10% TCA in acetone including 20 mM dithiothreitol (DTT)].
2. After incubation for 45 min at $-20°C$, the suspension is centrifuged for 15 min at 35,000g and 4°C.
3. The pellet is resuspended in washing solution [acetone including 20 mM DTT, 1 mM phenymethylsulfanyl fluoride (PMSF); volume: 10 times the weight of the pellet] and placed at $-20°C$ for 1 h, followed by a centrifugation for 15 min by 35,000g at 4°C.
4. The pellet is again resuspended in washing solution and centrifuged as described above.
5. After air drying, the weight of the pellet is determined.
6. Subsequently, the proteins are suspended in lysis buffer for IEF (8 M urea, 4% (v/v) Triton X-100, 40 mM Tris base, 50 mM DTT) at a concentration of approx 1 mg/40 µL.
7. Protein samples suspended in lysis buffer for IEF should be incubated for 1 h at room temperature with continuous shaking followed by centrifugation for 20 min by 18,000g to pellet insoluble material.
8. Afterward samples can be used directly for IEF (*see* below) or frozen and stored at $-80°C$ for later analyses.

2.2.2. Phenol Extraction and Ammonium Acetate/Methanol Precipitation

This method is a modification of the extraction procedure previously described by Hurkmann and Tanaka *(11)*.

1. Approximately 200 mg of homogenized potato leaves are suspended in 0.75 mL extraction buffer (700 m*M* saccharose, 500 m*M* Tris-HCl, 50 m*M* EDTA, 100 m*M* KCl, 2% β-mercaptoethanol, 2 m*M* PMSF) and incubated for 10 min on ice.
2. Afterward, an equal volume of water-saturated phenol is added.
3. The solution is incubated on a shaker for 10 min at room temperature.
4. Insoluble material, the aqueous phase, and the organic phase are separated by centrifugation for 10 min at 11,000*g* and 4°C.
5. The pellet and the aqueous phase (bottom) are discarded, and the phenolic phase (top) is recovered and re-extracted with an equal volume of extraction buffer.
6. Subsequently centrifugation for phase separation is repeated.
7. For protein precipitation, 5 vol of 0.1 *M* ammonium acetate in methanol is added to the recovered phenol phases.
8. After incubation for at least 4 h at –20°C, precipitated proteins are collected by centrifugation for 5 min at 18,000*g* and 4°C and afterward washed three times with 0.1 *M* ammonium acetate in methanol and finally with 80% acetone.
9. For collection of proteins, samples are centrifuged for 1–2 min at 18,000*g* and 4°C after each washing step.
10. Finally, the protein samples are air dried, suspended in lysis buffer for IEF [8 *M* urea, 4% (v/v) Triton X-100, 40 m*M* Tris base, 50 m*M* DTT] at a concentration of approx 1 mg/40 μL and treated as described in **Subheading 2.2.1., steps 7** and **8**.

2.3. Separation of Proteins by 2-DGE

2-DGE sorts proteins in two steps according to their isoelectric points (first dimension) and their molecular weights (second dimension). IEF was very much simplified by the introduction of immobilized pH gradient (IPG) gels, which are offered by several manufacturers in ready-made dried form *(11)*. These gel strips have very stable pH gradients and also offer sample loading during rehydration, which is more efficient than sample loading by electric forces. Different IEF units are commercially available and allow good results in protein separation (for comparison, *see* **ref. *12***). The following instructions are given for the IPGphor Isoelectric Focusing System of Amersham Pharmacia Biotech, which includes an integrated programmable 8000-V power supply and ceramic stripholders for rehydration and IEF in one step, but they can be easily transferred to the IEF units from other manufacturers.

1. For IEF of total protein from plant tissue, 18-cm-long ready-made IPG strips with a nonlinear pH gradient from pH 3 to 10 were chosen.
2. Protein samples (about 250 μg protein in 10 μL lysis buffer for IEF (*see* **Subheading 2.2.1., step 6** and **Subheading 2.2.2., step 10**) are combined with 340 μL rehydration solution [8 *M* urea, 2% (v/v) Triton X-100, 70 m*M* DTT, 0.5% IPG buffer, and a few grains of bromophenol blue] and afterward directly loaded into the IPG stripholders.
3. Subsequently the IPG strips are placed into the stripholders (gel side down), and an IPG cover fluid is added to minimize evaporation during IEF.
4. IEF is carried out under the following conditions: rehydration for 12 h at 30 V, initial focusing for 1 h at 500 V, and further focusing for 1 h at 500–1000 V (gradient), for 4 h at

1000–8000 V (gradient) and for 6 h at 8000 V (for further details and variations, *see* **ref. *13***).

5. Before separation on a second gel dimension, IPG strips are incubated with equilibration solution for two times 15 min [equilibration solution: 50 m*M* Tris-HCl, pH 8.8, 6 *M* urea, 30% (v/v) glycerol, 2% (w/v) SDS supplemented with 1% (w/v) DTT in the first equilibration and 2.5–4.8% (w/v) iodoacetamide in the second equilibration]. The second equilibration step reduces point streaking owing to DTT and other artifacts on silver-stained 2-D gels.

6. The gel strips are then horizontally transferred onto SDS polyacrylamide gels for protein separation in the second dimension.

7. SDS-PAGE is carried out according to Laemmli *(14)* or Schägger and von Jagow *(15)*. Gels are silver stained according to Heukeshoven and Dernick *(16)*.

2.4. Protein Identifications

Proteins separated on 2-D gels can be identified by several different procedures, which are generally applicable, no matter whether proteins are from plant, animal, or microbial sources. If antibodies are available, proteins can be identified by immunoblotting. Blotted proteins can also be directly sequenced by cyclic Edman degradation. However, in recent times the most powerful protocols for protein identification have been based on mass spectrometry. As the development of plant-specific procedures is not necessary, methods can be taken from the general literature (reviewed in **ref *17***).

3. Results

TCA/acetone precipitation is often used in plant proteomics to extract total proteins from different tissues such as pine needles *(18)*, maize leaves *(19)*, clover roots *(20)*, or rapeseed seedlings and flower buds *(21)*. This method is useful for (1) inactivation of proteases to minimize protein degradation, (2) the enrichment of alkaline proteins, and (3) the removal of interfering substances *(11)*.

In our experiments, TCA/acetone precipitation led to high-quality 2-D gels of total protein extracts from etiolated rapeseed seedlings, rapeseed flower buds, and potato seedlings (**Fig. 1A, B,** and **D**). However, it turned out to be more difficult to obtain high-quality 2-D gels of total protein from green tissues after TCA/acetone precipitation, especially if leaves were fully differentiated. Although the quality of 2-D gels of total proteins from rapeseed flower buds is satisfactory (**Fig. 1D**), there is more background and more vertical streaking compared with the 2-D gels for etiolated seedlings. Likewise, TCA/acetone precipitation of total potato leaf proteins led to 2-D gels with higher background and reduced sharpness (**Fig. 1C**). In contrast, phenol extraction and subsequent ammonium acetate/methanol precipitation of potato leaf protein gave much better results upon 2-D analysis (**Fig. 1E**).

In summary, TCA/acetone precipitation is the most common procedure for protein purification prior to 2-DGE. Contaminating cell wall and fiber components of the precipitate are removed by centrifugation after resuspension of the precipitate in lysis solution for IEF and subsequent centrifugation *(10)*. However, other contaminants are sometimes present and disturb IEF. In these cases, extraction of proteins in high-salt buffers and phenol and subsequent precipitation of proteins by the ammonium acetate/methanol seems to be a good solution, because most contaminating compounds are removed by the first centrifugation step. This method also led to good results for

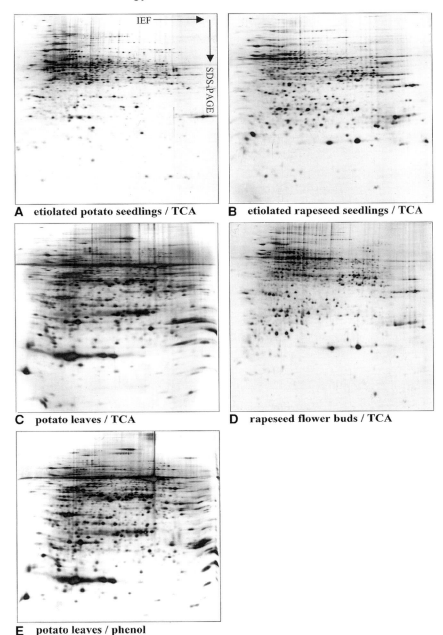

A etiolated potato seedlings / TCA

B etiolated rapeseed seedlings / TCA

C potato leaves / TCA

D rapeseed flower buds / TCA

E potato leaves / phenol

Fig. 1. Two-dimensional resolution of total protein from different plant tissues after TCA/acetone precipitation or phenol extraction and subsequent ammonium acetate/methanol precipitation. Horizontal gel dimension: isoelectric focusing on nonlinear IPG gel stripes between pH3 (left border of the gel) and pH10 (right side of the gel). Vertical gel dimension: SDS PAGE (top: ca 100 kDa, bottom: ca 5 kDa). All gels were silver-stained. TCA, TCA/acetone precipitation; phenol, phenol extraction and subsequent ammonium acetate/methanol precipitation.

total protein extraction from *Medicago truncatula* roots *(22)* and from potato stems, roots, and tubers (Mihr and Braun, unpublished data).

4. Concluding Remarks

Currently, 2-DGE on the basis of IEF and SDS-PAGE is the most important technique for protein separation in plant biology. However, not all proteins are visible on the gels for different reasons, e.g., hydrophobic proteins are poorly resolved by IEF. One solution is the use of alternative 2-D gel systems like two-dimensional blue-native (BN)-SDS-PAGE *(23)*. This procedure is based on the presence of Coomassie Blue during the separation on the first gel dimension, which introduces negative charges to proteins and protein complexes without denaturing them. Highly hydrophobic proteins are easily resolved by 2-D-BN-SDS-PAGE. This 2-D procedure is especially suitable for the analysis of protein complexes, like the photosystems of chloroplasts or the mitochondrial protein complexes of the respiratory chain *(24–26)*. For optimal resolutions, BN gel electrophoresis can be combined with IEF and SDS-PAGE, if proteins are purified from Coomassie Blue after the first gel dimension. This three-dimensional procedure was successfully used to separate isoforms of protein complexes from plant organelles *(27)*.

Proteins of very low abundance do not show up on either 2-D IEF-SDS or BN-SDS gels. One possible way to solve this problem is the preparation of cellular subfractions *(28)*. Alternatively, proteins can be separated by gel-free systems, which were developed recently and which are based on 2-D chromatographic separations *(29,30)*. In the future these procedures might become very important, because protein separation can be directly combined with protein identification by MS, allowing high-throughput analyses. In a pilot study based on this technology, Koller et al. *(31)* identified more than 2000 different proteins of the rice proteome.

Protein identification will be very much facilitated in the future by progress in genome-sequencing projects. Currently, the genome sequences for *Arabidopsis thaliana* and rice are available, but genome-sequencing projects have been started for several other plants. Completely sequenced genomes theoretically allow us to define entire proteomes. However, theoretical proteomes are based on computer predictions, which often turn out to be incorrect. Comparison of theoretical and "real" proteomes will be an important task of future research in the field of plant proteomics. Furthermore, comparison of proteomes of different plant tissues, different developmental stages of plant organs, differentially treated plants, different genotypes of plant species, or different mutants of plants will allow us to generate data on protein function, giving new insights into the molecular physiology of plants.

References

1. The *Arabidopsis* Genome Initiative (2000) Analysis of the genome sequence of the flowering plant *Arabidopsis thaliana*. *Nature* **408,** 796–815.
2. Yu, J., Hu, S., Wang, J., et al. (2002) A draft sequence of the rice genome (*Oryza sativa* L. ssp. *Indica*). *Science* **296,** 79–92.
3. Goff, S. A., Riche, D., Lan, T. H., et al. (2002) A draft sequence of the rice genome (*Oryza sativa* L. ssp. *Japonica*). *Science* **296,** 92–100.
4. Thiellement, H., Bahrman N., Damerval C., et al. (1999) Proteomics for genetic and physiological studies in plants. *Electrophoresis* **20,** 2013–2026.

5. Rossignol, M. (2001) Analysis of the plant proteome. *Curr. Opin. Biotechnol.* **12,** 131–134.

6. Darmerval, C., de Vienne D., Zivy M., and Thiellement H. (1986): Technical improvements in two-dimensional electrophoresis increase the level of genetic variation detected in wheat-seedling proteins. *Electrophoresis* **7,** 52–54.

7. Hurkman, W. J. and Tanaka, C. K. (1986) Solubilization of plant membrane proteins for analysis by two-dimensional gel electrophoresis. *Plant Physiol.* **81,** 802–806.

8. Komatsu, S., Muhammad A., and Rakwal R. (1999) Separation and characterisation of proteins from green and etiolated shoots of rice (*Oryza sativa* L.): towards a rice proteome. *Electrophoresis* **20,** 630–636.

9. Gallardo, K., Job, C., Groot, S. P., et al. (2002) Proteomics of *Arabidopsis* seed germination. A comparative study of wild-type and gibberellin-deficient seeds. *Plant Physiol.* **129,** 823–837.

10. Tsugita A. and Kamo M. (1999) 2-D electrophoresis of plant proteins. *Methods Mol. Biol.* **112,** 95–97.

11. Görg, A., Obermeier C., Boguth G., et al. (2000) The current state of two-dimensional electrophoresis with immobilized pH gradients. *Electrophoresis* **21,** 1037–1053.

12. Choe, L. H. and Lee, K. L. (2000) A comparison of three commercially available isoelectric focussing units for proteome analysis: the Multiphor, the IPGphor and the Protean IEF cell. *Electrophoresis* **21,** 993–1000.

13. Berkelman, T. and Stenstedt T. (1998) *2-D Electrophoresis Using Immobilized pH Gradients: Principles and Methods.* Amersham Pharmacia Biotech, Piscataway, NJ.

14. Laemmli, U. K. (1970) Cleavage of structural proteins during the assembly of the head of bacteriophage T4. *Nature* **227,** 680–685.

15. Schägger, H. and von Jagow, G. (1987) Tricine-sodium dodecyl sulfate-polyacrylamide gel electrophoresis for the separation of proteins in the range from 1 to 100 kDa. *Anal. Biochem.* **166,** 368–379.

16. Heukeshoven, J. and Dernick, R. (1986) Silver staining of proteins, in *Electrophoresis Forum* (Radola, B. J., ed.), Technische Universität München, Munich, pp. 22–27.

17. Lottspeich, F. (1999) Proteome analysis: a pathway to the functional analysis of proteins. *Angew. Chem. Int. Ed.* **38,** 2476–2492.

18. Costa, P., Barman, N., Frigerio, J.-M., Kremer, A., and Plomion, C. (1998) Water-deficit-responsive proteins in maritime pine. *Plant Mol. Biol.* **38,** 587–596.

19. Porubleva, L., Velden K. V., Kothari S., Oliver D. J., and Chitnis, P. R. (2001) The proteome of maize leaves: use of gene sequences and axpressed sequence tag data for identification of proteins with peptide mass fingerprints. *Electrophoresis* **22,** 1724–1738.

20. Morris, A. C. and Djordjevic, M. A. (2001) Proteome analysis of cultivar-specific interactions between *Rhizobium leguminosarum* biovar *trifolii* and subterranean clover cultivar *Woogenellup. Electrophoresis* **22,** 586–598.

21. Mihr, C., Baumgärtner, M., Dieterich, J.-H., Schmitz, U. K., and Braun, H.-P. (2001) Proteomic approach for investigation of cytoplasmic male sterility (CMS) in *Brassica. J. Plant Physiol.* **158,** 787–794.

22. Bestel-Corre, G., Dumas-Gaudot, E., Poinsot, V., et al. (2002) Proteome analysis and identification of symbiosis-related proteins from *Medicago truncatula* Gaertn. by two-dimensional electrophoresis and mass spectrometry. *Electrophoresis* **23,** 122–137.

23. Schägger, H. and von Jagow, G. (1991) Blue native electrophoresis for isolation of membrane protein complexes in enzymatically active form. *Anal. Biochem.* **199,** 223–231.

24. Jänsch, L., Kruft, V., Schmitz, U. K., and Braun, H. P. (1996) New insights into the composition, molecular mass and stoichiometry of the protein complexes of plant mitochondria. *Plant J.* **9,** 357–368.

25. Kügler, M., Jänsch, L., Kruft, V., Schmitz, U. K., and Braun, H. P. (1997) Analysis of the chloroplast protein complexes by blue-native polyacrylamide gelelectrophoresis. *Photosynthesis Res.* **53,** 35–44.

26. Kruft, V., Eubel, H., Jänsch, L., Werhahn, W. and Braun, H.-P. (2001) Proteomic approach to identify novel mitochondrial proteins in *Arabidopsis. Plant Physiol.* **127,** 1694–1710.

27. Werhahn, W. and Braun, H. P. (2002) Biochemical dissection of the mitochondrial proteome from *Arabidopsis thaliana* by three-dimensional gel electrophoresis. *Electrophoresis* **23,** 640–646.

28. Cordwell, S. J., Nouwens, A. S., Verrills, N. M., Basseal, D. J., and Walsh, B. J. (2000) Subproteomics based upon protein cellular location and relative solubilities in conjuction with composite two-dimensional electrophoresis gels. *Electrophoresis* **21,** 1094–1103.

29. Washburn, M. P., Wolters, D., and Yates, J. R. (2001) Large scale analysis of the yeast proteome by multidimensional protein identification technology. *Nat. Biotechnol.* **19,** 242–247.

30. Wolters, D. A., Washburn, M. P., and Yates, J. R. (2001) An automated multidimensional protein identification technology for shotgun proteomics. *Anal. Chem.* **73,** 5683–5690.

31. Koller, A., Washburn, M. P., Lange, B. M., et al. (2002) Proteomic survay of metabolic pathways in rice. *Proc. Natl. Acad. Sci. USA* **99,** 11969–11974.

IV

DATA ANALYSIS

29

Bioinformatics in Proteomics

Mauno Vihinen

1. Introduction

Now that several complete and draft genomes, including that for human, have been determined, the biggest task in functional genomics remains to identify the functions, reactions, interactions, localization, and structures of gene products. Proteomics is in the forefront of these studies of protein expression, function(s), activity, regulation, and post-translational modifications. As is evident from the wide scope of the term *proteomics*, a large variety of protein properties and technical approaches for their investigation is involved. The subdivision of proteomics into four categories by Ng and Ilag *(1)* is followed here.

Expression proteomics provides a view on protein production. Differential proteome analysis (comparison of different samples) forms the basis of this approach. In *interaction proteomics*, also known as cell-map proteomics, the aim is to investigate interactions of proteins and their subcellular localization. *Functional proteomics* involves analysis of effects of functional inactivation or perturbation of proteins of interest. *Structural proteomics*, also called structural genomics, refers to three-dimensional (3-D) structure determination in large scale.

Proteomics was started by the development of two-dimensional gel electrophoresis (2-DGE) technology. In recent years, several new approaches have been developed including protein identification methods, especially mass spectrometric methods, protein arrays, localization methods, and protein interaction studies including two-hybrid assay and phage display technology (**Fig. 1**). Proteomics analyses generate huge amounts of data. In a single experiment it is possible to obtain expression information for hundreds or thousands of proteins. For the analysis, storage, and retrieval of all the information, bioinformatics approaches are essential *(2,3)*.

Samples for proteomics, whether cells, organisms, tissues, organs, body fluids, or molecular complexes, are applied to either 2-DGE or protein arrays. 2-DGE technology has been supplemented by various other techniques. Protein arrays or chips are prepared by attaching suitable baits onto a solid support. Antibodies are the most prominent bait molecules, and there are already commercial antibody arrays available. Other feasible capture molecules include antigen-binding fragments, affinity-binding agents, synthetic DNA, RNA or protein aptamers, affibodies, ligands, or substrates of enzymes *(4,5)*. In the future, microfluidics-based arrays (lab-on-a-chip) may be available as well.

From: *Handbook of Proteomic Methods*
Edited by: P. Michael Conn © Humana Press Inc., Totowa, NJ

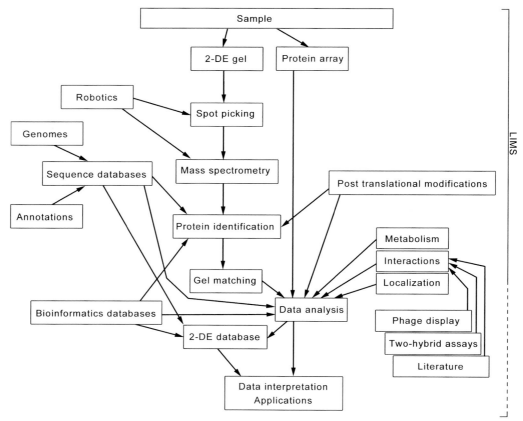

Fig. 1. General overview of proteomics research and types of bioinformatics databases and programs required.

In most proteomics studies, mass spectrometry (MS) is the method of choice for protein identification. Gel running, spot picking, and MS analysis can be automated with robotics. Integrated systems are essential for high-throughput proteomics. Protein identification is based on comparison of peptide fingerprints with proteins in databases.

To compare results from different gels and experiments, gel matching is the prerequisite for profound data analysis. It is important to record all information about the experiments to be able to repeat and compare the results. For DNA microarrays, Minimum Information About a Microarray Experiment (MIAME; http://www. mged.org/Workgroups/MIAME/miame.html) (*6*) has been developed to record essential data and to facilitate verification of information and data transfer between different analysis programs. In large centers, laboratory information management system (LIMS) packages are widely used not only for expression data like proteomics and microarrays, but for data from all kinds of experiments.

The ProteinChip system of Ciphergen provides another experimental approach to analyze protein interactions. The technique permits fast analyses of proteins that bind to a specific substrate. The method has been applied, e.g., to identification of biomarkers, toxicology screening, drug target validation, and so on. These studies require special software provided by the manufacturer and are not discussed here.

The bioinformatics steps of a proteome analysis project are discussed in the following, paying special attention to free internet-based services. Most of the bioinformatics tasks related to proteomics can be done in the Internet. Furthermore, all major databases are accessible in the web.

2. Protein Identification

Protein identification from spots on 2-DGE gels is one of the first steps in analysis. MS of the protein spots and their unique digestion fingerprints with matrix-assisted laser desorption/ionization/time-of-flight (MALDI-TOF) MS is the standard method. Proteins can be further characterized by sequence tagging with tandem mass spectrometry (MS/MS).

Protein identification relies on sequence databases to search for proteins with similar fingerprints. The more accurate and complete the sequence databases, the more accurate the identification. Genome-sequencing projects have remarkably increased the fidelity of identification. In principle, the complete genome for an organism facilitates the prediction of all the fingerprint patterns. However, alternative splicing and post-translational modifications are still poorly known and can still hamper the identification in some cases.

Spots from gels can be used directly for MS analyses. To identify proteins, peptides of characteristic molecular weights generated by the cleavage with proteases are analyzed. Fingerprints that consist of pI, M_r, peptide masses, sequence tags, or amino acid composition can be used for protein identification. For all proteomics-related information, the ExPASy service of the Swiss Institute of Bioinformatics (SIB; http://www.expasy.ch) provides the largest selection of Internet tools and links with the annotated SwissProt sequence database. Whenever possible, manually annotated sequence databases such as SwissProt (http://www.expasy.org/sprot/) or RefSeq of the NCBI (http://www.ncbi.nih.gov/RefSeq) should be used to obtain functional information on proteins instead of automatically generated entries. The Human Proteomics Initiative (HPI) (http://www.expasy.org/sprot/hpi/) of the SIB and the European Bioinformatics Institute (EBI) have attempted to annotate, describe, and distribute automatically highly curated information on human protein sequences *(7)*.

Dozens of web-accessible MS tools can be connected to services for protein molecular weight calculation and protein sequence tags, for peptide mass fingerprinting, for fingerprinting with sequence tag information, and for other tools, e.g., for amino acid composition, isoelectric points, post-translational modifications, and digestion information. For a detailed grouping, *see* **ref. 3**. CombSearch (http://www.expasy.org/tools/CombSearch/) and ProteinProspector of UCSF (http://prospector.ucsf.edu/) can launch a number of the aforementioned tools simultaneously.

3. Comparison of Gels

To gain qualitative and quantitative information about protein expression, gels have to be compared (**Fig. 2**). First, the gels are scanned, and often further processing is required to enhance the images, e.g., by increasing contrast or by subtracting the background. It is not possible to apply equal amounts of proteins to gels, and therefore the intensities in the images of individual images can differ and the data should be normalized. Then spots on the gels are detected and quantified. To compare gene expression in different conditions or, e.g., in healthy versus diseased samples, the gels have

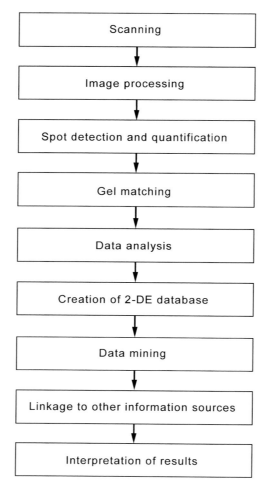

Fig. 2. Flow chart of analysis of 2-DGE gels and data mining.

to be matched. One has to be very careful when drawing conclusions on quantitative changes. For that purpose, the compared images ought to be for gels and samples that have been prepared, run, and stained in the same way. Even subtle variations in different steps can cause changes to the observed expression patterns.

Some freeware is available for proteomics image analysis, such as Flicker (http://www.lecb.ncifcrf.gov/flicker) *(8)* and WebGel (http://www-lecb.ncifcrf. gov/webgel) *(9)*, a gel database analysis system. However, these systems have not been updated for a while. Image comparison is one of the areas in proteomics in which commercial programs such as PDQuest (Bio-Rad), Melanie3 (GeneBio), and Progenesis (Perkin Elmer) are clearly much more advanced than programs in the public domain.

Gel matching has remained the bottleneck in 2-DGE analyses. Gels are not exactly identical, and inhomogeneities are caused by many factors including sample preparation, staining, unequal mobility in the different regions of gels, and variations in electrophoretic conditions. Therefore, gels cannot simply be superimposed. Gel matching has been a highly time-consuming manual process. Recently, automatic gel matching methods have been developed. Some commercial applications are available at high or very high prices. In Progenesis, the spot detection, gel warping, and auto matching are fully automated and do not allow any manual intervention. These programs are very useful, and they can save much time if resources allow the purchase.

From the matched gels, qualitative and quantitative differences in gene expression can be investigated. The matched gels and identified spots are the starting point for more detailed analysis and data mining, including statistical studies. Proteomics information can be used in many investigations and applications, including, e.g., identification of coexpressed proteins, synthesis rates, expression levels, determination of protein function and localization, effects of post-translational modifications, regulation, drug target identification and validation, diagnostics, search for characteristic markers for conditions such as tumor markers, follow-up of disease progression, differential proteome analysis, studies on microbial pathogenesis, responses to treatments and agents, and risk analysis. The expression of the components of metabolic pathways can be identified and followed with proteomics as well.

Nucleotide microarray analyses of transcriptomes generate vast amounts of information, which should be correlated to proteomes. This would allow the study of responses of cells, tissues, and organisms. Protein and mRNA expression levels do not always correlate *(10,11)*. It is evident that many methods developed for microarrays will also be applied to proteomics, especially to protein chip studies, which have the same format and principles as nucleotide arrays.

Detailed analysis of the gels also requires the use of other data sources, and therefore it is important that the program in use allow access to other databases, usually on the Internet, and supports integration with them.

4. 2-DGE Databases

2-DGE gels contain easily at least hundreds of protein spots. The identified protein spots have to be linked to the images of the gels. Several interactive, clickable images are already available on the Internet. These gels have been generated with widely diverse methods, and it may be difficult or even impossible to compare gels from one source with those from another laboratory. In particular, quantification of protein spots on gels produced with different methods is not possible, and gels stained with different methods give different results even for the very same sample. Furthermore, analysis of low-expression proteins is very error-prone. Despite all these cautions, data from other laboratories may be useful in many ways.

The web-accessible 2-DGE databases can be classified as general, organ/tissue-specific, or cell/organism-specific registries (for a full list, *see* **ref. 3**). A special meta-database, 2DWG (http://www-lecb.ncifcrf.gov/2dwgDB/) *(12)*, allows a flexible search for 2-DGE gels. WORLD-2DPAGE (http://www.expasy.ch/ch2d/2d-index.html) contains an index of 2-DGE databases and services. Multiple types of searches from Internet-based databases can be made with the 2DHunt program (http://www.expasy.ch/ch2d/2DHunt). 2-DGE databases can be used, e.g., for searching for a protein function, an expression pattern, or cellular localization along with annotations.

Certain guidelines have been developed for the distribution of images as databases in the web (http://www.expasy.ch/ch2d/fed-rules.html) *(13)*. Make2ddb software for generating federated databases is freely available from www.expasy.ch/ch2d/make2ddb.html *(14)*.

5. Interaction Proteomics

Proteomics is a method to describe biological events in cells (**Fig. 1**). Together with bioinformatics, the raw 2-DGE data can be converted into knowledge. Protein expression studies have numerous applications. Expression proteomics adds to the understanding of all the components in a proteome and of their relative abundances, but only interaction proteomics can describe the functional systems they form.

Interactions of proteins with each other and with other molecules are being extensively studied. Several experimental and computational methods are available. A (yeast) two-hybrid system *(15)* has been used to generate large datasets of interacting protein pairs. Systematic coprecipitation and MS studies have revealed very large protein complex networks *(16,17)*. Phage display facilitates the screening of a large number of interactions, e.g., with peptide ligands and epitopes, single-chain antibody fragments, or enzyme substrates. Also cDNA libraries can be displayed on phages *(18)*. Expression patterns in microarrays can also provide interaction information, although indirectly *(19,20)*.

Protein interaction networks have been calculated based on the expression information by applying reverse engineering based, e.g., on a linear system of equations *(21)*, Bayesian *(22)*, or Boolean networks *(23)*. Bioinformatics can further add to the interactions by analysis of genomes. In prokaryotes, genes for interacting proteins often appear in operons. Genes for interacting proteins have often co-occurrences in different genomes. Sometimes interacting proteins form fusion proteins in some genomes. Still another theoretical approach observes correlated mutations, based on the assumption that mutations in interacting parts in a binding partner will lead to a compensatory mutation in the other partner. Hence comutating pairs of residues are more frequent in physically interacting proteins than in a random pair of proteins *(24)*.

Much interaction information has been generated and published. Data mining of literature, e.g., with natural language processing to extract information automatically from Medline abstracts or full-text articles *(25)* has been utilized in several databases.

Different interaction analysis approaches provide complementary information. Recent large-scale comparative analyses of interaction networks in yeast obtained by several different ways indicated that only a small fraction of interactions have been detected with several methods *(26,27)*. The use of multiple datasets is highly recommended whenever possible.

Only experimentally verified interactions are listed in the following databases. Two-hybrid analysis information for yeast proteins is available in the Yeast Protein Linkage Map (http://depts.washington.edu/sfields/yp_project/) for the Uetz et al. *(15)* datasets along with published interactions in the literature *(28)*. Another yeast data source is the Yeast Interacting Proteins Database (http://genome.c.kanazawa-u.ac.jp/Y2H/) *(29)*.

Interactions of human proteins can be searched from the freely available part of Pronet (http://www.myriad-pronet.com/). PIMRider (http://pim.hybrigenics.com/pimrider/pimriderlobby/PimRiderLobby.jsp) provides two free services for academic scientists, namely, for *Helicobacter pylori* interactions and human immunodeficiency virus and human protein interactions. Molecular INTeraction (MINT) (http://cbm.bio.uniroma2.it/mint/) contains experimentally verified protein–protein interactions of both direct and indirect relationships *(30)*.

In addition to experimental interaction data, the following services also contain inferred information. The Biomolecular Interaction Network Database (BIND) (http://bind.mshri.on.ca/) describes interactions, molecular complexes, and pathways *(31)* for proteins, nucleic acids, and small molecules. The Database of Interacting Proteins (DIP) (http://dip.doe-mbi.ucla.edu) contains data on protein–protein interactions and interaction networks *(32)*.

In addition to yeast services, organism-specific interactions are catalogued for *Caenorhabditis elegans* (http://cmmg.biosci.wayne.edu/finlab/PIMdbv01.htm) and *Drosophila melanogaster* fruit fly (http://vidal.dfci.harvard.edu/interactome.htm).

The Kyoto Encyclopedia of Genes and Genomes (KEGG; http://www.genome.ad.jp/kegg/) includes all kinds of molecular interactions: metabolic pathways, regulatory pathways, and molecular assemblies *(33)*.

6. Functional Proteomics

This subsection of proteomics is based on the analysis of protein function by inactivating or perturbing protein function. Gene function is almost routinely probed by deleting or disrupting the gene from a model cell or organism, resulting in a knockout model.

For vertebrates, the mouse has been the most widely used animal model. TBASE (http://tbase.jax.org) is a large database on knockout models, mainly for mouse. In addition to experimental details of knockouts, there are also descriptions of phenotypes of heterozygous and homozygous knockouts. Knockouts can also be found in On-Line Mendelian Inheritance in Man (OMIM; http://www.ncbi.nlm.nih.gov/Omim/) in the section "Animal model" for each gene. Disease-causing mutations are in fact knockouts generated by nature. Locus-specific mutation databases are already available for about 250 genes (http://www.hgvs.org/).

In addition to gene knockouts, the function of proteins can be investigated in several other ways. Protein (enzyme)-specific inhibitors can be utilized. Recently, new approaches have been applied for gene silencing, namely, antisense and RNA interference techniques. Antisense oligonucleotides bind to complementary mRNA and thereby prevent transcription *(34)*. Short, double-stranded RNA molecules can be highly effective inhibitors of gene expression. Originally described for *Caenorhabditis elegans (35)*, the technique has subsequently been applied to many other organisms including human *(36)*. An RNAi database for *C. elegans* (http://www.rnai.org) has been constructed based on the analysis of Piano et al. *(37)*.

More than 95% of the yeast genes have already been knocked out in the *Saccharomyces* Deletion Project (http://www-sequence.stanford.edu/group/yeast_deletion_project/deletions3.html). In another project, double mutants have been generated *(38)*. There are several other consortia for investigating the effects of deletions in other organisms.

7. Structural Proteomics

Proteins are functional only when folded to their native 3-D structure. Therefore, structural information is crucial for understanding the function of a protein. Three-dimensional structures determined with either X-ray crystallography or nuclear magnetic resonance (NMR) spectroscopy have been collected in the Protein Data Bank (PDB; http://www.rcsb.org). There are plans for a structural genomics consortium *(39,40)*. To be able to model as many structures as possible based on determined structures, it is important to select crystallization targets so that they are representative for the sequence and structure space *(41)*.

Structures can be visualized, e.g., with the programs Chime or RasMol (http://www.umass.edu/microbio/chime/), Cn3D (http://www.ncbi.nlm.nih.gov/Structure/CN3D/cn3d.shtml), or WebMol (http://www.cmpharm.ucsf.edu/~walther/webmol.html). The PDB contains a versatile search engine. PDBsum (http://www.biochem.ucl.ac.uk/bsm/pdbsum/) lists a large number of structure-related features for each entry and provides links to other services.

Protein structures consist of a limited number of different architectures. The number of folds is not known; most estimates range from about 1000 to a few thousand folds. Protein fold classifications are available at CATH (http://www.biochem.ucl.ac.uk/bsm/cath_new/index.html) *(42)*, SCOP (http://scop.mrc-lmb.cam.ac.uk/scop/) *(43)*, and FSSP (http://www.ebi.ac.uk/dali/fssp/). Even the smallest organisms contain several hundred different proteins, and higher organisms have tens of thousands of genes and even more of proteins. Thus, there is a great demand for both experimental and theoretical structures and models.

When we consider all the different facets of proteomics and combine them with other databases for cell-, tissue-, and organism-wide data, we are entering the area of systems biology, which aims to study simultaneously the complex interactions of many levels of biological information.

Acknowledgments

Financial support was provided by the Finnish Academy, the National Technology Agency of Finland, the Sigrid Juselius foundation, and the Medical Research Fund of Tampere University Hospital.

References

1. Ng, J. H. and Ilag, L. L. (2002) Functional proteomics: separating the substance from the hype. *Drug Discov. Today* **7,** 504–505.
2. Wojcik, J. and Schächter, V. (2000) Proteomic databases and software on the web. *Brief. Bioinf.* **1,** 250–259.
3. Vihinen, M. (2001) Bioinformatics in proteomics. *Biomol. Eng.* **18,** 241–248.
4. Templin, M. F., Stoll, D., Schrenk, M., et al. (2002) Protein microarray technology. *Drug Discov. Today* **7,** 815–822.
5. Walter, G., Bussow, K., Lueking, A., and Glokler, J. (2002) High-throughput protein arrays: prospects for molecular diagnostics. *Trends Mol. Med.* **8,** 250–253.
6. Brazma, A., Hingamp, P., Quackenbush, J., et al. (2002) Minimum information about a microarray experiment (MIAME)—toward standards for microarray data. *Nat. Genet.* **29,** 365–371.
7. O'Donovan, C., Apweiler, R., and Bairoch, A. (2001) The human proteomics initiative (HPI). *Trends Biotechnol.* **19,** 178–181.
8. Lemkin, P. F. (1997) The 2DWG meta-database of two-dimensional electrophoretic gel images on the Internet. *Electrophoresis* **18,** 2759–2773.
9. Lemkin, P. F., Myrick, J. M., Lakshmanan, Y., et al. (1999) Exploratory data analysis groupware for qualitative and quantitative electrophoretic gel analysis over the Internet-WebGel. *Electrophoresis* **20,** 3492–3507.
10. Anderson, L. and Seilhamer, J. (1997) A comparison of selected mRNA and protein abundances in human liver. *Electrophoresis* **18,** 533–537.

11. Gygi, S. P., Rochon, Y., Franza, B. R., and Aebersold, R. (1999) Correlation between protein and mRNA abundance in yeast. *Mol. Cell. Biol.* **19,** 1720–1730.

12. Lemkin, P. F. (1997) Comparing two-dimensional electrophoretic gel images across the Internet. *Electrophoresis* **18,** 461–470.

13. Appel, R. D., Bairoch, A., Sanchez, J. C., et al. (1996) Federated two-dimensional electrophoresis database: a simple means of publishing two-dimensional electrophoresis data. *Electrophoresis* **17,** 540–546.

14. Hoogland, C., Baujard, V., Sanchez, J. C., Hochstrasser, D. F., and Appel, R. D. (1997) Make2ddb: a simple package to set up a two-dimensional electrophoresis database for the World Wide Web. *Electrophoresis* **18,** 2755–2758.

15. Uetz, P., Giot, L., Cagney, G., et al. (2000) A comprehensive analysis of protein-protein interactions in *Saccharomyces cerevisiae. Nature* **403,** 623–627.

16. Gavin, A. C., Bosche, M., Krause, R., et al. (2002) Functional organization of the yeast proteome by systematic analysis of protein complexes. *Nature* **415,** 141–147.

17. Ho, Y., Gruhler, A., Heilbut, A., et al. (2002) Systematic identification of protein complexes in *Saccharomyces cerevisiae* by mass spectrometry. *Nature* **415,** 180–183.

18. Zozulya, S., Lioubin, M., Hill, R. J., Abram, C., and Gishizky, M. L. (1999) Mapping signal transduction pathways by phage display. *Nat. Biotechnol.* **17,** 1193–1198.

19. Cho, R. J., Campbell, M. J., Winzeler, E. A., et al. (1998) A genome-wide transcriptional analysis of the mitotic cell cycle. *Mol. Cell* **2,** 65–73.

20. Hughes, T. R., Marton, M. J., Jones, A. R., et al. (2000) Functional discovery via a compendium of expression profiles. *Cell* **102,** 109–126.

21. Yeung, M. K., Tegner, J., and Collins, J. J. (2002) Reverse engineering gene networks using singular value decomposition and robust regression. *Proc. Natl. Acad. Sci. USA* **99,** 6163–6168.

22. D'Haeseleer, P., Liang, S., and Somogyi, R. (2000) Genetic network inference: from co-expression clustering to reverse engineering. *Bioinformatics* **16,** 707–726.

23. Friedman, N., Linial, M., Nachman, I., and Pe'er, D. (2000) Using Bayesian networks to analyze expression data. *J. Comput. Biol.* **7,** 601–620.

24. Pazos, F. and Valencia, A. (2002) In silico two-hybrid system for the selection of physically interacting protein pairs. *Proteins* **47,** 219–227.

25. Blaschke, C. and Valencia, A. (2001) The potential use of SUISEKI as a protein interaction discovery tool. *Genome Inform. Ser. Workshop Genome Inform.* **12,** 123–134.

26. von Mering, C., Krause, R., Snel, B., et al. (2002) Comparative assessment of large-scale data sets of protein-protein interactions. *Nature* **417,** 399–403.

27. Bader, G. D. and Hogue, C. W. (2002) Analyzing yeast protein-protein interaction data obtained from different sources. *Nat. Biotechnol.* **20,** 991–997.

28. Schwikowski, B., Uetz, P., and Fields, S. (2000) A network of protein-protein interactions in yeast. *Nat. Biotechnol.* **18,** 1257–1261.

29. Ito, T., Chiba, T., Ozawa, R., Yoshida, M., Hattori, M., and Sakaki, Y. (2001) A comprehensive two-hybrid analysis to explore the yeast protein interactome. *Proc. Natl. Acad. Sci. USA* **98,** 4569–4574.

30. Zanzoni, A., Montecchi-Palazzi, L., Quondam, M., Ausiello, G., Helmer-Citterich, M. and Cesareni, G. (2002) MINT: a Molecular INTeraction database. *FEBS Lett.* **513,** 135–140.

31. Bader, G. D., Donaldson, I., Wolting, C., Ouellette, B. F., Pawson, T., and Hogue, C. W. (2001) BIND—the Biomolecular Interaction Network Database. *Nucleic Acids Res.* **29,** 242–245.

32. Xenarios, I., Rice, D. W., Salwinski, L., Baron, M. K., Marcotte, E. M., and Eisenberg, D. (2000) DIP: the database of interacting proteins. *Nucleic Acids Res.* **28,** 289–291.

33. Kanehisa, M., Goto, S., Kawashima, S., and Nakaya, A. (2002) The KEGG databases at GenomeNet. *Nucleic Acids Res.* **30,** 42–46.
34. Dean, N. M. (2001) Functional genomics and target validation approaches using antisense oligonucleotide technology. *Curr. Opin. Biotechnol.* **12,** 622–625.
35. Fire, A., Xu, S., Montgomery, M. K., Kostas, S. A., Driver, S. E., and Mello, C. C. (1998) Potent and specific genetic interference by double-stranded RNA in *Caenorhabditis elegans*. *Nature* **391,** 806–811.
36. Elbashir, S. M., Harborth, J., Lendeckel, W., Yalcin, A., Weber, K., and Tuschl, T. (2001) Duplexes of 21-nucleotide RNAs mediate RNA interference in cultured mammalian cells. *Nature* **411,** 494–498.
37. Piano, F., Schetter, A. J., Mangone, M., Stein, L., and Kemphues, K. J. (2000) RNAi analysis of genes expressed in the ovary of *Caenorhabditis elegans*. *Curr. Biol.* **10,** 1619–1622.
38. Tong, A. H., Evangelista, M., Parsons, A. B., et al. (2001) Systematic genetic analysis with ordered arrays of yeast deletion mutants. *Science* **294,** 2364–2368.
39. Burley, S. K. (2000) An overview of structural genomics. *Nat. Struct. Biol.* **7 (Suppl.),** 932–934.
40. Williamson, A. R. (2000) Creating a structural genomics consortium. *Nat. Struct. Biol.* **7 (Suppl.),** 953.
41. Brenner, S. E. (2000) Target selection for structural genomics. *Nat. Struct. Biol.* **7 (Suppl.),** 967–969.
42. Orengo, C. A., Michie, A. D., Jones, S., Jones, D. T., Swindells, M. B., and Thornton, J. M. (1997) CATH—a hierarchic classification of protein domain structures. *Structure* **5,** 1093–1108.
43. Lo Conte, L., Brenner, S. E., Hubbard, T. J. P., Chothia, C., and Murzin, A. G. (2002) SCOP database in 2002: refinements accommodate structural genomics. *Nucleic Acids Res.* **30,** 264–267.

Quantitative Characterization of Proteomics Maps
by Matrix Invariants

Milan Randić

1. Introduction

In this chapter, we outline some theoretical schemes that attempt to summarize collective information on proteomes contained in a two-dimensional (2-D) gel, usually presented as a list of x, y coordinates of protein spots and their abundances. Visually, the information of a 2-D gel can be presented as a "bubble" diagram (**Fig. 1**) in which spots are shown as circles with the radius of the circle indicating the relative abundance. (In drawing the "bubble" diagram, we have included only the 20 proteins of **Table 1**, which are the spots having the greatest abundance in the data taken from **ref. 1**.) One can view a bubble diagram of a proteomics map as graphical *output* from an experimental laboratory. From the theoretical point of view, a bubble diagram (or the corresponding tables of x, y, z coordinates) represents *input* information on a map for theoretical considerations. What we would like to see as an *output* of theoretical studies of proteomics maps is a *list* of properties of the map as a whole, which are technically referred to as map *invariants* or map *descriptors*. An *invariant* in mathematics is a quantity that is independent of labels (such as numbers assigned to individual spots) used to record data.

We outline the construction of several map invariants, which one can order in a sequence that offers a characterization of a map. Any representation of an object (a molecule, DNA sequence, or 2-D map) by invariants is typically associated with some loss of information. This means that from a set of invariants, generally one cannot *reconstruct* a map, as is possible from a list of x, y, z coordinates. However, invariants may allow *classification* of maps, easy *comparison* of maps, and finding a degree of *similarity* among maps. The art in the construction of map invariants, as has been the case with construction of molecular descriptors *(2–5)*, is in finding invariants that have captured important structural features of a map leading to *useful* data reduction.

1.2. Strategy

Proteins have a constant charge and a constant mass. As a consequence they will appear with the same (x, y) coordinates on a proteomics map (providing the same protocol has been followed). Thus proteomics maps originating from the same kind of tissue or cell will differ among themselves only in the distributions of the third coordinate in 3-D space, all thus generating the same 2-D projection on the (x, y) plane. Our goal is now to combine N apparently independent points distributed in 3-D space into a single

From: *Handbook of Proteomic Methods*
Edited by: P. Michael Conn © Humana Press Inc., Totowa, NJ

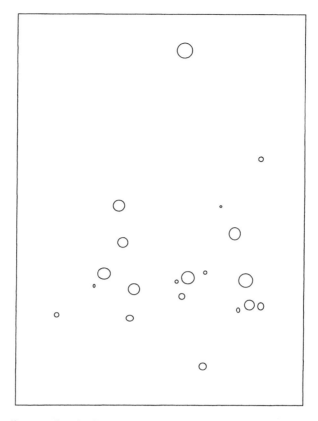

Fig. 1. A bubble diagram for the 20 most abundant proteins using data from the control group in **ref. *1***.

object. A way to do this is to connect points by lines, thus creating a recognizable geometrical object. We outline three such objects that have been considered recently in the literature.

In **Fig. 2** we have summarized the strategy for construction of map descriptors. We start with a list of *x, y, z* coordinates signifying the mass, the charge, and the abundance of individual proteins in a gel. The first step requires that we associate with a map certain well-defined geometrical objects, such as a zigzag line connecting various gel spots, or a graph connecting various gel spots. If we use only information on the location of spots, we obtain 2-D geometrical objects, but if we include abundance information, these geometrical forms become three-dimensional and represent objects in 3-D space. In **Table 1** we list coordinates for the 20 most abundant protein spots of rat liver cells for the control group, in descending order of proliferators, as reported in **ref. *1***. In **Figs. 3–5** we have illustrated for these data the following geometrical objects, respectively: (1) a zigzag curve obtained by connecting the 20 spots sequentially *(6–9)*; (2) a graph constructed after partial ordering of spots with respect to mass and charge *(10,11)*; and (3) a graph constructed by connecting neighboring spots at a specified distance or closer *(12)*.

After selecting a well-defined geometrical object, we associate it with a mathematical representation based on *matrices*. Matrices are numerical arrays that satisfy cer-

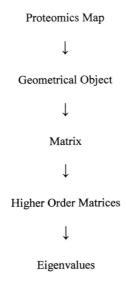

Fig. 2. Summary of the strategy for construction of map descriptors.

tain rules of algebra, including rules for multiplication of matrices. There are two kinds of matrix multiplications: (1) an individual matrix element is multiplied by a corresponding matrix element in the other matrix (or the same matrix; known as Kronecker and Hadamard multiplications); or (2) alternatively, one can consider the standard matrix multiplication. Of particular interest here are matrices in which an entry (i, j) is the euclidean distance between the spots i and j and matrices in which an entry (i, j) is the distance between spots measured along connecting lines. From such matrices we can construct additional matrices using matrix algebra, from which matrix invariants can be extracted.

1.3. Integrated Proteomics

The ultimate goal of proteomics studies is to identify proteins in a cell or tissue and to gain some understanding of their function and mutual inter-relationship. Our goal is less ambitious and focuses on the development of a tool that will allow *quantitative* characterization of maps. Although a list of spots and their abundances allows one to follow increases and decreases of particular proteins in a cell exposed to outside agents, we would like to characterize numerically an *overall change* of the proteome in a cell. Assuming that we know the nature of the outside agent (a chemical drug, toxin, and so on), we could try to relate the *biodescriptors* that characterize a map with the *chemodescriptors* that caused the observed changes in the map. Chemodescriptors represent mathematical invariants that describe molecules; they have been extensively explored in the last 25 years (*2–5*). Even when a cause for a change in a proteome is not known or cannot be numerically characterized, one can nevertheless build a library of 3-D maps based on characterization of maps by invariants such as those under review here. Such a library will then facilitate *comparisons* of maps, *prediction* of changes in a proteome, and possibly even *identification* of unknown external sources from the changes in a cell or tissue proteome. The underlying assumption here is that one can extend the dogma

Table 1
Gel Coordinates and Abundance for the 20 Most Intensive Protein Spots of Rat Liver Cells

	x	*y*	Abundance	Mass	Charge	Canonical
1	2111.7	2278.6	144,357	1	12	11
2	2804.3	903.6	143,630	15	8	17
3	1183.9	959.6	136,653	20	15	12
4	2182.2	928.8	127,195	7	2	13
5	2685.6	1196.1	118,581	5	17	18
6	1527.9	825.5	114,929	9	5	16
7	1346.0	1352.5	112,251	18	10	9
8	2868.5	778.0	108,883	3	20	1
9	1406.3	1118.1	98,224	4	18	10
10	2450.2	409.2	93,601	2	4	3
11	1474.0	665.1	90,004	16	1	5
12	2974.9	772.8	86,730	19	13	2
13	2068.4	823.1	84,842	6	16	15
14	642.2	669.8	82,492	13	6	6
15	2860.7	1649.9	81,965	8	11	4
16	2032.7	902.8	80,015	12	9	8
17	2752.7	765.6	79,847	17	7	20
18	2334.2	982.2	72,791	11	3	7
19	1053.6	864.3	72,173	14	19	14
20	2519.5	1365.9	69,452	10	14	19

of Emil Fisher that *similar compounds* will show *similar properties* from the structure-activity studies to the structure-proteome studies.

Invariants that have been found to be useful so far for characterization of proteomics maps include the *eigenvalues* of various matrices M, and in particular the *leading eigenvalue* of such matrices. Formally, this means solving the equation:

$$Det \mid M - \lambda I \mid = 0$$

Here *Det* is the determinant of a matrix $(M - \lambda I)$, I is an $N \times N$ unit matrix (having all elements zero except on the main diagonal, which are equal to 1, N being the number of vertices), and λ are the eigenvalues sought of the selected graph matrix M. There are several computer programs that calculate matrix eigenvalues. We found MATLAB *(14)* (an abbreviation for Matrix Laboratory, The MathWorks, Natick, MA) to be suitable for our needs in the proteomics map studies.

For a given map, we can construct many different geometrical objects, from a simple one like the zigzag curve to the most complex graph, the complete graph, in which all N spots are connected to one another. When N is large, or when one wants to compare large numbers of proteomics maps, the calculation of graph invariants for such graphs may be computationally intensive. Hence, it is of considerable interest to select geometrical objects that may contain sufficient structural information on a map but that are computationally friendlier.

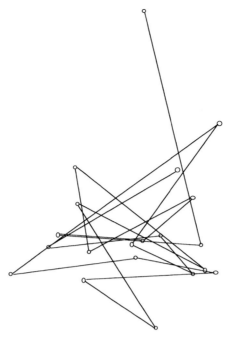

Fig. 3. A zigzag curve obtained by connecting the 20 most abundant spots (of **Table 1**) sequentially.

1.4. Graph Matrices

We have mentioned already the euclidean matrix *(E)*, in which the actual euclidean distance between vertices i and j are used as the matrix elements e_{ij}, and the path-distance matrix D in which elements d_{ij} are given by the shortest distance between vertices i and j taken along the existing lines connecting matrix vertices i and j. From these two distance matrices, we can construct the matrix E/D, in which the elements are given by the quotient of the corresponding distance between vertices i and j through space and the distance between vertices i and j through the connecting lines between the same vertices.

2. Methodology

2.1. Zigzag Curve

The shape of the zigzag curve is sensitive to the ordering of spots and can dramatically change. An alternative ordering of the same N points will generally produce a quite different zigzag curve. However, in comparative studies of proteomics maps, when gel preparations from different laboratories may differ and when different coloring agents may have been used, a very high sensitivity on a map of descriptors would be counterproductive. Hence, we also need alternative map representations that are less sensitive to variations among the input data. This, as will be seen, is the case with the graph based on partial ordering of spots. However, before calculating matrix elements, we should follow the recommendation of Kowalski and Bender *(13)* and scale the (x, y, z) coordinates in order to give equal relative weight to three experimental quantities expressed in different measurement units.

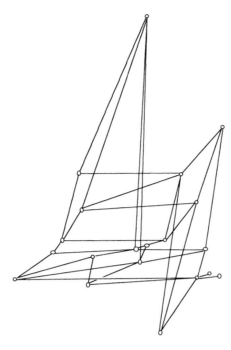

Fig. 4. Graph of the partial ordering of the same 20 spots with respect to the mass and the charge.

In **Table 2** we list the abundance of the 20 proteins of **Table 1** when rats have been exposed to four different proliferators, as reported by Anderson et al. *(1)*. Observe the apparent *chaotic* behavior of the cell proteome under different situations, showing no regularity in changes in the abundance of different proteins. A protein that has increased in abundance in one case may have dramatically decreased in abundance when a different agent is used, or it may remain unchanged when different toxins are introduced into the mouse diet. In order to *facilitate* comparisons of different 2-D maps or 3-D counterparts, one uses the following rules for ordering spots:

Rule 1: Select a set of *N* protein spots, the variations of abundance of which one wishes to examine more closely.
Rule 2: Order the set of selected *N* spots according to some adopted criterion (e. g., relative abundance).
Rule 3: Take the numbering of protein spots of the control group (e.g., healthy animals or tissue) and use it when considering other gels to be analyzed.

An advantage of the zigzag curve is the simplicity of its construction and the calculations that follow.

2.2. Graph of Partial Ordering

In **Table 1** we also list the rank order of the spots with respect to their masses and their charges *separately*. Partial order consists of ordering spots with respect to their masses and their charges *simultaneously*. As one can see in the corresponding columns

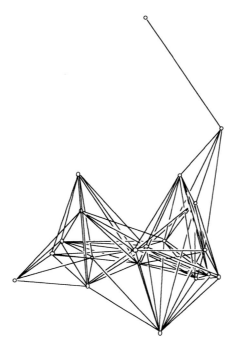

Fig. 5. Graph constructed by connecting neighboring spots within certain specified maximal distances.

for mass and charge, spot 1 is ahead of spot 7, which is ahead of spot 3, which is ahead of spot 19, which is finally ahead of spot 14. Hence, the ordering 1, 7, 3, 19, 14 is satisfied for *both* properties, mass and charge. Thus we have established a *dominance* of protein 1 over protein 7 over protein 3, and so on. A list of all such dominance sequences represents the partial ordering. Graphically, this is illustrated in **Fig. 4**, in which dominance is indicated by connecting lines. The sequence 1, 7, 3, 19, 14 is depicted as a path starting at the vertex 1 and ending at the vertex 14. Construction of the partial ordering graph over a map is not difficult if one observes a simple rule for construction: (1) all the connecting lines have a positive slope; and (2) one does not draw the connecting lines between the intermediate members in partially ordered sequence. Thus, in the above case it suffices to connect 1 to 7 and 7 to 3, but there is no need to connect 1 and 3, since their dominance has been already indirectly established. As we see from **Fig. 4**, the number of lines (edges) between points has increased by a factor of approx 2 in comparison with the number of line segments in the zigzag curve **(Fig. 3)**.

An advantage of the graph of partial ordering is that it can be constructed regardless of the data for the control group. This is important when there is no clear control. For example, proteomics maps of tumor cells (leukemia, melanoma, and so on) or proteomics of swollen tissues (e.g., synomial fluid) or proteomes of different bacteria may have no such obvious control group. The partial ordering graphs are not very dense, even if the number of spots has increased considerably; thus it is not difficult to find the shortest distances of interest for construction of *E/D* matrices.

Table 2
Unscaled Abundances for the 20 Proteins of Table 1

	Control	PFOA	PFDA	Clofibrate	DEHP
1	**144,357**	108,713	95,028	147,081	165,886
2	**143,630**	155,565	188,582	159,898	155,055
3	**136,653**	113,859	150,253	163,645	8111
4	**127,195**	99,160	73,071	76,642	112,096
5	**118,581**	112,790	49,769	109,856	138,795
6	**114,929**	192,437	221,567	166,080	180,590
7	**112,251**	58,669	38,915	73,159	77,075
8	**108,883**	26,105	50,735	45,923	116,849
9	**98,224**	91,147	82,963	84,196	92,942
10	**93,601**	83,172	62,934	79,870	109,381
11	**90,004**	129,340	112,361	112,655	119,402
12	**86,730**	70,746	78,691	105,760	116,281
13	**84,842**	73,814	45,482	71,911	97,444
14	**82,492**	73,974	74,466	84,703	88,545
15	**81,965**	16,137	16,501	60,077	148,992
16	**80,015**	77,314	80,072	76,027	100,836
17	**79,847**	20,782	13,103	38,816	53,830
18	**72,791**	76,369	52,749	55,599	77,432
19	**72,173**	77,982	60,376	46,808	78,121
20	**69,452**	37,838	16,129	57,167	71,274

PFOA, perfluorooctanoic acid; PFDA, perfluorodecanoic acid; DEHP, di(ethylhexyl)phthalate.

2.3. Graph of Clustering of Spots

In **Fig. 5** we show a graph obtained by connecting protein spots that are within a selected prescribed distance. A cluster graph can be based on 2-D euclidean distances between spots or on 3-D euclidean distances between spots that takes into account information on their relative abundance. As we see from **Fig. 5**, in the case of cluster graphs the number of lines (edges) between points has increased by an order of magnitude in comparison with the graph of partial ordering (**Fig. 4**). Clearly the matrices constructed for cluster graphs have more information on the underlying maps, which may be an advantage. In addition, this approach allows variations in the selected critical distance and gradual changes in the analysis using matrices of different density. A disadvantage of the approach is that the search for the shortest paths could be computer-intensive and may represent a slow step in construction of E/D matrices. We employed the Dijkstra algorithm for finding the shortest paths, which in the case of 100–200 spots may take a few seconds of central processing units (CPUs) (**12**).

3. Map Invariants

Given a matrix M (which in our case is the E/D matrix), numerous matrix invariants can be considered, but of interest for us are the eigenvalues, in particular the leading eigenvalue of a matrix. To increase the number of invariants for characterization of maps, we will consider the leading eigenvalue of matrices $M^{(n)}$ and $M^{[n]}$, the first of which represents the standard multiplication of a matrix by itself, whereas the second represents

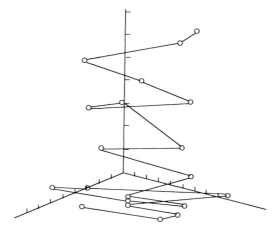

Fig. 6. 3-D zigzag line connecting the first 20 most intensive spots of **Table 1**.

Hadamard multiplication of matrices by itself, where n takes values $1 - N$. In the past, the $N = 80$ and higher values have been considered, in which case we have for each matrix 80 instead of a single descriptor. Finally, $N \rightarrow \infty$ gives the limiting value for the leading eigenvalue of a matrix when only the contributions from directly linked vertices remain, while all other disappear.

4. Illustrations

4.1. Use of the Zigzag Curve

In **Fig. 6** we show a 3-D zigzag line connecting the first 20 most intensive spots of **Table 1**, and in **Table 3** we have illustrated a 5×5 fragment of the E/D 20×20 matrix. As we see, the elements on the main diagonal of the E/D matrix are zero, and the adjacent elements are equal to 1, because the euclidean distance and the distance along the connecting line are the same for directly connected spots. All other elements are necessarily smaller than 1, unless it happens (which is possible but not likely) that three or more points are collinear. In **Table 4** we have listed the leading eigenvalues of 20×20 E/D matrices for a selection of exponents n. As the exponent n increases to infinity, we see a slow convergence of the leading eigenvalue to the limiting value $\lambda = 1.97766165245026$, which in this case is the leading eigenvalue of a path of length 19. For an $n \times n$ matrix, the limit would be the leading eigenvalue of a path of length $n - 1$, unless colinearity appears.

4.2. Use of Partial Ordering

In **Table 5** we have listed for the same proteomics map and the same 20 most abundant proteins the leading eigenvalues obtained from the graph of the partial ordering (**Fig. 4**). Here we used the weighted adjacency matrix AD, the elements of which are either zero (if spots are not directly linked by an edge) or are equal to the euclidean distance between a pair of spots (if directly linked). The leading eigenvalues for matrices AD obtained by the Hadamard product and for the standard matrix product, respectively are shown in the first row of **Table 5**. The remaining rows correspond to higher exponents, with $n = 10$, and $n = 15$, as the maximum for the Hadamard matrix product and

Table 3
A 5 × 5 Fragment of the 20 × 20 {*E/D*} Matrix

	1	2	3	4	5
1	0	1	0.5309	0.4728	0.5002
2	1	0	1	0.5582	0.6047
3	0.5309	1	0	1	0.9741
4	0.4728	0.5582	1	0	1
5	0.5002	0.6047	0.9741	1	0

Table 4
The Leading Eigenvalues of *E/D* Matrices for a Selection of Exponents *n*

n	Zigzag
1	8.0800
2	4.6107
3	3.3403
4	2.8182
5	2.5781
10	2.3811
20	2.3734
50	2.3587
100	2.3361
250	2.2760
500	2.1972
1000	2.0966
2500	1.9981
5000	1.9795
10,000	1.9776
25,000	1.977661652
50,000	1.97766165245026
Limit	1.97766165245026

for the standard matrix product, respectively. To curb the magnitudes of the leading eigenvalues as n increases, we used as a normalizing factor the reciprocal factorial ($1/n!$). As we have already seen in the case of the Hadamard matrix product, when $n = 10$ we have a small value for λ, whereas in the case of the standard matrix multiplication, at $n = 15$ the λ values have sufficiently decreased so that we could truncate the process of construction of higher power of matrices. The remaining columns of **Table 5** list the leading eigenvalues for the four proliferators considered.

One can view the eigenvalues in **Table 5** as sets of vectors having 10 and 15 components, respectively, which can be used to estimate the similarities among different proteomics maps. In **Table 6** we show the degrees of similarity/dissimilarity among the five proteomics maps, which have been designated as A, B, C, D, and E, corresponding to the control, PFOA, PFDA, clofibrate, and DEHP, respectively. An entry (X, Y)

Table 5
The Leading Eigenvalue of Distance-Weighted Adjacency Matrix and Corresponding Higher Order Matrices Obtained by Using the Hadamard Product (10-D) and Standard Matrix Multiplication (15-D), Respectively

	A	B	C	D	E
10-D					
1	5.142	5.497	6.038	5.700	5.420
2	5.014	5.473	6.060	5.477	5.274
3	4.179	4.549	4.844	4.427	4.338
4	2.841	3.100	3.229	3.011	2.945
5	1.589	1.748	1.810	1.699	1.653
6	0.749	0.834	0.864	0.809	0.783
7	0.304	0.343	0.367	0.332	0.320
8	0.108	0.124	0.130	0.120	0.115
9	0.034	0.040	0.042	0.038	0.037
10	0.010	0.012	0.012	0.011	0.011
15-D					
1	5.142	5.497	6.038	5.700	5.420
2	13.220	15.109	18.229	16.244	14.690
3	22.660	27.686	36.689	30.862	26.541
4	29.130	38.175	55.383	43.976	35.965
5	29.957	41.833	66.881	50.131	38.988
6	25.674	38.327	67.305	47.622	35.221
7	18.860	30.098	58.056	38.777	27.272
8	12.122	20.682	43.818	27.627	18.478
9	6.926	12.633	29.397	17.497	11.129
10	3.561	6.943	17.750	9.973	6.032
11	1.665	3.470	9.743	5.167	2.972
12	0.713	1.590	4.903	2.454	1.343
13	0.282	0.672	2.277	1.076	0.560
14	0.104	0.264	0.982	0.438	0.217
15	0.004	0.100	0.395	0.267	0.078

corresponds to the maps X and Y. The smallest entry corresponds to the most similar maps. The entries in the first numerical column correspond to the similarities based on the 10-D vectors, whereas the entries in the next column correspond to the similarities based on the 15-D vectors. A comparison of the two columns shows that the two vectors give somewhat different estimates of similarity. For instance, both vectors show as the most similar maps (A, E), (B, D), (B, E), and (D, E), although not in the same order. The same is again true if we look for the three most dissimilar pairs of maps, which are (B, C), (C, E), and (A, C), but again the order of the three is different for the two cases.

4.3. Use of Cluster Graphs

In **Fig. 5** we illustrate the cluster graph constructed for the same 20 protein spots of **Table 1**. As we see, the cluster graph is rather dense, that is, it has many lines and

Table 6
**Similarity/Dissimilarity for Each Pair of Maps Based
on 10-D and 15-D Vectors and Local Descriptors
Having 26 Compoments**

	10-D	15-D	26-D
(A, B)	0.758	25.9	4.83
(A, C)	0.864	85.7	5.76
(A, D)	0.832	44.6	5.27
(A, E)	0.367	19.3	5.14
(B, C)	1.599	52.9	2.98
(B, D)	0.259	18.8	3.18
(B, E)	0.356	6.6	3.09
(C, D)	0.830	41.2	1.87
(C, E)	1.171	66.5	2.35
(D, E)	0.367	25.3	1.99

involves more information on a map than has been the case with the zigzag curve or the partial order graph. In **Table 7** we list the normalized x, y coordinates and the abundances for the data of **Table 1**. As the critical distance we selected the 3-D distance between spot 1 and spot 15 (of **Table 7**), which is the maximal smallest distance between any pair of different spots, which is equal to 0.45447. By selecting the maximal smallest distance among spots, the cluster graph remains connected. If we had selected a still smaller distance, we would have obtained a disconnected graph. The distance-weighted adjacency matrix has zeros on the main diagonal and zero matrix elements for all pairs of protein spots that are not connected directly by a line. Otherwise the matrix elements are given by the euclidean distance between the corresponding points.

4.3.1. The Quotient Matrix Q

Having constructed the weighted adjacency matrix, which we denote as *ED*, we proceed to construct the quotient matrix $Q_{ij} = (ED_{ij})/(PD_{ij})$. Here PD_{ij} is the path-distance matrix, the elements of which are defined as the length of the shortest path between vertices i and j. The advantage of a quotient matrix in comparison with the euclidean distance matrix is that its elements are nondimensional, that is, they do not depend on the units chosen to measure distances. Moreover, all the elements are smaller than 1, except for the spots that are adjacent, in which case the matrix elements are equal to 1. Thus, when constructing the higher order matrices by using the Hadamard product, there is no need for additional normalization, because the magnitude of the elements of the higher order matrices will steadily decrease. If, however, we consider the higher order matrices obtained by using the standard matrix multiplication, we have to introduce a normalization factor $(1/n!)$ in order to secure convergence of the leading eigenvalues of corresponding matrices.

For construction of the path distance matrix, one has to find for every pair of vertices among numerous possible paths the shortest paths. In the case of dense graphs,

Table 7
Normalized Map Coordinates for 20 Proteins of *Table 1*[a]

	x	y	z	z/z_1
1	1.0230	2.2550	1.4446	1.0000
2	1. 3587	0.8941	1.4373	0.9949
3	0.5736	0.9496	1.3675	0.9466
4	1.0573	0.9191	1.2728	0.8811
5	1.3012	1.1836	1.1866	0.8214
6	0.7403	0.8169	1.1501	0.7961
7	0.6521	1.3384	1.1233	0.7776
8	1.3898	0.7699	1.0896	0.7543
9	0.6814	1.1064	0.9829	0.6804
10	1.1871	0.4049	0.9367	0.6484
11	0.7142	0.6581	0.9007	0.6235
12	1.4413	0.7647	0.8679	0.6008
13	1.0021	0.8145	0.8490	0.5877
14	0.3112	0.6628	0.8255	0.5714
15	1.3860	1.6326	0.8202	0.5678
16	0.9849	0.8934	0.8007	0.5543
17	1.3337	0.7576	0.7990	0.5531
18	1.1309	0.9719	0.7284	0.5042
19	0.5105	0.8553	0.7222	0.4999
20	1.2207	1.3516	0.6950	0.4811

[a]z = abundance.

this is generally a difficult task. However, the problem of searching for the shortest paths has been solved in graph theory by Dijkstra *(14)*. His algorithm has been implemented in **ref.** *12*.

After examining the leading eigenvalues of the Hadamard multiplication and the standard multiplication matrices, we found in this particular application that the leading eigenvalues of the standard multiplication give a better discrimination of the data for the four peroxisome proliferators. In **Fig. 7** we have plotted variations of the leading eigenvalues with the exponent n for clofibrate and the DEHP proliferators. As one can see, the two curves are different, but of a similar shape, having maximum values of approx 950 and 675 for clofibrate and DEHP curves, respectively. Plots for the remaining three maps have the same shape and fall in between, with the control group having the maximum at 925, PFOA at 875, and PFDA at 825. In order to see the differences better between the curves of **Fig. 7**, in **Fig. 8** we have plotted the quotient of eigenvalues for the four proliferators with the corresponding eigenvalues of the control group as the normalization factor. We have obtained much better visual differentiation for the four maps of the four proliferators. The data displayed in **Figs. 7** and **8** allow one to use the standard curve fitting procedure and reduce the number of essential descriptors from the currently used 20 for each map to at most half a dozen.

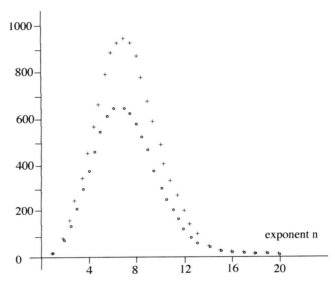

Fig. 7. Variations in the leading eigenvalues with the exponent *n* for the clofibrate (+) and DEHP data (o).

5. Quantitative Characterization of Map Differences

5.1. Characterization of Global Changes in Proteomics Map

So far we have used map descriptors (collected as components of vectors in *n*-dimensional vector space) to assess similarities and dissimilarities among set of maps. When the same toxic agent is administered to animals at different doses, one can test the current map descriptors to see how well they can characterize numerically variations of proteomics maps with variations in doses of toxin. We re-examined in **ref. 9** the data of Anderson et al. *(1)*, who reported data for mouse liver cells under a series of doses of peroxisome proliferator LY711883. Using the zigzag line based on the 20 most abundant proteins on proteomics maps having the following concentrations: 0 (the control), 0.003, 0.01, 0.03, 0.1, 0.3, and 0.6, we calculated the leading eigenvalues for each concentration (shown in **Table 8** for selected values of the exponent *n*).

By using multivariate regression analysis (MRA) with the eigenvalues of **Table 8** as a pool of descriptors, one obtains the following statistical parameters when using two, three, and four eigenvalues as descriptors:

$$\lambda_1, \lambda_{40} \qquad\qquad r = 0.923 \qquad\qquad s = 0.11$$

$$\lambda_1, \lambda_{40}, \lambda_5 \qquad\qquad r = 0.939 \qquad\qquad s = 0.11$$

$$\lambda_1, \lambda_{40}, \lambda_5, \lambda_{80} \qquad\qquad r = 0.970 \qquad\qquad s = 0.10$$

Here *r* is the regression coefficient, and *s* is the standard error of estimate. It turns out that the best single descriptor is the leading eigenvalue of the *E/D* matrix with the exponent *n* = 1. The order of selection of the leading eigenvalues of different exponents was based on the so-called greedy algorithm, in which one first finds the best single descrip-

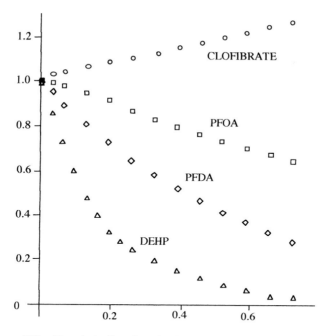

Fig. 8. The data of **Fig. 7** and similar data for PFOA and the PFDA plotted as a quotient of the leading eigenvalues with the corresponding leading eigenvalue of the control group. DEHP, di(ethylhexyl)phthalate; PFOA, perfluorooctanoic acid; PFDA, perfluorodecanoic acid.

tor, which was λ_1, and then keeps this descriptor and searches for the best second descriptor, which was λ_{40}, and so on. Better results can be expected from an exhaustive search of all possible combination of four descriptors. If we have to choose 4 descriptors from the 6 that we listed in **Table 8**, this would require examination of 15 regressions, but if we are to choose 4 from all 80 computed, the number of combinations to be examined would be over 1.5 million. Although this appears prohibitive, some recent developments in quantitative structure-activity relationship (QSAR) studies outlined a highly efficient algorithm for exhaustive search of the optimal combination of descriptors *(15)* that can also be applied here, if desired.

In **Fig. 9** we show the plot of the predicted concentrations (doses) against the experimental concentrations when using four descriptors, and in **Fig. 10** we show a similar plot based on the use of the four principal components of the principal component analysis (PCA) of the same data. Using three and four principal components we find the following statistical characterizations *(9)*:

$$PC_1, PC_2, PC_3 \qquad r = 0.374 \qquad s = 0.30$$

$$PC_1, PC_2, PC_3, PC_4 \qquad r = 0.993 \qquad s = 0.05$$

Observe that use of three principal components as descriptors does not give good regression even though three principal components account for 98.5% of the variance of data. The use of four principal components, however, gives slightly better regression that using four descriptors in multiple regression analysis (MRA).

Table 8
The Leading Eigenvalues for Zigzag Curves of Proteomics Maps
for Different Concentrations of LY1711883 in Mouse Diet

	0	0.003	0.01	0.03	0.1	0.3	0.6
1	9.781	9.691	9.745	9.764	9.718	9.668	9.642
5	3.027	2.766	2.865	2.927	2.773	2.087	2.857
10	2.731	2.649	2.689	2.678	2.661	2.755	2.466
20	2.692	2.551	2.616	2.597	2.573	2.501	2.350
40	2.648	2.425	2.414	2.483	2.461	2.394	2.302
80	2.580	2.279	2.364	2.319	2.327	2.339	2.233

5.2. Characterization of Local Changes in Proteomics Map

Once we have constructed a zigzag curve for a map, we can consider characterization of local map features. In **ref. *16*** we described one such characterization of local map features by considering a segment of a zigzag curve of length five. A zigzag curve involving N spots will generate $(N + 1 - k)$ local matrices of size $k \times k$, giving $(N + 1 - k)$ leading eigenvalues as local descriptors. The advantage of local descriptors is that local differences in abundances of proteomic maps are less likely to be averaged than the global differences. In addition, as presented in **ref. *16***, when large number of spots are considered one can use *all* eigenvalues of small $k \times k$ matrices, rather than considering only the leading eigenvalue. A close examination of the set of eigenvalues belonging to different fragments of the same or different zigzag curve has shown that eigenvalues of $k \times k$ matrices can be expressed as a combination of eigenvalues of the two unique matrices belonging to the complete graph having k vertices and the path graph (a chain) having k vertices, respectively. The coefficients of such expressions offer condensed information on all eigenvalues and can be used to evaluate the degree of similarity of different maps. In the last column of **Table 6**, we show the degree of similarity/dissimilarity of various pairs of maps using 26-component vectors based on the 30 most abundant spots and characterization of the local eigenvalues of 5×5 matrices. Not surprisingly, one can observe some differences, with the relative degree of similarity/dissimilarity based on global descriptors and local descriptors. It is interesting to see from **Table 6** that the four "perturbed" proteomics maps associated with the four proliferators are among themselves visibly more similar (the entries in **Table 6** being in the range 1.87–3.18) than they are when compared with the control proteomics map (the corresponding entries in **Table 6** being in the range 4.83–7.76).

6. Canonical Labeling for Proteomics Maps

We end this brief overview on mathematical characterization of proteomics maps by map invariants by addressing the problem of map storage and map retrieval, as well as map manipulations by computer. The problem is not dissimilar to data storage, data retrieval, and computer manipulation of molecules. Ideally this would be best accomplished if one could assign to objects of storage (proteomics maps) and retrieval *unique* labels, rather than conventional labels based on arbitrary labels for protein spots. It may be surprising that although there is no unique universal labeling of molecules that will

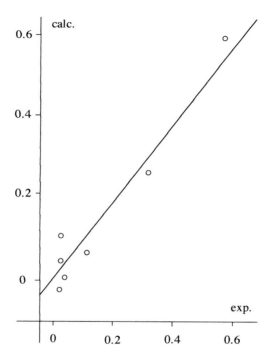

Fig. 9. Plot of the predicted doses against the experimental doses of LY711883 based on four eigenvalue descriptors.

follow directly from examination of molecular geometry, such unique labeling for proteomics maps is possible and has been proposed *(17)*. This was possible because unique labels are possible for graphs (the representation of which does not require information essential for chemistry: atom coordinates and the kind of atom present in molecules). In the last column of **Table 1**, we includ under the heading "Canonical labels" the unique labels for the 20 protein spots listed.

The canonical labels for proteins are based on the graph of the partial ordering of **Fig. 5** and are derived by requesting that the associated binary adjacency matrix A of the graph correspond to the smallest binary number *(18,19)*. This means that the first row of **A** should be the smallest binary number possible, which in this case is 1 (preceded by 19 zeros), because there are spots having but a single neighbor, which thus assumes label 20. In fact, there are two such spots, so both the first and the second row will have 19 zeros followed by 1. To resolve the ambiguity as to which of the two spots (which from a graph theoretical point of view are equivalent) should be given label 1 and which label 2, we will in such situations give preference to the spot belonging to a protein of *heavier mass*. One continues to search for the location of label 3 and so on. Although the search for canonical labels belongs to the category of so-called NP problems [which require nonpolynomial (NP) algorithms for solution] *(20)*, which are tedious and could be time-consuming, this problem, once solved, holds for all maps having the same set of proteins selected for characterization. Moreover, the problem is more tedious for regular and highly symmetrical graphs, but if graphs have many vertices of different degrees, as is typical for proteomics maps, finding canonical labels need not be so difficult.

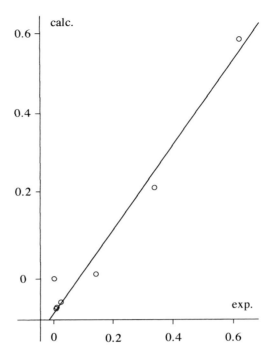

Fig. 10. Plot of the predicted doses against the experimental doses based on four principal components.

Once the labels have been assigned to protein spots, a list of abundances can serve as the map signature. Thus, from **Table 1** we can construct the canonical list of spots:

(8, 13, 11, 16, 12, 15, 19, 17, 7, 9, 1, 3, 4, 20, 14, 6, 2, 5, 10, 18)

from which we obtain (by replacing each spot number with the normalized abundance shown in the last column of **Table 8**) the following signature for the proteomics map of the control group:

(1.000, 0.779, 0.827, 0.735, 0.797, 0.753, 0.663, 0.733, 1.031, 0.902, 1.326, . . .)

For the other four maps of **Table 2** we obtain similarly:

PFOA   (1.000, 2.828, 4.955, 2.962, 2.710, 0.618, 2.987, 0.796, 2.247, 3.491, 4.164, . . .)

PFDA   (1.000, 0.896, 2.215, 1.578, 1.551, 0.325, 1.190, 0.258, 0.767, 1.635, 1.873, . . .)

Clofibrate (1.000, 1.566, 2.453, 1.656, 2.303, 1.308, 1.019, 0.854, 1.593, 1.833, 3.203, . . .)

DEHP   (1.000, 0.834, 1.022, 0.863, 0.995, 1.275, 0.669, 0.461, 0.660, 0.795, 1.420, . . .)

If one is interested only in recording the graph that defines a class of proteomics maps having the same partial ordering graph, this can be accomplished by listing the rows of the canonical adjacency matrix and converting the rows of the adjacency matrix into decimal numbers.

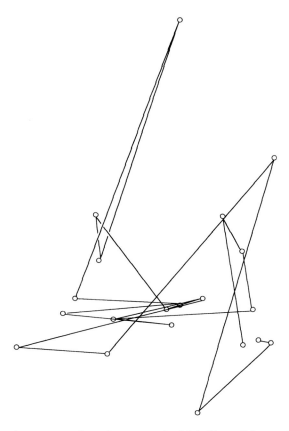

Fig. 11. Fox trail, a zigzag curve based on canonical labeling of the protein spots in **Fig. 4**.

6.1. Fox Trail Curves

Once the canonical labels have been assigned to the selected protein spots, one can construct a zigzag curve by using the canonical labels as a guide *(17)*. **Figure 11** shows such a zigzag curve, which we will call a *fox trail curve*, to distinguish it from similar curves constructed by using abundance as the guide in linking the protein spots. All the zigzag curves constructed over a map have a general shape resembling a fox trail, as documented in recordings of the movement of a fox at equal time intervals over a certain longer time interval *(21,22)*. An important advantage of the fox trail curves over ordinary zigzag curves is that they can be constructed *without* information on spot abundances, that is, they are independent of measured abundances. This may be of particular importance when one wishes to compare gels originating from different laboratories or different experimental conditions. Computationally there are no novelties, and one proceeds in the same way as outlined for the zigzag curve of **Fig. 3**.

7. Challenges Ahead

Before we end the present outline of currently available numerical procedures for obtaining quantitative characterizations of proteomics maps, we should point out that the same procedures described here are equally good for characterization of other kinds of 2-D maps, such as 2-D NMR maps, for instance. Among the challenges lying ahead, we should mention a systematic search for additional map invariants based on the existing methodologies as well as the search for alternative mathematical objects to be associated with maps. We should recall that 25 years ago there were at most a dozen various topological indices for the characterization of molecular structure, whereas today we have more than several hundred *(19,20)* and their numbers continue to grow. It is reasonable to anticipate that a similar situation will occur with the recently proposed sequence invariants [used for characterization of DNA sequences *(23–33)*] and map invariants to be used in proteomics studies. At the same time we will have to explore a number of arbitrary choices in the current schemes. This includes finding the optimal number of spots to be used for construction of map invariants, as well as finding optimal numbers of lines incorporated into geometrical objects used for building matrices to be associated with a proteomics map. We can summarize the present situation in quantitative proteomics, with the same words that Winston Churchill uttered when referring to events at the beginning of World War II: "This is not the beginning of the end but the end of the beginning." In the near future we can expect a doubling of map invariants and a doubling of the new methodologies—all designed to facilitate the use of computers to digest the enormous amount of data on the abundance of proteins in proteomes of different origins.

Acknowledgments

The author thanks Professor Jure Zupan for the hospitality enjoyed at the Chemometrics Laboratory of the National Institute of Chemistry, Ljubljana, Slovenia.

References

1. Anderson, N. L., Esquer-Blasco, R., Richardson, F., Foxworthy, P., and Eacho, P. (1996) The effects of peroxisome proliferators on protein abundance in mouse liver. *Toxicol. Appl. Pharmacol.* **137**, 75–89.
2. Randić, M. (1998) Topological indices, in *Encyclopedia of Computational Chemistry* (Schleyer, P. V. R., Allinger, N. L., Clark, T., et al. eds.), John Wiley & Sons, Chichester, pp. 3018–3032.
3. Balaban, A. T. (2001) A personal view about topological indices for QSAR/QSPR, in *QSPR/QSAR Studies by Molecular Descriptors* (Diudea, M. V., ed.), Nova Science Publishers, Huntington, NY, pp. 1–30.
4. Randić, M. (2001) The connectivity index 25 years after. *J. Mol. Graph. Modelling* **20**, 19–35.
5. Balaban, A. T. (1999) Historical developments of topological indices, in *Topological Indices and Related Descriptors in QSAR and QSPR* (Devillers, J. and Balaban, A. T., eds.) Gordon and Breach, Amsterdam, pp. 403–453.
6. Randić, M. (2001) On graphical and numerical characterization of proteomics maps. *J. Chem. Inf. Comput. Sci.* **41**, 1330–1338.
7. Randić, M., Zupan, J., and Novic, M. (2001) On 3-D graphical representation of proteomics maps and their numerical characterization. *J. Chem. Inf. Comput. Sci.* **41**, 1339–1334.
8. Randić, M., Witzmann, F., Vračko, M., and Basak, S. C. (2001) On characterization of proteomics maps and chemically induced changes in proteomics using matrix invariants: application to peroxisome proliferators. *Med. Chem. Res.* **10**, 456–479.

9. Randić, M., Novič, M., and Vračko, M. (2002) On characterization of dose variations of 2-D proteomics maps by matrix invariants. *J. Proteomic Res.* **1**, 217–226.

10. Randić, M. (2002) A graph theoretical characterization of proteomics maps. *Int. J. Quantum Chem.* **90**, 848–858.

11. Randić, M. and Basak, S. C. (2002) A comparative study of proteomics maps using graph theoretical biodescriptors. *J. Chem. Inf. Comput. Sci.* **42**, 983–993.

12. Bajzer, Ž., Randić, M., Plavšić, D., and Basak, S. C. Novel matrix invariants for characterization of toxic effects on proteomics maps. *J. Mol. Graphics Modelling*, in press.

13. Kowalski, B. R. and Bender, C. F. (1972) A powerful approach to interpreting chemical data. *J. Am. Chem. Soc.* **94**, 5632–5639.

14. Rade, L. and Wedergoen, B. (1990) *Beta Mathematics Handbook.* CRC Press, Boca Raton, FL.

15. Lučić, B., Trinajstić, N., Sild, S., Karelson, M., and Katritzky, A. R. (1999) New efficient approach for variable selection based on multiregression: prediction of gas chromatographic retention times and response factors. *J. Chem. Inf. Comput. Sci.* **39**, 610–621.

16. Randić, M., Zupan, J., Novič, M., Gute, B. D., and Basak, S. C. (2002) Novel matrix invariants for characterization of changes of proteomics maps. *SAR QSAR Environ. Sci.* **13**, 689–703.

17. Randić, M., Plavšić, D., Basak, S. C., and Gute, B. D. On canonical labeling of proteins of proteomics maps. *J. Chem. Inf. Comput. Sci.*, in press.

18. Randić, M. (1974) On recognition of identical graphs representing molecular topology. *J. Chem. Phys.* **60**, 3920–3928.

19. Randić, M. (1977) On canonical numbering of atoms in a molecule and graph isomorphism. *J. Chem. Inf. Comput. Sci.* **17**, 171–180.

20. Garey, M. R. and Johnson, D. S. (1979) *Computers and Intractability (A Guide to the Theory of NP-Completeness).* Freeman, San Francisco.

21. Hall, G. G. (1972) Modelling—a philosophy for applied mathematicians. *Bull. Inst. Math Applications* **8**, 226–228.

22. Siniff, D. H. and Jesson, C. R. (1969) Simulation model of animal movement patterns. *Adv. Ecol. Res.* **6**, 185–220.

23. Todeschini, R. and Consonni, V. (2000) *Handbook of Molecular Descriptors, Methods and Principles in Medicinal Chemistry*, vol. 11, (Mannhold, R., Kubinyi, R., and Timmerman, H., eds.), Wiley-VCH, New York.

24. Todeschini, R. and Consonni, V. (2001) *Dragon, rel. 1.12 for Windows.* Milano, Italy. Program for the calculation of molecular descriptors from HyperChem, Sybyl, and SD file formats.

25. Randić, M. (2000) Condensed representation of DNA primary sequences by condensed matrix. *J. Chem. Inf. Comput. Sci.* **40**, 50–56.

26. Randić, M. and Vračko, M. (2000) On the similarity of DNA primary sequences. *J. Chem. Inf. Comput.* **40**, 599–606.

27. Randić, M., Vračko, M., Nandy, A., and Basak, S. C. (2000) On 3-D graphical representation of DNA primary sequences and their numerical characterization. *J. Chem. Inf. Comput. Sci.* **40**, 1235–1244.

28. Randić, M. and Basak, S. C. (2001) Characterization of DNA primary sequences based on the average distance between bases. *J. Chem. Inf. Comput. Sci.* **41**, 561–568.

29. Guo, X., Randić, M., and Basak, S. C. (2001) A novel 2-D graphical representation of DNA sequences of lower degeneracy. *Chem. Phys. Lett.* **350**, 106–112.

30. Randić, M. and Balaban, A. T. (2003) On 4-dimensional representation of DNA primary sequences. *J. Chem. Inf. Comput. Sci.* **43**, 532–539.

31. Randić, M., Vračko, M., Lerš, N., and Plavšić, D. (2003) Novel 2-D graphical representation of DNA sequences and their numerical characterization. *Chem. Phys. Lett.* **368,** 1–6.
32. Randić, M., Vračko, M., Lerš, N., and Plavšić, D. (2003) Analysis of similarity/dissimilarity of DNA sequences based on novel 2-D graphical representation. *Chem. Phys. Lett.* **371,** 202–207.
33. Randić, M., Nandy, A., Basak, S. C., and Plavšić, D. On numerical characterization of DNA primary sequences. *J. Math. Chem.*, submitted.

31

Complexity of Protein–Protein Interaction Networks, Complexes, and Pathways

Danail Bonchev

1. Introduction
1.1. The Network Approach to Proteomics

The focus of proteomic research in developing experimental techniques for protein identification and interaction studies is shifting from individual proteins to their organization in reaction pathways, complexes, and networks, i.e., to the proteome—the large-scale network comprising all protein–protein interactions in a cell, tissue, or organism. The number of complete proteomes in accessible databases exceeds 100 *(1)*, thus making possible proteome-wide and across-proteomes analyses. Such a systemic approach offers a view of the biological machine as a whole, revealing important new details of its work. Thus, one could regard aging and diseases as specific patterns of protein network degradation and, vice versa, evolutionary beneficial factors as creating patterns of larger proteome complexity. Medicines' side effects could be analyzed in terms of the extremely high network connectivity, thus orienting the search for new medicines toward protein complexes, rather than individual compounds *(2)*. Potential drug and marker candidates could be identified proceeding from protein connectivity and centrality patterns.

1.2. Quantifying Networks

Network analysis and applications are necessarily related to numbers that uniquely characterize each network, making possible comparison, classification, structure-activity/toxicity relationships, and prediction. Regarding networks as graphs *(3)*, one can use graph theory *(4,5)* to generate such numbers, usually termed *topological indices (6,7)*, proceeding from different graph invariants. A variety of topological indices have been proposed in an area of theoretical chemistry called *mathematical chemistry*. From this rich arsenal of descriptors we select for characterizing proteomic networks only a few representatives that mirror patterns of *network topological complexity*. Information theory *(8,9)* is also used as a source of descriptors of *network diversity and composition*.

1.3. The Concept of Network Complexity

Two concepts of complexity as a general property of systems have been developed during the last half century. The concept of new phenomena emerging in a highly com-

From: *Handbook of Proteomic Methods*
Edited by: P. Michael Conn © Humana Press Inc., Totowa, NJ

plex system focuses on the nonlinearity of processes in complex dynamic systems *(10)*. In this chapter, we present the lesser known concept of *topological (or structural) and compositional complexity*. The roots of this concept can be traced back to the ideas of the information content of a system (a molecule, a cell, an organism) advanced in the 1950s *(11)*, to the first estimates of graph complexity *(12)*, and to the later hierarchical concepts of complexity *(13–15)*.

Definition 1. The larger the number and connectivity of the subnetworks, the larger the network topological complexity.

Definition 2. The more diverse the distribution of the network elements, the larger the network compositional complexity.

Definition 1 proceeds from counting all subgraphs and the links within each of them. It quantifies the idea that the whole is more than its parts and provides nonlinear quantitative measures of network structural complexity. Definition 2 accounts for the different aspects of network complexity, based on the element's distribution, properties, weights, interactions, and so on. When applied to dynamic evolutionary networks, the definitions presented might be viewed as a step toward unifying the two alternative approaches to systems complexity.

2. Some Basic Notions from Graph Theory

Protein–protein interaction networks are presented as graphs *(4,5)*. A *graph* is a structure composed of points (*vertices* or *nodes*), connected by lines (*edges* or *links*). A *subgraph* is a graph obtained from the parent graph by deleting at least one edge or vertex with its incident edges. A *loop* is an edge that begins and ends in the same vertex. A *multigraph* is a graph in which some pairs of vertices are linked by more than one edge. *Simple graphs* are graphs having no multiple edges and loops. In a *complete graph* any two vertices are connected by an edge. A *directed* graph is a graph having at least one directed edge. Directed edges are termed *arcs*. The graph is *connected* when there is a path between any pair of vertices in it; otherwise the graph is *disconnected*. A *path* in the graph is a sequence of adjacent edges without traversing any vertex twice. A *walk* is an alternating sequence of vertices and edges, each of which could be traversed more than once. The *walk length* is the number of edges in it. A *cycle* is a path that starts from and ends in the same vertex. *Trees* are graphs containing no cycles. Two vertices j and i are called *adjacent* when they are connected by an edge $\{i,j\}$. The adjacency relation is quantified by the term $a_{ij} = 1$. Graph *components* are connected subgraphs or vertices that are not connected to each other. Euler's theorem relates the number of vertices V, edges E, cycles C, and components K:

$$C = E - V + K \tag{1}$$

Illustrations of the notions introduced above are given in **Fig. 1**.

Definition 3. Vertex degree a_i is the number of edges $\{i,j\}$ connecting vertex i with its adjacent vertices j (denoted as $j \leftrightarrow i$):

$$a_i = \sum_{j \leftrightarrow i} a_{ij} \tag{2}$$

In a multigraph one may distinguish between *minimal* vertex degree, for which $a_{ij} = 1$, and *multiple* vertex degree, which accounts for all links to the nearest neighbors ($a_{ij} \geq 1$). In directed graphs, *in-degree* and *out-degree* are defined as the number of arcs entering into and emanating from the vertex, respectively.

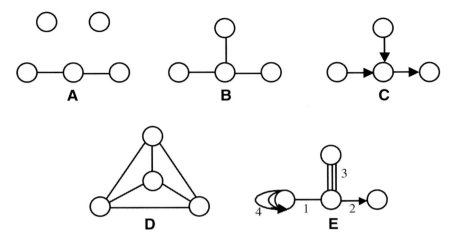

Fig. 1. (**A**) A disconnected graph with three components. (**B**) A simple connected graph. (**C**) A directed graph. (**D**) A complete graph with three cycles. (**E**) a multigraph with loops. 1, edge; 2, arc; 3, triple edge; 4, two loops.

Definition 4. Graph adjacency $A(G)$ is the sum of all vertex degrees a_i or all vertex adjacencies a_{ij}. In directed graphs, the total adjacency $A(DG)$ is the sum of the respective in- and out-degrees:

$$A(G) = \sum_{i=1}^{V} a_i = \sum_{i=1}^{V} \sum_{j \leftrightarrow i} a_{ij} \quad A(DG, out) = \sum_{i=1}^{V} a_i(out) = \sum_{i=1}^{V} \sum_{j \leftrightarrow i} a_{ij} (out) \qquad (3)$$

Graph invariants are numbers uniquely derived from the graph; they do not depend on the way the graph is drawn or on the manner in which the vertices are labeled. Graph adjacency is a graph invariant, whereas vertex degree is a vertex (or local) invariant.

Definition 5. (a) The *distance* d_{ij} between the vertices i and j is the number of edges along the shortest path connecting i and j. (b) The vertex distance degree d_i is the sum of distances between vertex i and all other vertices j in the graph:

$$d_i = \sum_{j \neq i} d_{ij} \qquad (4)$$

Definition 6. The *graph distance* $d(G)$ is the sum of distances between all pairs of vertices i and j; it is also a half sum of the distance degrees of all V vertices in the graph:

$$d(G) = \sum_{i=1}^{V} \sum_{j \neq i} d_{ij} = \frac{1}{2} \sum_{i=1}^{V} d_i \qquad (5)$$

In nondirected graphs G, $d(G) = 2W(G)$, $W(G)$ being the *Wiener number* of G (**16**). In directed graphs DG, $d(DG) = W(DG)$.

Definition 7. (a) *Vertex eccentricity* e_i is the maximum distance between vertex i and the remaining graph vertices. (b) *Graph center* is the vertex(es) having minimum eccentricity and minimum distance degree (*see* **ref. 17** for a more detailed definition):

$$e_i = min; \quad d_i = min \qquad (6)$$

Illustrations of Definitions 3–7 are given in **Fig. 2.**

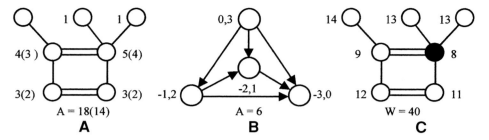

Fig. 2. **(A)** Vertex degrees and adjacency of a mutigraph; the minimal degrees are given in parentheses. **(B)** In- and out-degrees and adjacency of a directed graph; in-degrees are denoted by a minus sign. **(C)** Vertex distance degrees and the Wiener number (16) of a graph. The filled vertex having eccentricity 2 and distance degree 8 is the graph center.

3. Topological Complexity Measures

For a review of the entire area of complexity descriptors, the reader is addressed to **ref. *15***. Here, we focus on several complexity measures that satisfy the requirements to increase with the increase in network size and connectivity, to vary regularly with the topological patterns of branching *(18)*, cyclicity *(19)*, centrality *(17)*, and clustering *(18,19)*, and to be in compliance with the idea of biological complexity regarding the whole as more than the sum of its parts.

3.1. Connectedness (Connectance)

Definition 8. Connectedness *Conn* is the ratio between the number of edges E in the network and the number of edges in the complete graph having the same number of vertices V:

$$Conn(\%) = \frac{2E}{V(V-1)} \times 100 \qquad (7)$$

For simple graphs (graphs having no loops or multiple edges), *Conn* varies from zero for a totally disconnected graph having $E = 0$, to 100 for a complete graph. An example is given in **Fig. 3A**. For multigraphs, *Conn* could exceed 100. In such cases one may regard separately the multiple connectedness *Multiconn* from the simple one, which measures only whether the pairs of vertices are connected but not how many edges connect them. Similarly, for characterizing the local (or vertex) complexity one may use the vertex degrees and multiple vertex degrees introduced in **Subheading 2.** The factor 2 in **Eq. 7** is omitted for directed graphs. Connectedness can be used as a preliminary estimate of topological complexity; it is not sensitive to variations in topology of networks having the same number of vertices and edges.

3.2. The Substructure Count SC

See **refs. 20–22**.

Definition 9. (a) The *complexity index SC* is the total number of subgraphs in the graph (**Fig. 3**). (b) The *eth* order index ^eSC is the count of all subgraphs having e edges:

$$SC = {^0SC} + {^1SC} + {^2SC} + \ldots + {^ESC} = \sum_{e=0}^{E} {^eSC} \qquad (8)$$

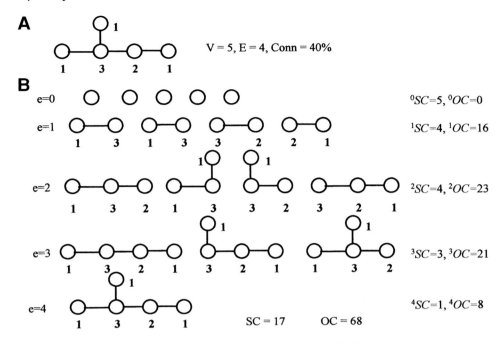

Fig. 3. **(A)** Calculation of connectedness according to **Eq. 7**. **(B)** Illustration of subgraph count *(20–22)* and overall connectivity indices *(22,23)* calculation (**Eqs. 8** and **9**).

where ^0SC, ^1SC, and ^2SC are the number of vertices V, edges E, and two-edge subgraphs, respectively, and $^ESC = 1$ stands for the graph itself (**Fig. 3B**).

3.3. Overall Connectivity

See **refs. 22** and **23**.

The overall connectivity concept (Definition 1) combines subgraph count with subgraph connectivity.

Definition 10. (a) The *overall connectivity index OC* is the sum of total adjacencies A_i of all subgraphs G_i belonging to the graph G. (b) The eth order overall connectivity term ^eOC is the sum of the total adjacencies ^eA_i of all subgraphs ^eG_i having e edges (**Fig. 3B**).

$$OC = {}^1OC + {}^2OC + \ldots + {}^EOC = \sum_{e=1}^{E} {}^eOC$$
$$^eOC = \sum_{i=1}^{^e SC} {}^eA_i\,({}^eG_i) = \sum_{i=1}^{^e SC} \sum_{j=1}^{V_i} a_j\,(j \in {}^eG_i) \tag{9}$$

3.4. The Graph Walk Count

Another realization of the idea to characterize complexity by regarding the structure as a whole is to account for all walks within it (*24–26*; *see* also Chap. 2 in **ref. 15**).

Definition 11. (a) The *graph walk count TWC* is the total number of walks ^lWC of length $l = 1$ to $V - 1$. (b) The *walk count of order l*, ^lTWC is the number of walks of length l. (c) The *vertex walk count vwc_i* is the total number of walks of length l that start in the vertex i.

$$TWC = {}^1WC + {}^2WC + {}^3WC + \ldots + {}^{V-1}WC = \sum_{l=1}^{V-1} {}^lWC = \sum_{i=1}^{V} vwc_i = \sum_{l=1}^{V-1}\sum_{i=1}^{V} {}^lvwc_i \tag{10}$$

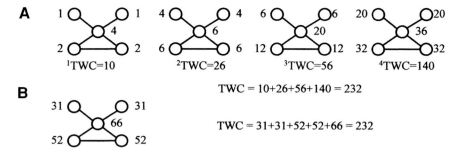

Fig. 4. (**A**) Calculation of the total walk count *TWC (24–26)* as a sum of walk counts of lengths 1 to V – 1. (**B**) Calculation of *TWC* from the vertex walk count sums (**Eq. 10**).

An example is shown in **Fig. 4**. A simple method for calculating vertex walk counts and total walk count is based on the Morgan algorithm. Starting with vertex degrees, one calculates the count of walks of length 1 as the sum of vertex degrees, and then an *extended degree* of *l*th order is calculated for each vertex by summation of the (*l* – 1)th order degrees of the vertex neighbors.

4. Compositional Complexity Measures

Diversity of network composition is another aspect of complexity (*see* Definition 2). Graph representation can incorporate such an aspect by assigning weights to graph vertices; however, additional insight can be gained by Shannon's information theory *(8)*.

4.1. Shannon's Basic Formulas as Applied to Finite Networks

See **refs. 8**, **9**, and **27**.

Consider a network composed of N elements, distributed into k classes, according to a certain equivalence criterion α. The elements could be network vertices, edges, or any other type of subgraphs, as well as distance-based invariants. Denote the number of elements in classes 1, 2, . . . , k as $N_1, N_2, \ldots, N_k$, respectively. The probability for a single randomly chosen network element to belong to the class i is $p_i = N_i/N$, where $N = \Sigma N_i$, and $\Sigma p_i = 1$.

Definition 12. *Shannon's entropy* $H(\alpha)$ of the network distribution $\{N_1, N_2, \ldots, N_k\}$ is

$$H(\alpha) = N\log_2 N - \sum_{i=1}^{k} N_i \log_2 N_i \text{ bits} \tag{11}$$

$$\bar{H}(\alpha) = -\sum_{i=1}^{k} p_i \log_2 p_i \text{ bits/element} \tag{12}$$

Here, base 2 logarithms are used for calculating entropy in binary digits (bits). Entropy is maximum when $N_i = 1$, and $p_i = 1/N$:

$$H_{max}(\alpha) = N \log_2 N; \quad \bar{H}_{max}(\alpha) = \log_2 N \tag{13}$$

Definition 13. The *information content* $I(\alpha)$ of network probability distribution, the *average* and the *normalized* information content, $\bar{I}(\alpha)$ and $I_n(\alpha)$, are respectively

$$I(\alpha) = H_{max}(\alpha) - H(\alpha) = \sum_{i=1}^{k} N_i \log_2 N_i \; ; \; \bar{I}(\alpha) = \frac{1}{N} \sum_{i=1}^{k} N_i \log_2 N_i$$

$$I_n(\alpha) = \frac{1}{N \log_2 N} \sum_{i=1}^{k} N_i \log_2 N_i \; ; \; 0 \le I_n(\alpha) \le 1 \tag{14}$$

Different information-theoretic indices can be introduced depending on the specific equivalence relation α. Such a relation for proteins in a proteome network means they belong to a certain protein complex or to a certain network component. The corresponding information indices of compositional complexity may be termed *information on proteome complexes*, *I*(complexes), and *information on proteome components*, *I*(component), respectively.

The Shannon entropy/information measures based on equivalence of system elements were previously shown *(27)* to fail in evaluating structural or topological complexity. Thus, proceeding from vertex equivalence, one obtains the same null information content for three graphs with very different complexity: a totally disconnected, a single-cycle, and a complete graph. A modification of the theory captured many of the complexity features *(9,28)*. According to this modification, each system element is assigned a weight, uniquely derived from its topology. Then, the distribution of the total graph weight M into weights M_i of individual elements i (vertices, edges, two-edge subgraphs, and so on) is also a probability distribution with the probability of a randomly chosen element to have weight M_i equal to $p_i = M_i/M$.

Definition 14. Weighted information content $^M I$, of the network elements distribution, average and *normalized* weight information content, $^M \overline{I}$ and $^M I_n$, are respectively

$$^M I = \sum_{i=1}^{m} M_i \log_2 M_i; \; ^M \overline{I} = \frac{1}{M} \sum_{i=1}^{m} M_i \log_2 M_i; \; ^M I_n = \frac{1}{M \log_2 M} \sum_{i=1}^{m} M_i \log_2 M_i \qquad (15)$$

Typical network "weight" distributed among its elements is the total adjacency A as partitioned into vertex degrees a_i (**Eqs. 2** and **3**). Similarly, the total network distance $d(G)$ (**Eq. 5**) is considered as composed of vertex distance degrees d_i or, alternatively, of the distance values $d(i) = 1, 2, \ldots, d_{max}$. The information indices thus constructed are termed *vertex degree information index*, I_{vd}, *distance degree information index*, I_{dd}, and *distance information index*, I_d, respectively.

5. Examples of Complexity Assessments of Protein–Protein Interaction Networks, Complexes, and Pathways

5.1. Networks Presented by Nondirected Graphs

Nondirected graphs are used to represent protein–protein interaction networks, the direction of interaction in which is not known or disregarded. Also included here are networks the nodes of which are protein complexes, and an edge between two nodes indicates the presence of a protein common to the two complexes. An example is the functional net of membrane biogenesis and traffic in the *Saccharomyces cerevisiae* proteome extracted from the data of Gavin et al. *(29)*, (**Fig. 5**). The network contains 147 proteins organized into 20 complexes. The corresponding compositional distribution is 147{20, 15, 13, 13, 10, 10, 5 × 7, 5, 5, 3 × 4, 3, 3 × 2}, and **Eq. 14** yields *I*(*complexes*) = 463.9 bits, $\overline{I}$ = 3.16, and I_n = 0.438. The 20 *protein complexes* are distributed into 10 components, 1 large including 11 complexes, and 9 single-complex ones: 20{11, 9 × 1}, from whence *I*(*component*) = 38.05, $\overline{I}$ = 1.903, and I_n = 0.440. The *protein* distribution into components 147{*94*, 10, 10, 3 × 7, 4, 4, 2, 2} differs considerably from the even distribution, which produces high values of information indices: *I′*(*component*) = 761.5, $\overline{I}$ = 5.18, and I_n = 0.720. Thus, the presence of a very large component, typical for biological networks, can be related to a trend toward higher information content.

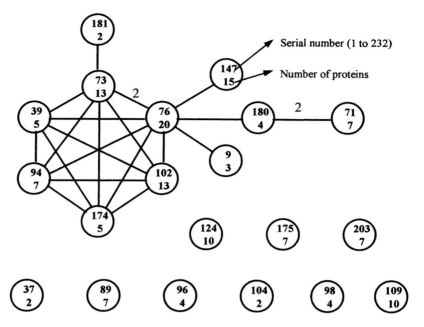

Fig. 5. The membrane biogenesis and traffic network in yeast proteome. Each node represents a protein-protein interaction complex, each edge stands for a protein shared between two complexes. Edge weight denotes the number of proteins shared. (Data from **ref. 29.**)

The large component of the network contains 20 edges, 2 of which are double edges. Connectedness is quite high: $(2 \times 22 \times 100)/(11 \times 10) = 40\%$, owing to the presence of a complete six-vertex subgraph. The vertex degree distribution $\{9, 7, 4 \times 5, 3, 2, 3 \times 1\}$ obeys a power law to a good approximation *(30,31)*. The central point *(17)* of the network is complex *76*, which shares nine proteins with eight other complexes and reaches to all other complexes via only one or two steps. Such central points are regarded as potential marker/drug candidates. They are of crucial importance for the stability of the network.

The Wiener number of the large component is $W = 102$, and the average distance in it is only $<d> = (102 \times 2)/(11 \times 10) = 1.85$. Thus, each pair of complexes in the functional set is connected on average by less than two links, a typical manifestation of a "small-world" network *(32)*, which is another feature of biological networks. The distance distribution $\{20 \times 1, 24 \times 2, 10 \times 3, 1 \times 4\}$ is not very different from the least complex even distribution, as demonstrated by the calculated low value of the information index $I_{d,n} = 0.152$. The distance degree distribution $\{28, 24, 21, 21, 19, 4 \times 16, 15, 12\}$ is considerably more diverse, placing the corresponding information index $I_{dd,n} = 0.554$ in the middle of the 0 to 1 complexity scale.

The three topological complexity measures, when applied to the connected component regarded as a simple graph, yield very high values ($SC = 729{,}449$; $OC = 26{,}594{,}270$; $TWC = 122{,}082{,}804$). It would suffice for characterizing the complexity of large-scale biological networks to use the first several terms of these indices, ^eSC, ^eOC, ^lTWC, with e or $l = 1$ to 3 or 4. Such terms are, for example, the number of two-edge subgraphs, known as the Platt index *(33)*, $^2SC = Pt$, and that of the three-edge subgraphs, ^3SC. For the large component of the membrane biogenesis and traffic network regarded as a multi-graph, one obtains $Pt = 99$ and $^3SC = 518$, respectively. Other examples are $^1OC = 246$, $^2OC = 1714$; $^2TWC = 248$, and $^3TWC = 1338$. All these descriptors can be used for quantitative comparative studies in proteomics, as well as for network structure-activity/toxicity relationships (NSAR). Another application is to quantify the degradation of the network as a result of a sickness or environmental effects. As an example, consider the degradation of the membrane biogenesis and traffic network resulted from the elimination of the central complex 76. The connected component of the network decomposes into a subgraph of six vertices, another subgraph of two vertices, and two isolated vertices. The dramatic loss of network complexity can be illustrated with the decrease in the corresponding complexity descriptors: *Conn*—from 40 to 28.9%; the Wiener index—from 102 to 20; the Platt index—from 99 to 35; ^1OC—from 246 to 98; and ^2OC—from 1724 to 422. The information index on network components also reduces strongly to half its initial value ($38.05 \rightarrow 17.51$).

5.2. Networks Presented by Directed Graphs

The direction of arcs in the graphs representing protein-protein interaction networks is assumed from the bait protein to the interacting partner. As an example, consider the DNA damage response network in yeast proteome (data taken from Ho et al. *(34)*; **Fig. 6**). The network includes 76 proteins organized in five components, a large one with 59 proteins, and four small ones having 7, 4, 4, and 2 proteins, respectively. This highly uneven distribution produces a normalized compositional complexity index of 0.810, close to the upper limit of 1.0. The directed interactions prohibit many paths between proteins thus diminishing network topological complexity. The network descriptors are calculated as the sum of the respective values for all components, their values remaining close to the values of the large dominant component. Thus, for the entire DNA damage response network and its large component, one obtains a connectedness of 3% versus 4.2%, a Platt index of 72 versus 57, and a Wiener number of 273 versus 230. The in and out first-order overall connectivities of the complex are 316 versus 274 and 316 versus 266, respectively; the second-order overall in-connectivities are 1503 versus 1381, and the out ones are 1302 versus 1152. The total walk count of lengths two and three is 72 versus 57, and 52 versus 35, respectively.

The vertex degrees in directed networks also could not reach very high values, 10 being the highest value of in-degree in the large component (however, this indicates that protein Dun1 could be a good potential marker/drug target) versus only 6 for the out-degree. This makes vertex degree distribution less complex, and closer to the even distribution, as witnessed by the low values (0.210 for out-degrees versus 0.177 for the in-degrees) of the corresponding information indices. In contrast, the distance degree distribution is considerably more uneven, owing to the hindered protein-protein communication. Thus, the in-degree distribution in the large component is $\{16 \times 1, 3 \times 2, 8 \times 3, 3 \times 4, 2 \times 5, 3 \times 6, 2 \times 7, 10, 11, 17, 23, 31, 38\}$. However, the presence of 16 distance degrees equal to unity reduces the complexity of the distribution, and the corresponding normalized information index $I_{dd}(in)$ has the medium range value of 0.429. For comparison, if the DNA damage response network were nondirected, the average distance in the large component would increase from 3.90 to 4.53 (still remaining a small-world network), the maximum distance would go up from 4 to 11, and distance

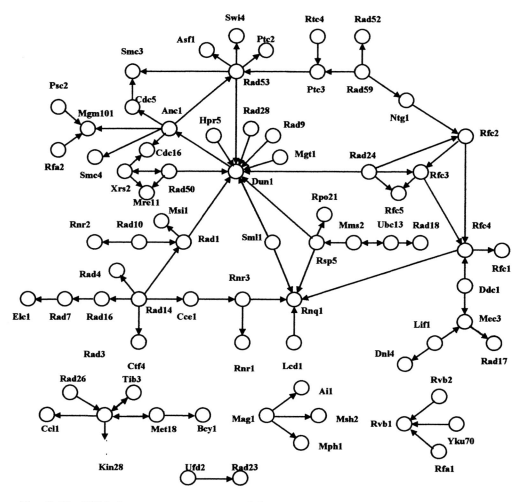

Fig. 6. The DNA damage response network in yeast proteome. Directed edges (arcs) stands for interacting pairs of proteins. Arrows point from the bait protein to the interacting protein. (Data from **ref. 34**.)

degree would span within the range of 159 to 442, with the average degree over 262. The average vertex degree of the nondirected graph is almost twice that of the undirected (2.34 versus 1.22), and the slightly more uneven distribution makes the corresponding complexity index $I_{vd,n}$ to increase in value from 0.177 (in) and 0.210 (out) to 0.220.

It should be mentioned that the examples given in this section are taken from publications dealing with incomplete yeast proteomes. Thus the specific conclusions made could change when the complete proteome is analyzed; they should be regarded only as an illustration of the manner in which the networks can be analyzed. However, the methodology presented here is not limited to protein-protein interaction networks and can be applied to any network, including metabolic, regulatory, and other biological networks, as well as to the pathways within the networks.

Acknowledgments

The comments and suggestions made by Dr. Christoph Rücker (Bayreuth, Germany) are gratefully acknowledged.

References

1. www.ebi.ac.uk/proteome/.
2. Vogelstein, B., Lane, D., and Levine, A. J. (2000) Surfing the p53 Network. *Nature* **408,** 307–310.
3. Kanehisa, M., Goto, S., Kawashima, S., and Nakaya, A. (2002) The KEGG Databases at GenomeNet. *Nucleic Acids Res.* **30,** 42–46; *see* also KEGG Encyclopedia, http://fire2.scl.genome.ad.jp/kegg/kegg2.html.
4. Harary, F. (1972) *Graph Theory.* Addison-Wesley, Reading, MA.
5. Trinajstić, N. (1992) *Chemical Graph Theory.* CRC Press, Boca Raton, FL.
6. Todeschini, R. and Consonni, (2000) V. *Handbook of Molecular Descriptors.* Wiley Europe, Weinheim.
7. Devillers, J. and Balaban, A. T. (eds.) (1999) *Topological Indices and Related Descriptors in QSAR and QSPR.* Gordon & Breach, Reading, UK.
8. Shannon, C. and Weaver, W. (1949) *Mathematical Theory of Communications.* University of Illinois Press, Urbana, IL.
9. Bonchev, D. (1983) *Information-Theoretic Indices for Characterization of Chemical Structures.* Research Studies Press, Chichester, U. K.
10. Nicolis, G. and Prigogine, I. (1989) *Exploring Complexity.* Piper, Munich.
11. Rashewsky, N. (1955) Life, information theory, and topology. *Bull. Math. Biophys.* **17,** 229–235.
12. Mowshowitz, A. (1968) Entropy and the complexity of graphs. I. An index of the relative complexity of a graph. *Bull. Math. Biophys.* **30,** 175–204.
13. Bertz, S. H. (1983) A mathematical model of molecular complexity, in *Chemical Applications of Topology and Graph Theory* (King, R. B., ed.), Elsevier, Amsterdam, pp. 206–221.
14. Bonchev, D. and Polansky, O. E. (1987) On the topological complexity of chemical systems, in *Graph Theory and Topology in Chemistry* (King, R. B. and Rouvray, D.H., eds.), Elsevier, Amsterdam, pp. 126–158.
15. Bonchev, D. and Rouvray, D. H. (eds.) (2003) *Mathematical Chemistry*, vol. VII, *Complexity in Chemistry.* Taylor and Francis, London, UK.
16. Wiener, H. (1947) Structural determination of paraffin boiling points. *J. Am. Chem. Soc.* **69,** 17–20.
17. Bonchev, D. (1989) The concept for the center of a chemical structure and its applications. *Theochemistry* **185,** 155–168.
18. Bonchev, D. (1995) Topological order in molecules. 1. Molecular branching revisited. *Theochemistry* **336,** 137–156.
19. Bonchev, D., Balaban, A. T., Liu, X., and Klein, D. J. (1994) Molecular cyclicity and centricity of polycyclic graphs. I. Cyclicity based on resistance distances or reciprocal distances. *Int. J. Quantum Chem.* **50,** 1–20.
20. Bertz, S. H. and Herndon, W. C. (1986) The similarity of graphs and molecules, in *Artificial Intelligence Applications to Chemistry* (Pierce, T. H. and Hohne, B. A., eds.). ACS, Washington, DC, pp. 169–175.
21. Bertz, S. H. and Wright, W. F. (1998) The graph theory approach to synthetic analysis: definition and application of molecular complexity and synthetic complexity. *Graph Theory Notes NY Acad. Sci.* **35,** 32–48.

22. Bonchev, D. (1997) Novel indices for the topological complexity of molecules. *SAR QSAR Environ. Res.* **7,** 23–43.
23. Bonchev, D. (2000) Overall connectivities/topological complexities: a new powerful tool for QSPR/QSAR. *J. Chem. Inf. Comput. Sci.* **40,** 934–941.
24. Rücker, G. and Rücker, C. (2000) Walk count, labyrinthicity and complexity of acyclic and cyclic graphs and molecules. *J. Chem. Inf. Comput. Sci.* **40,** 99–106.
25. Gutman, I., Rücker, C., and Rücker G. (2001) On walks in molecular graphs. *J. Chem. Inf. Comput. Sci.* **41,** 739–745.
26. Rücker, G. and Rücker, C. (2001) Substructure, subgraph and walk counts as measures of the complexity of graphs and molecules. *J. Chem. Inf. Comput. Sci.* **41,** 1457–1462.
27. Bonchev, D. and Seitz, W. A. (1996) The concept of complexity in chemistry, in *Concepts in Chemistry: A Contemporary Challenge* (Rouvray, D. H., ed.). Research Studies Press, Taunton, UK, pp. 348–376.
28. Bonchev, D. and Trinajstić, N. (1977) Information theory, distance matrix, and molecular branching. *J. Chem. Phys.* **67,** 4517–4533.
29. Gavin, A.-C., Böschke, M., Krause, R., et al. (2002) Functional organization of the yeast proteome by systematic analysis of protein complexes. *Nature* **415,** 141–147.
30. Barabasi, A.-L. and Albert, R. (1999) Emergence of scaling in random networks. *Science* **286,** 509–512.
31. Jeong, H., Tombor, B., Albert R., Oltval Z. N., and Barabasi, A.-L. (2000) The large-scale organization of metabolic networks. *Nature* **407,** 651–654.
32. Watts, D. J. and Strogatz, S. H. (1998) Collective dynamics of "small-world" networks. *Nature* **393,** 440–442.
33. Platt, J. R. (1947) Influence of neighbor bonds on additive bond properties in paraffins. *J. Chem. Phys.* **15,** 419–420.
34. Ho, Y., Gruhler, A., Heilbut, A., et al. (2002) Systematic identification of protein complexes in *Saccharomyces cerevisiae* by mass spectrometry. *Nature* **415,** 180–183.

32

Patchwork Peptide Sequencing

Protein Identification by Accurate Mass-to-Sequence Conversion of High-Resolution Q-TOF Tandem Mass Spectrometry Data

Andreas Schlosser and Wolf D. Lehmann

1. Introduction

The advent of new analyzer systems based on Fourier-transform ion cyclotron resonance (FT-ICR) and quadropole time-of-flight (Q-TOF) technologies allows protein digests to be analyzed under high-resolution conditions, resulting in mass spectometry (MS) and tandem MS (MS/MS) peptide mass spectra characterized by a resolution of 10,000 [full width half-maximum (FWHM) definition] or better. From mass spectra recorded under these conditions, the m/z values of peptides can be measured with mass errors between somewhat below 1 ppm and 50 ppm, depending mainly on the type of instrument used, the sample purity, the signal intensity, and the mode of calibration employed.

For the peptide mass fingerprinting (PMF) approach *(1,2)* accurate mass data generated, e.g., by delayed extraction matrix-assisted laser desorption/ionization (MALDI)-TOF or Q-TOF analyzers are of great benefit, since the specificity of PMF is directly related to the mass tolerance of the molecular weight data *(3,4)*.

In the alternative strategy of protein identification by electrospray (ESI) MS/MS, the benefits of using accurate mass data are not as obvious as for the PMF approach, since the specificity of the MS/MS approach is based on sequence information, which often can be extracted to a sufficient extent from tandem mass spectra recorded at low to medium mass resolution *(5)*. Nevertheless, peptide MS/MS spectra recorded at high resolution with high mass accuracy contain much more analytical information of higher specificity than those recorded at low resolution. For instance, discrimination between K and Q residues has been demonstrated by MS/MS with accurate mass measurement *(6,7)*. In addition, high-resolution conditions have been used to identify most of the low-mass fragment ions observed in tandem mass spectra of peptides as internal dipeptide b ions *(8)*. These dipeptide b ions can be regarded as sequence quanta, since they represent the smallest unit of sequence information. It has been demonstrated that these sequence quanta can be used for a highly efficient protein identification based on a direct accurate mass-to-sequence conversion, an approach termed *patchwork peptide sequencing (8)*. In the following, this new strategy is demonstrated by complex examples, and its relevance for the field of peptide sequencing is validated.

From: *Handbook of Proteomic Methods*
Edited by: P. Michael Conn © Humana Press Inc., Totowa, NJ

2. Materials and Methods

2.1. Sample Preparation

1. Ubiquitin was from Sigma Aldrich (Taufkirchen, Germany).
2. Recombinant protein kinase A catalytic subunit was kindly supplied by V. Kinzel and D. Bossemeyer (DKFZ Heidelberg, Germany).
3. Digestion with trypsin (Roche, Mannheim, Germany) was performed in 0.1 M ammonium hydrogen carbonate (pH 8) overnight at 37°C.
4. Prior to analysis by nanoESI MS, all samples were desalted with C_{18}-ZipTips (Millipore, Bedford, MA) using 2% formic acid for equilibration and washing and 50% acetonitrile, 2% formic acid for peptide elution.

2.2. Mass Spectrometry

1. Mass spectra were recorded on a Q-TOF mass spectrometer type Q-TOF2 (Micromass, Manchester, UK) equipped with a nanoESI source.
2. Spray capillaries were manufactured in-house using a micropipet puller type P-87 (Sutter Instruments, Novato, CA) and coated with a semitransparent film of gold in a sputter coater type SCD 005 (BAL-Tec, Balzers, Liechtenstein).
3. The mass spectrometer was calibrated over the mass range 100–2000 using a 20 mM solution of phosphoric acid (in 30% isopropanol).
4. Internal recalibration was performed using a single lock mass as described in the text.

3. Results and Discussion

The data presented in this study were generated by nanoESI MS/MS on a Q-TOF instrument. The MS/MS spectra were recorded by fragmenting a small tryptic peptide and a small intact protein. **Figure 1** shows the low-mass region of the nanoESI MS/MS spectrum of a doubly charged tryptic peptide (*m/z* 582.8) from recombinant protein kinase A catalytic subunit. The signal at *m/z* 147 was assigned as the y_1 ion of lysine and was used for internal recalibration. It is known from earlier studies that the *m/z* values of an internally recalibrated spectrum show a ±3 SD tolerance equivalent to 10 mDa *(8)*. On this basis, the *m/z* values of the spectrum shown in **Fig. 1** were assigned by comparison with the calculated accurate mass values of our reference mass list, which is available as a download file from the Internet *(9)*. This mass list contains the accurate mass values of 19 immonium ions, 19 y_1 ions, 38 tryptic y_2 ions, 190 dipeptide a ions, and 190 dipeptide b ions.

The result of this assignment is summarized in **Table 1**; 13 peaks could be assigned in this way. Generally, signals from immonium and single amino acids can be identified unambiguously. Among the dipeptide ions often more than one ion is compatible with an experimental accurate mass. The ambiguities involve composition, ion type, and sequence direction. Some of these ambiguities can be resolved by additional information. For instance, the fragment ion at *m/z* 225.0994 can be assigned either as a dipeptide a ion of DH or as a dipeptide b ion of SH. However, the accompaying loss of CO allows the clear assignment of this ion fragment as a b-type ion.

Following identification of the y_1 ion, the y_2 ion is seen to contain unidirectional sequence information. Unfortunately, this is not the case for b_2 ions, since b_1 ions are normally not observed in the MS/MS spectra of unmodified peptides. Accurate mass-based assignment of tripeptide ions is connected with even more ambiguities than assignment of dipeptides. As a rational method to restrict this diversity, we decided to

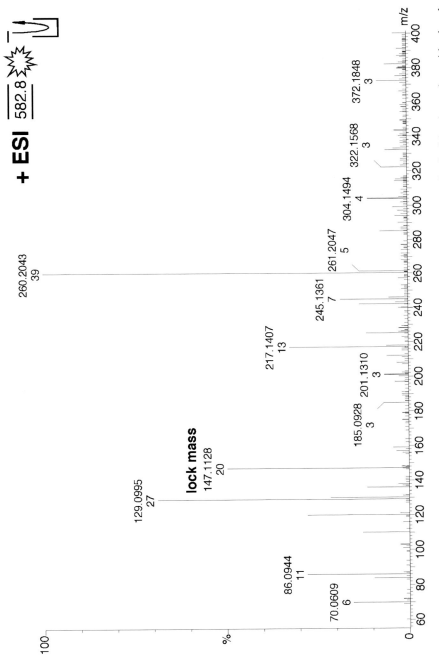

Fig. 1. Low-mass region of the nano-electrospray ionization (ESI) MS/MS spectrum of a doubly charged peptide signal observed in the tryptic digest of protein kinase A catalytic subunit. The ion at m/z 147 was assigned as the y_1 ion of lysine and used for internal recalibration. The numbers on top of the mass-annotated peaks indicate the intensity (counts).

Table 1
Accurate Mass-Based Assignment (Mass Tolerance ± 10 mDa) of Ion Signals from the Partial MS/MS Spectrum in Fig. 1[a]

	Intensity	*m/z*	Ion type	Neutral loss
Immonium ions	11	86.0944	**im-X**	
	11	120.0792	**im-F**	
	6	70.0609	**im-P**	
	5	110.0695	**im-H**	
	5	136.0752	**im-Y**	
Single-residue ions	27	129.0995	**y_1-K-H_2O**	+H_2O
	20	147.1128	**y_1-K**	–H_2O
Dipeptide ions	39	260.2043	**y_2-XK**	–H_2O
	13	217.1407	**a_2-FP**	+CO
			a_2-XM	
	7	245.1361	**b_2-FP**	–CO
			b_2-XM	
	5	225.0994	**b-SH**	–CO
			a-DH	
	3	185.0928	**b-PS**	–CO
			a-DP	
Tripeptide ions	3	322.1568	**b-PSH**	–CO

[a]The most abundant dipeptide b ion was assigned as the b_2 ion. The presence of –CO satellite ions was used as criterion for b ion assignment. The assignments are in bold. X = L or I

consider only those tripeptide combinations that contain a previously identified dipeptide. The sequence possibilities for the tripeptides were selected with the support of an Internet tool correlating tripeptide accurate mass values and their composition (tripeptide finder) *(10)*.

For performing a database search in a protein database such as SwissProt, all peptide sequences compatible with the data in **Table 1** have to be considered, owing to the lack of information on the direction of the sequence. These possibilities are summarized in **Table 2** in the nomenclature required for input into the search engine Mascot *(11)*.

The six sequence possibilities for the tripeptide ion with the composition PSH could be reduced to the two sequences PSH and HSP using the two identified birectional b ions PS and SH. The variety of sequences summarized in **Table 2** also takes into account the ambiguity and bidirectionality in b_2 ion assignment, which leads to four structures of the b_2 ion. The y_2 ion carries an unidirectional sequence information and a leucine/isoleucine ambiguity, which cannot be resolved on the basis of an accurate peptide mass. As listed in **Table 2**, the combination of the patchwork-type sequence information results in a set of eight different combinations, of which only one results in a significant search hit. The summary of the corresponding Mascot search result is displayed in **Fig. 2**.

Subsequently, the same strategy was applied to the MS/MS spectrum of the small intact protein ubiquitin. As the precursor ion, the $[M+12H]^{12+}$ ion at *m/z* 714.6 was selected. The low-mass region of the MS/MS spectrum obtained in this way is given in **Fig. 3**.

Table 2
**Permutations of all Patchwork-Type Sequence Information
Compatible with the Accurate Mass-Based Ion
Assignments Given in Table 1**[a]

Search input	Search result
c-[IL]K + n-FP + b-PSH	Single significant hit (score 90)
c-[IL]K + n-PF + b-PSH	No result
c-[IL]K + n-[IL]M + b-PSH	No result
c-[IL]K + n-M[IL] + b-PSH	No result
c-[IL]K + n-FP + b-HSP	No result
c-[IL]K + n-PF + b-HSP	No result
c-[IL]K + n-[IL]M + b-HSP	No result
c-[IL]K + n-M[IL] + b-HSP	No result

[a]Only one combination gives a significant hit by protein database searching
with the search engine Mascot *(11)*. The peptide FPSHFSSDLK of protein
kinase A catalytic subunit is identified. c-, C-terminal sequence; n-, N-terminal
sequence; b-, internal sequence; all sequences are given in the standard direction, N-terminal to C-terminal from left to right. Thus the information in the first
line is compatible with a peptide sequence FPSH—[IL]K)

{MATRIX}
{SCIENCE} **Mascot Search Results**

```
User          : Andreas Schlosser
Email         : andreas.schlosser@charite.de
Search title  :
Database      : SwissProt 40.27 (184988 sequences; 100860165 residues)
Timestamp     : 10 Sep 2002 at 13:57:59 GMT
Top Score     : 90 for P36887, cAMP-dependent protein kinase, alpha-catalytic subunit
```

Probability Based Mowse Score

Score is -10*Log(P), where P is the probability that the observed match is a random event.
Protein scores greater than 65 are significant (p<0.05).

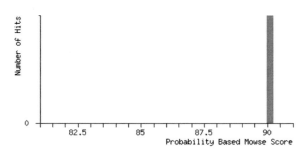

Fig. 2. Search result of the search engine Mascot using as input data the complete set of
sequence information summarized in **Table 2**. A single significant hit is obtained identifying a
cAMP-dependent protein kinase catalytic subunit on the basis of the peptide FPSHFSSDLK.

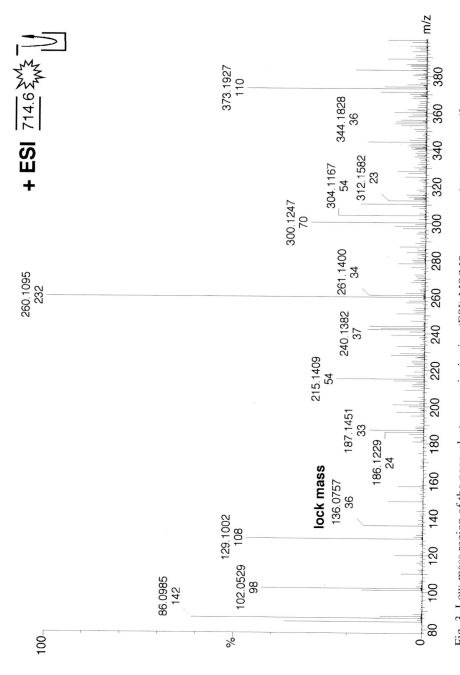

Fig. 3. Low-mass region of the nano-electrospray ionization (ESI) MS/MS spectrum of the [M+12H]$^{12+}$ ion of ubiquitin. The ion at m/z 136 was assigned as the immonium ion of tyrosine and used for internal recalibration. The numbers on top of the mass-annotated peaks indicate the intensity (counts).

Table 3
Accurate Mass-Based Assignment of Ion Signals from the Partial MS/MS Spectrum of Ubiquitin Given in Fig. 3[a]

	Intensity	*m/z*	Ion type	Neutral loss
Immonium ions	142	86.0985	**im-X**	
	98	102.0529	**im-E**	
	60	104.0509	**im-M**	
	36	136.0757	**im-Y**	
	17	120.0779	**im-F**	
	13	110.0697	**im-H**	
Single-residue ions	108	129.1002	**y₁-K-H₂O**	
Dipeptide ions	232	260.1095	**b₂-MQ**	−CO
			b₂-CR	
	54	215.1409	a-EX	−CO
			a-KN	
			b-TX	
	33	187.1451	**a-TX**	+CO
	24	186.1229	a-NV	
			b-GK	
	16	157.0937	**a-PS**	+CO
			b-GV	
	10	185.0959	a-DP	−CO
			b-PS	
Tripeptide ions	110	373.1927	**b₃-MQX**	−CO
			b₃-CRX	
			b-SVW	
			b-FPQ	
			b-ERS	
			b-DRT	
	36	344.1828	**b-TXE**	−CO

[a]The most abundant dipeptide b ion was assigned as the b₂ ion. The presence of −CO satellite ions was used as criterion for b ion assignment. The assignments are in bold.

The spectrum was recalibrated using the immonium ion of tyrosine at *m/z* 136, and the assignment of the ions was performed as described above for the MS/MS spectrum displayed in **Fig. 1**. The identifications of the low-mass fragment ions of ubiquitin are summarized in **Table 3**.

From the fragment ion assignments summarized in **Table 3**, the sequence information was extracted and assembled in all possible combinations, as listed in **Table 4**. These input data were used in a single search in combination with the molecular weight of 8564.5 ± 0.5 Daltons determined by the ESI survey spectrum. As experienced for the PKA-derived peptide described above, only one combination resulted in a significant hit that correctly identified ubiquitin with the sequence MQIFVKTLTGKTITLEVEPSDTIENVKAKIQDKEGIPPDQQRLIFAGKQLED GRTLSDYNIQKESTLHLVLRLRGG.

The search result is displayed in **Fig. 4**.

Table 4
Permutations of all Patchwork-Type Sequence Information
Compatible with the Accurate Mass-Based Ion Assignments
Given in Table 3[a]

Search input	Search result
n-MQ[IL] + b-T[IL]E + b-PS	Single significant hit (score 79)
n-QM[IL] + b-T[IL]E + b-PS	No result
n-MQ[IL] + b-[IL]TE + b-PS	No result
n-MQ[IL] + b-T[IL]E + b-SP	No result
n-QM[IL] + b-[IL]TE + b-PS	No result
n-QM[IL] + b-T[IL]E + b-SP	No result
n-MQ[IL] + b-[IL]TE + b-SP	No result
n-QM[IL] + b-[IL]TE + b-SP	Nonsignificant hit (score 48)
n-CR[IL] + b-T[IL]E + b-PS	No result
n-RC[IL] + b-T[IL]E + b-PS	No result
n-CR[IL] + b-[IL]TE + b-PS	No result
n-CR[IL] + b-T[IL]E + b-SP	No result
n-CR[IL] + b-[IL]TE + b-SP	No result
n-RC[IL] + b-T[IL]E + b-SP	No result
n-RC[IL] + b-[IL]TE + b-PS	No result
n-RC[IL] + b-[IL]TE + b-SP	No result

[a]Only one combination gives a significant hit by protein database searching with the search engine Mascot *(11)*. Sequence information is written as required for Mascot (for an explanation, *see* **Table 2**).

4. Summary and Outlook

The patchwork peptide sequencing approach outlined in this study uses similar information as the established sequence-tag approach *(12)* in that internal protein sequences are extracted from MS/MS spectra as a basis for protein identification. However, a sequence tag represents an internal continuous stretch of amino acids (from 3 up to about 10 residues), which is derived from the mass differences of an ion fragment series. In contrast, the patchwork peptide sequencing approach uses a set of discontinuous, sometimes bidirectional stretches of two or three amino acid residues derived from absolute mass values. The accurate mass-to-sequence conversion is supported by experience-based fragmentation rules and combinatorial considerations. The patchwork strategy uses previously neglected or unrecognized sequence information in the low-mass region of the MS/MS spectra, which has become accessible with the introduction of high-resolution mass analyzers. It is a highly useful new tool, which supports the established strategies, e.g., in cases in which MS/MS spectra contain only a minor amount of sequence information in their classical b and y ion series. The patchwork approach also has the potential to be applied for the identification of covalently modified peptides, as shown by the identification of an elastase-generated phosphopeptide *(8)*. All steps in this pilot study have been performed manually. However, an appropriate software might be developed for the complete patchwork sequencing procedure directly assessing the mass spectrometry raw data. The approach will gain further specificity in parallel to the ongoing accuracy improvement in mass determination.

{MATRIX SCIENCE} **Mascot Search Results**

User   : **Andreas Schlosser**
Email   : **andreas.schlosser@charite.de**
Search title :
Database   : **SwissProt 40.26x (184160 sequences; 100505185 residues)**
Timestamp   : **30 Aug 2002 at 14:05:00 GMT**
Top Score   : **79 for P02248, Ubiquitin**

Probability Based Mowse Score

Score is $-10*Log(P)$, where P is the probability that the observed match is a random event. Protein scores greater than 65 are significant ($p<0.05$).

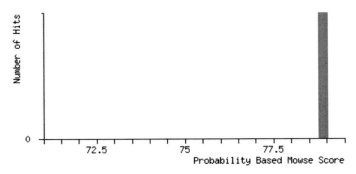

Fig. 4. Search result of the search engine Mascot using as input data the complete set of sequence information summarized in **Table 4**. A single significant hit is obtained identifying ubiquitin.

References

1. Henzel, W. J., Billeci, T. M., Stults, J. T., Wong, S. C., Grimley, C., and Watanabe, C. (1993) Identifying proteins from two-dimensional gels by molecular mass searching of peptide fragments in protein sequence database. *Proc. Natl. Acad. Sci. USA* **90,** 5011–5015.
2. Pappin, D. J. C., Hojrup, P., and Bleasby, A. J. (1993) Rapid identification of proteins by peptide-mass fingerprinting. *Curr. Biol.* **3,** 327–332.
3. Jensen, O. N., Podtelejnikov, A. V., and Mann, M. (1997) Identification of the components of simple protein mixtures by high-accuracy peptide mass mapping and database searching. *Anal. Chem.* **69,** 4741–4750.
4. Clauser, K. R., Baker, P., and Burlingame, A. L. (1999) Role of accurate mass measurement (±10 ppm) in protein identification strategies employing MS or MS/MS and database searching. *Anal. Chem.* **71,** 2871–2882.
5. Hunt, D. F., Yates, J. R. III, Shabanowitz, J., Winston, S., and Hauer, C. R. (1986) Protein sequencing by tandem mass spectrometry. *Proc. Natl. Acad. Sci. USA* **83,** 6233–6237.
6. D'Agostino, P.A., Hancock, J. R., and Provost, L. R. (1997) Analysis of bioactive peptides by liquid chromatography-high-resolution electrospray mass spectrometry. *J. Chromatogr. A* **767,** 77–85.

7. Bahr, U. and Karas, M. (1999) Differentiation of isobaric peptides and human milk oligosaccharides by exact mass measurements using electrospray ionization orthogonal time-of-flight analysis. *Rapid Commun. Mass Spectrom.* **13,** 1052–1058.

8. Schlosser, A. and Lehmann, W. D. (2002) Patchwork peptide sequencing—extraction of sequence information from accurate mass data of peptide tandem mass spectra recorded at high resolution. *Proteomics* **2,** 524–533.

9. http://www.dkfz-heidelberg.de/z_spek/MS/startseite.htm.

10. http://www.colby.edu/chemistry/NMR/scripts/tripeptides.html.

11. http://www.matrixscience.com.

12. Mann, M. and Wilm, M. (1994) Error-tolerant identification of peptides in sequence databases by peptide sequence tags. *Anal. Chem.* **66,** 4390–4399.

33

Estimation of Bias in Proteome Research

Ralf Mrowka and Hanspeter Herzel

1. Introduction

This chapter focuses on statistical analysis of large-scale yeast two-hybrid (Y2H) experiments compared with the data on protein–protein interactions obtained by single Y2H experiments in traditional laboratory setups. Reports on protein–protein interactions from large-scale experiments differ qualitatively from results obtained in traditional setups. We discuss possible reasons for the discrepancies and estimate the possible fraction of false-positive protein–protein interaction reports in the large-scale datasets.

The number of base pairs of DNA sequences in the public databases has increased rapidly, and for a large number of lower organisms, complete genomes are available *(1)*. The natural next step in analysis was annotation of the genomes by means of bioinformatics *(2)* and experimental tools *(3)*. Finding the DNA coding regions for proteins is not an easy task, as illustrated by the fact that the "correct" number of genes in the human genome is still unknown and the initial predictions of individual genes of the two sequencing projects show little overlap *(4)*. Having found the coding regions for a putative protein, the biological function of the protein has to be characterized. This task is even harder than finding the coding parts of the DNA. One important step in the functional analysis of proteins and in understanding their functions is the determination of possible interaction partners. Proteins often "act" in protein complexes, i.e., multiple proteins interact physically and then exhibit a common biological function. This observation leads to a demand for knowledge of the physical interaction of proteins. Considering the high number of proteins encoded in the DNA of each given organism, it becomes clear that single by single experiments done by hand are not feasible. Therefore, automated high-throughput methods have been developed.

2. The Yeast Two-Hybrid Assay

One underlying method to test for physical interaction is the Y2H assay *(5)*. The result of a Y2H experiment is a type of color reaction indicating the interaction, i.e., the answer falls into two categories: interaction and no interaction. The inherent property of any biological assay is that it may give false results, i.e., the result may therefore be classified as false positive, false negative, true positive, or true negative. At this point, it should be mentioned that it might be problematic to classify the result of the Y2H experiment into the categories of false and true. A classification of a physical interac-

From: *Handbook of Proteomic Methods*
Edited by: P. Michael Conn © Humana Press Inc., Totowa, NJ

tion into *present* or *not present* is the highest degree of simplification of the property—interaction—studied. In addition, the interaction may depend on many other biological factors not tested in the Y2H system. For simplicity, however, we assume that this classification is applicable.

3. Statistical Evaluation of Databases

We can use statistical methods to evaluate databases with the results of Y2H experiments. Each database for Y2H experiments may be regarded as a subset sample of all possible Y2H experiments. The idea behind the following analysis is very simple. If we randomly take two large samples from a population and compare a statistical feature of each sample, the features should not differ significantly. For instance, if two researchers sample randomly married couples in Europe and calculate the distribution of the arithmetic mean of the age of each couple, the two distributions should not differ. If the age distributions differ, then the sampling procedure was not the same for both datasets, i.e., it was biased. In the case of the married couples, one explanation could be that one researcher recruited his samples mainly from a region in Europe in which the typical age of marriage is above average. We note in passing that the finite sample size might also induce a bias of certain estimated statistical quantities. In this work, we focus on a bias caused by inappropriate sampling induced, for instance, by biased selection or false positives.

Focusing on protein–protein interactions, we can analyze two large Y2H protein–protein interaction datasets *(6,7)*, which represent the results of high-throughput approaches. In addition, a literature collection of single by single Y2H experiments is available (part of the MIPS *(8)* database). In theory, high-throughput approaches should not be biased by hypotheses caused by the large-scale design and are therefore regarded as hypotheses-free approaches.

With the protein–protein interaction elements of each database we have defined the sets of all possible interactions. Now we want to test whether the sampling procedure (experiments) was identical for each database. As described earlier, this requires at least one other characteristic biological property for each sample. For instance, the expression levels of the corresponding genes can be analyzed. As a byproduct, we can detect possible dependencies between the "interactome" and the "transcriptome" *(9,10)*.

Fortunately, a number of genome-wide expression-profiling experiments have been described in the literature. For example, one study refers to the mitotic cell cycle. At different time points, the mRNA concentration was measured for each gene of the yeast. The statistical feature we are referring to is the correlation of the mRNA concentrations of the genes of each protein–protein pair. To take the expression correlation as an independent measure only works if there is a nonrandom relationship between protein–protein interaction and expression. To assume that there is a relationship is biologically reasonable, since it makes sense that proteins that interact are also present in the cell at the same time. To test for positively or negatively correlated pairs compared with random controls, one can derive a histogram of correlation coefficients *(h)* between mRNA abundance levels for protein pairs. For example, in analyzing gene expression data on the yeast's cell cycle *(11)*, as well as Y2H *(6,7)* data, we find a significant shift toward positive *h* values ($p < 10^{-7}$, Kolmogorov-Smirnov test). This means that the amount of positively regulated protein–protein pairs exceeds the number we expect for randomly chosen controls, indicating that there is a nonrandom relationship between protein–protein interaction and gene expression.

Now, we can compare pooled single experiments from the literature (literature data) with the hypothesis-free genome-wide Y2H experiments. We find that the ratio of positively to negatively regulated genes is significantly higher for the literature data compared with the genome-wide experiments *(9)*. Protein–protein interactions from single experiments show more positively correlated gene expression profiles compared with interactions from the large-scale experiments. This means that interacting partners from the literature data are more closely related to each other with respect to gene regulation compared with genome-wide databases. At first glance, this seems to prove a selection bias in the "literature data." Reasons might be found in the failure to report on interaction, which cannot be understood from prior knowledge, or failure to perform experiments on such interactions in the first place. The *a priori* hypothesis of the researcher about protein function and other properties may have influenced the single-experiment design and interpretation, leading to a possibly biased protein–protein interaction data pool in the literature.

However, it must be noted that there might be a second major reason for the difference between the literature Y2H dataset and the genome-wide Y2H approaches. Because of the large-scale nature of the Y2H assays, it might be that they may contain many false-positive reports on protein–protein interactions. Again, a statistical procedure might give us some figures related to this hypothesis. Let us assume that the literature data constitute a nonbiased representative subsample (gold standard) of all possible true-positive protein–protein interactions and the genome-wide scans are affected by false positives. Because the number of all true-positive interactions is much less than all possible combinations of protein–protein interactions, it is reasonable to assume that a false-positive interaction would "behave" statistically like pairs of randomly chosen proteins with respect to other properties. Under these assumptions, we can estimate the number of false positives in the genome-wide scans by adding a fraction of random pairs to the defined gold standard data-set until the statistical properties of the genome-wide Y2H datasets are reached. By doing this we have to add between 47 and 91% false positives *(9)* to the genome-wide datasets to explain the difference between the literature data and the genome-wide Y2H datasets. Again, this figure holds only if we assume that the literature dataset is the "true" gold standard.

In a recent paper by von Mering et al. *(12)*, a similar method is described to estimate false positives and false negatives in the protein–protein interaction databases. The authors define a type of gold standard from trusted literature data and compute the coverage, that is, the fraction of the reference set covered by data, as well as the accuracy, that is, the fraction of data confirmed by the reference set. Consistent with previous studies *(9)*, the authors find that the different methods for protein–protein interaction detection differ considerably. A combination of more than one method will increase the accuracy of the prediction.

4. Conclusions

In summary, we have shown that genome-wide Y2H methods in protein-protein interaction detection differ quantitatively from Y2H "literature data" obtained by single experiments. Hypothesis-driven single experiments may harbor a selection bias toward correlated pairs, whereas false-positive reports in genome-wide scans may lead to a selection bias toward random pairs. It is important that the reasons for the differences in results of the different approaches using the identical underlying Y2H system be identified. A statistical analysis may give first clues to interpretations. However, a statistical analysis has to be complemented by an appropriate experimental design, to find the reasons for discrepancies in the databases. Independent data sources *(13)* such as structural information, pull-down experiments *(14)*, and functional assays may help to deal with this important question.

References

1. Benson, D. A., Karsch-Mizrachi, I., Lipman, D. J., Ostell, J., Rapp, B. A., and Wheeler, D. L. (2002) GenBank. *Nucleic Acids Res.* **30,** 17–20.
2. Hubbard, T., Barker, D., Birney, E., et al. (2002) The Ensembl genome database project. *Nucleic Acids Res.* **30,** 38–41.
3. Shoemaker, D. D., Schadt, E. E., Armour, C. D., et al. (2001) Experimental annotation of the human genome using microarray technology. *Nature* **409,** 922–927.
4. Hogenesch, J. B., Ching, K. A., Batalov, S., et al. (2001) A comparison of the Celera and Ensembl predicted gene sets reveals little overlap in novel genes. *Cell* **106,** 413–415.
5. Fields, S. and Song, O. (1989) A novel genetic system to detect protein-protein interactions. *Nature* **340,** 245–246.
6. Uetz, P., Giot, L., Cagney, G., et al. (2000) A comprehensive analysis of protein-protein interactions in *Saccharomyces cerevisiae*. *Nature* **403,** 623–627.
7. Ito, T., Chiba, T., Ozawa, R., Yoshida, M., Hattori, M., and Sakaki, Y. (2001) A comprehensive two-hybrid analysis to explore the yeast protein interactome. *Proc. Natl. Acad. Sci. USA* **98,** 4569–4574.
8. Mewes, H. W., Frishman, D., Gruber, C., et al. (2000) MIPS: a database for genomes and protein sequences. *Nucleic Acids Res.* **28,** 37–40.
9. Mrowka, R., Patzak, A., and Herzel, H. (2001) Is there a bias in proteome research? *Genome Res.* **11,** 1971–1973.
10. Jansen, R., Greenbaum, D., and Gerstein, M. (2002) Relating whole-genome expression data with protein-protein interactions. *Genome Res.* **12,** 37–46.
11. Cho, R. J., Campbell, M. J., Winzeler, E. A., et al. (1998) A genome-wide transcriptional analysis of the mitotic cell cycle. *Mol. Cell* **2,** 65–73.
12. von Mering, C., Krause, R., Snel, B., et al. (2002) Comparative assessment of large-scale data sets of protein-protein interactions. *Nature* **417,** 399–403.
13. Grünenfelder, B. and Winzeler, E. A. (2002) Treasures and traps in genome-wide data sets: case examples from yeast. *Nat. Rev. Genet.* **3,** 653–661.
14. Gavin, A. C., Bosche, M., Krause, R., et al. (2002) Functional organization of the yeast proteome by systematic analysis of protein complexes. *Nature* **415,** 141–147.

34

Scoring Functions for Mass Spectrometric Protein Identification

Robin Gras, Patricia Hernandez, Markus Müller, and Ron D. Appel

1. Introduction

The new reliability and availability of mass spectrometric instruments and of protein separation techniques associated with complete sequencing of several hundred genomes (www.ebi.ac.uk/genomes) allow us to carry out large gene and protein expression studies (proteomics) *(1,2)*. The need to produce, manage, and analyze a huge amount of data calls for the application of specific biological and bioinformatics techniques. A complete protein project is built in several steps. Proteins must be purified to make analysis by mass spectrometry (MS) feasible. Then the spectra must be analyzed, providing a list of peptide masses in the case of peptide mass fingerprinting (PMF spectra or MS spectra) or peptide fragment masses in the case of MS/MS spectra.

There are three main approaches for protein identification using MS data *(3,4)*: the PMF approach for a comparison between peptide masses, ion search for a comparison between fragment ion masses, and *de novo* sequencing for inference of the peptide sequence from the fragment ion masses followed by sequence matching. All these methods require a comparison between experimental data and theoretical data from a sequence database. The difficulty of these approaches comes from the large number of parameters included in the complete process (mixture of proteins in gel, calibration of spectra, chemical and biological modification of peptides, possible external contaminants, and so on) and the measurement uncertainties that have to be taken into account for an automatic identification. It is therefore essential to develop tools for handling MS data efficiently in spite of these difficulties. A fundamental aspect is the scoring schemes used for comparing experimental and theoretical data. In this chapter we present an overview of the standard scoring functions of the MS and tandem MS (MS/MS) identification tools. We particularly focus our discussion on the specificities and limitations of each method and show some recent innovative advances in this field.

2. MS Identification

Protein identification by PMF involves the comparison of a list of experimental peptide masses with a list of theoretical peptide masses obtained from a virtually digested protein sequence. There are two kinds of noise that affect the complexity of the identification. Theoretical noise corresponds to the size of the database, the variability in

From: *Handbook of Proteomic Methods*
Edited by: P. Michael Conn © Humana Press Inc., Totowa, NJ

sequence length, and the number of possible mass modifications, whereas experimental noise corresponds to the limited resolution of mass measurements, calibration errors, contaminants, and protein mixture. The identification programs must be able to deal with most of these properties to provide correct identifications.

2.1. Mono-Spectrum Identification

Different scoring schemes may be used for PMF. The simplest one defines the score of each protein in the database by the number of matches, that is, the number of experimental masses that differ from a theoretical mass by a value smaller than a certain threshold. Here, the noise intervenes in the scoring process in only a discrete way. The principal programs based on this scheme are PeptideSearch *(5)*, PepSea *(6)*, and PeptIdent/MultiIdent *(7,8)*. These programs can work efficiently with high-quality spectra but are limited by the fact that their scoring scheme has a low discrimination power, often allowing ambiguous identifications and also because the size of the protein database strongly influences the result. Post-translational modifications and mutations can be taken into account by these programs, especially by PeptIdent, which uses SWISS-PROT annotations to select the protein modifications consistent with the database information.

The second approach is based on a probabilistic or a statistical scoring scheme. Mowse *(9)* and MS-Fit *(10)* derive their scoring from a statistical analysis of the frequency of peptide masses in the OWL database. All proteins in the database are subdivided according to their mass into bins of 10 kDa. For each bin, the frequency of peptide masses is calculated for each 100-Dalton interval. The identification score of a protein is the product of the inverse frequencies of all matching peptides. Possible missed cleavages are taken into account by weighting the matches corresponding to missed cleaved peptides with a user-defined coefficient (Pfactor). This scoring scheme uses important information from the database composition, but no further experimental information (isoelectric point, molecular weight) and no further attributes of the candidate protein. MassSearch *(11)* from the Darwin library and ProFound *(12)* from the PROWL server use different probabilistic scoring schemes. MassSearch bases its identification score on the probability of a random match between n experimental and n theoretical peptide masses, given the range of the experimental peptide masses and the maximal mass deviation. This computation is repeated several times. At each step, a new mass is added by increasing the allowed maximal mass deviation until an optimal n is found. ProFound calculates the probability, given by a bayesian formula, for the identification of the correct protein using the distance between experimental and theoretical masses and the protein sequence length obtained from the NCBInr database. ProFound also uses a weighting factor to increase the influence of peptides that are found in consecutive regions in the candidate protein sequence. The authors also claim to consider experimental factors in the scoring formula, but no details have been published so far. Mascot *(13)* combines a statistical and a probabilistic approach. The Mowse frequency model is integrated with a score based on the probability of obtaining a given number of matches for a given size of database. The scoring scheme is not given explicitly, but it is claimed that information about match accuracy and peptide modifications is included.

Eriksson et al. *(14)* carried out a comprehensive statistical analysis of random matches in PMF. They calculated the score distribution for random matches as a means of finding α-significance thresholds for two scoring schemas: the shared peak count and the Profound scoring. The α-significance threshold S_C is defined by the condition that the probability to obtain a random match with a score higher than S_C is α. The authors investigated the influence on the critical score S_C of the number of masses in the peptide map, the mass accuracy, the number of missed cleavage sites, the protein mass, and the size of the database. The random tryptic peptide fingerprints used to calculate the score distribution were generated by digesting yeast open reading fragments and picking peptides randomly.

Machine learning approaches have also been used to infer statistical significant information from data. The tool SmartIdent *(15,16)* uses a genetic algorithm to discover an optimal scoring for automatic protein identification from most of the available experimental and theoretical information. The idea is to develop a tool for high-throughput PMF identification that is totally automatic. The scoring scheme is composed of a set of weighted parameters including peak intensity, peptide hydrophobicity, peptide C-terminal amino acid, occurrence of chemical modification and missed cleavage, global correspondence between experimental and theoretical masses, and percentage of sequence covering. The genetic algorithm learned the weights of these parameters in the global score and also the optimal values of filter parameters, such as allowed peptide mass error or the minimal number of peptide matches. The training dataset consisted of real spectra from proteins already identified by MS and confirmed by at least one other method (amino acid composition, microsequencing, or gel matching). The optimal weights provide that the scoring scheme discovered is maximally discriminating between the true protein and the rest of the database. One particularly important feature is the global alignment between theoretical and experimental peptide masses using a linear regression. This regression is also useful to recalibrate the masses from spectra without internal standards and allows one to correct a large mass error (± 1 Dalton). The test performed with this tool showed a very good discrimination power for a large set of proteins using a unique set of parameters, which is essential for a high-throughput process.

The same linear regression approach was also used in a more recent report *(17)* (it seems that the authors were not aware of its prior publication since they called it "a new strategy"). The authors used a simple score taking account of the mass standard deviation after linear regression correction, sequence coverage by matching peptides, and the number of matches. They showed that this score has a much better selectivity than the shared peak count and that the results are less dependent on whether chemical noise is included in the fingerprints or not. The ChemApplex program *(18)* makes the most comprehensive use of chemical sequence information so far. Like SmartIdent, it calculates a peptide score (ChemScore) considering whether the peptide contains arginine or lysine [for matrix-assisted laser desorption/ionization (MALDI) experiments], the occurrence of modifications, sequence-specific missed cleavage rules, peptide signal intensity, and mass error. The peptide scores are then combined into a final score using the contribution of the matching peptides to the total ChemScore and the total intensity of all theoretical peptides as well as an intensity-weighted mass error. As a special feature, it is possible to use a dominant protein for recalibration of experimental masses to refine the search for minor components.

2.2. Contextual Identification

Since measuring mass spectra is a complex and probabilistic process, each measurement will produce different results. A peptide mass might be detected in one experiment but not in another, or its intensity might be very different. Several spectra are acquired at different positions of the sample and compiled into one spectrum to enhance the signal-to-noise ratio. Contextual information might also be used to purge noise from the mass lists. The molecular scanner was devised as a high-throughput identification method for proteins separated by one-dimensional (1-D) or 2-D gel electrophoresis. Circumventing time-consuming spot excision and expensive robotic sample handling, peptide digests from proteins are simultaneously transferred from the gel onto a collecting polyvinylidene fluoride (PVDF) membrane, which is scanned by a MALDI-time-of-flight (TOF) mass spectrometer *(19)*. The measured PMFs reveal the presence and identity of protein spots without prior knowledge of their position and generate virtual 2-D maps of the identified spots *(20)*. The multidimensional PMF data can be well represented by intensity distributions. For each mass found in the experiment, they describe the intensity of the mass signal as a function of the position on the membrane. These intensity distributions reveal which masses are expressed in similar regions and therefore belong together and which belong to chemical noise *(21)*. It was also realized that false positives in PMF identification often have matching peptide masses with dissimilar intensity distributions. Including the similarity of the matching masses into a PMF identification score could therefore help to detect false positives. This idea was investigated in a recent publication *(22)*. To define a similarity score, the similarity between the intensity distributions of two masses had to be defined. This measure was then used to calculate the similarity table of all matching masses of a protein at a certain position on the membrane. The correlation score S_{Corr} between all n matching masses had to be robust against a small number of outliers, and the average of the n strongest similarities (out of $n(n-1)/2$ similarities) was taken as such a robust measure. Subsequently, the original PMF score S_{PMF} was multiplied by S_{Corr}, yielding a combined score $S_C = S_{Corr}S_{PMF}$. It was shown that incorporation of S_{Corr} improved the signal-to-noise ratio and helped to identify proteins, which would have been difficult to detect otherwise.

$S_C(P,i)$ could then be calculated for every candidate protein P at every scanned point i on the membrane, and a criterion had to be found to decide whether protein P is really present or not. This criterion should be locally adaptive, since the score S_C of the highest scoring proteins showed a strong dependence on the region on the membrane. To reduce the dynamic range of S_C, the scanned points were partitioned into regions, where each region represented a set of masses with strongly correlated intensity distributions. It was shown that these regions corresponded well to protein spots. Then the average combined score was calculated for every region, and a local statistical decision criterion, based on the distribution of the average S_C of all candidate proteins, was applied to identify proteins within each region. These methods allowed identification of several proteins in a human plasma experiment and the results corresponded very well to results obtained by previous identification experiments carried out with gels of higher resolution and spot excision.

3. MS/MS Identification

A more informative technique for protein identification is tandem or MS/MS mass spectra. These spectra contain a set of masses corresponding to peptide fragments and provide information about the peptide amino acid sequence. Unfortunately, in addition to the difficulties of identification by PMF spectra, the MS/MS process provides new sources of noise. The peak detection is more difficult than with MS spectra because of the possibility of multiply charged fragments. Each peptide fragment can be seen as several ion fragments in an MS/MS spectrum, and the probability of seeing each kind of ion fragment is dependent on the spectrometer, the peptide sequence, and the mass of the fragment. Thus, it is particularly essential to develop efficient scoring schemes to be able to exploit the very rich information of these data.

There are two principal ways to use MS/MS spectra for protein identification. First, one can try to interpret the spectra directly in order to infer a peptide amino acid sequence by the so-called *de novo* sequencing process. The sequence is then parsed against a peptide sequence database (with the Blast or Fasta programs, for example) to find the most similar sequence. *De novo* sequencing is a difficult task, because inferring a sequence from an MS/MS spectrum is a highly combinatorial task. It is often performed by an expert who "by hand" infers a small tag from the spectrum that can be used with or without PMF information to identify the protein. The tools MS-seq, PepSea *(5)*, PeptideSearch, Mascot, and TagIdent allow submission of a sequence tag for peptide identification. Automatic *de novo* sequencing has also been widely studied, leading to several tools: SHERENGA *(23)*, SeqMS *(24,25)*, and LUTEFISK97 *(26,27)*. Unfortunately, they are not reliable enough to be used in a fully automated mode. Furthermore, all these identification methods have the drawbacks that transforming spectral information into a sequence neglects much relevant information and that they lead to difficulties in handling possible modifications and mutations.

The second way of using MS/MS spectra for protein identification relies on a direct comparison of spectra with virtual spectra calculated from a sequence database. Its simplest form is an extension of the PMF approach, called shared peak count (SPC), which consists in counting and scoring the matches of ion fragment masses with theoretical fragment masses of a peptide sequence. Most of the PMF tools propose a MS/MS version of their program, adapting their scoring function to take into account specific information from ion fragments. Here we can mention MS-Tag, PepFrag *(28)*, and Mascot *(13)*. SEQUEST *(29–31)* and its extension TurboSequest, which is based on a peptide mass indexation, use the SPC as a first pass to select the best peptides from the database. Then they construct a simplified virtual spectrum for each of these peptide sequences and calculate the cross-correlation score between these virtual spectra and the real ones by means of a fast Fourier transform. SPC methods generally include in the database all modified peptides to be considered, which requires prior knowledge of the mass differences associated with the corresponding modifications. One of the current important difficulties in identification is related to biological modifications or mutations. Including all possible modifications of peptides in the database is unrealistic owing to the combinatorial explosion it implies. As a result, SPC methods usually take into account only a few very common modifications occurring on specific amino acids, such as methionine oxidation or cysteine carbamidomethylation. In addition to the combinatorial problem, SPC algorithms have two other limitations. First, they consider the peaks independently of each other, thereby wasting important information contained in MS/MS spectra. Second, SPC algorithms need to allow a large error tolerance when they are used with badly calibrated spectra. As a result, the high intrinsic accuracy of current mass spectrometers is basically lost.

A new approach, sonar, is specifically dedicated to automation of protein identifcation *(32)*. The similarity score between the experimental spectrum and theoretical peptides from a database is based on the comparison of their intensity distributions. The procedure implies reduction of the experimental and theoretical spectra into two multidimensional peak intensity vectors and calculation of the inner product between the two vectors. The identification is then validated using a statistical method *(14)* in order to discriminate false positives from true positives in the list of potential hits. There is actually no complete publication of the scoring function, so it is difficult to know whether this method is able to handle the classical identification difficulties.

Other reports propose an extension of the matching concept by spectral convolution and spectral alignment methods *(33,34)*. They have been implemented in PEDANTA *(35)*, which is claimed to be very efficient in dealing with modifications and mutations, including unpredictable modifications. Indeed, the spectral alignment has a major advantage over SPC methods, because it uses logical constraints imposed by the spectrum peak composition to limit the number of considered modifications and mutations. One obvious tradeoff of these two approaches is that one must parse the whole peptide database without using the parent mass as a filter. In addition, the combinatorial problem grows with the number of contemplated mass shifts. Accordingly, the number of modifications/mutations considered must be kept rather low in order to allow identifications that are sufficiently discriminating.

A new approach, called Popitam, exploits a graph structure in association with database searching *(36)*. The graph allows one to store all information coming from the source peak list, in addition to the relationships between the peaks. The graph is then compared with theoretical sequences, leading to a similarity score for each peptide sequence. As for the *spectrum graph* currently used for *de novo* sequencing *(23)*, the vertices are computed from the spectrum mass/charge ratios and represent masses of potential "ideal" fragments (corresponding to a b-ion-type charged one). Since one cannot know what ionic type is a peak, it is necessary to make several ionic hypotheses for each peak (for example, a, b, b $-$ H$_2$O, yy++), each of them being at the origin of one vertex. Then, vertices are clustered into families. The idea behind the concept of family is that a given fragment can be represented by several different ionic peaks in the MS/MS spectrum. In such a case, the computed masses corresponding to these peaks will be almost equal. Since vertices are built under assumption, it is necessary to define for each of them a "credibility" score, which is computed from the occurrence probability of the ionic hypothesis used to build the vertex and from the family size of the vertex. The latter step of the graph construction consists of connecting the graph: vertices whose masses differ by the mass value of one or two amino acids are connected by an edge.

Sequences can be constructed by moving from one vertex to another one following existing edges. Thus all complete sequences and subsequences that can possibly be built from the experimental MS/MS spectrum are represented in the graph. *De novo* sequencing methods typically try to find the most probable, or higher, scoring paths in the graph. In Popitam, the aim is to find sections in the graph that are well suited to theoretical peptides from the database. This implies parsing of the graph for each theoretical peptide from the database whose mass matches the parent mass observed in the spectrum (filter on the parent mass). For this highly combinatorial task, the authors use an ant colony optimization algorithm (ACO) *(37)*. ACOs are multiagent systems inspired by the foraging behavior of real ant colonies. They have been applied to a wide range of difficult combinatorial optimization problems, like vehicle routing, job-shop scheduling, graph coloring, and so on. Indirect communication between the ants is mediated by environmental modifications and leads to an emergent collective behavior. Environmental modifications are caused by the deposit by the ants of given amounts of pheromone on edges. As in the real world, ants are attracted on the pheromone paths. When they move, they take into account three pieces of information: (1) the visibility, represented by the vertex scores; (2) the desirability, that is, the amount of pheromone deposited on the edges; and (3) the theoretical sequence being compared, which is stored in a collective memory. The quantity of pheromone deposited depends on the correspondence level between the path and the current theoretical peptide. The latter is evaluated with a scoring function, defined using Genetic Programming *(38)*, that combines different subscores computed from the intensity, the ionic probabilities, and the masses of the parsed vertices. Because information coming from the database is used, the ACO iterative process is not very time consuming.

In this method, mutations and modifications (included unsuspected modifications) can be simulated by combining different sections of the graph in order to cover the theoretical peptide maximally. In this case, it is no longer possible to filter on the parent mass, but the search space can be reduced using tags highlighted from a first identification step to extract candidate peptides from the database by alignment.

4. Conclusions

Automatic protein identification using MS data is a central task of a proteomics project and requires efficient tools dedicated to these specific data. Many tools for MS and MS/MS spectra already exist, and some innovative techniques are in development. There are still some difficulties to be taken into account to achieve a completely automatic identification process, such as the possibility of identifying peptides with unknown modifications or the definition of scoring schemes able to discriminate the right protein even in large genomic databases. However, the new techniques aim to manage some or all of these objectives, and preliminary implementations already show promising results.

References

1. Wilkins, M. R., William, K. L., Appel, R. D., and Hochstrasser, D. F. (2000) *Proteome Research: Mass Spectrometry.* Springer-Verlag, New York.
2. Bennington, S. R., and Dunn, M. J. (2001) *Proteomics from Protein Sequence to Function.* BIOS Scientific, Springer-Verlag, New York.
3. Fenyo, D. (2000) Identifying the proteome: software tools. *Curr. Opin. Biotechnol.* **11,** 391–395.
4. Larsen, M. R. and Roepstroff, P. (2000) Mass spectrometric identification of proteins and characterization of their post-translational modifications in proteome analysis. *Anal. Chem.* **366,** 677–690.

5. Mann, M. and Wilm, M. (1994) Error-tolerant identification of peptides in sequence databases by peptide sequence tags. *Anal. Chem.* **66,** 4390–4399.

6. Mann, M., Hojrup, P., and Roepstorff, P. (1993) Use of mass spectrometric molecular weight information to identify proteins in sequence databases. *Biol. Mass Spectrom* **22,** 338–345.

7. Wilkins, M. R., Gasteiger, E., Bairoch, A., et al. (1999) Protein identification and analysis tools in ExPASy server. *Methods Mol. Biol.* **112,** 531–532.

8. Wilkins, M. R., Gasteiger, E., Wheeler, C. H., et al. (1999) Multiple parameter cross-species protein identification using Multident—a world-wide web accessible tool. *Electrophoresis* **19,** 3199–3206.

9. Pappin, D. D. J., Hojrup, P., and Bleasby, A. J. (1993) Rapid identification of proteins by peptide-mass finger printing. *Curr. Biol.* **3,** 327–332.

10. Clauser, K. R., Hall, S. C., Smith, D. M., et al. (1995) Rapid mass spectrometric peptide sequencing and mass matching for characterization of human melanoma proteins isolated by two-dimensional PAGE. *Proc. Natl. Acad. Sci. USA* **92,** 5072–5076.

11. Gonnet, G. H. (1992) *A Tutorial Introduction to Computational Biochemistry Using Darwin.* E.T.H., Zurich, Switzerland.

12. Zhang, W. and Chait, B. T. (2000) ProFound: an expert system for protein identification using mass spectrometric peptide mapping information. *Anal. Chem.* **72,** 2482–2489.

13. Perkins, D. N., Pappin, D. D. J., Creasy, D. M., and Cottrell, J. S. (1999) Probability-based protein identification by searching sequence databases using mass spectrometry data. *Electrophoresis* **20,** 3551–3567.

14. Eriksson, J., Chait, B. T., and Fenyo, D. (2000) A statistical basis for testing the significance of mass spectrometric protein identification results. *Anal. Chem.* **72,** 999–1005.

15. Gras, R., Gasteiger, E., Chopard, B., Müller, M., and Appel, R. D. (2000) New learning method to improving protein identification from peptide mass fingerprinting, in *4th Siena 2D Electrophoresis Meeting.*

16. Gras, R., Muller, M., Gasteiger, E., et al. (1999) Improving protein identification from peptide mass fingerprinting through a parameterized multi-level scoring algorithm and an optimized peak detection. *Electrophoresis* **20,** 3535–3550.

17. Egelhofer, V., Gobom, J., Seitz, H., Giavalisco, Lehrach, H., and Nordhoff, E. (2002) Protein identification by MALDI-TOF-MS peptide mapping: a new strategy. *Anal. Chem.* **74,** 1760–1761.

18. Parker, K. C. (2002) Scoring methods in MALDI peptide mass fingerprinting: ChemScore, and the ChemApplex program. *J. Am. Soc. Mass Spectrom.* **13,** 22–39.

19. Bienvenut, W. V., Sanchez, J. C., Karmime, A., et al. (1999) Toward a clinical molecular scanner for proteome research: parallel protein chemical processing before and during western blot. *Anal. Chem.* **71,** 4800–4807.

20. Binz, P. A., Muller, M., Walther, D., et al. (1999) A molecular scanner to automate proteomic research and to display proteome images. *Anal. Chem.* **71,** 4981–4988.

21. Muller, M., Gras, R., Bienvenut, W., Hochstrasser, D., and Appel, R. D. (2002) Visualization and analysis of molecular scanner peptide mass spectra. *J. Am. Soc. Mass Spectrom.* **13,** 221–231.

22. Muller, M., Gras, R., Binz, P.-A., Hochstrasser, D. F., and Appel, R. D. (2002) Molecular scanner experiment with human plasma: improving protein identification by using intensity distributions of matching peptide masses. *Proteomics* **2,** 1413–1425.

23. Dancik, V., Addona, T., Clauser, K., Vath, J., and Pevzner, P. A. (1999) De novo peptide sequencing via tandem mass spectrometry. *J. Comput. Biol.* **6,** 327–342.

24. Fernandez-de-Cossio, J., Gonzalez, J., Betancourt, L., et al. (1998) Automated interpretation of high-energy collision-induced dissociation spectra of singly protonated peptides by 'SeqMS,' a software aid for de novo sequencing by tandem mass spectrometry. *Rapid Commun. Mass Spectrom.* **12,** 1867–1878.

25. Fernandez-de-Cossio, J., Gonzalez, J., Satomi, Y., et al. (2001) Automated interpretation of low-energy collision-induced dissociation spectra by SeqMs, a software aid for de novo sequencing by tandem mass spectrometry. *Electrophoresis* **21**, 1694–1699.

26. Johnson, R. S. and Taylor, J. A. (2002) Searching sequence databases via de novo peptide sequencing by tandem mass spectrometry. *Methods Biotechnol.* **22(3)**, 301–315.

27. Taylor, J. A. and Johnson, R. S. (1997) Sequence database searches via de novo peptide sequencing by tandem mass spectrometry. *Rapid Commun. Mass Spectrom.* **11(9)**, 1067–1075.

28. Fenyo, D., Qin, J., and Chait, B. T. (1998) Protein identification using mass spectrometric information. *Electrophoresis* **19**, 998–1005.

29. Gatlin, C. L., Eng, J. K., Cross, S. T., Detter, J. C., and Yates, J. R. III (2000) Automated identification of amino acid sequence variations in proteins by HPLC/microspray tandem mass spectrometry. *Anal. Chem.* **72**, 757–763.

30. Yates, J. R. III, Eng, J. K., and McCormak, A. L. (1995) Mining genomes: correlating tandem mass spectra of modified and unmodified peptides to sequences in nucleotide databases. *Anal. Chem.* **67**, 3202–3210.

31. Yates, J. R. III, Eng, J. K., Clauser, K., and Burlingame, A. L. (1996) Search of sequence databases with uninterpreted high-energy collision-induced dissociation spectra of peptides. *J. Am. Soc. Mass Spectrom.* **7**, 1089–1098.

32. Field, H. I., Fenyo, D., and Beavis, R. C. (2002) RADARS, a bioinformatics solution that automates proteome mass spectral analysis, optimises protein identification, and archives data in a relational database. *Proteomics* **2**, 36–47.

33. Pevzner, P. A., Dancik, V., and Tang, C. L. (2000) Mutation-tolerant protein identification by mass-spectrometry. *J. Comput. Biol.* **7**, 777–787.

34. Pevzner, P. A. and Sze, S.-H. (2000) Combinatorial approaches to finding subtle signals in DNA sequences, in *Proceedings of the Eighth International Conference on Intelligent Systems for Molecular Biology, San Diego*, pp. 269–278.

35. Pevzner, P. A., Mulyukov, Z., Dancik, V., and Tang, C. L. (2001) Efficiency of database search for identification of mutated and modified proteins via mass spectrometry. *Genome Res.* **11**, 290–299.

36. Hernandez, P., Gras, R., Frey, J., and Appel, R. D. (2002) Automated protein identification from tandem mass spectrometric cata using ant colony optimization algorithms. *Proteomics*, in press.

37. Bonabeau, E., Dorigo, M., and Theraulaz, G. (1999) *Swarm Intelligence. From Natural to Artificial Systems.* Oxford University Press, New York.

38. Koza, J. R. (1992) *Genetic Programming: On the Programming of Computers by Means of Natural Selection.* MIT Press, Cambridge, MA.

Index

DATE DUE

DEMCO INC 38-2971